AF572723

Faulting, Fault Sealing and Fluid Flow in Hydrocarbon Reservoirs

Geological Society Special Publications

GEOLOGICAL SOCIETY SPECIAL PUBLICATION NO. 147

Faulting, Fault Sealing and Fluid Flow in Hydrocarbon Reservoirs

EDITED BY

G. JONES, Q. J. FISHER

AND

R. J. KNIPE

Rock Deformation Research
Department of Earth Sciences
The University of Leeds
UK

1998

Published by

The Geological Society

London

THE GEOLOGICAL SOCIETY

The Society was founded in 1807 as The Geological Society of London and is the oldest geological society in the world. It received its Royal Charter in 1825 for the purpose of 'investigating the mineral structure of the Earth'. The Society is Britain's national society for geology with a membership of around 8500. It has countrywide coverage and approximately 1500 members reside overseas. The Society is responsible for all aspects of the geological sciences including professional matters. The Society has its own publishing house, which produces the Society's international journals, books and maps, and which acts as the European distributor for publications of the American Association of Petroleum Geologists, SEPM and the Geological Society of America.

Fellowship is open to those holding a recognized honours degree in geology or cognate subject and who have at least two years' relevant postgraduate experience, or who have not less than six years' relevant experience in geology or a cognate subject. A Fellow who has not less than five years' relevant postgraduate experience in the practice of geology may apply for validation and, subject to approval, may be able to use the designatory letters C Geol (Chartered Geologist).

Further information about the Society is available from the Membership Manager, The Geological Society, Burlington House, Piccadilly, London W1V 0JU, UK. The Society is a Registered Charity, No. 210161.

Published by The Geological Society from:
The Geological Society Publishing House
Unit 7, Brassmill Enterprise Centre
Brassmill Lane
Bath BA1 3JN
UK
(*Orders*: Tel. 01225 445046
Fax 01225 442836)

First published 1998

The publishers make no representation, express or implied, with regard to the accuracy of the information contained in this book and cannot accept any legal responsibility for any errors or omissions that may be made.

British Library Cataloguing in Publication Data
A catalogue record for this book is available from the British Library.

ISBN 1-86239-022-3

Typeset & printed by the Alden Group, Oxford.

Distributors

USA
AAPG Bookstore
PO Box 979
Tulsa
OK 74101-0979
USA
(*Orders*: Tel. (918) 584-2555
Fax (918) 560-2652)

Australia
Australian Mineral Foundation
63 Conyngham Street
Glenside
South Australia 5065
Australia
(*Orders*: Tel. (08) 379-0444
Fax (08) 379-4634)

India
Affiliated East-West Press PVT Ltd
G-1/16 Ansari Road
New Delhi 110 002
India
(*Orders*: Tel. (11) 327-9113
Fax (11) 326-0538)

Japan
Kanda Book Trading Co.
Cityhouse Tama 204
Tsurumaki 1-3-10
Tama-Shi
Tokyo 206-0034
Japan
(*Orders*: Tel. (0423) 57-7650
Fax (0423) 57-7651)

Contents

It is recommended that reference to all or part of this book should be made in one of the following ways:

Jones, G., Fisher, Q. J. & Knipe, R. J. (eds) 1998. *Faulting, Fault Sealing and Fluid Flow in Hydrocarbon Reservoirs*. Geological Society, London, Special Publications, **147**.

Townsend, C., Firth, I. R., Westerman, R. *et al.* 1998. Small seismic-scale fault identification and mapping. *In:* Jones, G., Fisher, Q. J. & Knipe, R. J. (eds) *Faulting, Fault Sealing and Fluid Flow in Hydrocarbon Reservoirs*. Geological Society, London, Special Publications, **147**, 1–25.

Faulting, fault sealing and fluid flow in hydrocarbon reservoirs: an introduction

R. J. KNIPE, G. JONES & Q. J. FISHER

Rock Deformation Research Group, The University of Leeds, Leeds LS2 9JT, UK

Abstract: A predictive knowledge of fault zone structure and transmissibility can have an enormous impact on the economic viability of exploration targets and generate considerable benefits during reservoir management. Understanding the effects of faults and fractures on fluid flow behaviour and distribution within hydrocarbon provinces has therefore become a priority. To model fluid flow in hydrocarbon reservoirs, it is essential to gain a detailed insight into the evolution, structure and properties of faults and fractures. Generation of realistic flow models also requires calibration with data on the fluid distributions and flow rates from hydrocarbon fields. Most hydrocarbon geologists at one time or another have asked the question 'What is the behaviour of this fault?'. This question, as emphasized by the contributions to this volume, should more fundamentally be phrased; 'What is the geometry of this fault zone, what are the nature and petrophysical properties of any fault rocks developed and how are they distributed in the subsurface?'. An additional important question is 'What impact could the fault zone have on fluid flow through time?'. The properties and evolution of fault zones can be evaluated using the combined results of structural core and down-hole logging, microstructural and physical property characterization, together with analysis of faults from seismic and outcrop studies and well test data. Successful fault analysis depends upon the amalgamation of these data and incorporation into robust numerical flow models.

Compared to many other areas of petroleum geoscience, studies on the structural controls on fluid flow in hydrocarbon reservoirs are in their infancy. As hydrocarbon reserves have become depleted and the oil industry has become more competitive, the importance of cutting costs by minimizing well numbers, optimizing production and predicting the occurrence of subtle traps has highlighted the need for information on the way in which faults and fractures affect fluid flow. Structural geologists are now becoming increasingly expected to provide answers to questions such as:

- Are hydrocarbons likely to have migrated into (or out of) the trap?
- What is the likely height of hydrocarbons that a fault can support?
- Is it likely that compartments exist within a field which have not been produced and will therefore require further drilling?

Early research laid a firm foundation with which to address many of these questions. For example, papers such as Smith (1966, 1980), Schowalter (1979), Watts (1987), Allan (1989) and Bouvier *et al.* (1989) have presented and reviewed many of the fundamental principles which control fault sealing within hydrocarbon reservoirs. Knipe (1992*a*,*b*, 1994) also highlighted the importance of two long-recognized observations. Firstly, the petrophysical properties of deformation features can vary significantly depending upon factors such as the host rock composition, deformation mechanisms and the stress history, etc. Secondly, fault zones can have highly complex geometries, with strain being accommodated not just on a single fault plane but within a complex array of faults known as a damage zone. A particular implication of this is that existing reservoir models could not incorporate the real complexity of faults and even if they could, fundamental gaps existed in our understanding of fault structure and fault properties.

Recent progress in understanding faulting processes (Underhill & Woodcock 1987; Scholz 1989; Cowie *et al.* 1993; Sibson 1994), fault rock development (Knipe 1989), fault geometry (Peacock & Sanderson 1994), fault populations (Gillespie *et al.* 1993; Cowie *et al.* 1996) and improved analysis of reservoir hydrodynamics as well as new core recovery techniques and the capabilities of 3D seismic, all provide a platform for improving fault analysis. In particular, in the last six years new data has become available to allow the problem of fault and fracture related fluid flow to be addressed in a more sophisticated manner (Knipe 1992*a*; Gauthier & Lake 1993; Berg & Avery 1995; Møller-Pedersen & Koestler 1997; Coward *et al.* 1998). For example, the availability of 3D seismic data has provided a wealth of information on the macro-structure of faults. Integration of seismic, field and core studies has allowed the detailed internal structure of faults to be investigated. There has also been a

Knipe, R. J., Jones, G. & Fisher, Q. J. 1998. Faulting, fault sealing and fluid flow in hydrocarbon reservoirs: an introduction. *In*: Jones, G., Fisher, Q. J. & Knipe, R. J. (eds) *Faulting, Fault Sealing and Fluid Flow in Hydrocarbon Reservoirs*. Geological Society, London, Special Publications, **147**, vii–xxi.

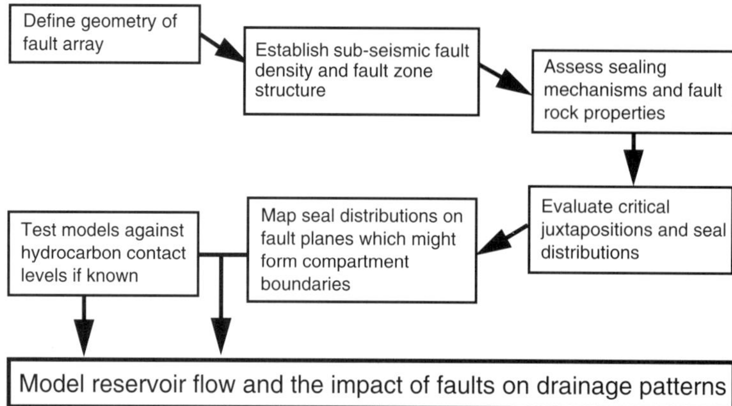

Fig. 1. Outline of the important stages in fault analysis and seal evaluation.

large increase in the characterization of the microstructural and petrophysical properties of fault rocks (Knipe 1989; Knipe *et al.* 1997; Gibson 1998).

The data now available to the geoscientist, geophysicist and reservoir engineer allow a new level of integration needed to develop and test different methodologies of fault evaluation. The challenge is to make use of these advances and to identify the best practices which lead to a more cost-effective and accurate prediction of faulting related influences on hydrocarbon reservoir behaviour.

Figure 1 outlines the important components needed for fault evaluation. The evaluation route emphasizes that a range of different elements has to be combined in order to assess the fault sealing or leaking potential. Each of these components carries its own resolution limits and sources of error. One of the limitations of fault analysis has been the complex nature of the variables involved and the difficulty in accurately defining each of the important factors. Calibration of any method is dependent upon the ability to separate the impact of individual components in the analysis and the need to include all the critical variables. The lack of data on the accurate characterization of fault zones and fault properties has resulted in the adoption of a number of assumptions about faults which are not always applicable and have resulted in the exclusion of a number of important factors from risk evaluations. Such omissions have reduced the success of fault seal/leak analysis and generated a perception that risk evaluation is impossible rather than difficult.

The present volume provides a broad sample of the areas critical to fault analysis and resulted from a conference entitled ‘Faulting, Fault Sealing and Fluid Flow in Hydrocarbon Reservoirs’, which was held at the University of Leeds in September 1996. The volume has been structured into four sections:

- Fault zone mapping, geometry and evolution.
- Faulting processes and fault seal characterization.
- Experimental and numerical modelling of deformation and fluid flow.
- Structure and seal analysis of hydrocarbon fields.

These sections are intended to cover the critical areas that are important to advancing our understanding of fault-related fluid flow, i.e. those where there is the need to understand the structure and flow properties of fault rocks, the need to model fluid flow, and finally the need to apply and calibrate these findings against information on the distribution and rate of fluid flow in existing hydrocarbon reservoirs. Each of these themes forms a sub-section of this introduction. A brief review of the recent advances in these areas, the requirements for future advances in these areas and the likely limitations to present solutions are discussed in the context of the volume contents. It should be noted that part of this introduction is based on a recent review written by the authors (Knipe *et al.* 1997), published by the Norwegian Petroleum Society.

Fault zone mapping, geometry and evolution

That fault zones are composed of deformation clusters which surround (or form halos to) large offset faults has been recognized for some time (e.g. Engelder 1974; Aydin 1978; Chester & Logan 1986; Wallace & Morris 1986). Damage zones represent the accommodation of strain

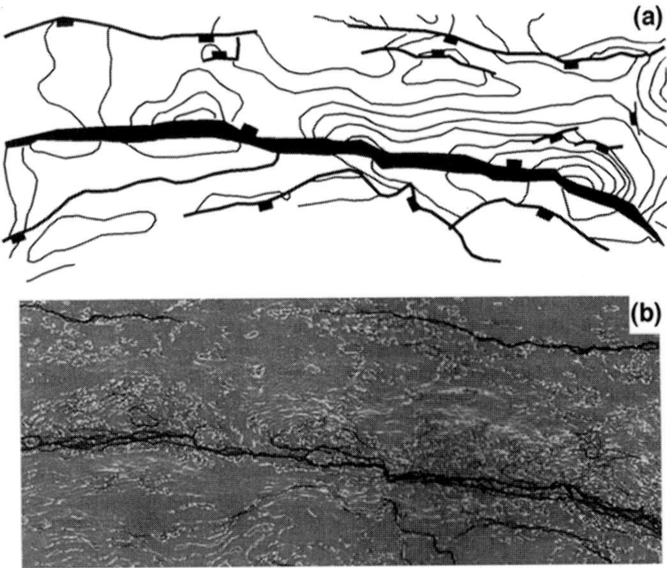

Fig. 2. An example of the complex architecture of a fault damage zone as imaged on a seismic attribute map (dip magnitude). Note that the fault zone is composed of linked segments and domains where different sub-structures are present. The simple interpretation of the fault zone (**a**) would be misleading if used as a basis for fault seal analysis.

around large faults, and are the products of fault propagation, displacement and linking processes operating during the life of the fault zone. It is important to recognize that the damage zone is the final product of the total history of strain accumulation in the volume around a large fault and should therefore be separated from fault process zones (Cowie & Scholz 1992) which develop at fault tips during propagation.

An example of the potential complexity of a fault zone is shown in Fig. 2. The illustration is of a seismic attribute map (dip magnitude) and shows that the fault zone is composed of linked segments and domains where different sub-structures are present. The internal structure of the northern and southern segments is made up of anastomozing faults which enclose lenses of more intact reservoir. The central segment appears to be composed of a smaller number of faults with larger throws. It is also interesting to note that a concentration of low-dip, short discontinuities occur on the eastern side of the fault that probably represent the accommodation zone of small structures. The width of this zone increases towards the central (high displacement) portions of individual segments.

Figure 2a illustrates the type of simple interpretation of the fault structure often used as a basis of fault analysis. This interpretation is based on the representation of the fault as a single fault plane, where the offset is assumed to be equal to the seismic (cumulative) offset. This is clearly not always valid as a representation of throw distributions for use in fault seal analysis. Figure 2 also illustrates that sub-seismic fault populations may be clustered around larger faults with extensive areas of low fault densities away from the large faults. This has important implications for the spatial distribution of sub-seismic faults and emphasizes that uniform distributions of small faults are not always applicable (except perhaps in areas where more uniform straining is associated with doming). The information contained in Fig. 2 reinforces the results of other recent studies which have noted different fault architecture associated with fault segment structures or domains (Cartwright *et al.* 1996), relay zones (Peacock & Sanderson 1994), and tip zones (McGrath & Davison 1995). It is also clear that attribute mapping from high-quality 3D seismic surveys offers an important direction for the future characterization of fault segment and damage zone geometries (see Jones & Knipe 1996).

The critical elements of fault damage zones which are needed for fault seal evaluation and for input into reservoir behaviour simulation include: (i) the dimensions of the damage zone; (ii) the fault clustering characteristics; (iii) the fault offset populations, which can control the

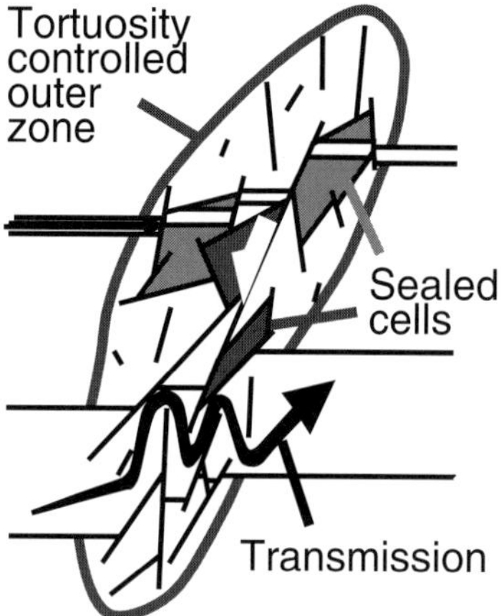

Fig. 3. Cartoon of the main structural elements of a fault damage zone. The zone is composed of a cluster of deformation features around a large offset fault. Note that the juxtapositions seen differ from those which would be occur if only a single fault was present and that the presence of an array of deformation features can induce the development of micro-compartments or sealed cells in the fault zone.

distribution of fault rocks and juxtapositions; (iv) the orientation distributions of deformation features present within damage zones; and (v) the total thickness of fault rocks. Each of these aspects are reviewed in Knipe *et al.* (1997).

The importance of damage zones to fault seal analysis in hydrocarbon reservoirs was highlighted by Knipe (1992*b*, 1994). The impact of the damage zones on fault seal are reviewed in Fig. 3. The primary influences of damage zones on fault seal and reservoir behaviour analysis are:

- The juxtapositions inferred by using a single fault model are different from those associated with a cluster of smaller faults.
- The volume of deformed rocks around faults can affect the volume of reservoir with recoverable hydrocarbons.
- The presence of an array of deformation features rather than a single fault can influence the changes in cross- or along-fault communication induced by reactivation events.

The development of an increased frequency of structural features in the volume around large faults is a ubiquitous characteristic of the cores studied from the North Sea (Fig. 4). The frequency of deformation features can increase from background levels of <50/100 m of core to values exceeding 600/100 m of core, close to faults. The dataset reveals that the structural frequency increases with increasing offset and that

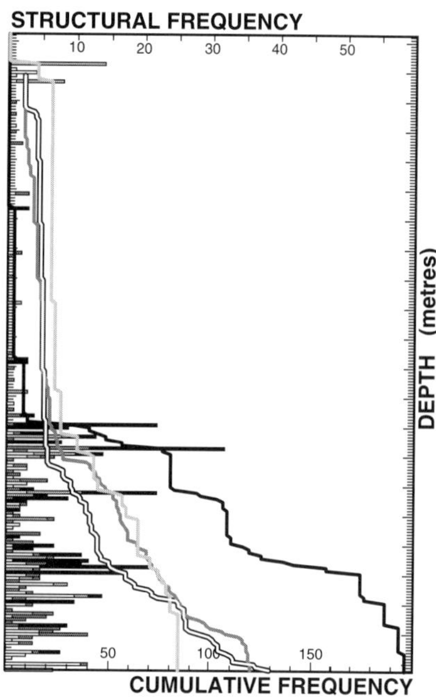

Fig. 4. Structural log of core entering a fault damage zone. The high structural frequency, faults and fractures concentrated at the base of the core, are arranged in clusters which define steps in the cumulative frequency curves.

faults with offsets of >75 m have damage zones which can extend for ~150 m. The edge of the damage zone is taken here as twice the background structural density. An example of the variation in the frequency of structural features away from one fault (the Ninety Fathom fault exposed in Whitley Bay, Tynemouth) is shown in Fig. 5. These data highlight the degradation of the reservoir properties close to the faults.

The size of the damage zone is also dependent upon the lithologies which have been faulted, the deformation conditions and the distribution of strain between the hanging wall and footwall. Figure 6 is a cartoon which reviews the main factors which may control the dimensions and shape of damage zones and illustrates how the observed concentrations of deformation in either the hanging wall and the footwall can arise. It should be noted that these variations form part of the 3D variation in fault zone structure likely to occur along larger faults.

Most fault offset population analyses (see Walsh & Watterson 1992; Cowie *et al.* 1996) have concentrated on the prediction of the number of sub-seismic faults over large areas (>1 km^2) rather than the distribution of the faults within the fault damage zones themselves.

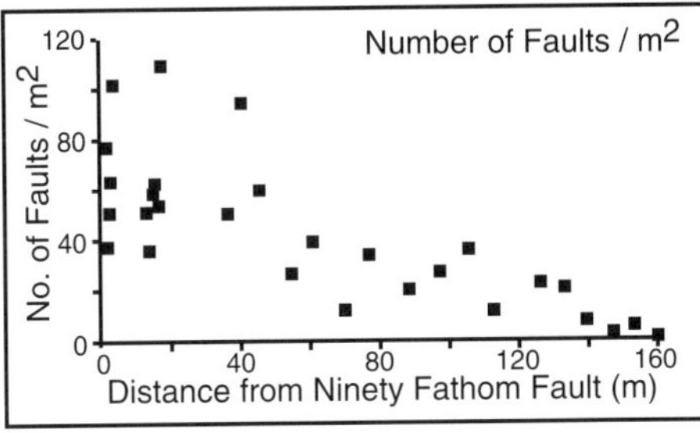

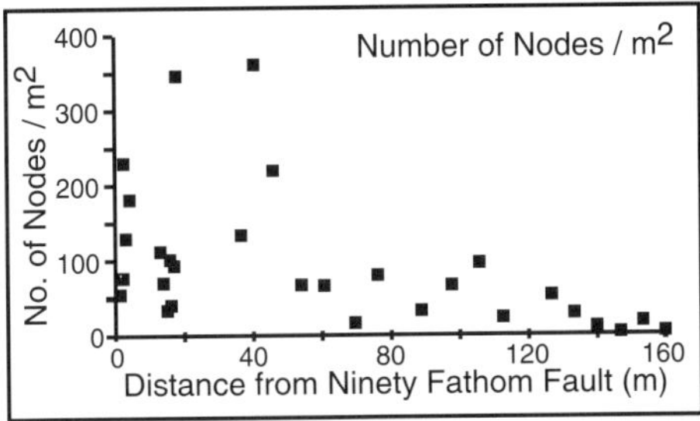

Fig. 5. Example of the increased structural frequency and number of fault intersection nodes present in a damage zone associated with the Ninety Fathom Fault, Whitley Bay, Tynemouth, U.K.

In many cases a uniform distribution of faults across an area is assumed. The data from the structural logging of North Sea wells illustrate that the characteristics of faults found on a field scale are also present within individual fault zones identified on seismic or from well data. As the population of small faults around larger structures will control the distribution of juxtapositions and fault rocks, detailed characterization of the offsets is important to seal analysis. Figure 7 illustrates the population characteristics of three fault zones with different offsets. The plot illustrates the increasing population and a similar slope (a power-law, fractal relationship) for the central part of the measured populations.

Figure 8 shows a number of fault offset population data sets from cores through different faults and presents information on the growth of fault populations. The figure shows that with increasing fault development (increasing total population) there is a change in the fractal number or slope towards higher values (i.e. fault populations are not simple fractal systems where one fractal number can describe the fault population characteristics). This indicates that as the level of deformation increases (either with increasing fault offset magnitude or proximity to a fault) the fault population grows to contain a larger proportion of small offset faults.

The 3D orientation distributions of structural features within damage zones are also important for modelling of fluid flow as the non-parallel members of the arrays induce an intersection network which will control the connectivity of barriers. It is insufficient to characterise the average fault orientations or to identify average fault trends or families. A more detailed statistical analysis of fault orientations is needed in order to evaluate the 3D distribution and connectivity of flow paths and barriers. An analysis of fault dips from North Sea wells yields an average dip of 59°. However, the standard deviations of these datasets, which will control the density and pattern of intersections, is typically between 15° and 26°. It is clear that the small scale structures exhibit more variation in orientation than the larger scale structures, and demonstrates that seismic fault orientation distributions cannot be used in a simple way to predict sub-seismic fault patterns.

The model of fault damage zones which emerges from the data presented above is of a volume of deformed reservoir surrounding fault

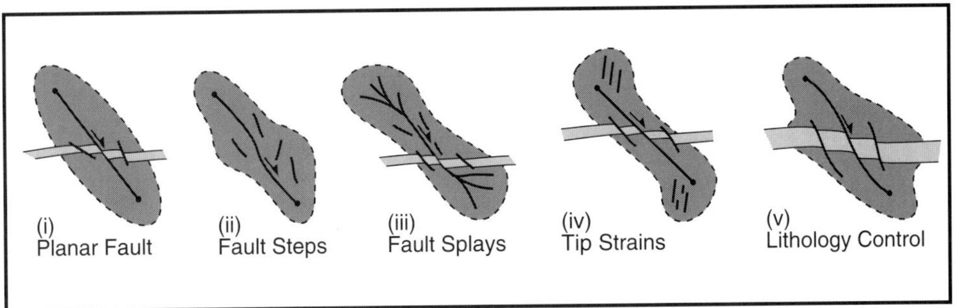

Fig. 6. Cartoon which reviews variations in the geometry of fault damage zones. The simple elliptical geometry of the planar fault (i) is likely to be modified in reality by strain fields caused by: (i) fault steps, (ii) deviations away from a plane, (ii) at fault splay points, (iv) from extensional tip strains and (v) fault offset populations in damage zones.

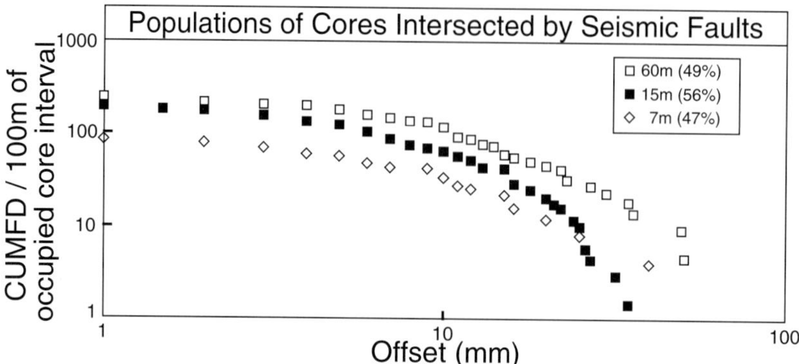

Fig. 7. The fault offset populations of three fault zones with different cumulative offsets. The plot illustrates the increasing population and a similar slope (or power-law, fractal relationship) for the central part of the measured populations.

zones, where with increasing proximity to the main fault, the structure of the fault zone alters (see also Knipe *et al.* 1997). The damage zone can be viewed as being composed of 'onion skins' each with different densities, architectures and connectivities between potential barriers. Damage zones can be considered to be made up of two main domains with different flow properties: an outer zone and an inner zone. The outer zone of the damage zone will be composed of a volume with a higher structural density (minor faults and fractures) than outside where, if the fault rocks are effective barriers, tortuosity controls the flow behaviour. The inner zone is where the structural density and architecture (i.e. fault population, clustering and orientation characteristics present) generate a linked 3D array, which together form a continuous barrier, where flow is controlled by the fault rock properties. The challenge for fault analysis is to be able to define the geometry and location of these inner and outer regions of damage zones.

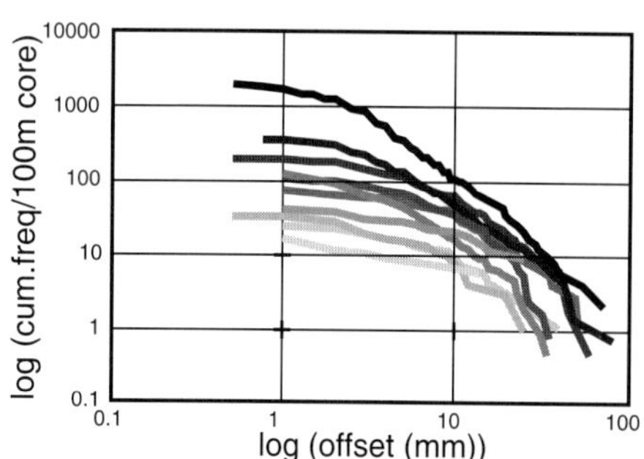

Fig. 8. Fault offset populations from cores through different faults. Note that with increasing fault development, there is a change in the fractal number or slope towards higher values.

The first section of this volume deals with the imaging, geometry and temporal and spatial evolution of fault arrays. A feature of the papers in this section is the wide range of approaches taken in this area of study. Geometrical characterization of fault arrays in the subsurface is still a major challenge, especially in basins that are often seismically indistinct, such as the North Sea.

At the seismic scale, **Townsend *et al.*** discuss a number of methods for the imaging and characterization of small scale faults in reservoirs, emphasizing the sensitivity of a number of seismic attributes, notably seismic amplitude, for resolving structures at the limits of the data. These authors also address the important issues concerning the future directions to be taken in the structural analysis of seismic data, and the tools that will be needed to undertake these tasks both rapidly and efficiently.

Steen *et al.* have attempted the difficult approach of linking the detailed fault geometries obtained from field mapping with seismic attributes, to produce predictive models for the densities and distributions of small faults in the subsurface. A number of synthetic horizon maps were generated based on outcrop and analogue model data, and compared to those obtained from the Snorre Field in the northern North Sea.

Marchal *et al.* use X-ray tomography of scaled physical models to elegantly illustrate how fault arrays could initiate and propagate in three dimensions. Such analysis is an invaluable aid to the interpretation of fault geometries in, for example, 3D seismic data. The paper suggests a means of reconciling presently recognized models of fault propagation (radial tip zone propagation model, Cowie & Scholz 1992; segment linkage model, Cartwright *et al.* 1995).

Faulting processes and fault seal characterization

A detailed understanding of the fundamental processes which control the evolution of fault rocks and their properties is required in order to predict their effect on fluid flow in hydrocarbon reservoirs. The deformation processes which result in the development of fault-related permeability barriers or pathways have been reviewed by Mitra (1988) and Knipe (1989, 1992*a*, 1993*a*,*b*) and include: deformation-induced porosity collapse by disaggregation, mixing and grain boundary sliding without large scale cataclasis; diffusive mass transfer; cataclasis; cementation and clay/phyllosilicate smearing. Studies reporting the detailed physical properties or microstructural evolution of fault rocks in hydrocarbon reservoirs include the analysis of clay smears (Knipe 1992*a*, 1994; Knipe & Lloyd 1994; Berg & Avery 1995), cataclasites within clean sandstones (Pittman 1981; Antonellini & Aydin 1994; Fowles & Burley 1994), and faults within impure sandstones (Sverdrup & Bjørlykke 1992; Gibson 1994, 1998). Despite these studies, the lack of data on the petrophysical properties of deformed reservoir rocks has placed major constraints on the ability of geologists to not only model fluid flow, but also to interpret fluid and pressure distributions, as well as fluid flow rates, within hydrocarbon reservoirs. The lack of a detailed link between the petrophysical properties of deformed rocks and factors such as the sediment composition and lithification at the time of deformation or the stress and thermal history experienced has also prevented the prediction of fault rock properties prior to drilling.

Detailed evaluation of fault-rock properties requires the integration of microstructural information on the deformation mechanism history of fault rocks with porosity, permeability and capillary entry pressure data. Such an analysis should involve the use of electron microscope based techniques (especially BSEM and CL) for detailed microstructural analysis (see Knipe 1992*a*). In addition, equipment capable of accurately measuring low ($\gg$0.01 mD) permeabilities is required. Some of the studies reported in the literature have not used techniques which allow clear resolution of the important microstructural elements or have been restricted by the measurement ability of equipment used for petrophysical property determination. Without such information, identifying the origin of the petrophysical properties may be impossible, because the type of (and timing of) deformation processes which control the pore characteristics and the fault rock strength, of the fault rocks remain poorly defined.

In the second section of this volume, the identification and characterization of faults and fault seals, including the microstructural, diagenetic and petrophysical properties of fault rocks, are considered in a series of papers.

Adams & Dart review the characteristics of likely sealing faults in the sub-surface, as determined from borehole imaging techniques. This dataset bridges the important scale gap between core and seismic data, and is of significant value in that it provides *in situ* data regarding the length, dip, orientation and fluid retardation properties of subsurface structures. The authors emphasise the largely qualitative nature of the sealing data generated and recommend an integrated approach to interpretation which incorporates core, conventional wireline log, pressure and hydrocarbon production datasets.

Foxford *et al.* describe the results of their detailed field study of the geometry of the Moab Fault zone in Utah and have discussed the implications that the observed fault geometry would have for the sealing properties of similar structures in the subsurface. The authors conclude that the fluid transmissibility properties of such a fault zone would be essentially impossible to predict, due to the heterogeneous nature of the fault zone over short distances as a result of fault propagation processes. However, their analysis has also revealed that shaley gouge is present all along the fault, except where the mudstone content of the faulted sequence is less than 20%, and note that this conforms to existing predictive models used for clay smear analysis in the subsurface.

The contribution of **Peacock *et al.*** links detailed outcrop work on fault geometries and propagation processes in carbonate rocks with microstructural analysis of the resultant fault rocks. They conclude that pressure solution features may be important fluid retardation barriers in contractional regions of fault arrays in such sequences. They also note that phyllosilicate material becomes concentrated along faults in fine-grained carbonates, which again may result from enhanced pressure solution during or after faulting.

Fisher & Knipe focus on the microstructural controls on the petrophysical properties of deformation features obtained from cored reservoir units from the North Sea. These authors provide the first comprehensive detailed classification of fault rock and fault seal types based on factors such as the clay content at the time of deformation, the amount of cataclasis experienced and the extent of post-deformation lithification. The paper also outlines many of the key factors which influence the petrophysical properties of fault rocks.

Two experimental and analytical papers in this volume consider the efficiency of shales as membrane seals. **Krooss *et al.*** have considered in detail fluid transport in faulted and unfaulted pelitic rocks and conclude that no systematic relationship could be found between permeability and microfault frequency, the distance of a sample to a microfault plane, or to sonic velocity anisotropy in the sampled mudrocks. Meanwhile, **Faulkner & Rutter** have measured the permeability of fine-grained fault gouge from a compressional fault zone and found that the gouge exhibited permeability anisotropies of up to three orders of magnitude.

Experimental and numerical modelling of deformation and fluid flow

The data presented above highlight the complexity of fault zones and that a detailed analysis of fault sealing can be of limited value if a simple fault zone structure is assumed. Despite these complexities, it is also clear that a more constrained analysis of faulting and flow behaviour is possible if the detailed fault zone architecture is considered.

Two variables are particularly fundamental to assessing the flow across complex fault zones. The first variable is the cumulative fault-rock thickness across the flow path through the fault zone, i.e. the total thickness of fault rock from all faults along the flow path. This depends upon the fault frequency along the flow path and is not equivalent to the fault damage zone thickness (cf. Knott 1993), unless the fault zone is invaded by cements. The second variable is the connectivity of the faults or deformation features with low permeabilities in the fault zone. In the case of a completely connected array with no windows of undeformed material along possible flow paths, the flow is controlled by the permeability of the fault rocks. Where a more open network of faults is present, then the flow will depend upon the tortuosity associated with flow around the low permeability zones and the ratio of matrix to fault-rock permeability. The interaction of these two factors will control the effective transmissibility of the zone. The development of reservoir simulation packages which can evaluate the flow through complex fault networks in damage zones is an important component of future fault analysis (see Knipe *et al.* 1997).

There are a number of papers in this volume which assess specific aspects of fault evolution and the relationship of faulting to fluid flow. **Matthai *et al.*** use extensive outcrop data from faulted sandstones to construct a numerical reservoir model that describes how joints and deformation bands could affect radial drawdown under production conditions. They show that normal faults with highly permeable slip planes can compartmentalize fluid pressure in the model over timespans greater than years and also focus fluid flow.

The experimental approach of **D'Onfro *et al.*** for identifying conductive fracture flow paths has used the extensive test facilities available to these researchers to attempt a complete geological and geophysical characterization of the flow characteristics of a subsurface fractured carbonate unit. The results obtained from high resolution seismic surveys (pre- and post-air injection) and interference tests, were then tested by drilling. This validated the presence of a previously predicted open fracture in the sub-surface. The paper of **Maillot *et al.*** describes a fully tensorial 3D model of fault growth in an elastic medium. These authors relate their findings to the assessment of anisotropic stresses around faults on strain accumulation and consequently to the spatial organization of crustal deformation.

Fleming *et al.* tackle the problem of fluid flow both in and around open fault zones, and describe the associated thermal effects within sedimentary basins. They suggest that large-scale buoyancy-driven circulation can occur within fault damage zones in a normally pressured basin situation, which is associated with thermal anomalies. **Henderson** takes a slightly different slant on the problem of fluid flow in fault zones by discussing how compaction affects the degree and rate of fluid flow, using a finite element model. When compaction is slow, a power-law relationship of earthquake event sizes occurs, whereas under rapid compaction, non power-law statistics apply. He suggests that the non power-law scaling of fault offsets often observed may be a real feature of the datasets, rather than a result of poor sampling, as is often assumed. **Leary** observes that borehole logs of rock properties in the metre to kilometre scale range have power-law Fourier spectra that scale inversely with spatial frequency to a power near unity, with a narrow range of scaling components. He suggests that if such 'long range' correlated random 'structures' control fluid flow, reservoir management cannot be accurately achieved by using flow models constrained by small-scale sampling of the reservoir rock. Lastly, **Lesnic *et al.*** discuss a mathematical solution for locating and assessing the hydraulic conductivity of a fault in a rock mass.

Structure and seal analysis of hydrocarbon fields

The final section of this volume deals with the use of structural geological techniques in reservoir characterization, and emphasizes with specific examples, the practical and applied aspects of fault and fracture characterization in the production of hydrocarbons.

The basic requirement for fault seal analysis is the generation of a realistic, maximum probability map of sealing capacities along individual fault zones. This involves evaluation of the possible juxtaposition patterns within the zone as well as an assessment of the variance of fault rock properties. The method most commonly used in evaluating fault seal distributions is the construction of Allan maps (Allan 1989) which illustrate stratigraphic geometries of horizon/fault plane intersections and are either drawn by hand or used in conjunction with fault mapping software (such as FAPS) (Freeman *et al.* 1989; Needham *et al.* 1996; Yielding *et al.* 1997). Construction of these maps can be an essential but difficult task, and the important limitations of these maps (e.g. seismic resolution, interpretation error) are not always considered in detail. The assumption made in constructing these fault plane maps is that the throw indicated across a fault identified on seismic represents a single fault plane.

Two factors are therefore critical in the evaluation of the geometrical distribution of seals in fault zones. The first relates to the accuracy of mapping, the stratigraphic horizons, the fault zone location and the cumulative throw distributions. This is usually controlled by the resolution of seismic data and is generally of the order of 20–30 m for high quality datasets in the North Sea. Such a resolution of both surface and fault mapping can introduce a large range of possible juxtaposition patterns, even if a fault zone with a single fault plane is present. The second limitation centres around the problem of characterizing the sub-seismic fault damage zone architecture along the fault zones and in tip areas. A previous section of the paper on fault structure reviewed how characterization and modelling of the critical parameters of sub-seismic faults can reduce the risk associated with fault analysis. How many fault seal analyses or reservoir modelling attempts have produced poor history matching because the potential variation in fault juxtapositions arising from the combined impact of resolution limitations and fault zone sub-structure have not been considered?

The evaluation of fault-seal potential along individual faults can be time-consuming if each fault has to mapped in enough detail to allow accurate definition of reservoir and fault intersections. It is often more efficient to divide the evaluation process into two phases.

(a) Phase 1: involving a rapid assessment of the impact of fault throw, sedimentary architecture and fault zone structure on the juxtapositions and sealing properties of faults in the field. This can be achieved by the use of simple juxtaposition/fault seal diagrams (Knipe 1992*b*, 1997) without the need for detailed seismic mapping of either stratigraphic horizons or faults. The advantage of this procedure is that critical fault throws, which create 'leaky' windows across faults, can be identified and then used to help locate areas of the field where more detailed analysis of the seismic data is needed.

Figure 9a, b illustrates the basis of the juxtaposition and fault seal diagrams. The details of the construction of these diagrams are presented by Knipe (1997). The figure plots the reservoir stratigraphies in the footwall and hanging wall along the vertical axes and increasing fault throw along horizontal axis. The diagrams can be considered as a horizontal view of a 'transparent' fault where the stratigraphy in the footwall is a horizontal and the hanging-wall inclined. The juxtapositions between the hanging-wall and footwall are represented as either triangles or parallelograms on the fault surface. The range of juxtapositions along a fault with constant or variable throw can be assessed from the range of triangular and parallelogram areas intersected by either a vertical or inclined line. The impact of seismic resolution can be assessed by considering the intersection of a band of throws with a thickness which accounts for the resolution. Damage zones can be evaluated by considering an array of faults, where the combined throw is equal to the cumulative throw interpreted from seismic. It is also useful to include 'side-wall' plots of the critical depth property data which help in the delineation and classification of fault rock types and properties. These side-wall plots can include porosity, permeability, net gross ratios, and various downhole-tool log data sets (see Fig. 10a).

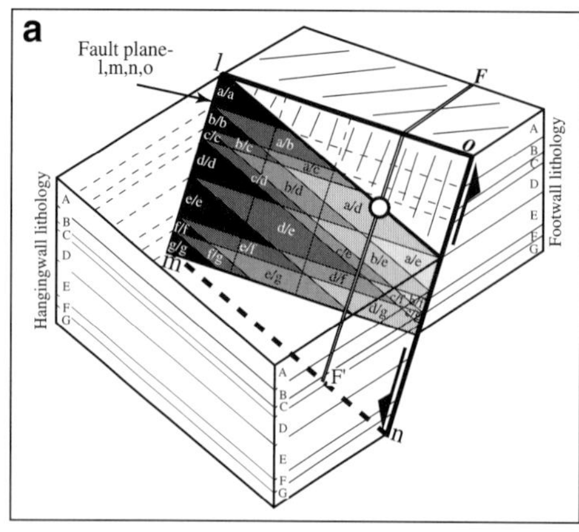

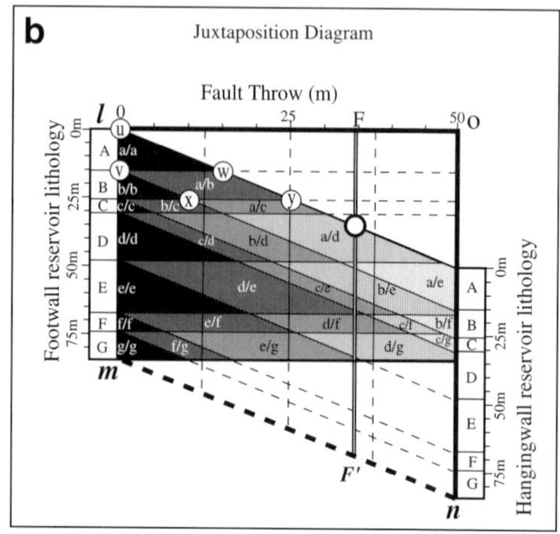

Fig. 9. Juxtaposition diagrams for use in fault seal analysis. See text for details.

Fault seal types, based on the analysis of fault rock properties derived from the different host rocks as well as on the throw and juxtaposition history, can be generated on a separate (fault seal type) diagram. Areas on the diagram with different properties can then be identified (see Fig. 10b). this procedure allows the mapping or contouring important sealing properties on the diagrams. The fault properties which can be mapped onto these diagrams include permeabilities, shale smear potential, transmissibilities, sealing capacities, and seal strengths. The diagrams can also be used to correlate these properties with well test or production data to validate the analysis.

(b) Phase 2: seismic based mapping (in selected areas) of detailed fault and reservoir horizon geometries using Allan diagrams (see Fig. 11) to constrain the depth and location of 'leaky' windows and to provide a platform for analysis of the potential controls on hydrocarbon/water contacts, pressure differences, production and drainage patterns as well as field communication.

Phase 2 involves the detailed mapping of both reservoir horizons and faults and requires assembling fault plane maps, as well as integration with well data on hydrocarbon/water contacts, pressure distributions and production data (e.g. Jev *et al.* 1993). The following is a list of the important considerations which should be involved in this stage of the analysis:

(1) Evaluation of the coherency of fault offset patterns and gradients (Walsh & Watterson 1991) to identify fault intersections, erosion of fault tops at unconformities, the substructure of tip areas and connectivity of fault zones (tip-tapping) into stratigraphic horizons able to generate cements.
(2) Assessment of data on the presence of fault damage zones and structural variation along fault zones, from both individual seismic lines as well as attribute data, if available (see Jones & Knipe 1996). The aim is to identify segments or domains, where different fault patterns may be present. For example, early relay zones, which mark the ends of linked fault segments, have a high probability of developing from overlapping faults and often represent areas where accommodation of displacement is distributed on a number of faults, i.e. these areas will be characterized by low throw juxtapositions (see Knipe 1997).
(3) Analysis of the location and heights of potential leaky fault juxtaposition windows on Allan diagrams which arise from variations in: (i) the possible depths, geometries and locations of stratigraphic horizons and faults; (ii) the difference between cumulative throws on individual fault zones, indicated from seismic and the most likely size of the throw on the largest real fault in that zone; and (iii) the sediment architecture and continuity. The end result should be a probability map of the distribution of sealed and leaky windows along the critical fault zones which can be compared and evaluated using pressure or production differences across faults (see also Yielding *et al.* 1997).
(4) The generation of communication and drainage maps for potential compartments and

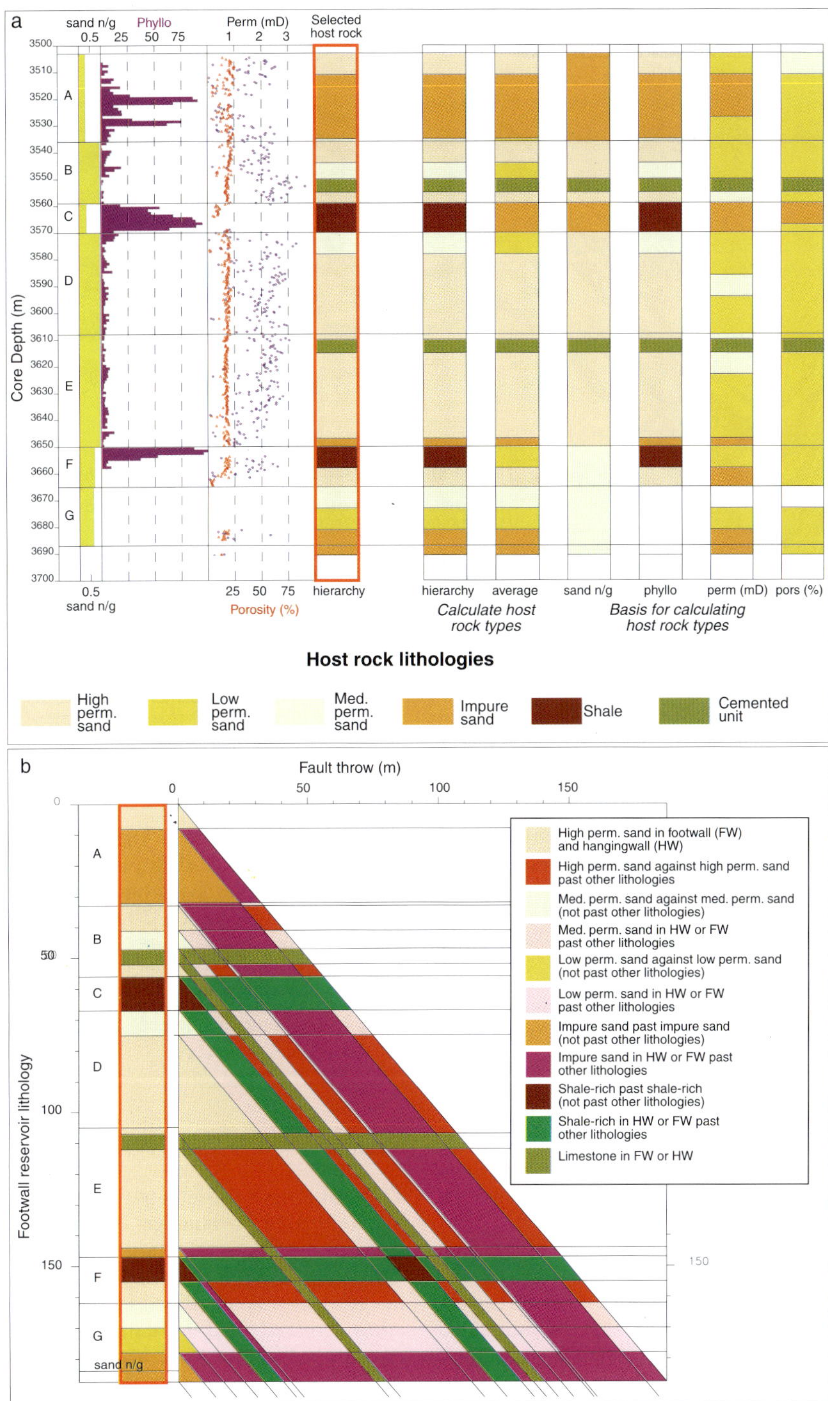

Fig. 10. Juxtaposition diagrams illustrating the modelling of different litho-classification scheme (**a**) and areas of different juxtapositions (**b**).

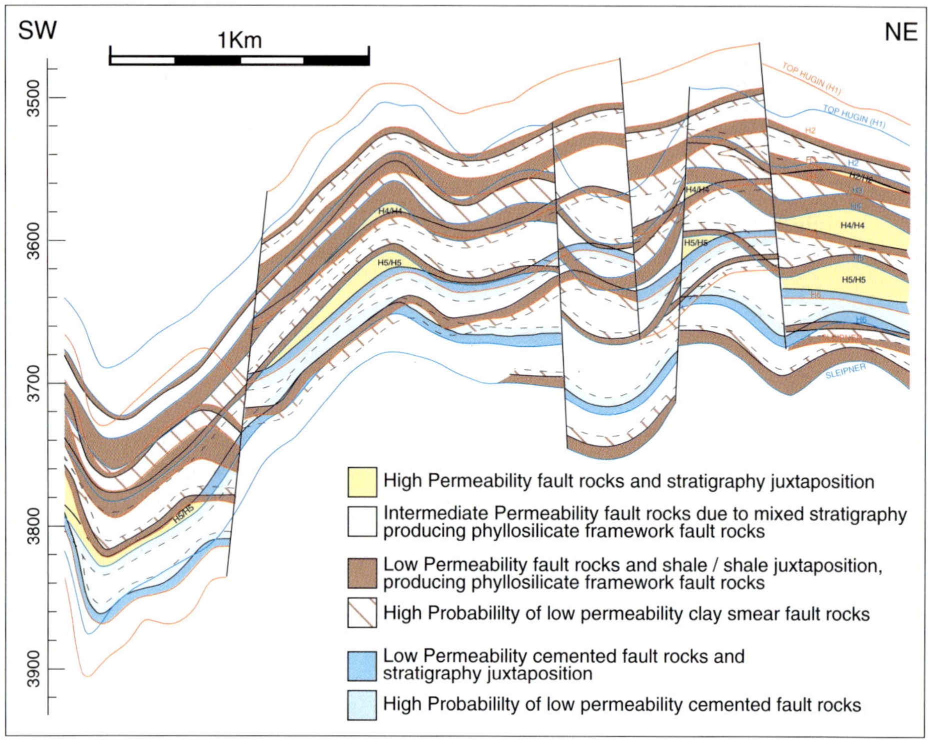

Fig. 11. Example of an Allan diagram illustrating the juxtaposition and fault seal types likely on a single fault.

the correlation and integration of these with hydrocarbon/water contacts, pressure test and production data.

(5) Input of the most robust reservoir characterization geometries and properties into simulation models.

Note that in the exploration situation, as opposed to the field development case, the different levels of data availability will dictate which of the above analytical procedures are possible and which of the missing (or poorly defined) elements represent the high risk factors.

There are three papers in the volume which describe the application of different fault analysis techniques to specific fields. Two of these hydrocarbon fields are located on the Norwegian Continental Shelf and are addressed in the papers of **Knai & Knipe** and **Ottesen *et al.*** These emphasize the impact that an integrated programme of fault rock property and fault zone characterization can have when used in reservoir prediction and management. A significant finding of these studies was the ability to use core-based studies as input for upscaling in reservoir simulations. The work undertaken has had an important impact on planning early in the production cycle of these two fields. The paper by **Ericsson *et al.*** draws on a large seismic and well database available from a producing field in the Arabian Gulf to illustrate how an understanding of tectonic fracture systems can significantly improve production and extend field life. Detailed characterization of the fractures in this field has successfully been linked to the geological structure.

Summary

This paper and those contained in this volume highlight that a number of components are important to the evaluation of fault flow behaviour. Some of these variables are often not included or not quantified in sufficient detail to allow a robust fault evaluation. The main components which are not always considered in detail are: (i) the errors in throw patterns which arise from seismic resolution and fault damage zone structures; (ii) the assumption that juxtaposition of reservoir against low permeability units and shale smear are the only sealing mechanisms; and (iii) that fault behaviour data

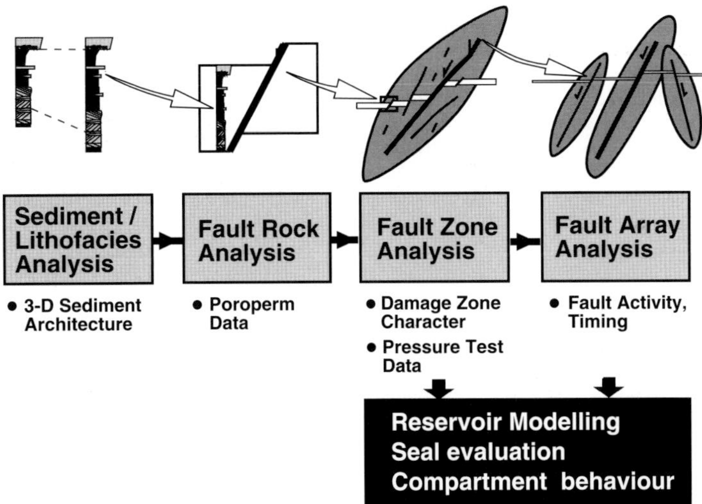

Fig. 12. Reviews of the critical factors needed for an integrated seal analysis. These include: (i) data on the 3D sediment architecture, (ii) the petrophysical properties of the fault rocks present, (iii) the architecture of individual fault zones and (iv) the fault array evolution.

from one area is directly applicable to any other sealing problem, i.e. that the geohistory is not critical to the seal evaluation. Each of these factors can have a major impact on seal analysis, and if not considered will induce a poorly constrained, high risk model and render detailed reservoir modelling of little value.

A number of elements stand out as being important directions for future studies. These include incorporation of the geomechanical properties of fault rocks into analysis of reservoir behaviour and integration with present day stress; assessment of the continuity, spatial distribution and petrophysical properties of fault rocks within fault zones and how these data may be integrated into flow models; an assessment of whether fault arrays are ordered or chaotic and finally the validation of the applicability of fault seal risk assessment procedures.

The analysis presented here has highlighted the need to integrate datasets from different scales into an amalgamated seal analysis (e.g. Leveille *et al.* 1997). Figure 12 reviews the four critical factors needed from the different scales. These include: (i) data on the 3D sediment architecture; (ii) the petrophysical properties of the fault rocks present; (iii) the architecture of individual fault zones; and (iv) the fault array evolution. It is the combined resolution and characterization level of each of these which defines the risk level of the seal analysis. There is an important geohistory component in each of these factors. This emphasizes the problems associated with transferring data or results from areas with different geohistories, without consideration of the different geohistories involved.

Despite the common assumption of fault sealing in hydrocarbon fields, very few faults have been characterized in the degree of detail which categorically allows identification of the sealing mechanism (s) or controls. Without the construction of a robust set of case histories from such analysis, future seal evaluation will remain a high-risk venture. These case histories are also needed to integrate seal behaviour with pressure test, production and *in situ* stress analysis. This paper has highlighted the importance of an integrated approach from micro to macro and stressed the value of core-based studies to quantify fault rock properties, sub-seismic fault populations and sealing mechanisms. The aim of this review has been to demonstrate that although a complex problem, there are techniques which can be, and should be, applied to fault seal analysis as they allow a clearer understanding, quantification and therefore predictability of the factors involved.

Many people helped to make the conference a success, particularly Jayne Harnett, Ned Porter, Liz White, Andy Farmer, Ewart Edwards and Kevin Leahy of RDR and the Conference Secretariat of the University of Leeds. We also thank the staff of the Geological Society and Bob Holdsworth of the Tectonic Studies Group for their contributions. Special thanks are due to the long list of people who generously spent time reviewing the manuscripts in this volume and of course to the authors for their efforts and patience. We particularly thank the following companies for financial and logistical support for the conference: Arco British, BP Norge UA, British Gas E & P, Chevron U.K., Conoco U.K., Elf Caledonia, Mobil U.K., Phillips Petroleum, Shell U.K. and Texaco U.K. Finally Alan Roberts is thanked for reviewing this introduction and, as editor of The Geological Society Special Publication Series, for suggesting inclusion of parts of our early review to help set the scene for the contents of this volume.

References

ALLAN, U. S. 1989. Model for hydrocarbon migration and entrapment within faulted structures. *American Association of Petroleum Geologists Bulletin*, **73**, 803–811.

ANTONELLINI, M. & AYDIN, A. 1994. Effect of faulting on fluid flow in porous sandstones: petrophysical properties. *American Association of Petroleum Geologists Bulletin*, **78**, 335–377.

AYDIN, A. 1978. Small faults formed as deformation bands in sandstone. *Pure and Applied Geophysics*, **116**, 913–942.

BERG, R. B. & AVERY, A. H. 1995. Sealing properties of Tertiary growth faults, Texas Gulf coast. *American Association of Petroleum Geologists Bulletin*, **79**, 375–393.

BOUVIER, J. D., KAARS-SIJPESTEIJN, C. H., KLUESNER, D. F., ONYEJEKWE, C. C. & VAN DER PAL, R. C. 1989. Three-dimensional seismic interpretation and fault sealing investigations, Nun River field, Nigeria. *American Association of Petroleum Geologists Bulletin*, **73**, 1397–1414.

CARTWRIGHT, J. A., MANSFIELD, C., & TRUDGILL, B. 1996. The growth of normal faults by segment linkage. *In*: BUCHANAN, P. G. & NIEUWLAND, D. A. (eds), *Modern Development in Structural Interpretation, Validation and Modelling*. Geological Society, London, Special Publications, **99**, 163–177.

CHESTER, F. M. & LOGAN, J. M. 1986. Implications for mechanical properties of brittle faults from observations of the Punchbowl fault zone, California. *Pure and Applied Geophysics*, **124**, 77–106.

COWARD, M. P., DALTABAN, T. S. & JOHNSON, H. (eds) 1998. *Structural Geology in Reservoir Characterisation*. Geological Society, London, Special Publications, **127**.

COWIE, P. A. & SCHOLZ, C. H. 1992. Displacement-length scaling relationships for faults: data synthesis and discussion. *Journal of Structural Geology*, **14**, 1149–1156.

—— VANNESTE, C. & SORNETTE, D. 1993. Statistical physics model for the spatio-temporal evolution of faults. *Journal of Geophysical Research*, **98**, 21 809–21 821.

—— KNIPE, R. J. & MAIN, I. G. 1996. Introduction to the Special Issue. Scaling Laws for fault and fracture populations – analyses and applications. *Journal of Structural Geology*, **18**, 135–383

ENGELDER, J. T. 1974. Cataclasis and the generation of fault gouge. *Geological Society of America Bulletin*, **85**, 1515–1522.

FOWLES, J. & BURLEY, S. D. 1994. Textural and permeability characteristics of faulted, high porosity sandstones. *Marine and Petroleum Geology*, **11**, 608–623.

FREEMAN, B., YIELDING, G. & BADLEY, M. 1989. Fault correlation during seismic interpretation. *First Break*, **8** (3), 87–95.

GAUTHIER, B. D. M. & LAKE, S. D. 1993. Probabilistic modelling of faults below the limit of seismic resolution in Pelican Field, North Sea, Offshore United Kingdom. *American Association of Petroleum Geologists Bulletin*, **77**, 761–777.

GIBSON, R. G. 1994. Fault zone seals in siliclastic strata of the Columbus Basin, Offshore Trinidad. *American Association of Petroleum Geologists Bulletin*, **78**, 1372–1385.

GIBSON, R. G. 1998. Physical character and fluid flow properties of sandstone-derived faults. *In*: COWARD, M. P., DALTABAN, T. S. & JOHNSON, H. (eds) *Structural Geology in Reservoir Characterisation*. Geological Society, London, Special Publications, **127**, 83–98.

GILLESPIE, P. A., HOWARD, D. W., HOLLOWAY, S. & HULBERT, A. G. 1993. Measurement and characterisation of spatial distribution of fractures. *Tectonophysics*, **226**, 113–141.

JEV, B. I., KARS-SIJPESTEIJN, C. H., PETERS, M. P. A. M., WATTS, N. L. & WHITE, J. T. 1993. Akaso field, Nigeria: use of integrated 3-D seismic, fault slicing, clay smearing, and RFT pressure data on fault trapping and dynamic leakage. *American Association of Petroleum Geologists Bulletin*, **77**, 1389–1404.

JONES, G. & KNIPE, R. J. 1996. Seismic attribute maps; application to structural interpretation and fault seal analysis in the North Sea Basin. *First Break*, **14** (12), 449–461.

KNIPE, R. J. 1989. Deformation mechanisms – recognition from natural tectonites. *Journal of Structural Geology*, **11**, 127–146.

—— 1992*a*. Faulting processes and fault seal. *In*: LARSEN, R. M., BREKKE, H., LARSEN, B. T. & TALLERAAS, E. (eds) *Structural and Tectonic Modelling and its application to Petroleum Geology*. NPF Special Publication **1**, Elsevier, Stavanger, 325–342.

—— 1992*b*. Faulting processes, seal evolution and reservoir discontinuities: an integrated analysis of the Ula Field, Central Graben, N. Sea. *Abstracts of the Petroleum Group Meeting on Collaborative Research Programmes in Petroleum Geoscience between U.K. Higher Education Institutions and the Petroleum Industry*. Geological Society, London.

—— 1993*a*. The influence of fault zone processes and diagenesis on fluid flow. *In*: HORBURY, A. D. & ROBINSON, A. G. (eds) *Diagenesis and Basin Development*. AAPG Studies in Geology, Tulsa, OK, **36**, 135–154.

—— 1993*b*. Micromechanisms of deformation and fluid behaviour during faulting. *In*: HICKMAN, S., SIBSON, R. & BRAHN, R. (eds) *The Mechanical Involvement of Fluids in Faulting*. USGS Open-File Report, **94–228**, 301–310.

—— 1994. Fault zone geometry and behaviour; the importance of damage zone evolution. *Abstracts of Meetings Modern Developments in Structural Interpretation*. Geological Society, London.

—— & LLOYD, G. E. 1994. Microstructural analysis of faulting in quartzite, Assynt, NW Scotland: Implications for fault zone evolution. *Pure and Applied Geophysics*, **143**, 229–254.

—— 1997, Juxtaposition and seal diagrams to help analyze fault seals in hydrocarbon reservoirs. *American Association of Petroleum Geologists Bulletin*, **81** (2). 187–195.

—— FISHER, Q. J., JONES, G., CLENNELL, M. R., FARMER, A. B., HARRISON, A., KIDD, B., MCALLISTER, E., PORTER, J. R. & WHITE, E. A. 1997. Fault seal analysis: successful methodologies, application and future directions. *In*: P. Moller-Pedersen & A. G. Koestler (eds), *Hydrocarbon Seals: Importance for Exploration and Production*. NPF Special Publication, **7**, 15–40.

KNOTT, S. D. 1993. Fault seal analysis in the North Sea. *American Association of Petroleum Geologists Bulletin*, **77**, 778–792.

LEVEILLE, G. P., KNIPE, R. J., MORE, C., ELLIS, D., DUDLEY, G., JONES, G. & FISHER, Q. J. 1997. Compartmentalisation of Rotliegendes gas reservoirs by sealing faults, jupiter area, southern North Sea. *In*: ZIEGLER, K., TURNER, P. & DAINES, S. R. (eds) *Petroleum Geology of the Southern North Sea: Future Potential*. Geological Society, London, Special Publications, **123**, 87–104.

MCGRATH, A. & DAVISON, I. 1995. Damage zone geometry around fault tips. *Journal of Structural Geology*, **17**, 1011–1024.

MITRA, S. 1988. Effects of deformation mechanisms on reservoir potential in central Appalachian overthrust belt. *American Association of Petroleum Geologists Bulletin*, **72**, 536–554.

MØLLER-PEDERSEN, P. & KOESTLER, A. G. (eds) 1997. *Hydrocarbon Seals: Importance for Exploration and Production*. NPF Special Publication, **7**, Elsevier, Singapore.

NEEDHAM, D. T., YIELDING, G. & FREEMAN, B. 1996. Analysis of fault geometry and displacement patterns. *In*: BUCHANAN, P. G. & NIEUWLAND, D. A. (eds) *Modern Developments in Structural Interpretation Validation and Modelling*. Geological Society, London, Special Publications, **99**, 189–200.

PEACOCK, D. C. P. & SANDERSON, D. J. 1994. Geometry and development of relay ramps in normal fault systems. *American Association of Petroleum Geologists Bulletin*, **78**, 147–165.

PITTMAN, E. D. 1981. Effect of fault-related granulation on porosity and permeability of quartz sandstones, Simpson Group (Ordovician), Tulsa, OK. *American Association of Petroleum Geologists Bulletin*, **65**, 2381–2387.

SCHOLZ, C. H. 1989. Mechanics of Faulting. *Annual Revue of Earth and Planetary Sciences*, **17**, 309–334.

SCHOWALTER, T. T. 1979. Mechanisms of secondary hydrocarbon migration and entrapment. *American Association of Petroleum Geologists Bulletin*, **63**, 723–760.

SIBSON, R. H. 1994. Crustal stress, faulting and fluid flow. *In*: PARNELL, J. (ed), *Geofluids: Origin, Migration and Evolution of Fluids in Sedimentary Basins. American Association of Petroleum Geologists Bulletin*, **78**, 69–84.

SMITH, D. A. 1966. Theoretical consideration of sealing and non-sealing faults. *American Association of Petroleum Geologists Bulletin*, **50**, 363–374.

—— 1980. Sealing and non-sealing faults in Louisiana Gulf Coast salt basin. *American Association of Petroleum Geologists Bulletin*, **64**, 15–172.

SVERDRUP, E. & BJØRLYKKE, K. 1992. Small faults in sandstones from Spitsbergen and Haltenbanken. A study of diagenetic and deformational structures and their relation to fluid flow. *In*: R. M. LARSEN, H. BREKKE, B. T. LARSEN & E. TALLERAAS (eds), *Structural and Tectonic Modelling and its Application to Petroleum Geology*. NPF Special Publication **1**, Elsevier, Amsterdam, 507–518.

UNDERHILL, J. R. & WOODCOCK, N. H. 1987. Faulting mechanisms in high porosity sandstones; New Red Sandstone, Arran, Scotland. *In*: JONES, M. E. & PRESTON, R. M. F. (eds), *Deformation of Sediments and Sedimentary Rocks*. Geological Society Special Publications, **29**, 91–105.

WALLACE, R. E. & MORRIS. H. T. 1986. Characteristics of faults and shear zones in deep mines. *Pure and Applied Geophysics*, **124**, 107–125.

WALSH, J. J. & WATTERSON, J. 1991. Geometric and Kinematic coherence and scale effects in normal fault systems. *In*: ROBERTS, A. M., YIELDING, G. & FREEMAN, B. (eds) *The Geometry of Normal Faults*. Geological Society, London, Special Publications, **56**, 193–203.

—— & WATTERSON, J. 1992. Populations of faults and fault displacements and their effects on estimates of fault-related regional extension. *Journal of Structural Geology*, **14**, 701–712.

WATTS, N. L. 1987. Theoretical aspects of cap-rock and fault seals for single and two-phase hydrocarbon columns. *Marine and Petroleum Geology*, **4**, 274–307.

YIELDING, G., FREEMAN, G. & NEEDHAM, B. 1997. Quantitative fault seal prediction. *American Association of Petroleum Geologists Bulletin*, **81**, 897–917.

Small seismic-scale fault identification and mapping

C. TOWNSEND[1], I. R. FIRTH[2*], R. WESTERMAN[3†], L. KIRKEVOLLEN[2], M. HÅRDE[2] & T. ANDERSEN[2]

[1] *Statoil, Statoil's Research Centre, Postuttak, N-7005 Trondheim, Norway*
[2] *Statoil, Hovedkontoret, N-4035 Stavanger, Norway*
[3] *Seistran Ltd, 29 St David's Drive, Doncaster, DN5 8PW, UK*
[*] *Present address: Gaffney, Cline & Associates Ltd, Bentley Hill, Blacknest, Alton, Hants GU34 4PU, UK*
[†] *Present address: Department of Petroleum Engineering, Heriot-Watt University, Enterprise Oil Building, Riccarton, Edinburgh EH14 4AS, UK*

Abstract: The primary focus of this paper is to emphasise the large volumes of information related to faults and fault systems that are present in, and extractable from, 3D seismic data. During most interpretations, this information is seldom included and transferred to the reservoir model so that their effects can be accounted for during reservoir simulation. Reasons why they are not included are generally related to the time constraints imposed on studies when commercial considerations are often given precedence above any scientific justification. The inadequacies of what are presently considered acceptable models are highlighted and methodologies that could lead to improved reservoir models are proposed. These methodologies are derived from an investigation of how fault systems will manifest themselves in seismic data, based on both theoretical concepts and the use of synthetic models. From these information sources, principles are derived for the identification of faults and fault systems in 3D seismic volumes. These principles are then tested in two case studies selected to emphasize the limitations imposed on seismic resolution by both target depth and seismic frequency content. After highlighting the diversity of fault related information, which is accessible and currently under-utilized by current reservoir modelling techniques, potential methods for automatic mapping and digital extraction of fault information are proposed. When these automated methods are implemented into seismic interpretation software, commercial reasons for ignoring small seismic-scale faults will be significantly reduced and as a result, reservoir performance predictions will become more realistic.

Seismic data often contain large amounts of geological information that are rarely utilized during normal interpretation studies (see for example Henriquez & Jourdan 1995). This not only includes indications of fluid content, lithology and sedimentary heterogeneities, but also small seismic-scale faults. Traditionally, faults within hydrocarbon reservoirs have been divided into 2 types (see for example Gauthier & Lake 1993); seismically resolvable faults and sub-seismic faults. The seismic-scale faults should be identified during seismic interpretation, providing time allows, whereas little information about the sub-seismic-scale faults can be collected (except perhaps for a few fault intersections seen in wells).

The exact location of the boundary between these two fault types cannot be defined in absolute terms, but varies between seismic datasets because it is dependant on frequency content and signal to noise ratio. In most cases, however, there are many more seismically resolvable faults within a 3D dataset than are normally mapped during the 'seismic interpretation'. In addition seismic data also contains indications, which are rarely utilized, of faults whose displacements lie close the limits of seismic resolution.

A note of caution is required at this point; the magnitude of a fault's displacement changes across its surface and as it approaches the limits of seismic resolution, the criteria which are used to recognize seismic anomalies generated by faults become increasingly difficult to apply. Therefore, the degree of confidence that can be placed in the interpretation of the fault is also reduced. The exact location of the limit of seismic fault resolution for a given dataset must therefore be defined in two ways:

(1) on theoretical grounds, dependent on the frequency content, signal to noise ratio and acquisition-processing characteristics of the data;
(2) on interpretative grounds related to the criteria that are required for classifying features as potential faults.

Faults commonly give rise to problems when predicting reservoir performance. This may be because their sealing capacities are not fully

Townsend, C., Firth, I. R., Westerman, R. *et al.* 1998. Small seismic-scale fault identification and mapping. *In*: Jones, G., Fisher, Q. J. & Knipe, R. J. (eds) *Faulting, Fault Sealing and Fluid Flow in Hydrocarbon Reservoirs.* Geological Society, London, Special Publications, **147**, 1–25.

understood, their geometry has been poorly described, or the smaller sizes of the fault populations have not been accounted for in the simulation. The first mentioned problem can rarely be solved using seismic data and will not be considered further in this paper. The latter two problems can be addressed if faults can be mapped more accurately. To achieve more accurate representations of fault systems from interpretation studies, it is important for interpreters to develop a sound understanding of the geometry of fault systems. Furthermore, in order to recognize fault systems within a seismic dataset it is also necessary to understand how they manifest themselves within the data and how they influence it. If the second of these two constraints can be fully understood, then it should be possible to design a number of fault mapping principles. These same guidelines should also be adaptable for the mapping of small-scale seismic faults that lie close to the data resolution limits. Ideally, if these principles are sufficiently robust, then it may also be possible to design automatic mapping tools that can follow faults within the seismic data cube.

At present, a common method for accounting for the smaller sizes of the fault populations is modelling. This usually includes the use of stochastic techniques, which produce statistically realistic fault objects, but the generated fault patterns usually diverge significantly from observed examples. Poorly constrained fault locations can generate large variations between realizations of a single stochastic model. There are two obvious ways of reducing this problem, by either:

(i) improving the understanding of fault patterns in order that stochastic models can become more realistic; or
(ii) extracting and utilising the maximum amount of fault information from within seismic data cubes.

The latter option will only be effective for those faults lying close to the limits of seismic resolution. If these resolution limits can be lowered by improving our understanding of both faults and seismic data, this will lead to a significant reduction in the number of fault objects to be modelled using stochastic techniques.

A number of recently published papers have attempted to address the question of how seismic attributes can be used to identify faults. Hesthammer & Fossen (1997*a*,*b*) have used seismic attributes (mainly illuminated dip maps) to identify faults from the Gullfaks Field. They have put a significant effort into highlighting the possibility that 'seismic noise' could be responsible for some of the linear features found within their dataset, arguing that there were significant pitfalls in interpreting them as faults. Although Hesthammer & Fossen (1997*b*) presented convincing evidence for the presence of linear noise trains within their dataset they failed to explain what criteria they had used to uniquely differentiate faults from noise. This is a serious issue, since many of the features they had classified as 'noise' exhibit the same attribute characteristics as the features which had been interpreted as faults. A further area of controversy, with the suggestions of Hesthammer & Fossen, lies in the absence of an explanation of why the noise generating artefacts within their dataset should result in both amplitude anomalies and offsets along seismic reflectors. Jones & Knipe (1997) have shown how attribute maps can be used in some cases to decipher some of the complexities associated with fault zones. They implied that most seismic fault 'surfaces' have similar complexities at a scale beneath the resolution of data. In this volume, Steen *et al.* (1998) have analysed dip, azimuth and dip change from a faulted outcrop example and have applied the techniques to the analysis of real seismic attribute maps. In this paper, the use of dip change has not been considered, but it appears to be a promising attribute for fault identification, as it tends to highlight faults and dampens smaller background features.

This paper considers a number of methods for fault identification, which when pushed to their limits, demonstrate that fault resolvability from seismic data is often considerably below that of the smallest faults which are usually mapped during seismic interpretation. The approach taken is to consider:

(i) how faults manifest themselves in seismic data;
(ii) from this evaluation, develop principles for fault identification; and
(iii) maximize the chance of a fault being identified by optimizing visualization techniques.

This is carried out by first considering seismic theory and then testing the validity of the proposed fault identification principles by synthetic seismic experiments. These tested principles are subsequently used to identify small faults in two seismic datasets with contrasting characteristics. The final part of the paper discusses how methods for fault information extraction can be automated for fast and accurate mapping.

Methods for fault identification in section

During the 1980s, techniques were developed in the coal mining industry for the identification of small seismic-scale faults on 2D seismic sections;

unfortunately very little of this work was ever published (Fairburn *et al.* 1988 is a rare example). The faults being mapped lay close to the limits of seismic resolution, therefore the assumptions regarding how faults would manifest themselves in seismic data had to be pushed towards their limits in order to resolve the smallest possible structures. The assumptions made (out-lined below) were, however, uncontroversial, as they were based on structural geological observations and sound geophysical theory.

Faults displace geological layers, therefore the most obvious method for their identification is to observe offsets of reflection events. This is often helped when several strong reflection events occur close together allowing a fault's continuity to be recognized. In addition to displacing horizons, faults frequently deform the surrounding strata, which often results in changes to the orientation of the layering. This will be manifested in cross-section as a change in apparent dip.

Faulted reflectors will generate an amplitude anomaly at fault locations (Sheriff 1981; Badley 1985; Fairburn *et al.* 1988). A full discussion of the physics of seismic wave propagation, which would be required to fully explain why fault terminations are associated with amplitude anomalies, is beyond the scope of this paper. However, the following simplistic description may be of help to the less geophysically minded reader.

The process of multi-channel seismic imaging is usually visualized as a one-dimensional incident ray travelling from the source to a reflector and a single reflected ray travelling up to the receiver (Fig. 1a). Although this is a good conceptual representation of the process, in reality the seismic source generates a spherical zone of disturbance that propagates out in all directions from the source. This phenomenon is known as a wavefront. When this wavefront interacts with a reflector at the location of a CDP, a complex waveform, propagating three dimensionally away from the reflector returns (and not the simple one-dimensional reflected ray). As the incident waveform is three-dimensional, the area across which it interacts with the reflector is not a single point, but a circular area known as a Fresnel zone. A simple analogy to a Fresnel zone is the circular print left when a wet ball is bounced off a smooth, dry, absorbent surface.

The largest contributor to the energy of the reflected wavefront is the first Fresnel zone, the radius of which can be approximately estimated from the following expression (see McQuillian *et al.* 1984):

$$f = \left(\frac{\lambda h}{2}\right)^{1/2}$$

where, f is the radius of the first Fresnel zone, λ is the wavelength of the incident wavefront and h is the distance from the source to reflector (Fig. 1b). From the above expression it is clear that the area with which a reflected wavefront has interacted increases with either depth or lower frequency content (i.e. longer wavelength). This principle is the primary theoretical control on the lateral resolution of seismic data.

At fault locations, where the reflector is abruptly terminated, the interaction between the propagating wavefront and the reflector is further complicated. The segment of the Fresnel zone located along the continuous reflector behaves normally as described above. However at the location of the fault, a further wavefront is generated which moves spherically away from the point of termination; this is called the diffracted wave. Such waves are commonly observed producing hyperbolic diffraction patterns on stacked seismic sections. The diffracted waves undergo more rapid amplitude and frequency decay with increasing travel time than the normally reflected wavefront. Although the migration process attempts to collapse the diffracted energy back to its point of origin, the amplitudes recorded at the fault location are lower than those recorded from the continual reflector (Fig. 1c & d). The width of the amplitude anomaly will increase with increasing Fresnel zone radius. As stated above, this invariably increases with depth due to the progressive attenuation of higher seismic frequencies. When faults with different throws are compared, the amplitude reduction should be greatest for larger faults (Badley 1985; see Fig. 2). For individual faults with variable displacements, the magnitude of the amplitude anomaly should decrease towards the fault tip (Fig. 2).

The smallest fault that can be identified from an associated amplitude anomaly will generate a reflection strength perturbation, which can be statistically separated from the background variability of a particular reflection event. Fault resolution using amplitude anomalies will therefore be dependent upon the strength of the reflection event, the geological variability of the surface it represents and the amount of ambient noise.

The recommended method for the identification of small scale faults using attribute anomalies has 3 basic components:

(1) Generate maps in the area of interest by automatically tracking reflection events by following the maximum deflection of the peak or trough. Routines for autotracking

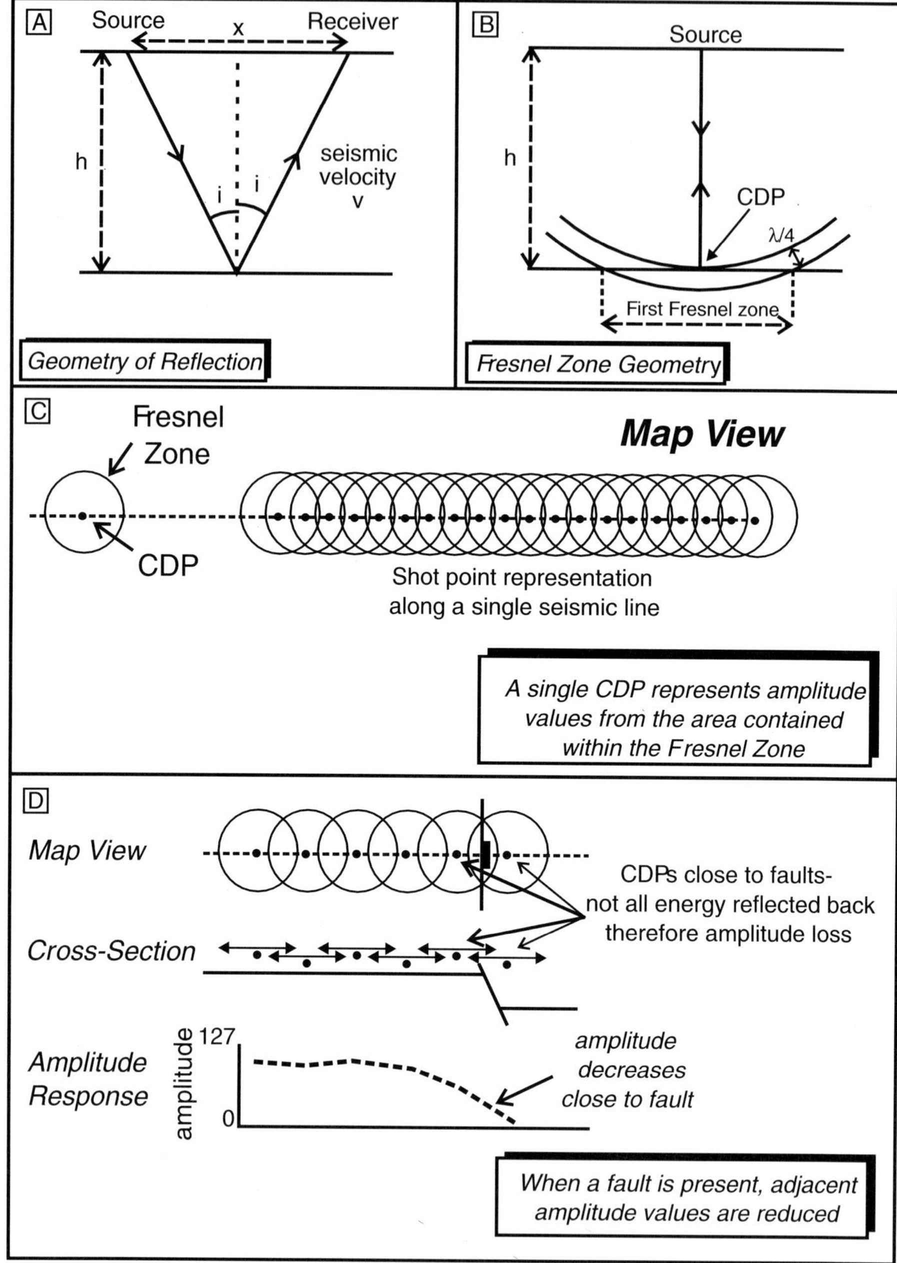

Fig. 1. Theory behind how amplitude values are affected by faults: (**a**) Ray path model for multi-channel seismic acquisition, a one dimensional ray travels from the source to the reflector, where it is reflected back to the receiver at the same angle as the incident ray (i). (**b**) The wavefront model for seismic reflection, illustrated as a two dimensional cross-section through a zero-incident wavefront, the radius of the first Fresnel zone is a function of frequency (a quarter of the wavelength) and the vertical distance between the source and reflector (h). (**c**) Map view of a seismic line, each CDP collects energy from the circular region defined by the Fresnel zone; the size of the Fresnel zone usually increases with depth. (**d**) When a fault is present, not all the energy from the Fresnel zones which span both sides of the fault, is reflected back to the receivers. This leads to a reduction in amplitude values around the fault: the width of the zone of reduced amplitudes will increase with increasing Fresnel zone size.

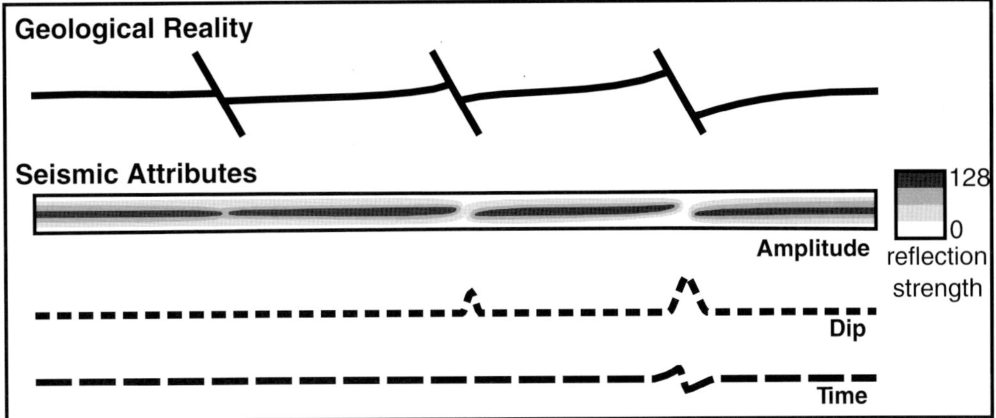

Fig. 2. An illustration of how amplitude reduction is expected to increase with increasing fault throw (adapted from Badley 1985). The figure also illustrates how dip values are expected to increase and how time values are expected to change. From experience, and as demonstrated later in this paper, for larger faults a change in time, dip and amplitude is expected, but as fault throw decreases observable time changes disappear. This is followed by dip changes, leaving only amplitude anomalies present when throws are at their smallest.

events are available on all commercial interpretation workstations. The tracking must be completed for every CDP within the area of interest. Attribute maps (e.g. amplitude, dip and correlation) can then be generated over the area of interest and linear anomalies should be selected for further examination.

(2) Attribute anomalies identified during step 1 as potential fault locations can then be further analysed on the seismic sections. Extreme vertical exaggeration should be used. When a fault is present, this will highlight the slightest offset of a reflector and emphasize subtle changes in apparent dip.

(3) Amplitude changes along reflections on seismic sections can be identified by the use of a colour palette that will react to small changes in reflectivity. A good colour scheme will be case-dependent and may also be subject to personal preference. However, a large number of colours are desirable and avoiding hard edges between them will help lead to an optimum palette. An optimum palette design should be based upon a histogram of amplitude values for the reflector of interest where the colours are concentrated around the most commonly occurring amplitude values.

Experience from the interpretation of 2D coalfield seismic data has shown that amplitude anomalies resolve the smallest faults, followed by apparent dip changes and then small offsets (see Fig. 2). When considering a fault with variable displacement, where the displacement is largest, all three components will be present (providing the maximum displacement is large enough). Moving towards the fault tip, the offset will first disappear, followed by dip change, leaving just an amplitude anomaly at the seismically resolvable tip.

It is by no means intended that the criteria proposed above for fault recognition are anything other than industry established, principles of best practice. Indeed it is fully accepted that the use of sophisticated 3D visualization and image processing techniques are currently becoming 'run of the mill' fault identification procedures. What this paper emphasizes is that despite certain developments in the methodologies for visualization and identification of faults, no real advances have been made in digitally capturing and exporting this information. Moreover, when pushed to their limits, some techniques can resolve many small features that could be seismically detectable faults (although with an uncertainty attached). The characteristics of faults that are focused upon in this paper are properties that are inherently associated with seismic data, which with minor modifications of existing software, could be automatically tracked and subsequently included in reservoir models.

Seismic modelling

2D seismic modelling has been carried out in order to confirm that the proposed methodologies for fault identification using seismic attributes are consistent with seismic theory and that the assumptions are reproducible. Geological models were generated using stochastic modelling software packages which are capable of producing complex, but realistic 3D models. These models were used as geological input to sophisticated finite difference seismic modelling which was used to produce synthetic seismic sections. Although the input to the experiments presented here were relatively simple, the system was designed to account for increasing levels of geological complexity.

The input for the seismic modelling was a simple layered model generated using the STORM™ package. It comprised of a homogeneous, *c.* 1 km thick, overburden overlying 6 'reservoir' layers; these 6 layers were thick enough to be clearly resolved (30–50 m), internally homogeneous and had sufficient density and velocity contrasts to generate clear reflections. The layered model was faulted using the HAVANA™ stochastic fault modelling software (Munthe *et al.* 1993, 1994). A single fault was placed at the same location in each model with its centre lying within (close to horizon 3) and extending through the 'reservoir' sequence. The fault had a dip of 60°. Three experiments were carried out where the maximum displacement of the fault was varied between 5, 10 and 15 m; in each case the maximum displacement of the fault lay at its centre and decreased linearly towards its tips. Faults of this size are generally considered to lie at the limits of seismic resolution.

The synthetic seismic sections were generated using a zero incidence finite difference simulator. The input geological model and its velocity–density characteristics are shown in Table 1. The simulator used in this instance was restricted to the use of symmetrical grid cells. In order to simulate faults with throws of less than 5 m, a 2 m vertical grid cell size was chosen. Therefore the CDP spacing was also defined as 2 m due to the symmetrical grid cells. A CDP spacing of 2 m is clearly incompatible with conventional 3D seismic data (usually 12.5 m); however, this does not detract from the validity of the results. As discussed above, the horizontal resolution of a seismic experiment is not defined by the CDP spacing but by the radius of the Fresnel zone. The latter is dependent upon the depth to the reflectors and the frequency content of the seismic source. As can be seen from Table 1, the frequency content of the input wavelet (40 Hz) is comparable to commercial seismic, as are the depths to the reflectors. The initial output of the model was a simulated zero-offset stacked section with a CDP spacing of 2 m. The stacked section was then re-sampled to 2 ms.

Table 1. *Synthetic seismic modelling input and processing parameters*

Layer	Thickness (m)	Velocity (m/s)	Density (kg/m^3)
1	900	1825–2154	1835–2134
2	50	2 350	2 080
3	30	2 400	2 150
4	50	2 500	2 130
5	30	2 600	2 200
6	90	2 750	2 180
7	350	2 800	2 250

Initial simulated stack	
Source:	40 Hz Zero phase Ricker wavelet
Reciever and source interval:	2m
CDP interval:	2m
Record length:	1.5 s
Sample rate:	0.25 ms
Resampled	2 ms
FD Migration:	Using input velocity field

and migrated using a finite difference migration algorithm and the input velocity model. No coherent or random noise was introduced during the finite difference modelling because the aim of the experiments was to find out how the very smallest seismic-scale faults manifest themselves in seismic data. The premise being that if faults cannot be seen without noise then they are unlikely to be observed with noise present.

A CHARISMA™ workstation was used for interpretation of the seismic models; the 2D sections were visualized using the colour manipulation techniques described above for highlighting amplitude anomalies, and the 6 horizons were picked using autotracking tools. Amplitude and time values for the autotracked horizons were extracted and plotted separately as line-graphs. A dip angle line-graph was derived from the time values. The advantages of using line-graphs include the easy comparison of adjacent values and the identification of subtle changes by using extreme extension of the vertical axis.

An unfaulted model was run as a control to make sure that there was no variability generated by the seismic method and that no fault-like features were present. This unfaulted model showed that there was less than 1% variation, in both time and amplitude, along each of the reflectors and obvious fault-like features were absent.

The faulted models showed that displacement could only be resolved in seismic section for the 15 m model, whereas an amplitude anomaly is present for the 10 and 15 m faults (see Fig. 3). For the 5 m fault, changes occurred on the seismic section, however, the identification of a fault is very difficult. The line graphs, displayed with extremely exaggerated vertical axes also show the anomalies (Fig. 4), but they are much clearer. The response of the 5 m fault is also better distinguished as a small offset on the time plot and as a trough on the amplitude plot. However, it is unlikely that such small anomalies will be distinguished from the variability in amplitude of a reflector from a real dataset. The dip plot also shows a significant anomaly at the location of the fault for displacements of 10 and 15 m (Fig. 4), whereas for the 5 m fault, it is indistinguishable from the noise generated by the fault.

It is notable from the line graphs that the amplitude anomaly is much wider than the sharply defined dip anomaly (Fig. 4). This has significant implications for attribute maps as a fault related amplitude anomaly will generally form a broad linear feature. On the other hand the dip anomaly will be much narrower, which should make it appear as a clearly defined lineation on dip maps. Therefore in general terms, dip should define faults more easily on attribute maps because of the sharper definition, whereas amplitude will resolve smaller faults.

These experiments, although simple geologically, show that the basic practices proposed here are consistent with observations from seismic modelling and predictions based on seismic theory. They also show that amplitude anomalies can resolve smaller faults in seismic section than either displacement or apparent dip. Moreover, the line-graphs proved to be extremely useful in resolving small offsets when used with an exaggerated vertical axis.

Fault identification on attribute maps

Having established the characteristics expected of small-scale faults on 2D seismic sections, these can now be extrapolated into the third dimension to see how they are portrayed on horizon attribute maps. Any such extrapolation should take into consideration that faults are near-planar structures which are expected to manifest themselves as linear features on maps when they intersect an horizon; the principle of 2 intersecting planes (see Hobbs *et al.* 1976). In addition, faults can be identified on maps by considering that they possess geometrical properties which can be observed from oucrops, analogues and other seismic data. These features include variable displacement (usually a maximum displacement close to the fault centre and gradually decreasing towards its tip line) and interaction with adjacent faults such as fault splitting, fault truncation and fault linkage (soft and hard).

(1) A fault-related amplitude anomaly should show up as a linear feature of low values on the amplitude anomaly map. In order to display faults on an amplitude map, low value linear features need to be highlighted. One drawback to use of amplitude maps is that if the Fresnel zones are large then fault generated anomalies can be wide, resulting in poor definition. Consequently, faults can sometimes be difficult to identify on amplitude maps. Another problem lies in areas of closely spaced faults where anomalies overlap; in such cases their identification as individual structures is especially difficult and the whole faulted zone tends to be a blur of low amplitude values. However, despite these shortcomings, the amplitude map can be used in a positive manner. Areas of continuous and relatively high amplitude values devoid of linear features can be considered to be unfaulted with respect to faults of a size large enough to generate amplitude anomalies.
(2) Apparent dip changes across a fault that can be seen in cross-section will be difficult to identify on dip attribute maps because they are invariably subtle. However, a dip attribute map should generate sharply defined high dip zones where faults intersect an horizon. This occurs because of slight offsets in the horizon, which in turn leads to an increased angle between adjacent CDP's. This results in a sharp linear high-dip anomaly being formed along the length of a fault. In order to visualize these high dip zones, a display should be used which emphasizes high values and dampens the lower background values. High dip features can also be highlighted by the use of directional illumination (see Fig. 5); although this is effective for identifying features perpendicular to the illumination direction it will introduce a bias into any map as it dampens those features lying parallel to this direction. Rotating the illumination direction and generating several maps will partially solve this problem; however, at the same time it will significantly increase the required interpretation time.

It is not unusual for a linear dip anomaly to be present in the dip attribute map that has not previously been identified as a fault on seismic sections. This is because of its high degree of continuity, which attracts the

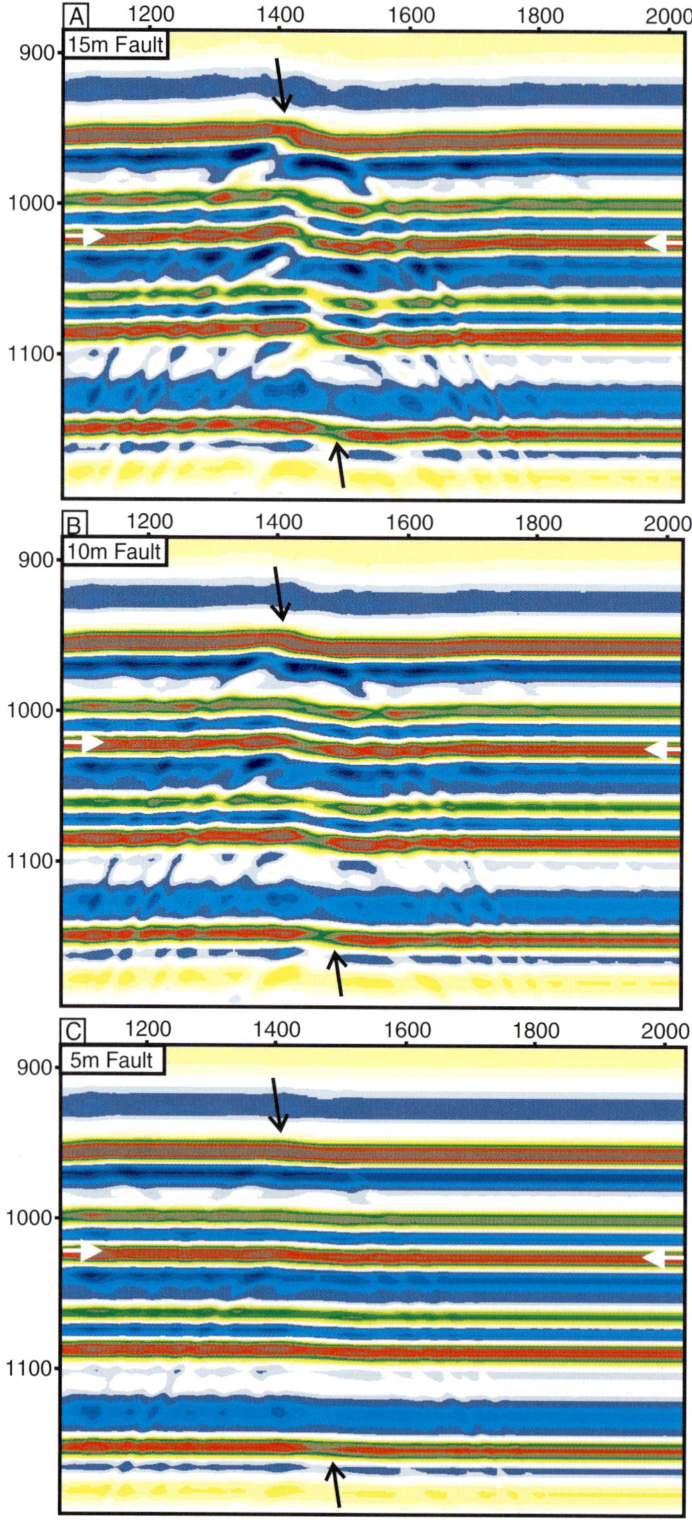
A
1200
1400
1600
1800
2000
15m Fault
900
1000
1100
B
1200
1400
1600
1800
2000
10m Fault
900
1000
1100
C
1200
1400
1600
1800
2000
5m Fault
900
1000
1100

human eye. However, in our experience once a linear dip anomaly has been identified, a second examination of the seismic sections shows that a previously unrecognized sharp amplitude reduction is present at the same location.

(3) Fault displacements should also manifest themselves on time structure maps. However, time maps are rarely used for fault identification because any single time structure map will have a relief in the order of one to two orders of magnitude greater than the throw on the smallest resolvable faults. When a colour palette is attributed to a time structure map it will usually cover the whole range of the relief and therefore subtle variations created by faults will not be recognized. This can be improved by using shaded relief in the display; however, this will only highlight features perpendicular to the lighting direction, therefore multiple map generation is required.

Other attribute maps, which can help with structural interpretation, are the azimuth and correlation maps. The azimuth map is good for highlighting significant changes in dip direction, which can sometimes be caused by faults. However, a faulted surface with azimuth changes is often indistinguishable from one which is folded. The azimuth map also suffers from a large range (i.e. 360); a colour palette distributed over this range will not be able to highlight subtle changes. A more obvious, but rarely used, structural map is the strike attribute map; as the range is half that of the azimuth map, subtle changes are somewhat easier to identify using a colour display. An alternative display of the strike attribute map would use lines to represent directions rather than colour, in some cases this will attract the human eye to subtle changes caused by faulting.

CHARISMA's™ correlation map compares adjacent seismic traces over the area covered by a reflection event. Where faults occur, adjacent traces will have a poor correlation. This is because faults generate amplitude anomalies at their points of dislocation and therefore traces close to faults will be subject to change. Although this highlights linear anomalies which are often fault related, it rarely highlights features which cannot otherwise be seen on either the dip or amplitude attribute maps. Like the amplitude map, it also suffers from a lack of definition because it is essentially related to the Fresnel zone effect as outlined earlier. However, the main advantage of the correlation map is that it can also be used on a time-slice without first having to interpret reflection events.

Application of techniques

The techniques outlined earlier have been applied to two separate 3D seismic datasets from the North Sea: the Siri Fault Zone and the Gullveig Field. The Siri Fault Zone is a very recent, salt-tectonic related normal fault system that extends near east–west through Block 5604/20. This block is located in the centre of the Norwegian–Danish Basin and is close to the northern margin of the Danish sector of the North Sea. The fault zone extends upwards and almost reaches the sea floor. Therefore in its shallower section, it is imaged by very high frequency seismic data. The faulted section considered from this area is at *c.* 0.9 s TWT and is of latest Miocene age.

The Gullveig Field is a satellite to the Gullfaks Field. It is situated in an area referred to as the Tampen Spur that forms the western margin to the North Viking Graben on the Norwegian continental shelf. The reservoirs are of Mid Jurassic age, belonging largely to the Brent Group. Ness1 is one of the prominent reflectors in this region. It lies within the Brent Group and is the event concentrated on in this example. In the area of interest, the Ness1 reflector (an event related to lower Ness Fm coals) lies at *c.* 2.5 s TWT.

The two datasets have been chosen to illustrate the influences of frequency content on seismic fault resolution. The Siri dataset has been zero phased and spectrally whitened, resulting in a flat amplitude spectrum at 2 s TWT with a bandwidth of 8 to 70 Hz. Wavelets extracted at 2 s TWT have a wavelength of *c.* 50 ms. The Gullveig dataset is mixed phase and has an

Fig. 3. Synthetic seismic sections from faulted models with varying fault displacements (see text for details): (**a**) 15 m displacement, (**b**) 10 m displacement, and (**c**) 5 m displacement. One fault is present in each model (marked by black arrows), it is has the same location in each and it dips towards the right. The displacement is at a maximum at horizon 3 (marked by white arrows) and decreases linearly towards its tips. For all 3 models the seismic has clearly been altered by the presence of the faults. Only subtle amplitude changes can be observed in relation to the 5 m fault, whereas for the 10 m fault the amplitude reduction is much clearer and for the 15 m fault both amplitude reduction and reflector offset can be seen.

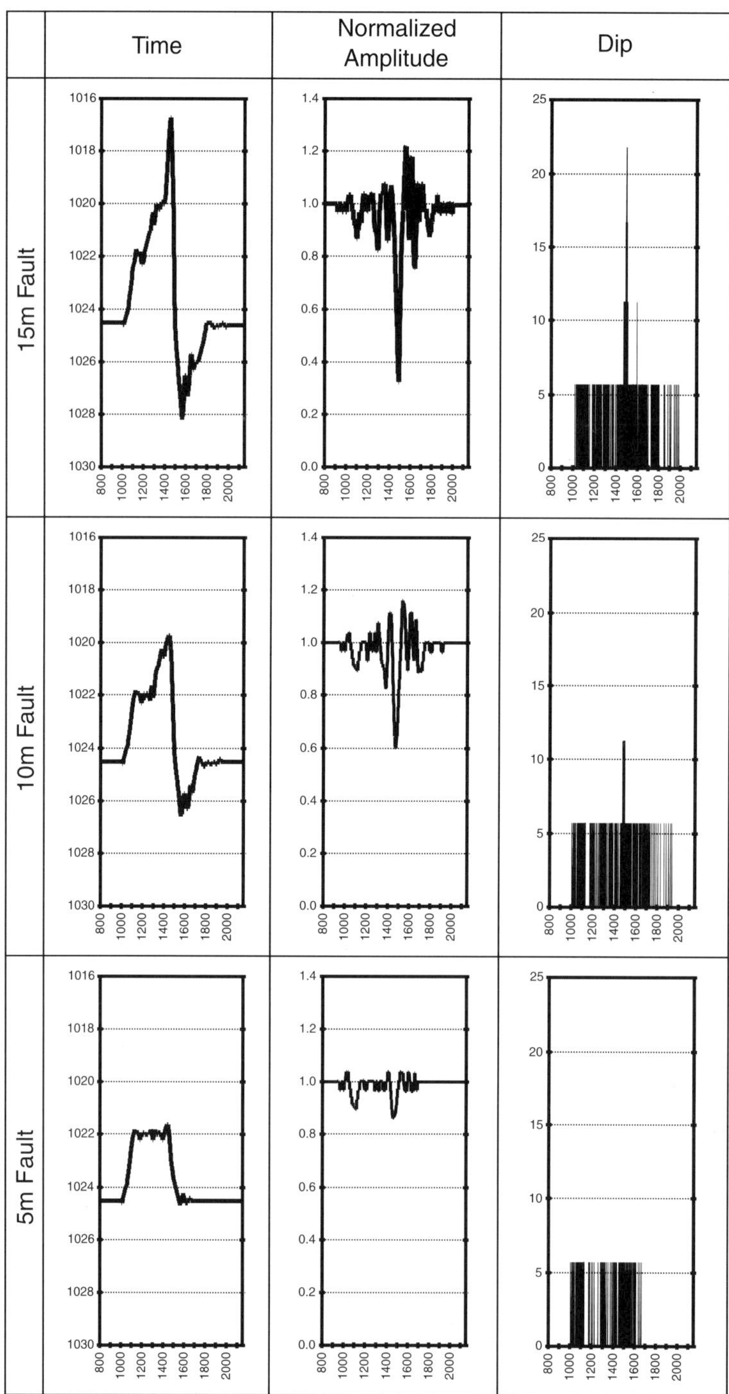

Fig. 4. Line graphs plotting time, normalized amplitude and dip values for horizon 3 from the 3 synthetic seismic sections in Fig. 3. These line graphs help to locate the subtle faults in the models, especially when extreme vertical exaggeration is used.

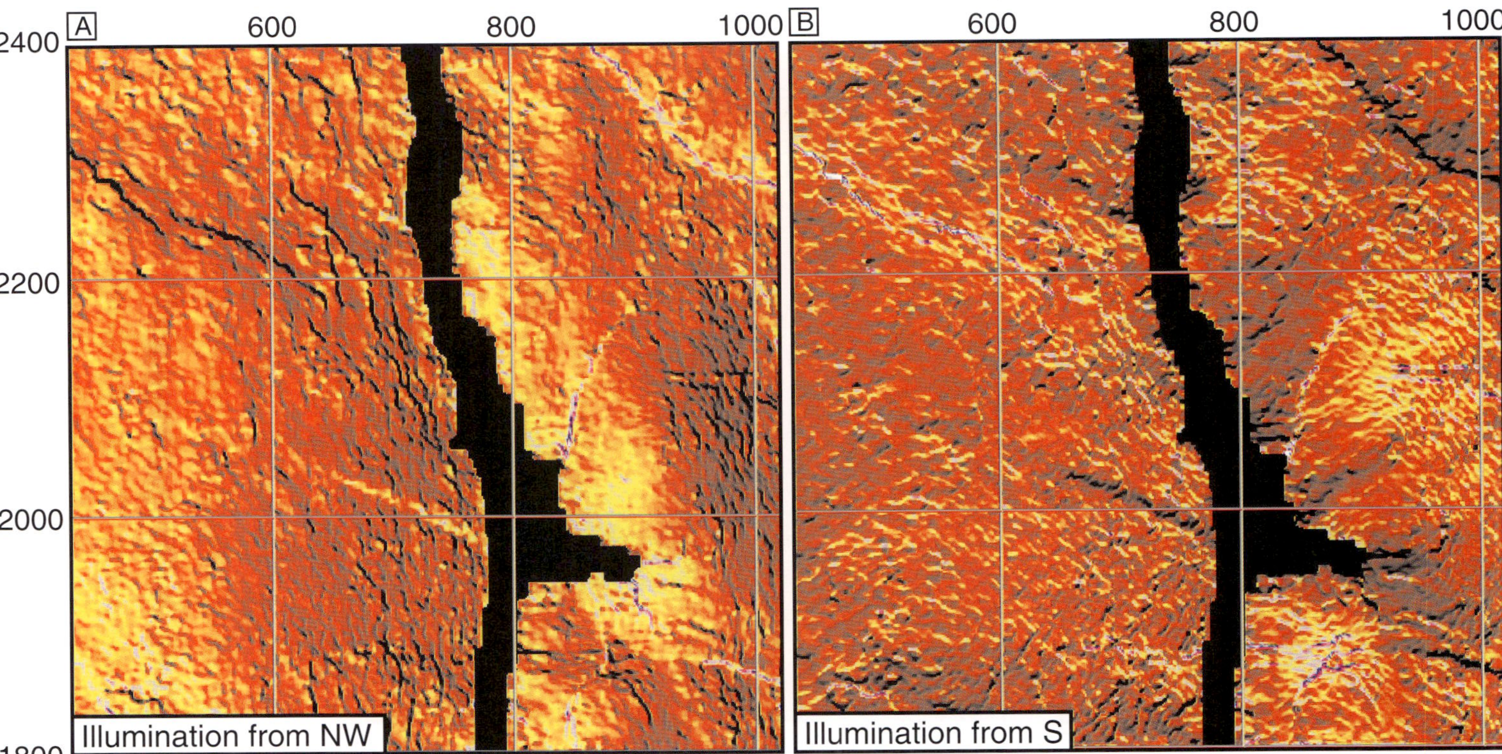

Fig. 5. Illuminated dip maps generated from the Ness1 reflector of the Gullveig Field with a light source from (**a**) the northwest and (**b**) the south. Dark lineaments indicate steep dipping zones (dipping away from the light source) whereas bright lineaments indicate high dip values towards the light source. The 2 maps illustrate how the observed linear features can vary significantly simply by changing the visualization parameters. In some cases, such visualization techniques may introduce bias into an interpretation.

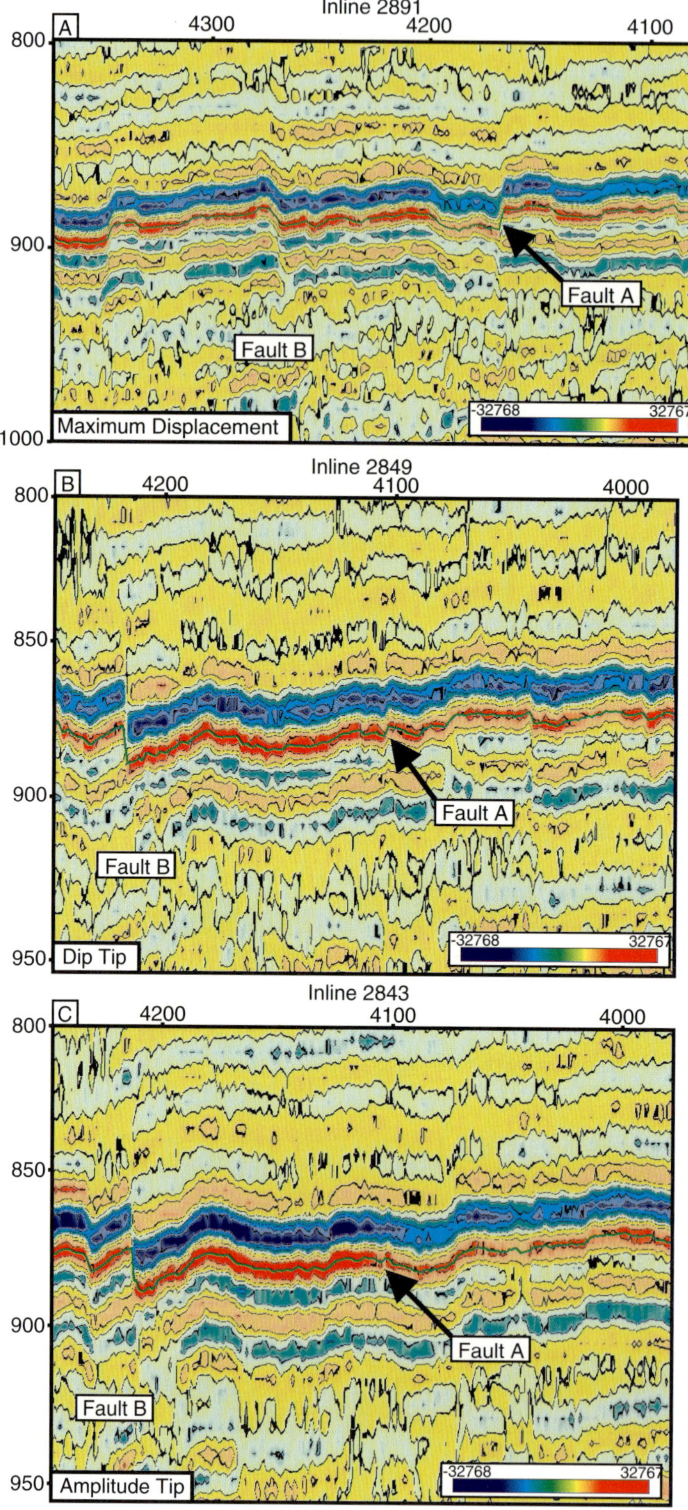

Inline 2891
A
4300
4200
4100
800
900
1000
Fault A
Fault B
Maximum Displacement
-32768
32767
Inline 2849
B
4200
4100
4000
800
850
900
950
Fault A
Fault B
Dip Tip
-32768
32767
Inline 2843
C
4200
4100
4000
800
850
900
950
Fault A
Fault B
Amplitude Tip
-32768
32767

effective bandwidth of 10 to 35 Hz. Wavelets extracted from the Ness1 level have a wavelength of *c.* 80 to 90 ms.

Siri fault zone

The frequency content of the seismic data from the Siri Fault Zone area is sufficiently high enough for the radius of the Fresnel zone to approach the same order of magnitude as the CDP spacing (12.5 × 12.5 m). In such an instance, it should be expected that faults of very small throw should be seismically detectable. The near top Miocene event (the reflector interpreted in green on Fig. 6) was autotracked over the entire dataset for structural evaluation of the faulting history. Dip and amplitude maps were then generated (see Fig. 7). The structure that is to be considered (marked as A on Fig. 7) forms the eastern bounding fault of a small graben system bounded to the west by a slightly larger fault (marked B on Fig. 7). Both of these faults are situated in the hanging wall of the main Siri Fault (marked C on Fig. 7). On the amplitude map, fault A appears continuous for some 4 km. However, when the dip map is analysed, it is clear that the feature is much more complex, as it is composed of a series of amalgamated dip anomalies that display characteristics similar to relay structures and transfer zones.

When considered in cross-section (see Fig. 6), the feature can be clearly identified as a fault on inline 2891 (arrow on Fig. 6a) where it has a displacement of 11 ms (*c.* 11 m). The displacement can be traced some 40–50 ms above and below the interpreted surface and can be observed cutting through much weaker events. In addition to the clear offset of the reflectors, amplitude and dip anomalies are also observed at all of the displaced events. These characteristics are also displayed by a number of other features on the same seismic line, about which there is little doubt that they too are faults.

Inline 2849 (Fig. 6b) is situated 525 m south of inline 2891 and represents the displacement tip of fault A. At this location the deflection of the interpreted surface is *c.* 5 ms (5 m). In this example, although the displacement or dip anomaly can be traced vertically above and below the interpreted horizon, it soon becomes difficult to follow through the weaker events. Again, an amplitude anomaly is also observed.

The amplitude anomaly observed near the tip of fault A on inline 2849 can be traced in section for a further six lines to inline 2843 (75 m south; Fig. 6c). No displacement or dip anomaly is observed at this location, however, the amplitude anomaly can still be traced above and below the interpreted surface for *c.* 20 ms.

The conclusion that can be drawn from this example is that, given sufficiently high frequency content within the seismic data, faults with throws of the order of *c.* 5 m (and possibly less) can be identified and mapped using existing attribute analysis techniques. What is also clear from Fig. 7 is that there are a very large number of dip anomalies of the same type present within the area of the study, nearly all of which also display amplitude anomalies. (It should be noted here that some of the lineaments seen on Fig. 7 are artifacts of the autotracking process, these can generally be identified by their grid parallel trends.) Individual inspection of a number of these anomalies revealed that without exception, they display all the same characteristics as Fault A. Therefore, they too are most likely to be faults. What is of concern regarding these minor faults is that their location and complex pattern is significantly different from that of the larger faults. If the larger fault pattern had been used to predict the smaller 'sub-seismic' faults then the complexities of the smaller faults observed on the dip and amplitude attribute map would not have been predicted. Our initial interpretation of these minor structures is that they may form a polygonal fault system similar to those described by Cartwright & Lonergan (1996).

Gullveig

Two possible faults have been selected on which to demonstrate the application of the techniques (see Figs 8, 9 & 10). Fault 1 (Figs 8 & 9) trends N–S, has a length of 1.9 km and a maximum throw of *c.* 40 ms ($\approx$40 m) at its centre which decreases towards both terminations. Although this is one of the smallest faults mapped during the interpretation of this field, there is little doubt that it is a real structure because of the size of its throw and that it forms a linear feature

Fig. 6. Seismic sections from the Siri Fault zone displayed using a colour palette which highlights subtle amplitude anomalies: (**a**) crossing Fault A at its maximum displacement, (**b**) at its displacement tip and (**c**) 75 m beyond its displacement tip, where the amplitude anomaly dies out. Arrow points to Fault A at the top Miocene level (horizon interpretation coloured green).

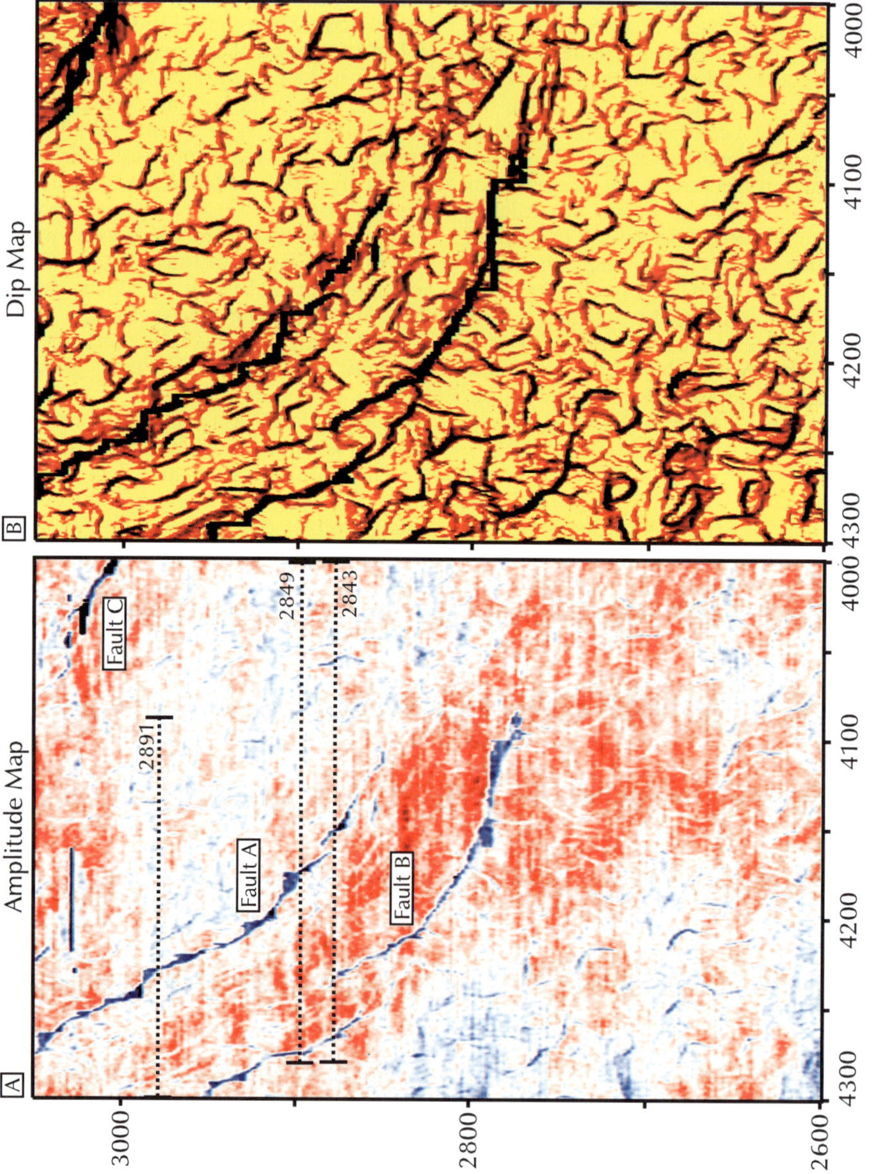

Fig. 7. Attribute maps from the top Miocene reflector: (**a**) amplitude map (red = high amplitude, white = intermediate and blue = low); and (**b**) dip map (yellow = low dip, red = intermediate and black = high). Faults A and B are referred to in the text and Fault C = Siri Fault zone.

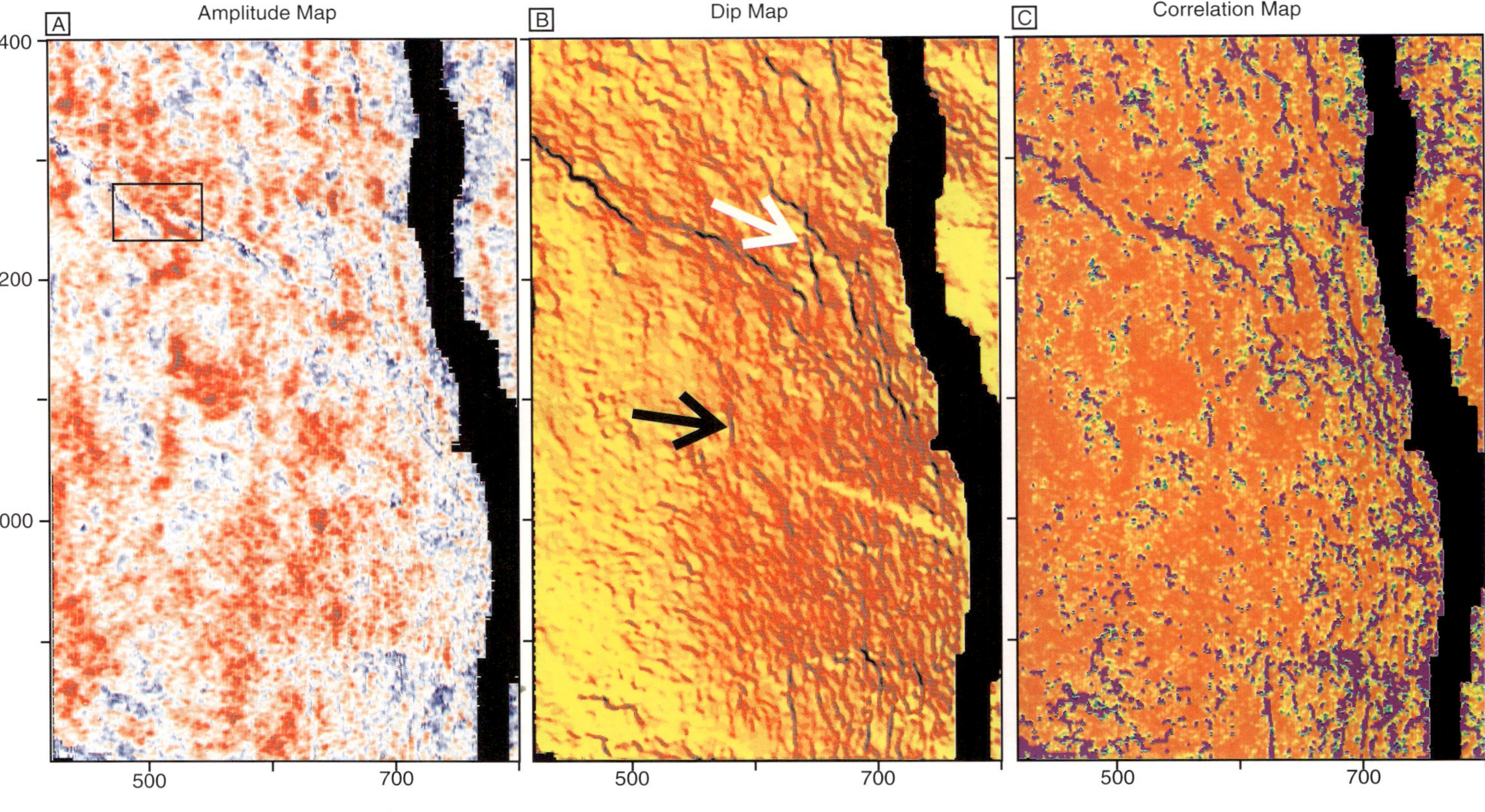

Fig. 8. Seismic attribute maps of the Ness1 reflector from the Gullveig field: (**a**) amplitude map (red = high amplitudes, white = medium and blue = low); (**b**) dip map illuminated from NW (yellow = areas dipping towards NW, red = areas dipping away from NW and black = steep dips away from NW); and (**c**) correlation map (red = areas of good correlation and purple = poor correlation). A number of linear features occur on all 3 maps, many of which can be interpreted as faults; generally the shorter the lineament the less confident the interpretation. Faults 1 and 2 are located on the dip map by arrows (white = Fault 1 and black = Fault (**2**) and they are examined in more detail (Figs 8–11). The box on the amplitude map shows the location of Fig. 16.

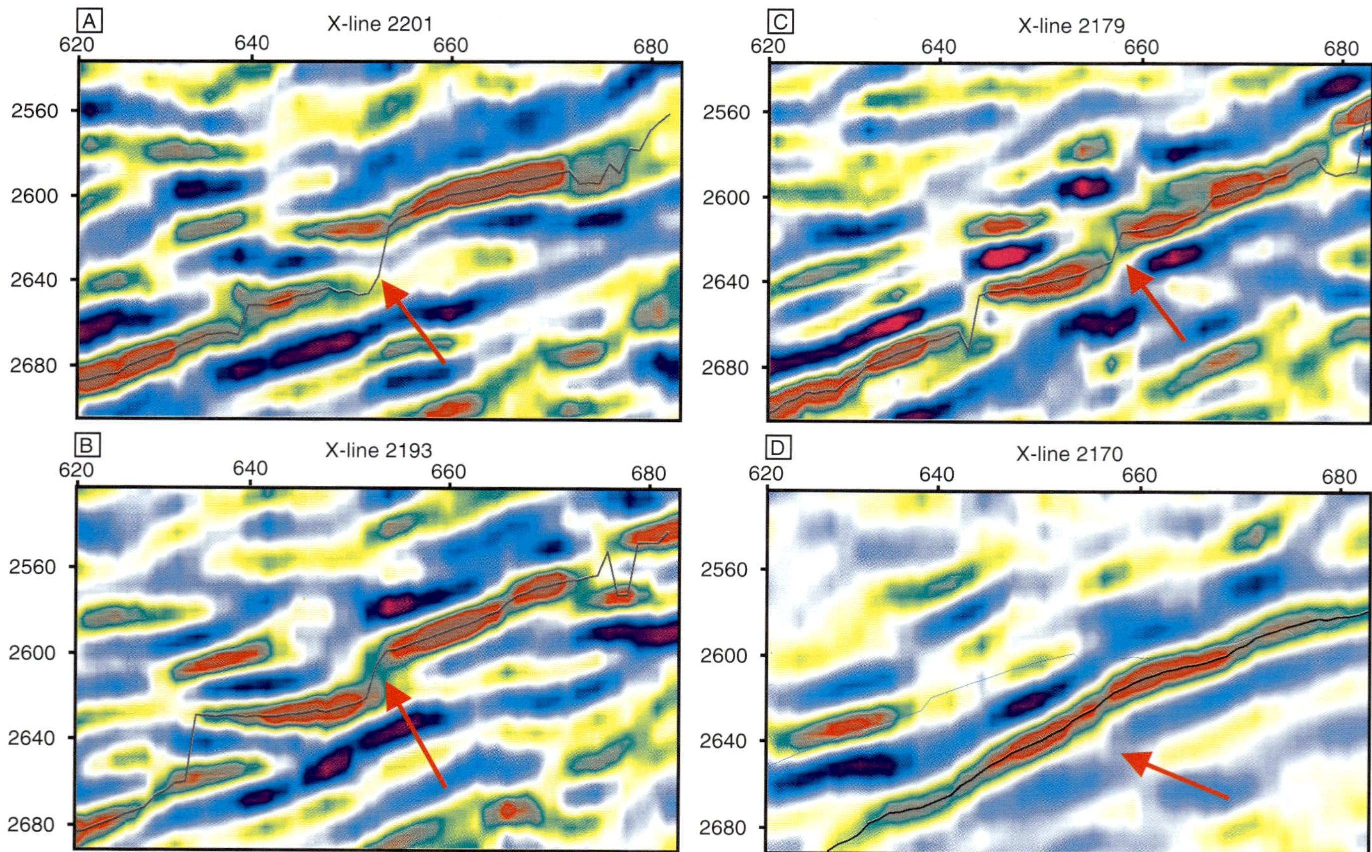

Fig. 9. Seismic sections through Fault 1 (red arrow) where it cuts the Ness1 reflector (red event): **(a)** at centre of fault where throw is largest, **(b)** half way to the visible displacement tip; **(c)** at the visible displacement tip; and **(d)** *c.* 110 m beyond the visible displacement tip, but where the amplitude anomaly is still present. The seismic sections have been displayed using a colour palette which reflects subtle amplitude changes caused when faults are present. Note that a number of amplitude anomalies occur along the Ness1 reflector, these should be investigated to see if they also caused by faults.

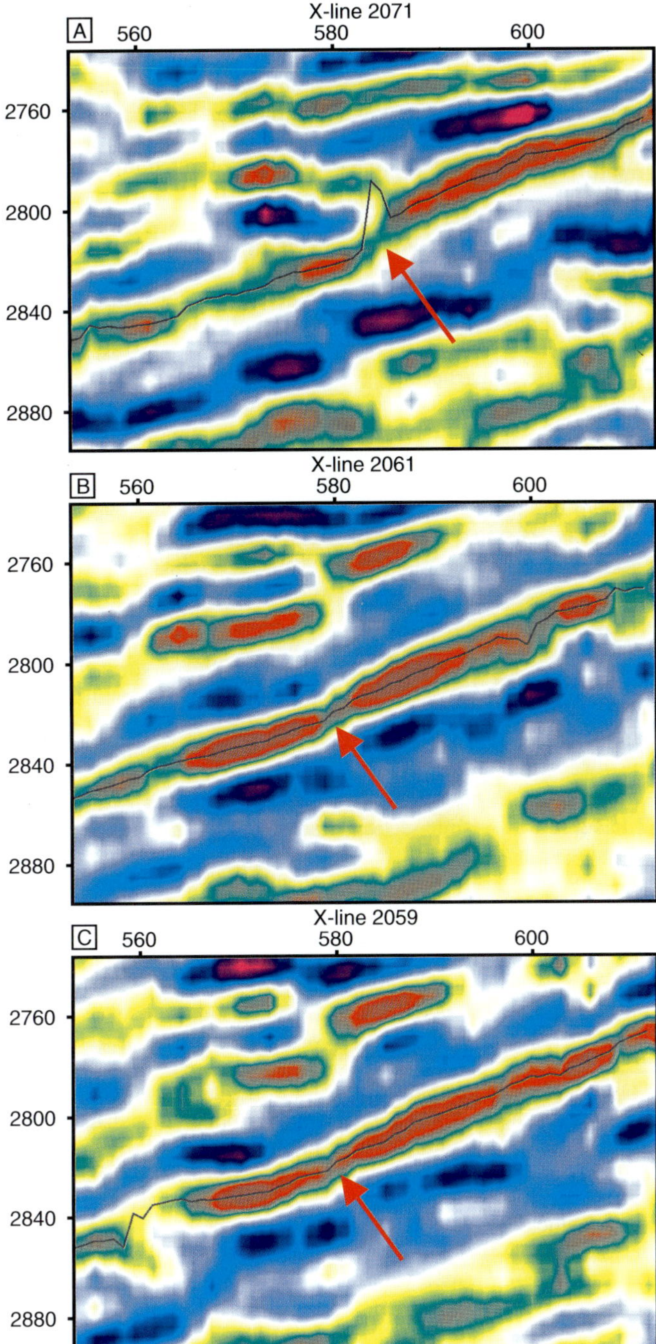

Fig. 10. Seismic sections through Fault 2 (red arrow) where it cuts the Ness1 reflector (red event), this fault is much smaller than Fault 1 and consequently its interpretation is less certain, however, the same characteristics are present: (**a**) at the centre of the fault where throw is largest; (**b**) at the visible displacement tip; and (**c**) 25 m beyond the visible displacement tip.

on the attribute maps. Confidence in its interpretation as a fault is helped by its location close to other faults; the relationship between these adjacent lineaments conforms to what is expected of faults (i.e. similar strike directions and truncation to the north by a NW–SE trending fault). Fault 2 (Figs 8 & 10) is less obvious as a real structure; it also has a N–S trend, it is located away from any other visible structures, it has a much shorter length (*c.* 700 m) and its maximum throw is smaller (20 ms $\approx$ 20 m).

The aim of the exercise is to identify the characteristics associated with Fault 1, to demonstrate that they are consistent with the fault identification techniques outlined earlier and to see how far a fault can be seen beyond its seismically resolvable offset. The same methods are then applied to the smaller less obvious feature (Fault 2) in order to see how far the interpretation of possible faults can be taken.

The two faults are displayed using a number of seismic sections (Figs 9 & 10). These cross the linear features:

(i) close to their centre's where the displacement is largest;
(ii) at their visible displacement tips (i.e. the point of displacement resolution) and;
(iii) beyond their displacement tips at the point where they are last seen using seismic attributes.

In addition, Fault 1 has a fourth seismic section which crosses approximately half way between the centre and visible displacement tip. The seismic sections (Figs 9 & 10) have been displayed using a colour palette specially designed to highlight the subtle amplitude changes that are expected when faults occur. Red and brown indicate very high amplitudes along the Ness1 peak, green represents moderate values, yellow indicates low amplitude and white is close to zero. The colour scheme has been scaled so that most of the Ness1 reflector has a red–brown colour. However this is expected to change sharply to either green or yellow (and sometimes white) when a fault is present. In addition to the seismic sections, the line graphs, which were used to display the results from the seismic modelling, are also used here to show amplitude values (Figs 11 & 12).

Fault 1 generates a linear, N–S trending anomaly on the dip, correlation and amplitude maps (Fig. 8). This anomaly is not so sharply defined on the amplitude map as it is on the other 2 maps. To the north, Fault 1 links with a NW–SE trending structure and southwards it dies out close to cross-line 2170; this coincides with the termination of a prominent NW–SE trending fault, which extends towards the northwest. The N–S trend is parallel to the main fault cutting through the data. It is also parallel to a number of other fault-like features, which occur to the north and southeast of Fault 1. All of these smaller structures lie in the footwall to the main fault bounding the Gullveig structure to the east.

On all of the seismic sections through Fault 1 (Fig. 9) a clear amplitude reduction can be observed at the location of the fault when compared to the local amplitude values. The size of the amplitude anomaly decreases with decreasing displacement; this is highlighted by the line-graphs (Fig. 11). This confirms that as displacement decreases along a fault, the size of the amplitude anomaly does likewise. The structure can be identified on the seismic sections with a small throw; *c.* 40 m at the centre, *c.* 15–20 m half way to the visible displacement tip, and no resolvable displacement beyond the tip. The visible displacement tip coincides almost exactly with the termination of the linear anomaly on the dip and correlation maps. The amplitude anomaly observed in the cross-lines can be followed some 9–12 lines (*c.* 110–150 m) beyond the visible displacement tip. In this case, the amplitude anomaly has the best resolution for smaller displacements along the fault.

It should be stated that although not all of the seismic sections crossing Fault 1 are presented here, panning through the lines shows that the structure is continuous. Furthermore, on all the cross-sections, several amplitude anomalies occur, in addition to that created by Fault 1 at the Ness1 level. These anomalies have not been investigated during this study; however, in a detailed mapping exercise they should be, in order to determine if they are also possible faults.

The feature labeled Fault 2 is less obvious than Fault 1. This is not only because of its shorter length and smaller displacement, but it also lies in an area devoid of other similar looking structures. Fault 2 has a N–S trend (Fig. 8), it forms a linear high dip feature on the dip map and a low correlation zone on the correlation map; its length on both these maps is *c.* 700 m. It is also represented as a low value anomaly on the amplitude map, but this is not easily distinguished from the background amplitude variations. The maximum throw on the fault lies close to its centre and is *c.* 15 m; this decreases towards the visible displacement tips in both directions.

In the seismic sections, an amplitude anomaly can be observed at the location of Fault 2 (Fig. 10). The size of this anomaly decreases towards the south (in the direction of the

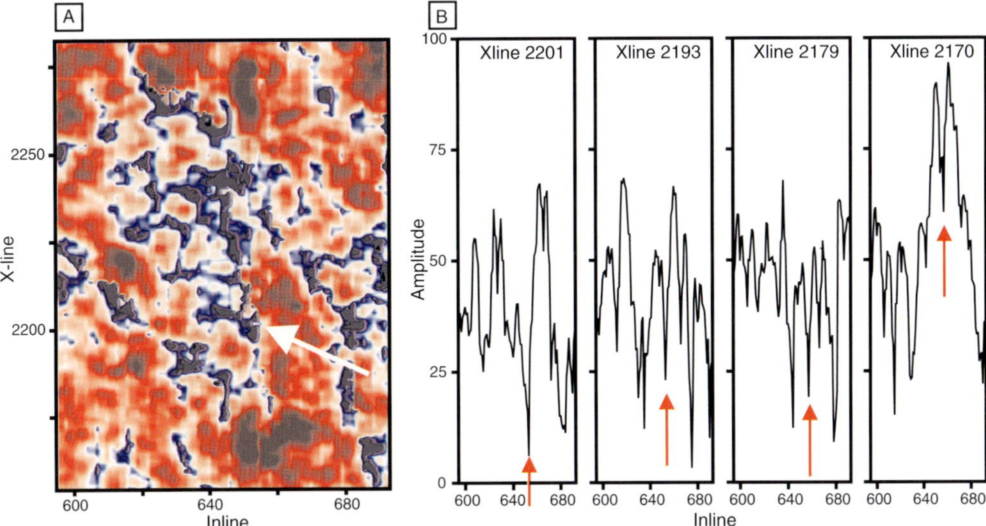

Fig. 11. (**a**) Detailed amplitude attribute map around Fault 1 and (**b**) amplitude line graphs following inlines which correspond to the sections in Fig. 9. Fault 1 is located with the white arrow on (**a**) and red arrows on (**b**). The amplitude reduction caused by Fault 1 (arrow) is easily followed scanning through adjacent line-graphs.

expected displacement decrease). The amplitude anomaly can be traced using line-graphs (Fig. 12) where it continues for some 5 to 6 cross-lines (62.5 to 75 m) to the south of the visible displacement tip. As with Fault 1, the structure is visible over a greater length as an amplitude anomaly than it is using any of the other available fault indicators.

The features seen associated with both Faults 1 and 2 are entirely consistent with what would be

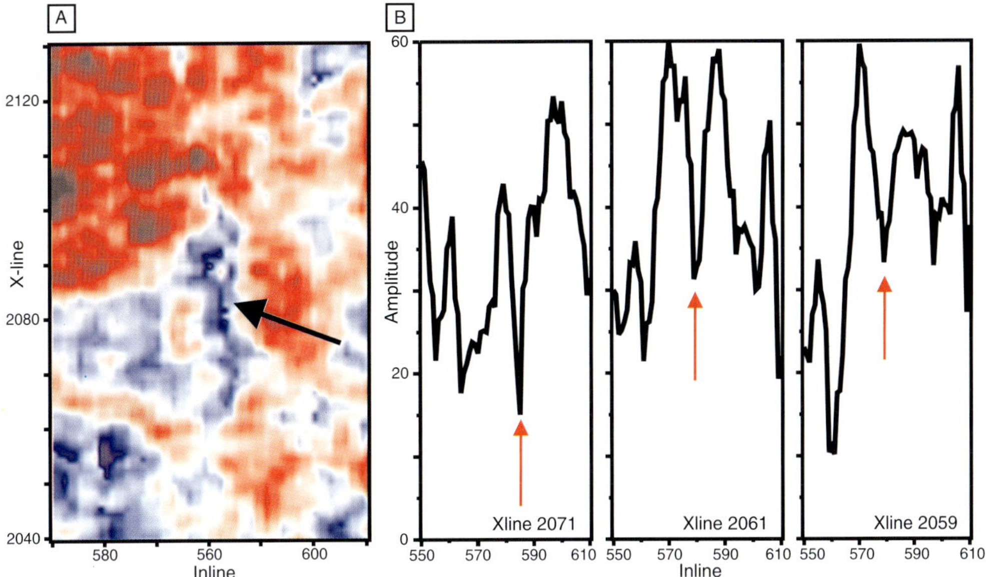

Fig. 12. (**a**) Detailed amplitude attribute map around Fault 2 (black arrow) and (**b**) amplitude line graphs following inlines which corresponding to the sections in Fig. 10. Fault 2 is located with the red arrows. The amplitude reduction caused by Fault 2 (arrow) can be followed by scanning through adjacent line-graphs.

expected if a fault were present. The following features are observed:

(i) an observable time offset;
(ii) the throw is largest at its centre and decreases towards the tips;
(iii) an amplitude anomaly is clearly observed in cross-section but is not so well defined on maps;
(iv) a well defined linear anomaly on the dip map; and
(v) a linear correlation map anomaly.

No observations have been made which would suggest that the features are not faults. Therefore, these two structures should be considered as likely faults.

Although the features examined are consistent with the way in which faults are observed in outcrop, geological analogues and their expected expression within 3D seismic data, the possibility should not be overlooked that they could have been generated by other geological structures, or even result from artifacts generated by seismic acquisition or processing. If the observed features are due to something other than faults, then the geological structure or seismic artifact must be capable of generating all the observations recorded here. However it should be emphasized that, even if an alternative interpretation of these features can be demonstrated, the fault interpretation cannot be discounted.

Uncertainty associated with mapping faults in 3D

Having shown that at least some of the linear features seen on seismic attribute maps have properties consistent with faults, consideration will now be given to how these structures manifest themselves in three dimensions. Faults are known to form 2D surfaces that extend a considerable distance above and below a linear feature identified on a good seismic reflector.

The properties that have been used to identify the faults can be quantified and attributed to individual fault surfaces (see Fig. 13). One such property is displacement (or throw). Displacements have previously been plotted and contoured on individual fault surfaces to help quality control and improve fault correlation between seismic lines (Rippon 1985; Barnett *et al.* 1987; Freeman *et al.* 1990). Other fault properties, e.g. amplitude, dip or correlation anomalies could also be treated in the same way (Fig. 13). The anomalies would have to be measured relative to the local background values and for some properties, values would have to be normalized to each reflector. Clearly there may be additional fault properties which have not been discussed here, which could also be plotted for fault plane analysis.

Fault surfaces could be displayed in 3D with the properties given by either intensity or colour. This will allow all the available fault properties to aid quality control. It will also increase confidence in any interpretation if more than one property exists for any individual fault. As it has been shown that the amplitude anomaly is often more sensitive to smaller fault displacements than any of the other properties, it will therefore be expected to cover a larger area of a given fault surface (see Fig. 13).

Deterministic mapping of faults in 3D is fine for seismic datasets with several good reflection events or for larger faults. However, for many of the smaller seismic-scale faults in the Tampen Spur region of the North Viking Graben, extrapolating between reflectors can generally be described as difficult and uncertain (see Fig. 14). Fault-like features can be clearly identified at the Ness1 level; however, when it comes to mapping these small faults above and below this reflector, any interpretation has a low confidence.

There are a number of possible reasons for this:

(1) Most of the other reflection events are rarely as well developed as the Ness1 event. The weaker events often do not allow small faults to be identified in the same way as the Ness1 event.
(2) Although the base Cretaceous unconformity often forms a good reflection event, the number of faults that can be interpreted is considerably less than on the Ness1 event. This may be due to; (i) most of the faults pre-dating the unconformity; (ii) other features overprinting the fault subtleties; or (iii) the faults at Ness1 level being confined to a narrow stratigraphic interval.
(3) Where a fault can be observed cutting the base Cretaceous unconformity, the correlation with an individual fault at Ness1 level can be uncertain, especially when a number of faults occur close together.

In areas such as the North Viking Graben, where fault-like features can be identified on a limited number of horizons (e.g. Fig. 14) and are difficult to map out in the third dimension, stochastic modelling could be used to take account of any uncertainty associated with the

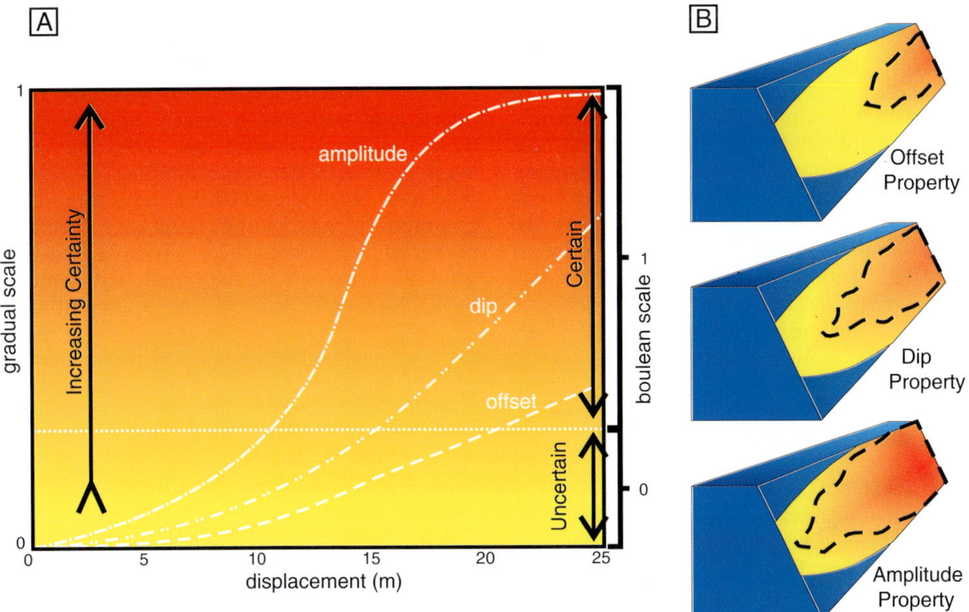

Fig. 13. An illustration of how various fault properties can be used to help improve the quality of fault interpretation. (**a**) Graph outlining how the effect of amplitude, dip and offset increase with increasing displacement showing that amplitude has the best resolution. The uncertainty associated with interpreting faults using these properties increases with progressively smaller faults. (**b**) Fault properties displayed on idealized fault planes where displacement decreases towards the tip line. Amplitude property will cover a larger area as it has the best resolution.

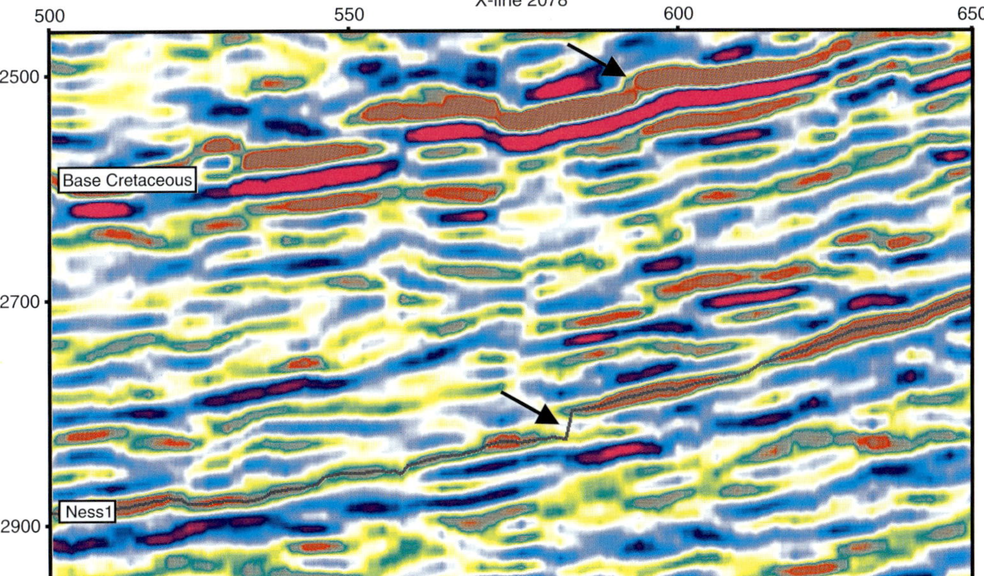

Fig. 14. Seismic section from the Gullveig field demonstrating the difficulties of correlating faults between different levels. A number of amplitude anomalies (some with offsets) can be observed at the Ness1 level. However, only one can be correlated with a similar feature at the base Cretaceous level (highlighted by arrows); even this has its problems as it is difficult to follow the fault through the intermediate weaker reflectors.

interpretation. For this to be as deterministic as possible, all available fault properties should be extracted from the linear features identified on each reflector and used as input parameters for the modelling routines. Such properties would act as control points for fault location, dip angle, dip direction, local maximum displacement and local length. These properties could be used to condition programs, such as the HAVANA fault modelling software (Munthe *et al.* 1993, 1994), to the seismic data. This would reduce some of the large uncertainties in sub-seismic fault modelling associated with the spatial distribution of modelled faults.

Automatic fault mapping

Today, seismic workstations have several tools for automatic tracking of seismic reflection events. These generally save a considerable amount of time for the interpreter and are capable of using all the available information (i.e. every CDP). More importantly, they tend to reduce any personal preferences in an interpretation, thus making them more objective.

When horizon-tracking tools were first introduced they were not especially effective. They have improved significantly over the years because of a better understanding of seismic reflectors. Having realized that reflectors follow a number of rules, workstations have been programmed to follow them. Today's tracking tools have become so advanced that some of them are even effective on weak, non-continuous reflectors.

If it can be determined how faults are likely to manifest themselves on seismic data, then it should, in theory, be possible to map them in an automatic (or at least in a semi-automatic manner). As with seismic reflectors, faults follow a number of rules and these need to be fully understood, such that interpretation systems can be developed to extract fault information efficiently. As faults are generally more complex geologically than seismic reflectors, the

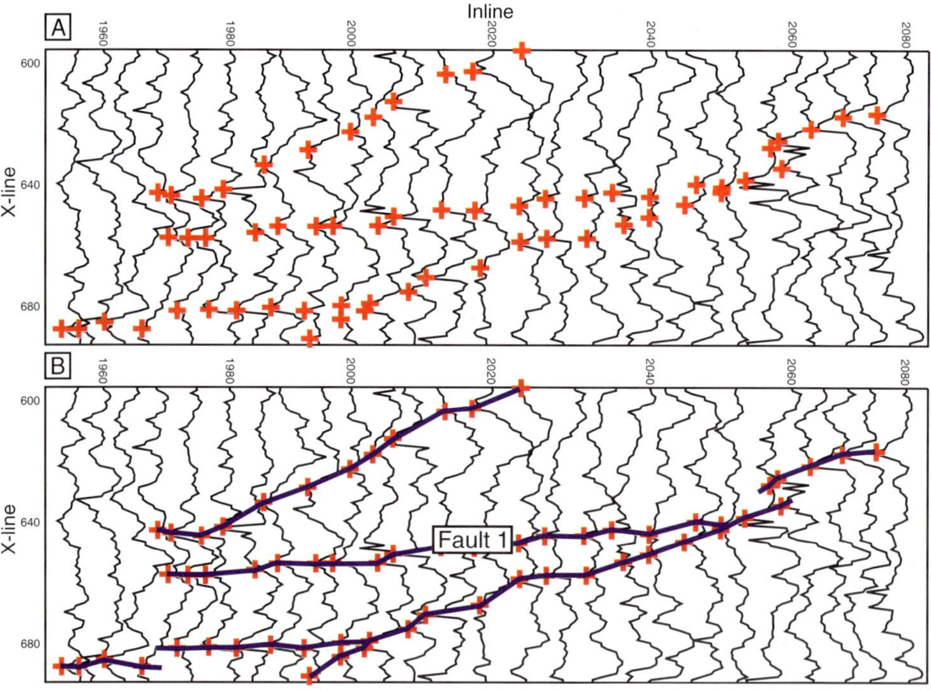

Fig. 15. Line graphs extracted every 5 cross-lines from the amplitude map around Fault 1 in Fig. 11 (amplitude increases to right). The line graphs have been interpreted interactively using both the line graphs and the amplitude map; the amplitude map indicates where the fault lies, whereas the trough on the line graph pin-points its exact position. This method of interpretation revealed a number of previously unrecognized relay structures. (**a**) Line graphs with fault locations and (**b**) with faults interpreted between locations. These interpretation methods could be easily automated using peak/trough-tracking software.

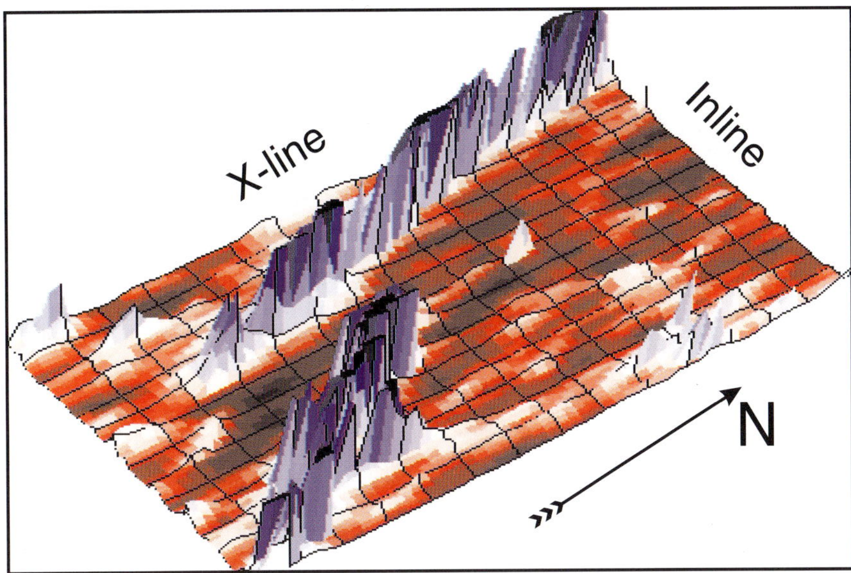

Fig. 16. An alternative way of visualizing the amplitude anomaly associated with a fault. The reciprocal of the amplitude values has been taken (making the amplitude anomaly into a peak) and then squared (this makes the larger values (i.e. faults) relatively larger and reduces any variation among the smaller values). The location of the 3D view is marked on Fig. 8, it is viewed looking northeast and shows the relay structure between 2 faults segments. The colour scheme has been inverted compared to Fig. 8 such that blue also indicates possible fault locations. One could easily imagine these ridges being tracked in an automatic manner.

programming of workstations for automatic fault mapping is likely to reflect this.

The first and most obvious design feature of an automatic fault mapping system should be to extract information from reflection events, such as those presented earlier in this paper (Figs 6 & 8). These horizons contain a significant number of linear features which have characteristics similar to those expected of faults. Different maps visualized using varying constraints (e.g. Fig. 5) can lead to a large amount of data to be interpreted. To map these by hand would normally take a considerable amount of time, much more than is often available during an interpretation project. Therefore, the task is rarely carried out.

Automatic extraction of fault-like features occurring on any horizon could be carried out either by the use of lineament analysis techniques used in remote sensing, or manually selecting each linear feature (i.e. by using a mouse) allowing it to be tracked until a given set of properties are no longer present (these properties could, for example, be a dip or amplitude anomaly size). This principle is illustrated in Fig. 15, where amplitude value line graphs from every 5th inline, across Fault 1 have been assembled together. The x- and y-axes on the line graphs have been transposed (compared to the way in which they are displayed in Figs 11 & 12), so that they are orientated in a similar way to other data where correlations are commonly performed (e.g. well data and seismic traces). Faults show up on these line graphs as troughs and in order to map a fault, troughs should be correlated through the data. In addition to correlating the troughs for Fault 1, a number of other faults have also been mapped. This is akin to autotracking an horizon through a 3D seismic dataset where it is represented as a trough.

To further emphasize the point that attributes anomalies created by faults form trackable features, the amplitude anomalies related to 2 faults separated by a relay ramp are plotted in 3D in Fig. 16. For visualization purposes, and in order to emphasize the fault related features, the reciprocal of amplitude values has been taken to transform the trough into a peak and these values have been squared; this reduces differences between the smaller background values and increases the effect of the amplitude anomaly. Note that the colours have also been inverted such that blue still represents a likely fault. The faults are represented as well defined and continuous ridges (Fig. 16) which can easily be traced by a peak following algorithm.

In addition to tracking the fault-like features, workstations should also be programmed to extract properties along each fault. These would be calculated at a number of points along each fault. These properties could be used either to help correlate faults between horizons or as input parameters for stochastic modelling.

The more challenging problem is that of mapping faults as surfaces in 3D space using auto-tracking techniques. Fault-tracking tools will have to be designed to cope with:

(i) steeply dipping structures rather than relatively flat lying reflectors;
(ii) extracting fault information which may be held by the weak reflection events occurring between the main events;
(iii) zones of poor reflectivity with low amplitudes, as opposed to good continuous events.

The latter may be accounted for by turning the seismic data cube *inside out* so those zones of poor reflectivity (which includes faults) become positive features. This is essentially how the coherence cube works (Bahorich & Farmer 1995). This uses only one of a number of possible methods for undertaking this task. Clearly, further investigation in this area would be beneficial for fault mapping.

The development of a reliable 3D auto-tracking fault tool probably lies a long way into the future. It requires not only a better understanding of fault systems, but also an increased comprehension of how seismic data is effected by faulting. Once the impact of faults on seismic data is understood, then tools can be designed to map their various characteristics. Although it will take a considerable amount of effort to develop a robust tool for fault mapping, the main benefits will include fast and accurate fault mapping from which a considerable part of the interpreter's subjectivity will be removed.

Conclusions

(1) Many features exist within seismic data that have characteristics consistent with those expected to develop from small seismic-scale faults.
(2) Amplitude anomalies caused by faults are not generally fully utilized in fault identification; they are probably the most robust fault indicators in cross-section. Amplitude anomalies are consistently better at resolving the smaller fault displacements than either the dip or time attributes.
(3) Fault identification using any seismic attribute is dependent upon understanding how each attribute will be effected by the presence of a fault and by using a visualization procedure which best displays the expected effects.
(4) Fault surfaces have properties; displacement is only one of a number of measurable properties. These can be used for both quality control of seismic fault interpretation and input to stochastic fault modelling.
(5) Automatic fault mapping routines need to be developed which will follow the properties of an individual fault. These routines should not only follow a fault but at the same time, extract its properties.

We are grateful to Statoil for permission to publish this paper. Kes Heffer (BP) is thanked for his encouragement at the start of this work. This paper has benefited from the reviews of G. Jones and an anonymous reviewer. CHARISMA is a trademark of Schlumberger–GeoQuest, STORM is a trademark of Smedvig Technologies and HAVANA is owned by the Norwegian Computing Centre.

References

BADLEY, M. 1985. *Practical seismic interpretation.* IHRDC press, Boston, USA.

BAHORICH, M. & FARMER, S. 1995. The coherence cube. *The Leading Edge*, **14**, 1053–1058.

BARNETT, J. A. M., MORTIMER, J., RIPPON, J. H., WALSH, J. J. & WATTERSON, J. 1987. Displacement geometry in the volume containing a single normal fault. *American Association of Petroleum Geologists, Bulletin*, **71**, 925–937.

CARTWRIGHT, J. A. & LONERGAN, L. 1996. Volumetric contraction during the compaction of mudrocks: a mechanism for the development of regional-scale polygonal fault systems. *Basin Research* **8**, 183–193.

FAIRBURN, C. M., GREEN, C. M. & WARD, W. A. 1988. Detection of small scale faulting from detailed investigation of reflection amplitudes. *50th meeting European Association of Geoscientists & Engineers*, The Hague, The Netherlands.

FREEMAN, B. G., YIELDING, G., & BADLEY, M. E. 1990. Fault correlation during seismic interpretation. *First Break*, **8**, 87–95.

GAUTHIER, B. D. M. & LAKE, S. D. 1993. Probabilistic modelling of faults below the limit of seismic resolution in Pelican Field, North Sea, Offshore United Kingdom. *American Association of Petroleum Geologists, Bulletin*, **77**, 761–777.

HENRIQUEZ, A. & JOURDAN, C. 1995. Challenges in modelling reservoirs in the North Sea and on the Norwegian Shelf. *Petroleum Geoscience*, **1**, 327–336.

HESTHAMMER, J. & FOSSEN, H. 1997a. Seismic attribute analysis in structural interpretation of the Gullfaks Field, northern North Sea. *Petroleum Geoscience*, **3**, 13–26.

—— & —— 1997b. The influence of seismic noise in the structural interpretation of seismic attribute maps. *First Break*, **15**, 209–219.

HOBBS, B. E., MEANS, W. D. & WILLIAMS, P. F. 1976. An outline of structural geology. John Wiley & Sons, New York, USA.

JONES, G. & KNIPE, R. J. 1997. Seismic attribute maps; application to structural interpretation and fault seal analysis in the North Sea Basin. *First Break*, **14**, 449–461.

MCQUILLIN, R., BACON, M. & BARCLAY, W. 1984. An introduction to seismic interpretation, reflection seismics in petroleum exploration. Graham & Trotman Limited, London, UK.

MUNTHE, K. L., OMRE, H., HOLDEN, L., DAMSLETH, E., OLSEN, T. & WATTERSON, J. 1993. Subseismic faults in reservoir description and simulation. **SPE 26500**, *68th Annual Technical Conference and Exhibition of the SPE, Houston, 3–6 October 1993.*

——, HOLDEN, L., MOSTAD, P. & TOWNSEND, C. 1994. Modelling sub-seismic fault patterns using a marked point process. Proceedings of the 4th European conference on the Mathematics of Oil Recovery, Røros, Norway, June 199, topic B: hetergeniety, description and assessment of uncertainty.

RIPPON, J. H. 1985. Contoured patterns of the throw and hade of normal faults in the Coal Measures (Westphalian) of north-east Derbyshire. *Proceedings of the Yorkshire Geological Society*, **45**, 147–161.

SHERIFF, R. E. 1981. Structural interpretation of seismic data. *American Association of Petroleum Geologists, Education Course Notes Series 23.*

STEEN, Ø., SVERDRUP, E. & HANSEN, T. H. 1998. Prediting the distribution of small faults in hydrocarbon reservoirs by combining outcrop, seismic and well data. *This volume.*

Predicting the distribution of small faults in a hydrocarbon reservoir by combining outcrop, seismic and well data

Ø. STEEN[1*], E. SVERDRUP[2] & T. H. HANSSEN[2]

[1] *Department of Geology, University of Oslo, P.B. 1047 Blindern, 0316 Oslo, Norway*
[2] *Saga Petroleum ASA, Kjørboveien 16, P.B. 490, 1320 Sandvika, Norway*
[*] *Present address: Statoil, Research Centre, Arkitekt Ebbels veg 10, 7005 Trondheim, Norway*

Abstract: This paper investigates the possibility of identifying small faults in a hydrocarbon reservoir from spatial derivatives of seismic horizons (dip, azimuth, rate of change of dip), combined with well log structural data and analogue models. The analogue models include examples from the literature and a faulted surface constructed from Kilve beach, Bristol Channel, which represent typical surface attributes associated with small faults and artefacts produced by modelling. Well log structural data of good quality were integrated with the seismic data to study the 1D spatial distribution of subseismic faults and their expression on seismic horizons. The results suggest that dip and azimuth modelling can be useful, but the methods need to be combined to achieve reliable interpretations of the individual features. The rate of dip change has less potential to trace individual structures, unless the signal/noise ratio is very high. The ductile strain observed on seismic sections is, to a large degree, assumed to be produced by subseismic faults. The amount of ductile strain may be quantified by calculating the rate of dip change across curved seismic horizons. Exact predictions of fault densities are difficult because of the inaccuracies related to the modelling technique and the variability in fault and fold styles that occurs on a subseismic scale. Detailed dipmeter and core interpretation can be used to calibrate structural dip and spatial frequency of faults.

Quantification and characterization of small ('subseismic') faults in reservoirs are topics which have been discussed by several authors during recent years (e.g. Yielding *et al.* 1992; Gauthier & Lake 1993; Jones & Knipe 1996; Knott *et al.* 1996). The validity of using scaling laws for fault population prediction is still debated as field examples show that gaps or breaks in fault size distribution may occur (e.g. Cowie *et al.* 1996 and references therein). Also, the spatial distribution of faults will be controlled by factors, such as structural history, lithology and fluid–rock interactions (Peacock 1996; Sverdrup & Bjørlykke 1997). Alternative ways to predict fault patterns in subsurface reservoirs which may provide new information and understanding of the deformation below seismic resolution are therefore needed (see Townsend *et al.* this volume). 3D seismic data have been increasingly used by oil companies for defining horizons as well as for describing reservoir architecture. In particular, the understanding and methods of such datasets to identify subtle faults and their impact on the horizon grids has accelerated the use of these maps for reservoir characterization. Seismic attribute mapping techniques comprise analyses of dip, azimuth, reflectivity and amplitude (e.g. Bouvier *et al.* 1989; Dalley *et al.* 1989; Hoetz & Watters 1992; Mondt 1993; Jones & Knipe 1996; Hesthammer & Fossen 1997*a*,*b*) which are commonly included in commercial software. The curvature (rate of change of dip) is another surface attribute which can be easily calculated and displayed. Ductile strain can be defined as a change in horizon shape produced by structures below the lower limit of resolution (Walsh *et al.* 1996). Discontinuities which are too small to be detected individually on a map or cross-section may be responsible for ductile strain which can be analysed on horizon maps. Outcrop data or analogue models are useful in predictive analyses, since the exact nature and geometry of small and mesoscale faults can be observed. If analogue data are digitized, processed and displayed alongside the seismic data, they can be efficiently used to calibrate seismic interpretations.

The main objectives of this paper are to:

(1) evaluate the potential to map individual structural features, either faults or folds, from geophysical attribute maps by construction of synthetic attribute maps;
(2) assess the possibility of predicting small fault density in a hydrocarbon reservoir on the basis of field observations and horizon attribute modelling;
(3) demonstrate and discuss the use of observational data from analogue outcrops and wells in characterizing the spatial distribution of subseismic faults.

STEEN, Ø., SVERDRUP, E. & HANSSEN, T. H. 1998. Predicting the distribution of small faults in a hydrocarbon reservoir by combining outcrop, seismic and well data. *In*: JONES, G., FISHER, Q. J. & KNIPE, R. J. (eds) *Faulting, Fault Sealing and Fluid Flow in Hydrocarbon Reservoirs*. Geological Society, London, Special Publications, **147**, 27–50.

We aim to show that by combining outcrop data, seismic and well log structural data, it is possible to improve the predictions of both the frequency and spatial distribution of faults in a hydrocarbon reservoir, the Snorre Field in the North Sea. The first two sections of this paper focus on a review of extensional fault deformation, as observed from outcrops and published experiments. These 2D examples aim to illustrate typical surface geometries associated with faults, and the expression they obtain along gridded curves. The main focus of this paper is to describe examples of synthetic attribute maps as modelled from outcrop and seismic data. From these maps, an assessment of the distribution of faults is made in a reservoir where detailed well log structural data are available for validation.

Dip changes and minor faults related to major faults

Rotation of fault blocks is common in several tectonic models for crustal extension, including the domino-model either with soft or hard-linked faults (Walsh & Watterson 1991), and the listric fault model (Gibbs 1983; Dula 1991; White 1992). The models of these authors indicate that block rotation increases with increasing displacement along the master faults. It is widely known that block rotation associated with major faults is accommodated by numerous minor faults. The geometry and spatial distribution of minor faults will depend on the style of fault-related folds. Several examples of fault-related folds from the laboratory and field illustrate this point (Figs 1a–e).

It has been demonstrated that minor faulting is important in adjusting roll-over folds above listric faults (Fig. 1a) (Dula 1991; Higgs *et al.* 1991). Models of rollover deformation suggest that minor faults can cause an inclined simple shear distributed within the hanging wall (Dula 1991; White 1992). Deformation structures found in roll-overs include bedding-parallel faults (flexural slip), synthetic and antithetic faults (Higgs *et al.* 1991).

Forced folds are likely to develop in a sedimentary cover above a basement especially if soft

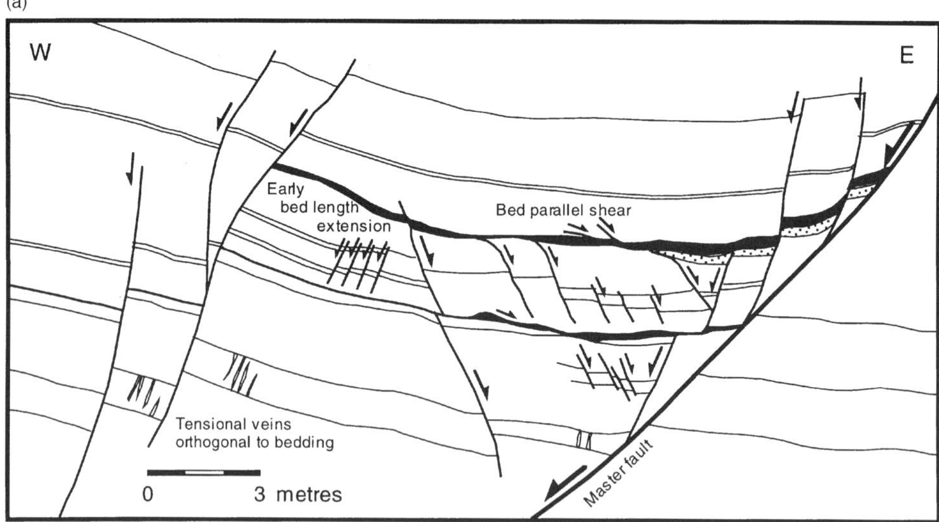

Fig. 1. Examples of extensional fault geometries from outcrops and laboratory. (**a**) A drawing from an outcrop showing a deformed hanging wall of a listric fault, from Higgs *et al.* (1991). Brittle structures include synthetic, antithetic and bedding parallel faults. Redrawn with permission from GSA. (**b**) Drawing of an extensional forced fold in sand and clay made in an experiment. Redrawn from Withjack *et al.* (1990) with permission of the AAPG. (**c**) A drawing showing the spatial distribution of small-scale faults in a forced fold system from Colorado National Monument. The small-scale faults, dominated by synthetic sets, increase in intensity as higher dip is recorded. Redrawn from Jamison & Stearns (1982) with permission from AAPG. (**d**) A relay zone mapped in Kilve beach (labeled F6 on Fig. 8a). The major south-dipping faults (marked by a thick line) show vertical separation of up to 65 m. In the overlap zone, bedding dip is high and minor antithetic faults are common. (**e**) Line-drawing of a fault zone exposed in the cliffs between Lilstock and Kilve (labeled F5 on Fig. 8a). Thin lines are bedding surfaces between limestones and shales. The fault zone consists of several major slip surfaces with variable internal deformation. Bedding planes are locally rotated by normal drag or cut by a web of minor faults.

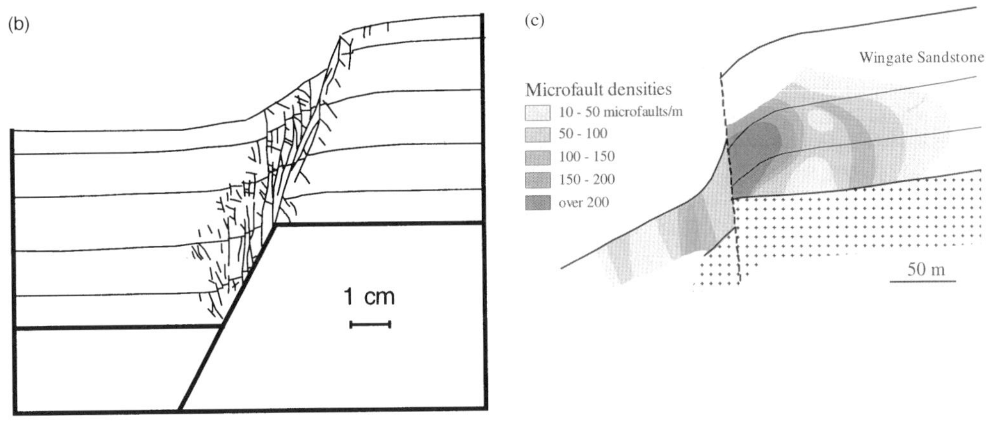
(b)
1 cm
(c)
Wingate Sandstone
Microfault densities
10 - 50 microfaults/m
50 - 100
100 - 150
150 - 200
over 200
50 m

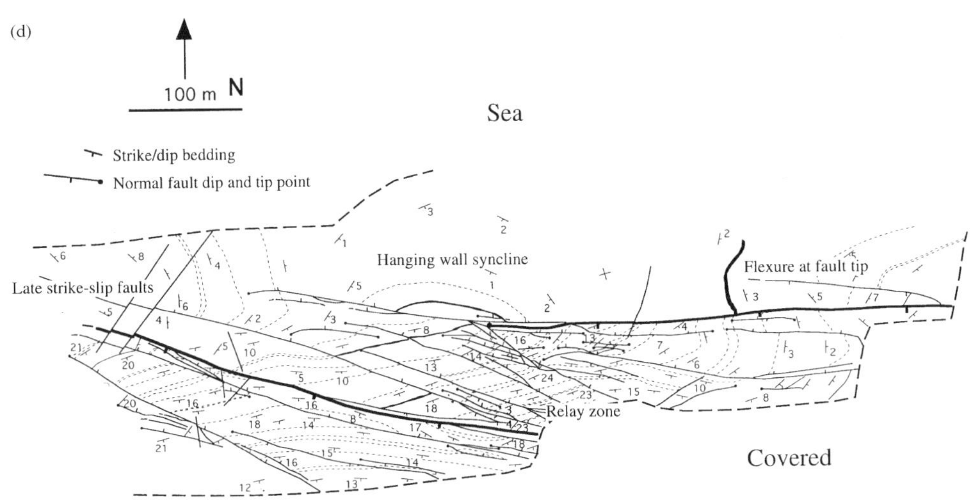
(d)
100 m
N
Sea
Strike/dip bedding
Normal fault dip and tip point
Hanging wall syncline
Late strike-slip faults
Flexure at fault tip
Relay zone
Covered

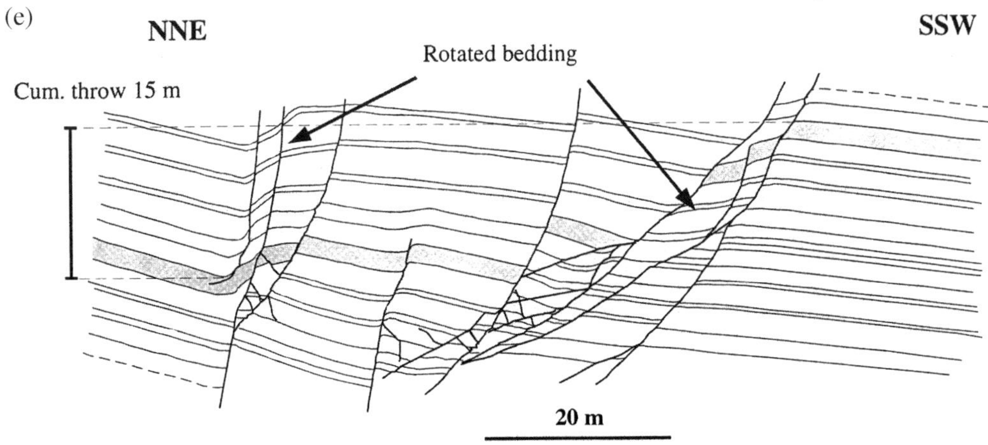
(e)
NNE
SSW
Rotated bedding
Cum. throw 15 m
20 m

layers are present to decouple folded strata from faulted strata. Field and laboratory studies indicate that small-scale synthetic faulting is important to accommodate extensional forced folding (Figs 1b & c) (Jamison & Stearns 1982; Withjack *et al.* 1990). Fault propagation folds can be compared with forced folds in terms of fault growth, and may contain minor faults related to a high stress field.

Relay zones often show minor faults developed in response to high shear strains in planes parallel to and also normal to the bounding faults (Morley *et al.* 1990; Peacock & Sanderson 1994). Field and offshore studies indicate that relay ramps break up by both synthetic and antithetic faults, which may be necessary to adjust the rotation of the ramp (Trudgill & Cartwright 1994). Figure 1d shows an outcrop scale relay zone within an extensional fault system. Hanging wall closures and flexures related to minor faults can be found outside the overlapping area of the two major S-dipping faults. Towards the overlapping zone, bedding dip increases up to 35° and oblique to the major faults. Antithetic faults dominate in the overlapping zone with high bedding dip. The apparent folds within the relay zone show no observable faults cutting the brittle limestone beds. Close observations show that the interbedded shales have variable thickness throughout the fold structures, apparently adjusted by small-scale shear fractures or non-observable structures.

It has long been noted that fault zones are often complex and consist of several slip surfaces (e.g. Antonellini & Aydin 1995; Little 1996). The total fault displacements can be partitioned between two or more slip planes with limited deformation in the intervening rock volume. Figure 1e shows a normal fault zone in which the cumulative displacement is distributed along several synthetic faults with minor throws. Subseismic scale faults may have significance for horizon geometry if they cluster in the rock volume. Complex fault zones, or fault zones consisting of several minor faults can be represented on a seismic section by a curved, and possibly less defined horizon (Jones & Knipe 1996).

Subseismic scale faults are commonly integral parts of larger structures, such as major faults and their related fold structures, which are seismically resolvable (or deterministic). To predict probable fault and bedding geometries on subseismic scales it seems useful to describe the fault and fold styles that appear on a large scale, such as listric faults with rollover folds, domino-style fault blocks, drag folds etc. (Figs 1a–e).

Methods to identify minor faults from surface geometry

Definition of dip, dip azimuth and dip change

The horizon grid (S) used in most commercial software is represented by a function of two perpendicular horizontal directions, x and y. The dip and azimuth attributes can be measured at each sample of the grid and represent the magnitude and direction, respectively, of the gradient vector (Mondt 1993). Dip magnitude is the tangent of the dip angle and is expressed in milliseconds per metre. Dip azimuth is expressed in degrees between 0° and 360°. Surface curvature has been quantified in terms of two orthogonal principal curvatures (maximum and minimum curvature) at each grid point; the product of the two principal curvatures is the Gaussian curvature (Lisle & Robinson 1995). Calculation and display of the Gaussian curvature were not available from the programs used in this study. An approximation of the maximum curvature at each grid point is obtained by calculating the dip magnitude twice from the horizon grid (S):

$$\text{Dip change} = \text{Magnitude (Magnitude (S))} \qquad (1)$$

expressed in milliseconds per metre^2. In practice, this calculation was made by transforming the dip magnitude map into a new 'surface grid', thereby repeating the dip derivation. This method is not adequate to characterize the true curvature of surfaces with various geometries. If dip azimuth is constant from one computation template to the next, the dip change approximates the maximum directional curvature. However, if there are changes in dip azimuth, the approximation becomes inaccurate. These relationships are indicated on Figs 2a and b, which show an ideally shaped cylinder and a cone. Natural folds are expected to show geometries intermediate between these end members. Using dip change to accentuate folds therefore requires a knowledge of the orientation and geometry of the folds.

Identification of faults using dip and dip change: examples in 2D

In this section we illustrate the possibility of identifying small faults on the basis of surface shape, using examples from 2D experiments and simple considerations of subsurface deformation. Whether faults occur individually or in groups is critical, as the lower observable limit determines the minimum size of an individual fault

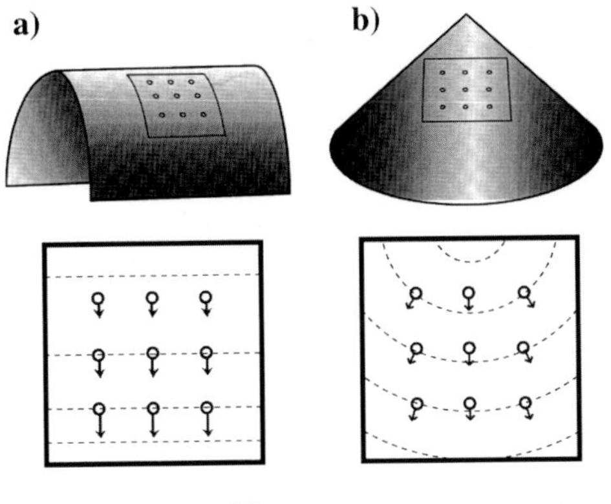

Fig. 2. Principal sketches showing the rate of dip change as calculated from equation (1). Stippled lines on the datagrid are contours. Arrows show the dip magnitude and dip direction, respectively. (**a**) Dip change derived from an ideally shaped cylinder. Dip change equals the maximum directional curvature as dip direction is constant throughout the grid area. (**b**) Dip change derived from an ideally shaped cone. Dip change is zero as the dip magnitude is constant throughout the grid area.

but not necessarily the minimum sizes of individual faults within a fault cluster. Both these cases are treated here.

Identification of individual faults. Figures 3a–d show a line-drawing of a faulted surface which was made experimentally by Cloos (1968). The faulted horizon was digitized from a photograph enlarged to a scale twice that of the performing experiment (Figs 3a–b). A grid was made to snap points at every single mm, which is the lower limit of resolution. The gridded surface is continuous through the faults between the footwall and hanging wall cut-offs. The mean dip (absolute value of $\delta y/\delta x$ on Fig. 3) and dip change (absolute value of $\delta^2 y/\delta x^2$ on Fig. 3) of the surface are calculated at variable horizontal intervals (Figs 3c–d). In the example with small difference intervals (Fig. 3c) a minor fault can be displayed by a peak, if the fault spacings are sufficiently large. Two closely spaced faults may be represented in one peak. A small fault can produce a steep gradient on the dip curve but even steeper gradients are produced on the dip change curve because of the 'kinks' digitized at fault cut-offs. This pattern leads to extremely variable values, ranging from zero to more than five times larger than the dip value. In the example with long difference intervals (Fig. 3d), minor faults are smoothed. Here the major fault and the general bedding geometry (rollover) are well displayed by the dip. The extreme values of dip change are very reduced as the difference interval is increased; the curve is flattened below the dip curve.

Figures 3c–d show that the choice of grid size is crucial for analysing dip and dip changes related to faults. If the difference intervals are close to the lower limit of resolution, each of the smallest features are detected. However, these features may be confused by artifacts produced by mispicks. It is important to note that dip change calculations are very sensitive to small curvatures at short difference intervals. For example, the unfaulted part of the horizon on the left on Fig. 3c shows variations in dip change due to mispicks along the surface. On a seismic horizon map, the areas containing small faults often show poorly defined reflectors and such mispicks are likely to occur. Dip change can be useful to display subtle features if the signal to noise ratio is extremely high, but this is rarely the case for complexly faulted lines that are commonly of interest to a structural geologist. The surface grid can be resampled at a grid spacing larger than the wavelength of artificial curvature. However, as the grid size is increased, the representation of curvatures will be less accurate on the derivated curve. Figure 4a shows that curvatures smaller than the resampling grid will be represented on a larger interval on the derivated curve; two single peaks appear on the plot. Dimensions and shapes of larger curvatures, however, can be more correctly represented (Fig. 4b). The bias caused by the resampling of small curvatures will be highlighted further by the second derivation of the surface. As shown by these examples (Figs 3 & 4), the first derivative (dip) seems more reliable for correct delineation of individual structures.

Groups of small faults. The examples in Figs 1a–d as well as other studies (e.g. Antonellini & Aydin 1995) show that displacements in a rock volume are often accommodated by groups of small faults. Studies of seismic and well data indicate that fault zones often are characterized by poorly distributed and/or folded reflectors (e.g. Jones & Knipe 1996; Hesthammer & Fossen 1997b). Small faults can be characterized by an amplitude anomaly (Townsend *et al.* this volume). Due to the difficulty in resolving individual structures by seismic data, large faults which may consist of numerous individual structures, are commonly interpreted as a single fault on seismic lines. Likewise, in regions with curved seismic horizons no faults are interpreted unless a visible offset is shown by the reflector.

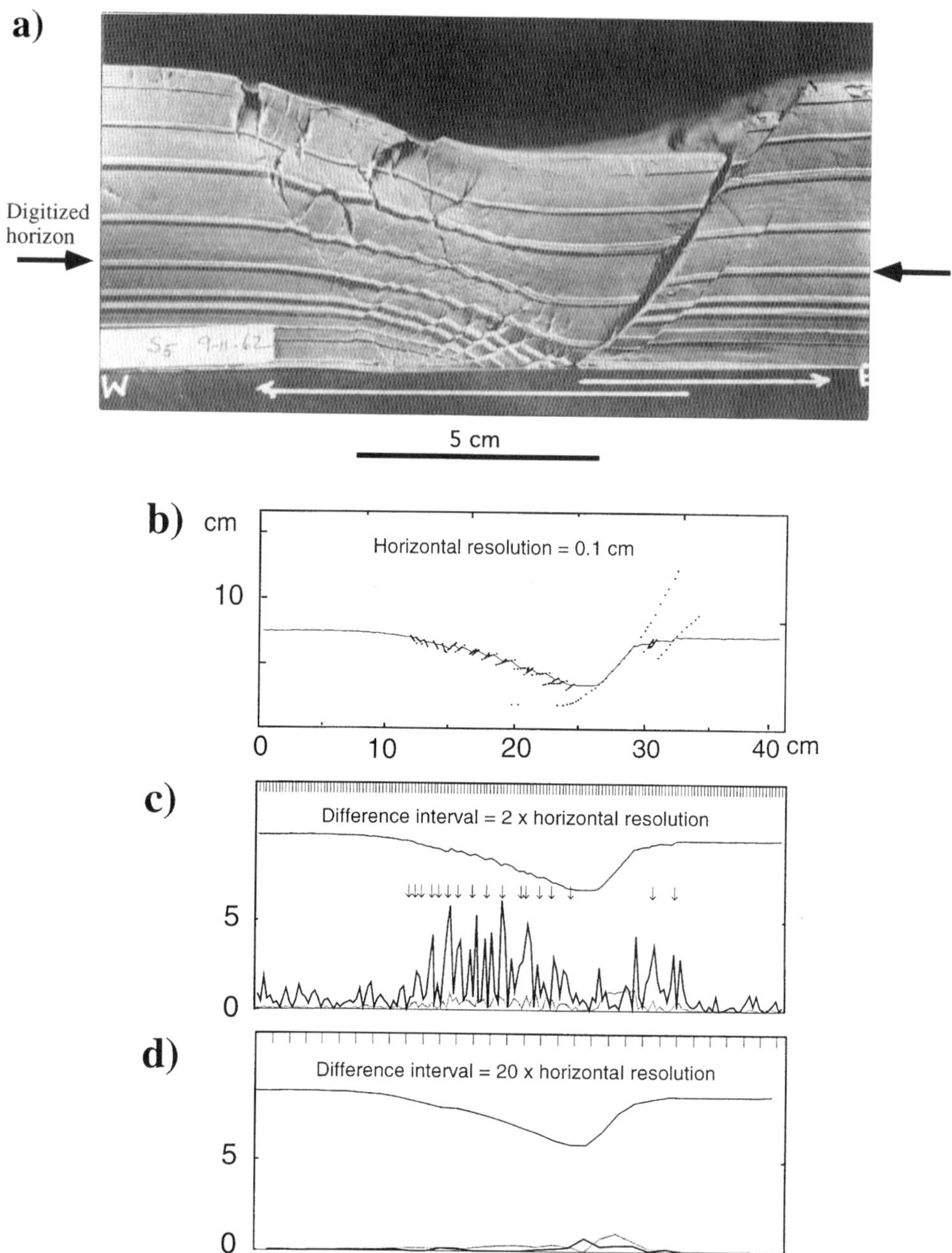

Fig. 3. (**a**) A photograph of a listric fault made in a clay. Numerous synthetic faults are developed in the roll-over to the master fault. From Cloos (1968), reprinted with permission of AAPG. (**b**) A digitized surface cut by the fault. The surface is continuous and connected through the faults between hanging wall and footwall cutoffs. The faults are marked by crosses. Digitizing interval was 1 mm. (**c–d**) First and second derivatives calculated from the digitized surface. Shaded line is the dip (1st derivative = $\delta y/\delta x$) and black line is the curvature (2nd derivative = $\delta^2 y/\delta x^2$). The positions of small faults are marked by arrows in (**c**). High absolute values can be seen when difference intervals are small and correspond to cut-offs of the small faults (**c**). In (**d**) small faults are smoothed and the rollover related to the master fault is highlighted.

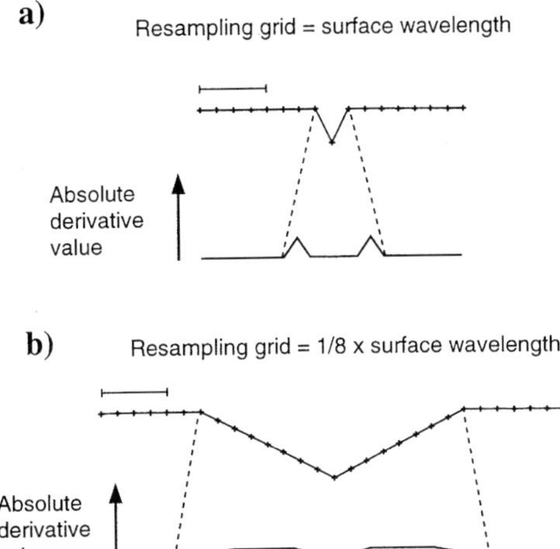

Fig. 4. A simplified sketch showing the expression of curved horizons on a derivated grid. (**a**) When the resampling grid size equals the wavelength of the curvature two, peaks appear on the derivated grid. If the wavelength is more than twice the grid size, a single peak is produced. (**b**) Wavelengths several times more than the resampling grid are more accurately represented.

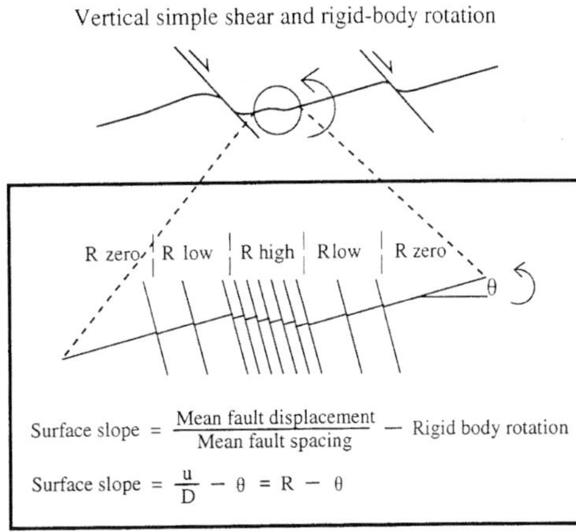

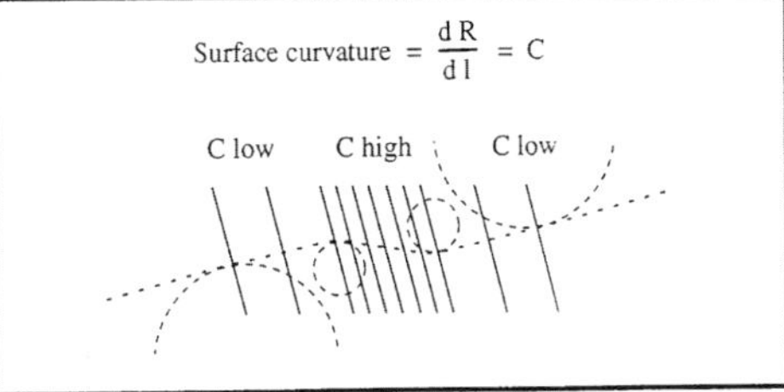

Fig. 5. Principal sketches aiming to illustrate the relationship between small-scale fault density, surface dip, and surface curvature. Deformation mechanisms are small-scale faulting, vertical simple shear and rigid-body rotation. Surface dip equals the ratio of mean fault displacement to fault spacing added by a component of rigid-body rotation. Surface curvature equals the change in ratio of mean fault displacement to fault spacing.

Several subseismic faults can produce curvatures on seismic lines if the ratio of fault displacement to spacing is sufficiently high (Fig. 5). The horizon geometry is then formed by heterogeneous simple shear similar to fold structures described in outcropping shear zones (Ramsay & Huber 1983; fig. 3.6). It must be mentioned that other folding mechanisms are likely to occur. Faults may develop as a result of folding, such as in fault-bend folds or in diapiric-induced folds. In faulted beds, bedding rotation can take place by complex interactions of rigid-body rotation and internal deformation. Information of bedding dip obtained by well logs and/or cores is useful when the rotation mechanism (heterogeneous simple shear, buckling or bending, rigid-body rotation) is to be assessed.

Calculation of the surface dip may be relevant for fault prediction in geological systems where we know the component and the direction of rigid-body rotation (Fig. 5). Rigid-body rotation about horizontal axes often takes place either by rotation of major fault blocks or by differential subsidence unaccompanied by tectonic deformation. These rotation components, however, are generally not determinable from the geological record and it is necessary to use the curvature to quantify subseismic faults. In a cylindrical fold (i.e. Gaussian curvature is zero), fault-related ductile strain can be described in 2D and relates to the maximum directional curvature. If the fold is non-cylindrical, where the Gaussian curvature is non-zero, we expect three dimensional fault strain. Before using dip change to estimate the density of small faults it is necessary to evaluate the fold and fault style in the field. Whether the folds are characterized by domes and basins, or are elongated with a fairly consistent orientation, is important to know for characterization of ductile strain.

Description of study areas

Northern North Sea (Snorre Field)

The Snorre Field is located at the Tampen Spur in the northernmost North Sea between the Viking Graben in the east and northeast and the Møre Basin in the north and northwest (Fig. 6a). The Tampen Spur consists of a series

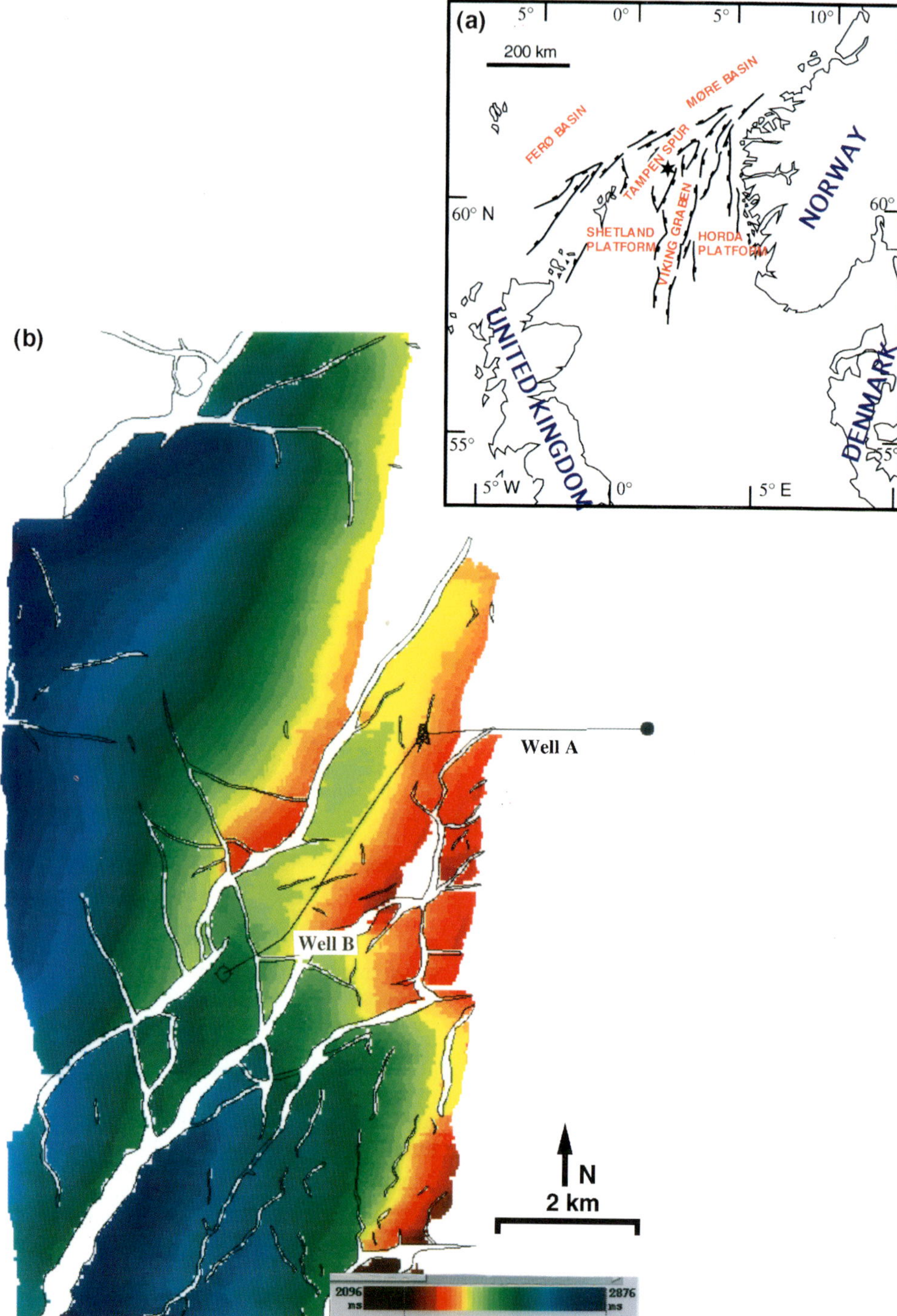

Fig. 6. (**a**) Geographical map showing major fault elements in the northern North Sea and the structural position of Tampen Spur. Study area is marked by a star. (**b**) Time map of the top reservoir horizon in the study area. The horizon is truncated by an erosional unconformity to the east. Fault polygons are shown in white. The paths of two wells described in this paper, labelled A and B, are indicated.

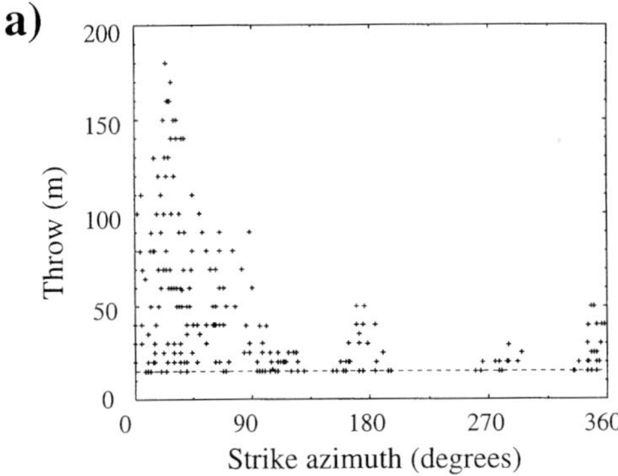

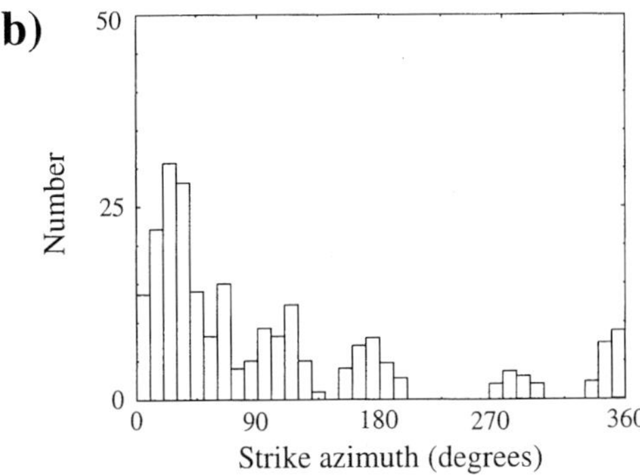

Fig. 7. (**a**) Strike of interpreted faults plotted against fault throw. The lowest resolvable limit is marked by the stippled line at 15 m throw. (**b**) Histogram of fault strikes. Histogram interval is 10.

of fault blocks, which were formed in the Late Jurassic to Early Cretaceous as a complex footwall uplift between the Viking Graben and Møre Basin (Badley *et al.* 1988). The study area displayed on Fig. 6b is located in the south-southeast of the Tampen Spur. Lower Jurassic fluvial reservoir rocks form structural traps in the westerly rotated fault blocks, one of which is the 'Snorre Fault Block'. Internally, the 'Snorre Fault Block' shows major, NNE–SSW trending normal faults and SE-NW to E–W trending cross faults. All three fault sets are present in the study area, although the NNE–SSW trending normal faults dominate (Fig. 7). The domino-style fault blocks characteristic of the Snorre Field were largely developed during the Middle to Late Jurassic rift phase (Badley *et al.* 1988). The Jurassic time horizons dip between 5 and 18 towards west and form major, elongated folds trending N–S to NNE–SSW. The throws of the seismically interpreted faults are typically in the range between 20 and 400 m.

Onshore Bristol Channel Basin (Kilve coast)

The onshore study area is located in Kilve at the southern coast of Bristol Channel (Fig. 8a). The beach between Kilve and Lilstock covers an area of *c.* 0.4×2.5 km on the wavecut platforms below the coastal cliffs (Fig. 8b). Lower Jurassic limestones and shales on these platforms are part of the southern margin of the E–W trending Bristol Channel Basin. These rocks are cut by a series of E–W trending normal faults probably developed during the Late Jurassic to Early Cretaceous (Dart *et al.* 1995 and references therein). Inversion took place during a Tertiary N–S directed shortening event and resulted in reverse reactivation of normal faults and development of conjugate strike-slip faults (Dart *et al.* 1995). These later structures are poorly developed east of Kilve stream where several normal fault zones are well preserved. From east to west the normal fault structures are divided into 6 distinct fault zones (Figs 8b & c). F1 to F5 are north-dipping and typically linked with several minor synthetic faults with individual throws less than 10 m. The cumulative vertical separation across each fault zone varies along strike within *c.* 10 to 60 m. F6 is composed of two overlapping, south-dipping faults of up to 65 m vertical separation. In the overlap zone, a series of closely spaced north-dipping faults dominate. Bedding dips commonly from 5 to 15° towards south between F1 and F5, but dips up to 35° are noted towards southeast in the overlapping zone of F6 (Fig. 1d). F6 is geometrically similar to some relay ramps mapped on a smaller scale on Kilve Beach (Peacock & Sanderson 1994). Lineations on fault planes indicate dip-slip movement.

The folds in the Kilve study area are dominated by anticlines and synclines trending parallel with the north-dipping fault array (F1 to F5). The fault system occurs within the hanging wall of a major north-dipping fault which strikes at Blue Ben (Fig. 8a) (Dart *et al.* 1995). The major fault and the domino-style fault blocks bear geometrical similarities with the NNE–SSW normal fault system in the southeastern part of the Tampen Spur (Badley *et al.* 1988). Although there are differences in scale, the Kilve outcrops can be useful for a qualitative interpretation of the seismic data from the Snorre Field.

Attribute map modelling and structural interpretation

An important tool in the interpretation of 3D seismic data is the powerful horizon processing

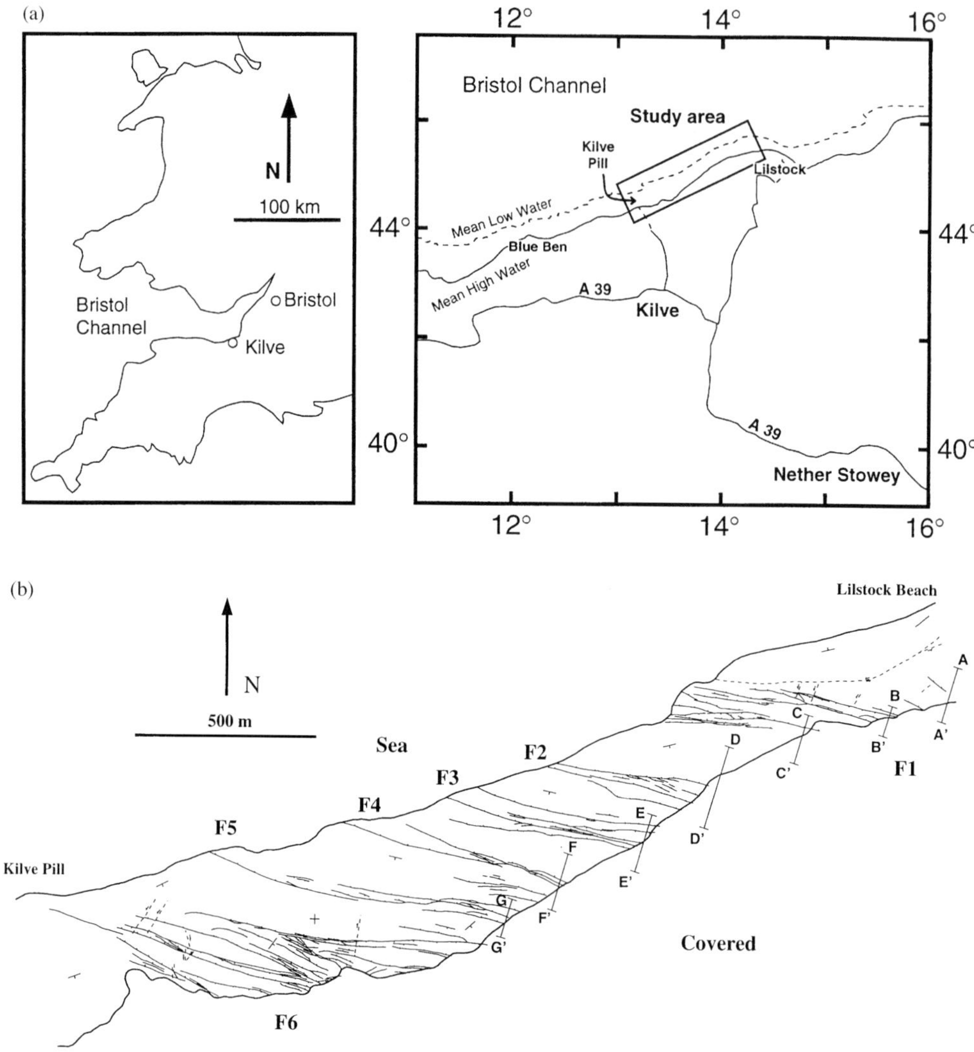

Fig. 8. (**a**) Simplified geographical maps showing the location of Kilve and the study area. The car road A39 is shown. (**b**) Fault traces mapped in the study area in Kilve. The minimum dip separation mapped on the surface is *c.* 0.05 m. Extensional fault zones are labelled F1 to F6. (**c**) Profile line-drawings constructed from the beach cliffs. Profile-lines are indicated in (**a**). The rocks are interbedded limestone and mudstone.

methods which can display and calculate certain spatial derivatives of horizons, such as dip and azimuth. The use of such maps can enhance detailed interpretation of structures in three dimensions and directs attention to certain anomalous features which may be seen on individual seismic sections but are difficult to interpret in three dimensions (Mondt 1993; Jones & Knipe 1996). We have investigated the possibility of delineating small faults and also assessing the variation in small fault frequency from dip, azimuth and curvature modelling. To achieve this, data from the excellent exposed bedding surfaces on Kilve foreshore were combined with seismic data through attribute modelling.

As illustrated on Fig. 2, the calculation of dip change gives a positive value with no information on the rate of azimuth change. This is a drawback of present commercial programs which only take one derivation operation at a time. The main argument for using dip change to characterize curvature in the described examples is that the fault patterns in both areas are dominated by subparallel faults showing the same

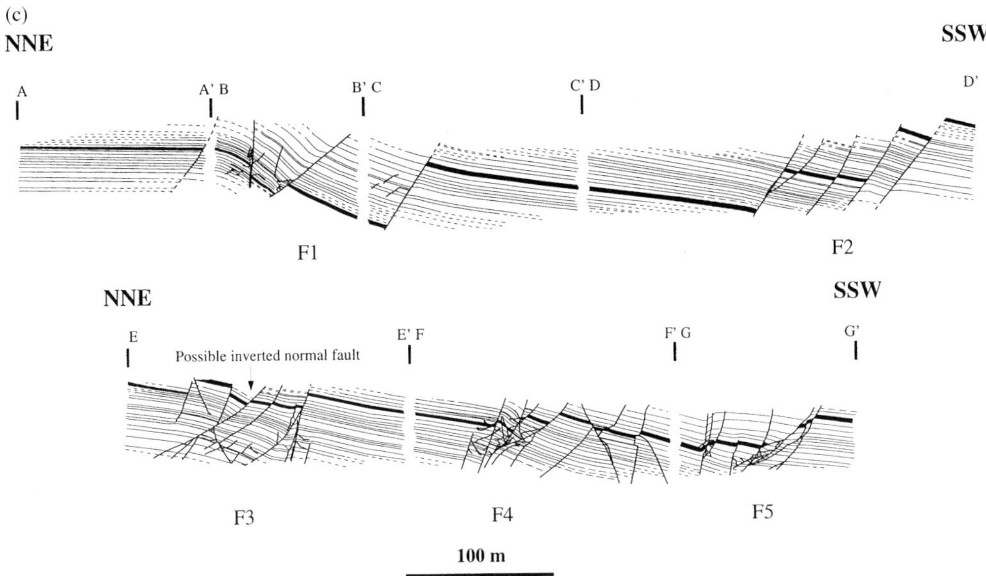

sense of shear, and the horizon in between shows elongated folds with axes subparallel to the faults (Figs 6b, 7d & 8b).

Onshore data

Construction of a faulted bed surface from an outcrop is difficult and must be treated with some reservations. Beds which are not directly observed must be constructed by extrapolation to form a surface. The advantage with the Kilve outcrops is that the limestone bed sequences can be recognized in vertical cliff sections as well on the tidal platforms. The platforms at Kilve display a mean seaward dip of 1.5° and is nearly 100% exposed. Bedding orientation were measured by using a supported wooden sheet with a 360° compass and clinometer (Suunto). The lateral persistence of the limestone beds can be tested by correlations of stratigraphic logs measured on different parts of the platform. A limestone bed that was easily recognizable was traced from the footwall of F2 and across F3, F4, F5 and F6. The following procedure was used to construct a horizon map of this surface:

(i) Orientations of bedding and faults were measured and mapped on aerial photographs that were enlarged to a scale of about 1:1000. The photographs allowed for accurate mapping (±1 m) of fault and bedding traces on the platform. Throws were estimated by calculating the vertical stratigraphic separation across the faults. This calculation assumes dip-slip movements along faults, no vertical variation in fault orientation and constant bed orientation at depth.

(ii) From the map, the faulted bedding surface was hand contoured at 1.5 m intervals. Contouring was performed throughout a 70 m vertical interval. However, the construction becomes more uncertain as the bedding is contoured away up or down from the platform level. Some simplifications were therefore made with respect to the displacement distribution along faults. At greater depth, the displacement along the major faults were systematically decreased and the small faults (throws <0.3 m) were ignored.

(iii) The contoured map was digitized on a workstation and gridded in IRAP with $0.4 \times 0.4\ m^2$ gridsize. The surface grid was in turn imported into a Charisma workstation wherein the attribute modelling was performed (Figs 9a–d). Figures 9b,c and d show the results from modelling of dip, azimuth and dip change, respectively, on the horizon map from Kilve. To better visualize the effects from small faults, the computation template was minimized to $0.8 \times 0.8\ m^2$. Extreme values of dip and dip change were clipped to enhance the resolution between major faults. The most important fault attributes can be identified on each

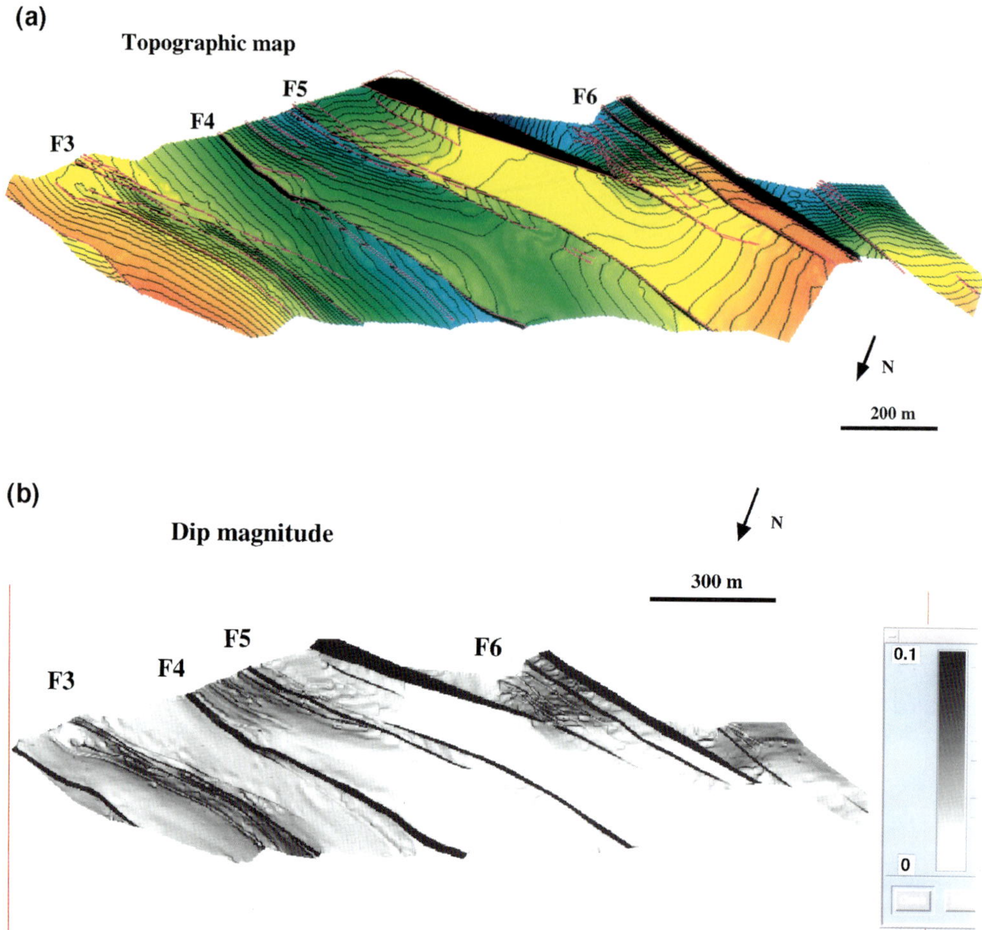

Fig. 9. (**a**) Horizon grid of the mapped bedding surface from Kilve. Contour interval is 1.5 m. Colour contours are from dark blue (at low level) to red (at high level). Fault-bedding cutoffs are indicated by pink lines. (**b**) Dip map; Data were clipped at 6.25 to give better resolution in low dip areas. Major faults are indicated by high dips (black). Also some domains between the major faults show high dips and contain minor faults. (**c**) Azimuth map; (**d**) Dip change map; extreme values were clipped. Edges along major faults show high dip change. Those areas containing high dip are also highlighted by the dip change. Cross-going lines and some sporadic patches of anomalous values are produced by gridding.

synthetic attribute map by comparisons with the topographic map (Fig. 9a).

Interpretation of individual structures: On the dip map faults are marked by dark grey/black polygons, which are fault planes that dip much more steeply than the bed surface (Fig. 9b). Some of the smaller faults present are also displayed, however, they are not well delineated in the areas containing high bedding dip, such as between F4 and F5, and in F6. Some of this problem may be overcome by the transformation of the dip magnitude into dip angle, which gives poorer resolution in low dip areas but may increase the potential to trace individual features with high dip.

Azimuth maps highlight faults which dip in the opposite direction to the faulted surface (Fig. 9c). These faults are well delineated by the blue and red colour contrasts on Fig. 9c. However, colour bar manipulation is important to get the best results from azimuth maps. Faults which dip oblique to the surface may be displayed by trial and error selections of colours and gradients. Several small faults in F3 and between F4 and F5, however, are not displayed. These are

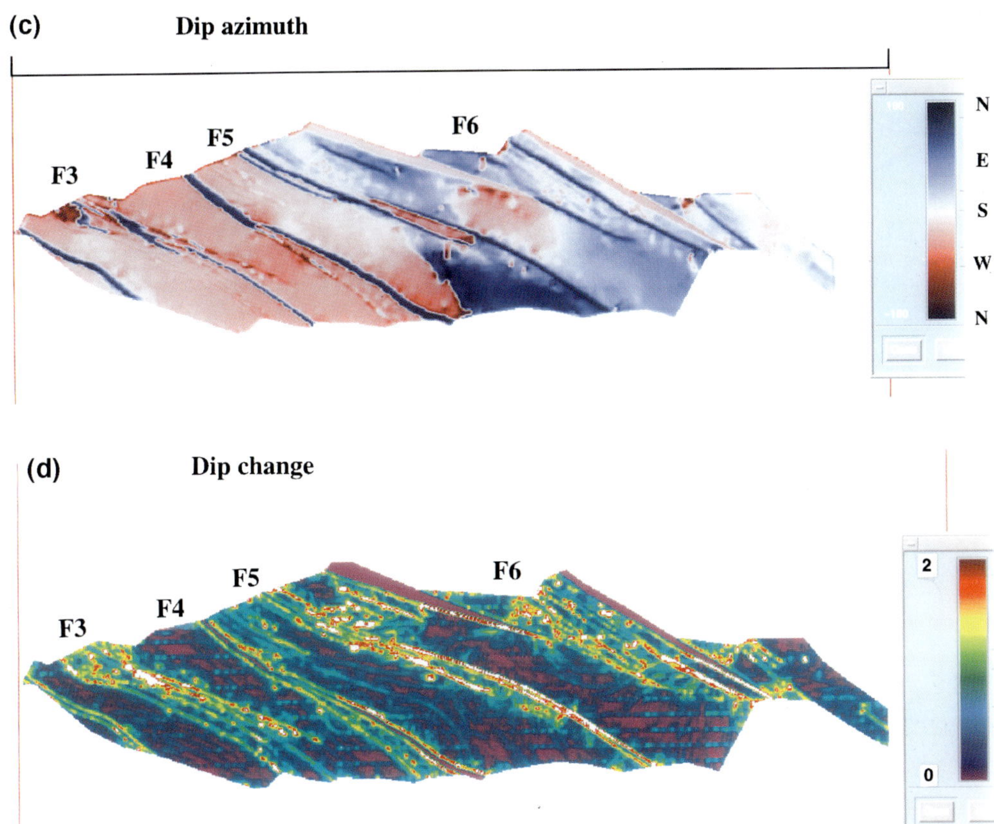

better displayed by the dip (Fig. 9b). In addition to several faults, some flexures and transverse folds are highlighted by azimuth. These include hanging wall synclines and flexures at the tip line of faults. Such folds are not well displayed by the dip.

On the dip change map, major faults are accentuated by the high curvature at fault cutoffs (Fig. 9d). There are, however, some limitations in the tracing of individual faults. Poor sampling and mispicks from the gridding procedure can greatly affect the dip change map. This is indicated by the lines going across faults in areas with low dip change and the patches with anomalous high 'curvature' which are not related to faults. As shown on Fig. 2, such artifacts are typically associated with the calculation of dip change at the lower limit of resolution. To avoid the noise produced from modelling, the resampling grid must be increased and consequently small features, such as faults, will be ignored.

Figures 10b–d show fault maps interpreted from each of the attribute maps on Figs 9b–d. Fault polygons were drawn by hand at places where obvious linear features were defined by a colour change. Non-linear features or features defined by a very weak colour change were not included. The construction clearly depends on the subjective criterias of how faults should appear on the surface. Nevertheless, this approach gives an assessment of how faults can be quantified from different attribute maps. Faults which are put into modelling may be partially represented on the interpreted maps (Figs 10b–d). A single fault may then be interpreted as two or several faults. Also note that azimuth can display the direction of dip of fault surfaces which are not interpreted from other maps. The interpreted maps also show misinterpreted faults which are the result of map generation (stars on Figs 10b–d). Results from map contructions are given in Table 1. Dip map gives the most positive results as more faults are correctly interpreted from this map compared with the others. Table 1 also shows that dip change has a poorer potential to trace individual faults.

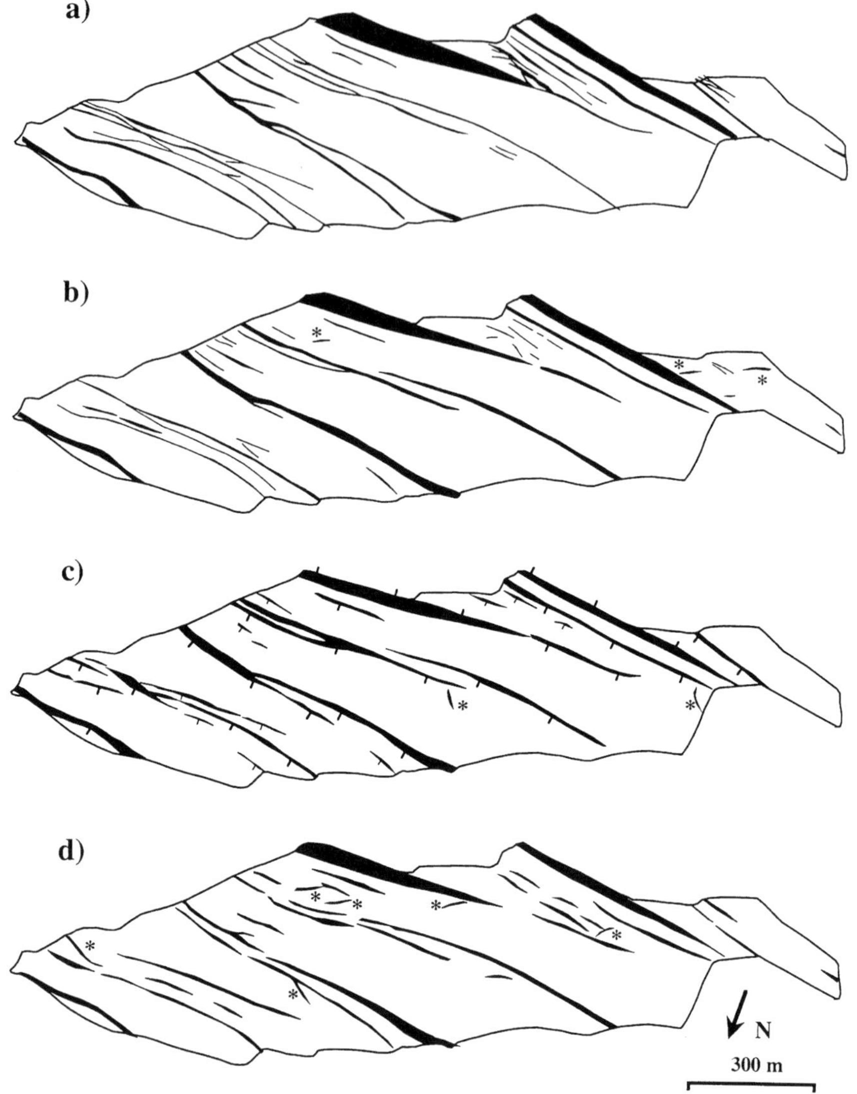

Fig. 10. Fault maps interpreted from the attribute maps modelled from the horizon grid from Kilve (cf. Fig. 9). (**a**) Fault polygons constructed on the original horizon grid. (**b**) Interpreted fault polygons from the dip map (cf. Fig. 9b). (**c**) Interpreted fault polygons from the azimuth map (cf. Fig. 9c). Also indicated is the dip direction of faults. (**d**) Interpreted fault polygons from the dip change map (cf. Fig. 9d). Stars indicate features on the attribute maps which may be misinterpreted as faults.

Fault density related to dip change: The theoretical considerations above suggest that the curvature of a gridded horizon cut by subparallel faults is a function of the fault density and fault displacement (Fig. 5). The fault array exposed in the cliff sections on Kilve shows dominantly north-dipping faults (Fig. 8c). To see if a relationship exists between dip change and small fault density, further modelling was performed on the horizon from Kilve. The original topographic map from Kilve (Fig. 9a) was transformed into a 'time map' by a multiplying factor of 0.66, so that values were comparable with those obtained from offshore maps. Dip change values were resampled at grid sizes ranging from $1 \times 4\,m^2$ to $12.5 \times 50\,m^2$, and domains of dip change were displayed in five colour ranges on separate maps. Sampling errors in areas containing many faults of limited length may be apparent if the resampling interval

Table 1. *Interpretation of faults from the attribute maps shown in Figs 10b–d*

Attribute type	Dip	Azimuth	Dip change
Number of faults represented on the map[1]	28 (65)[2]	20 (47)	22 (51)
Total trace length of interpreted faults[3] (m)	8190 (72)	6080 (53)	6310 (55)
Artifacts misinterpreted as a fault	3	2	6

[1] Includes fault segments which are fully or partially represented on the map.
[2] Numbers in parentheses refer to percent of total.
[3] Does not include wrongly interpreted faults.

is changed. Also, data were sampled from gridded cross-sections (Fig. 8c) to see if there are consistent results obtained from 2D and 3D maps (Fig. 11a). The fault density plotted versus dip change appears to follow a logarithmic function (Fig. 11b). This relationship results from the spatial distribution and the relative displacement distribution of faults and is difficult to explain. The systematics of these characteristics are still not well documented. However, the relevance of the results shown in Fig. 11 can be tested by complementary data from offshore reservoirs.

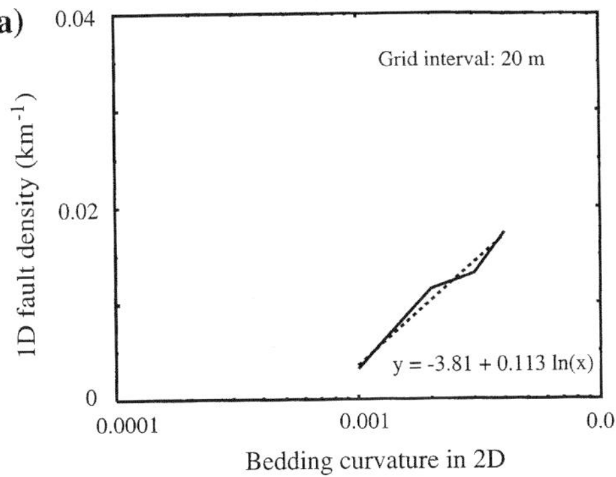

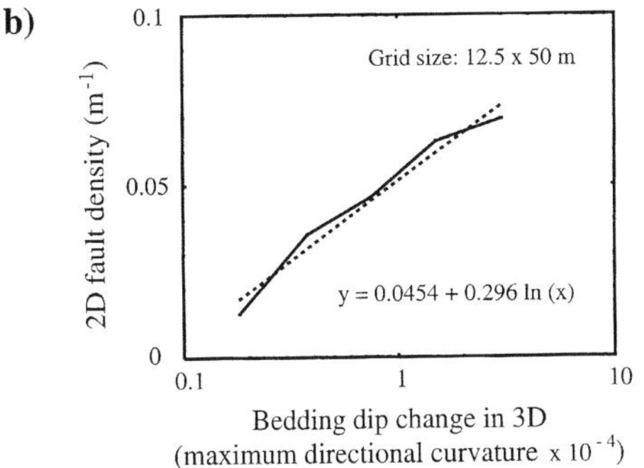

Fig. 11. (**a**) 1D fault density plotted against surface curvature as calculated from the composite profile constructed from Kilve (Fig. 5b). Grid spacing was 20 m. (**b**) 2D fault density (i.e. cumulative length of fault traces divided by km^2) plotted against dip change from the horizon in Kilve. Grid size was 12.5×50 m. Dashed lines are graphs of the logarithmic equations.

Offshore data

Automatic tracking methods within seismic datasets have been successfully used for interpretation of subtle features, such as minor faults and flexures (Bouvier *et al.* 1989; Mondt 1993). The quality of such methods will depend on the signal/noise ratio and also the presence of faults. This can be checked by counting the number of gaps and irregularities along the traced line (e.g. Mondt 1993). On the present seismic sections, a number of seed lines were interpreted as control grids, and subsequently the interpretation was expanded by automatic tracking. Dip and azimuth data were resampled on grids of 12.5×50 m size, whereas the dip change was resampled every 25×100 m. A median filter was used on the input grid to reduce the curvature from the smallest features which are often produced by mispicks. By filtering and increasing the resampling interval, resolution of the dip change decreases but the representation of the larger tectonic curvatures is improved.

Attribute map interpretation. Attribute maps were modelled from the top reservoir horizon shown on Fig. 6b. The northwestern parts of the maps show the highest signal/noise ratio and have the best potential for more detailed structural interpretation. On the dip map (Fig. 12a), features probably representing small faults are located at the end of seismically-interpreted faults (A and B). Azimuth displays some south-dipping flexures (C) and (D), which are not displayed by the dip (Fig. 12b). Gentle folds trending east–west are apparent (E).

Several features represented on the dip and azimuth maps (A–E), are also indicated on the dip change map (Fig. 12c). However, these features are not very well delineated and may

(a)

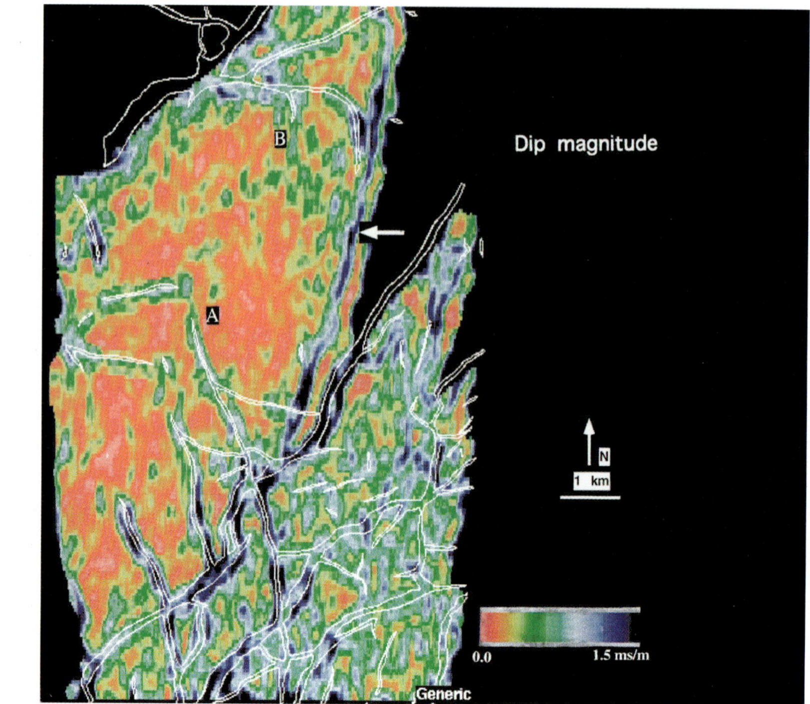

Fig. 12. Attribute maps modelled from the top reservoir horizon in Snorre Field. Interpreted fault polygons are indicated by white lines. The southeastern area of the maps has a poorer signal/noise ratio, shown by anomalous high attribute values. (**a**) Dip map; anomalous values can be found at the tip lines of some faults and between the two fault sets (A) and (B). Arrow shows a major feature which is an erosional truncation line against the overlying base Cretaceous. This feature is displayed on all three maps. (**b**) Azimuth map displays sharp flexures (C) and (D) not apparent on the dip map. Azimuth also highlights gentle open folds and flexures trending E–W (D). The features by (A) and (B) in (**a**) are not displayed by the azimuth. (**c**) Dip change map displays smaller features that are also shown on the dip and azimuth map (A–D) but they are not well delineated. Broader areas of anomalous dip change are found by (B) and between (D) and (E).

be subjects of ambigous interpretation. More distinct are the broader areas of high dip change (by (B) and between (D) and (E)). With reference to Figs 1, 3, and 11, and the earlier discussion above, such areas may be important in the mapping of subseismic faults.

The change in dip magnitude was modelled from several horizons in the Snorre Field (shown on Fig. 13a). The grid size used was 25×100 m. Values were subdivided into *low*, ranging up to 0.185×10^3 m/s^2, *intermediate* ranging from 0.185×10^3 m s^{-2} to 0.75×10^3 m s^{-2} and *high*, from 0.75×10^3 m s^{-2} to 1.5×10^3 m s^{-2}. From the relation on Fig. 11b, it is proposed that the regions with greater dip change represent higher small fault densities. High dip change can be found close to seismically interpreted faults probably associated with their damage zones. Such damage zones often contain numerous faults of different sizes as documented from field studies (e.g. Jamison & Stearns 1982) and seismic data (Jones & Knipe 1996). Areas of high dip change are also found in between the faults but these areas cannot be predicted from the spatial distribution of faults.

Well data. The use of structural information from wells may be of crucial importance for calibrating detailed structural interpretations from seismic data, as well as for validation of methods for estimating small-scale fault density. In particular, horizontal wells with high resolution downhole sampling tools gives valuable information, as fault data are sampled in 1D transects along seismic horizons. Two wells are presented here (well path labelled A and B on Fig. 6b) which are horizontal and dip *c.* 30°, respectively. The length of these wells (more than 1000 m) allows for data being compared with features which appear on the seismic attribute maps.

(b)

Dip azimuth
B
C
A
E
D
N
1 km
N
S
N

(c)

Dip magnitude change
B
C
A
E
D
N
1 km
0.0015 ms/m²

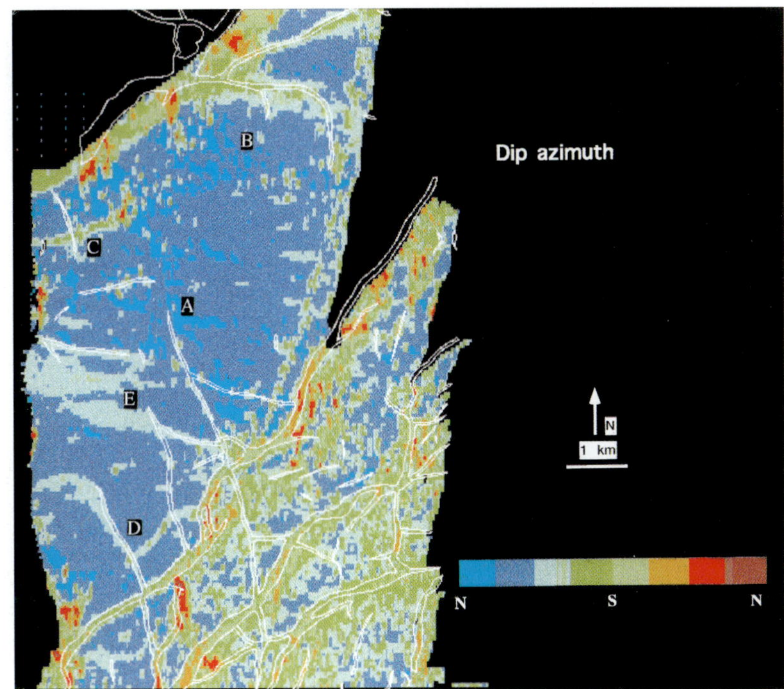

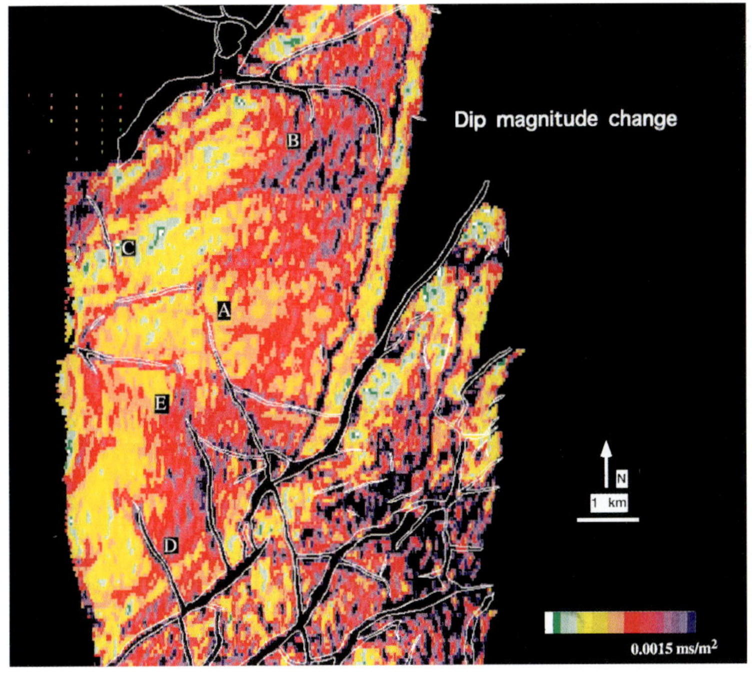

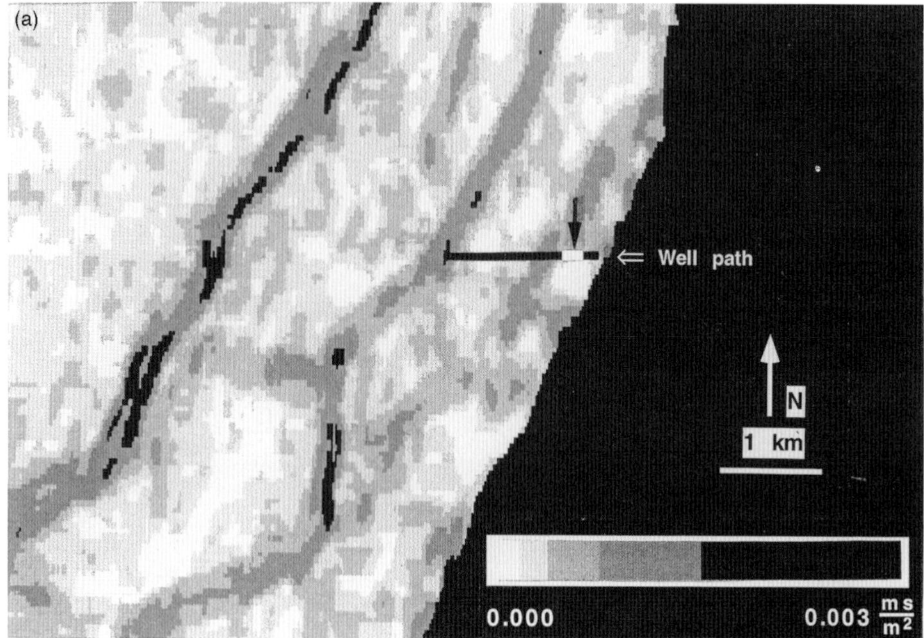

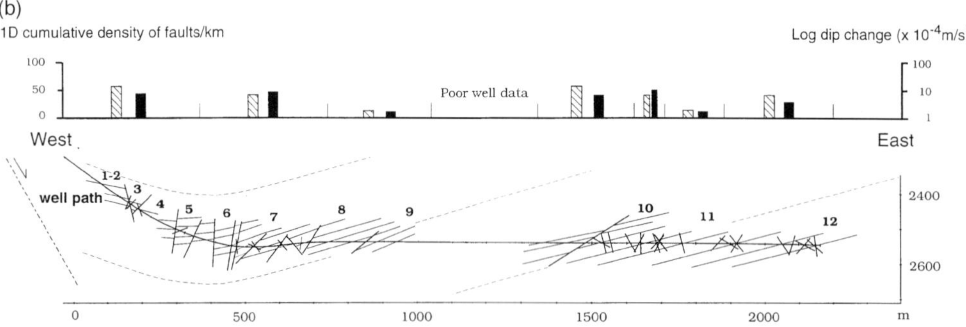

Fig. 13. (**a**) Dip change map modelled from an intra-reservoir horizon (cf. horizon 6 in (**b**)). Well A (black line) penetrates through domains with dip change ranging from low (white), via intermediate (lighter grey) to high (darker grey). From the relationship shown on Fig. 11b white and dark grey areas should show fault densities ranging from *c.* 40% to 145% about an average value. The intersection point between the well and the horizon is shown by the arrow and white spot. See Fig. 6a for the location of the well. (**b**) Interpreted section from an Ultrasonic Borehole Imaging (UBI) logging tool in well A. Filled bars are 1D fault density observed from the UBI log (vertical scale on the left). Shaded bars are dip change values modelled from horizon 2 (from 0–1000 m on the horizontal scale) and from horizon 6 (from 1350–2200 m on the horizontal scale) Vertical scale shown on the right.

Fault frequencies and bedding orientation were obtained by an Ultrasonic Borehole Imager (UBI) tool. Under ideal conditions, this tool permits us to observe faults with throws down to the borehole diameter (*c.* 0.1 m) if no borehole breakouts occurs. However, the interpretation of such tools usually exclude fault planes oriented perpendicular to the well path and faults striking parallel to the well. To compare fault frequency data from the wells with the predicted data, fault data need to be corrected for the variation in fault strike:

$$K = \int_{\theta_{\min}}^{\theta_{\max}} |\sin(\theta - \psi)| N(\theta)\,\mathrm{d}\theta \qquad (2)$$

where K is the correction factor, $N(\theta)$ is the relative number of faults with strike θ, and ψ is the direction of the well. Strike distribution determined for the interpreted faults (Fig. 7b) was used for this equation.

The significance of variable dip change for subseismic fault prediction can be tested from

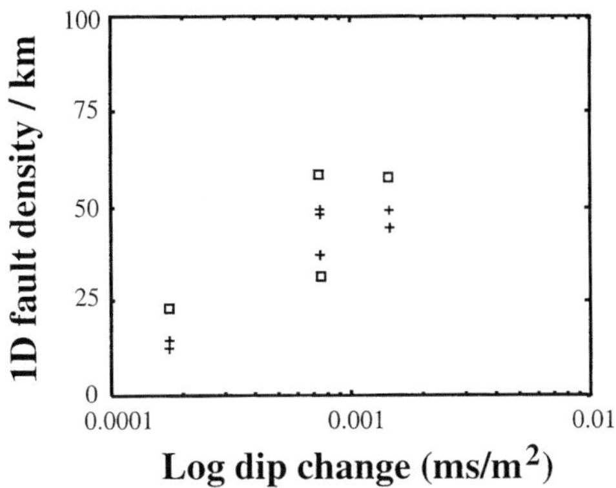

Fig. 14. 1D fault density observed from the wells plotted against modelled dip change. Values have been corrected for the strike variation of faults for the particular wells. Crosses are data from well A and squares are data from well B.

the well data. Figure 13a shows a dip change map modelled from an intra-reservoir horizon indicating the path of well A and its intersection point with the horizon. The well penetrates through regions showing low, intermediate and high dip change around this intersection point. As the well penetrates through higher stratigraphy towards the west, an overlying horizon was used in dip change modelling to compare with well data. Fault densities observed from the well are estimated in intervals with different dip change (Fig. 13b). Figure 14 shows the observed fault density plotted versus modelled dip change. The plot shows that the observed fault frequency is highest in regions with intermediate and high dip change whilst the lowest frequencies are in regions with low dip change. A similar procedure was used for well B, although fewer data were obtained because of the shorter lateral extent of this well. This plot also shows that the highest fault frequencies ($>50\,km^{-1}$) are observed in regions with intermediate and high dip change (Fig. 14).

Well data combined with data from Vertical Seismic Processing (VSP) can give useful information about the fault distribution and its association with seismic horizon geometry. A seismic section parallel with well B and a seismic section obtained by VSP are shown on Fig. 15a. On the seismic section, the reflector representing horizon 1 dips towards the west and is offset by one fault. An interpreted cross-section from the UBI log indicates that this horizon is cut by a number of west-dipping faults (Fig. 15b). Throw across these faults cannot be detected by the UBI log, but one or more faults must have produced a seismic offset. Bedding is almost horizontal or gently east-dipping around this fault (arrows on Fig. 15b). A plausible interpretation for the westward dip of the seismic horizon is the occurence of many small faults which are not detected individually, but produce a significant strain to affect the horizon geometry.

Discussion

Outcrop models presented in this paper show that dip and azimuth maps efficiently define structural trends and help to delineate faults on surfaces. Recent studies have demonstrated the efficiency of such maps in structural interpretation of larger fault zones and detection of tectonic disturbance in apparently unfaulted areas (Jones & Knipe 1996). Mapping of faults from dip and azimuth maps has, however, some limitations. If faults dip in the same direction as the faulted surface they will not be visible on the azimuth map. This may often be the case in relay ramps as described from field studies (Peacock & Sanderson 1994) and in heavily drag-folded beds close to major faults. Likewise, if bedding is folded by ductile drag at the tip line of the fault, the variation in dip across the fold limbs may be too small to be displayed by the dip map. If the wavelength of the fold is several times larger than grid size, it may be displayed by the azimuth map (D on Fig. 12b). These limitations should be considered when maps are to be implemented in further analysis. The lateral continuity of individual features on dip and azimuth maps may depend on variable dragfolding along faults, rotation of small faults in large fault blocks etc. Identification of linked and unlinked faults, juxtaposition analysis, and definition of structural compartments must therefore be carefully assessed from both dip and azimuth. Figure 10 shows that a single fault can be interpreted as two individual faults by dip or azimuth, depending on the displacement distribution and dip direction of faults. The second derivatives, either the change in dip or azimuth, are more sensitive to mispicks at the lower limit of the resolution and have poorer potential to trace individual features (Figs 9d and 10).

Modelling of dip change can be useful for rapid determination of the areal distribution of small faults which contribute to curvature of seismic horizons. Such methods also allow for manual tracing in regions with poorly defined reflectors, such as in complexly faulted areas. When dip change is to be calculated, attention should be drawn to possible noise interferences in order to choose appropriate grid sampling

(a)

(b)

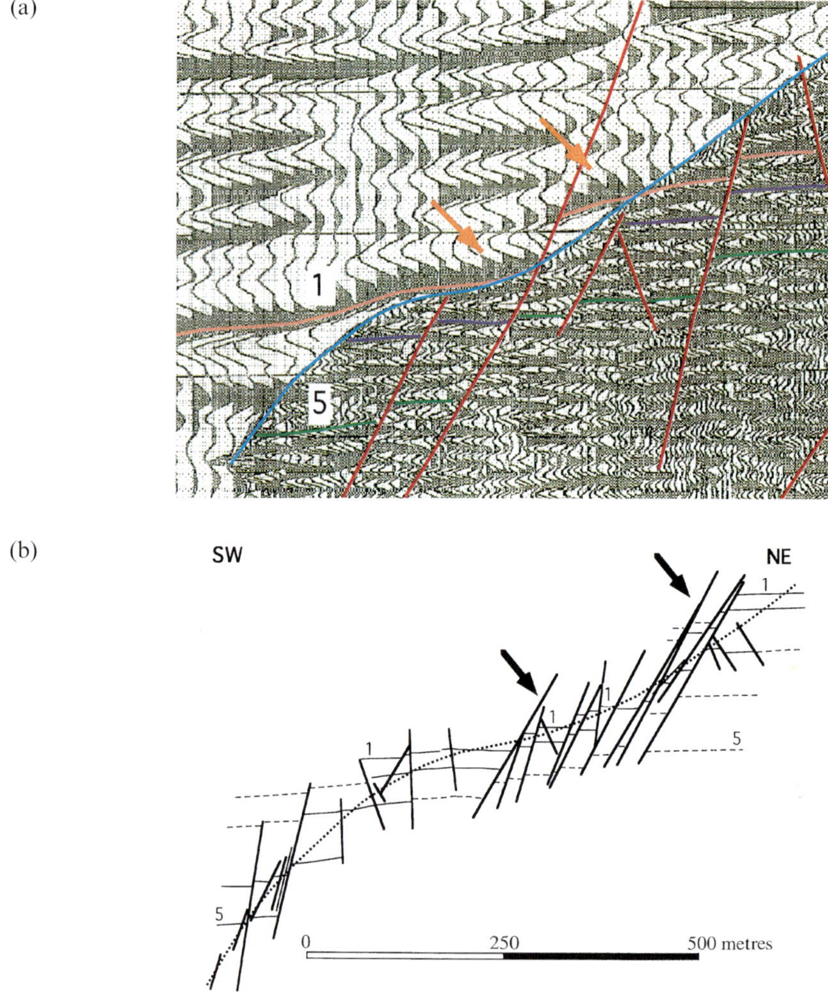

Fig. 15. (**a**) An interpreted seismic section showing the path of well B and a seismic section produced by Vertical Seismic Processing (VSP) from the well. VSP section are shown below the well path (blue line). Horizons are, from top to bottom, labelled 1 to 5. Horizon 1 (top reservoir horizon) is a strong reflector easily traced on the seismic sections. The horizon is offset at one place by a fault (between the arrows). See the horizontal scale in (**b**). (**b**) Interpreted section from the UBI log from well B. Well path is indicated by the stippled line. UBI data reveal a gently east-dipping or horizontal bedding cut by a series of west-dipping normal faults. VSP section in (**a**) shows gently horizontal to west-dipping reflectors. The apparent westward dip of the seismic horizon by the arrows in (**a**) is likely produced by the small normal faults.

sizes. The calculation of dip change is a good approximation to the true maximum curvature if the system contains elongated troughs and ridges which approach cylindrical shapes. If the horizon shows variation in dip azimuth but minor variation in dip magnitude, this calculation becomes inaccurate. If folds are characterized by domes and basins it may be useful to display the derivative of dip azimuth. Domes are typical in regions with salt intrusions (Lowell 1985). Outcrop studies of bending folds around salt diapirs show that bedding curvature can be correlated with the intensity of tension fractures or deformation bands depending on the dominant mode of deformation at the outcrop scale (Underhill 1988; Antonellini & Aydin 1995). Azimuth change across domes may therefore highlight areas of probable tectonic disturbance. The use of dip change to locate areas of probable faults will benefit from a step-by-step approach. After an assessment of the folding style within the basin and the selection of high-amplitude reflectors, the different attribute maps can be interpreted. Dip change maps can be rapidly evaluated by comparisons with other types of seismic attribute maps (reflection amplitude, dip, azimuth etc.) and with interpreted fault maps. When major faults are identified on the dip change map, it may be possible to locate other areas with anomalous high dip change. Abrupt variations of dip change from one grid to the next should be treated with care, since they may be produced by noise. The interpreter must also bear in mind other geological factors which can contribute to curvature, such as sedimentary channels, large-scale dunes, syndepositional deformation and erosional surfaces. It is therefore useful to integrate other geological knowledge about the region, e.g. the depositional style and spatial distribution of different lithologies to evaluate the seismic horizon geometry. Differential compaction has been suggested by Kerr & White (1996) for the development of hanging wall synclines but are rarely documented elsewhere. Non-geological factors contributing to curvature of seismic horizons may be interferences with artifacts such as multiples and diffraction lines occuring at fault terminations, by sedimentary channels etc.

Predicting fault density from horizon curvature will rely on some *a priori* assumptions about the spatial distribution of faults and their expression on seismic horizons. The Kilve outcrops have been useful for surface modelling because of the outcrop qualities but, the lithology and structural style are different from many hydrocarbon reservoirs in North Sea. Quantitative use of empirical relationships must therefore be treated with caution. The relationship between dip change and fault density shown on Fig. 11 implies that small and large faults are unevenly distributed in the rock volume and they cluster in patterns corresponding to the modelled dip change. The first assumption is of little controversy and is well documented by field studies (e.g. Jamison & Stearns 1982; Knott *et al.* 1996), whereas the second is critical and needs to be investigated by further studies. Data can be obtained from line construction from fault zones in 2D sections (e.g. Fig. 11a). However, such data give a poor resolution of both curvature and fault density, which are measured in one dimension. 3D construction of faulted surfaces may provide quantitative relationships (Fig. 11b), which in turn can be used to predict small faults in the subsurface. The problem in using analogue models is the

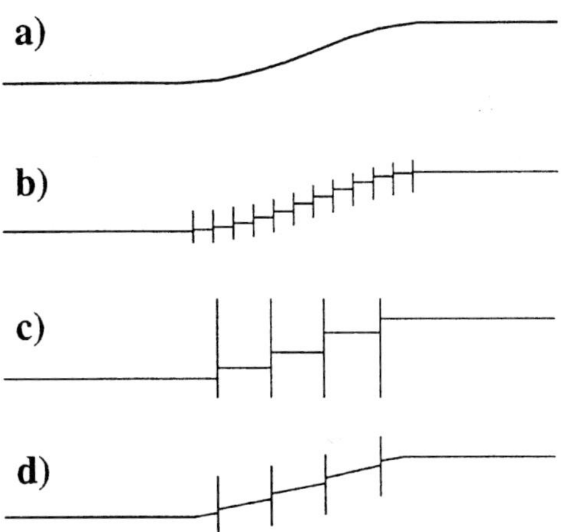

Fig. 16. Principal sketches of seismically ductile strain accommodated by fault arrays of different geometries. (**a**) Seismic horizon curvature. In (**b**) and (**c**) strain is entirely accommodated by faulting. In (**d**) strain is accommodated by faulting and small-scale ductile deformation (particulate flow). Note the variable fault density in (**b**) and (**c**) related to the displacement distribution of the fault array.

need for exceptional exposures and easy access of data. Alternatively, outcrop models may be constructed from small-scale examples, or one may obtain layer information from mechanical models.

Empirical relationships between dip change and fault density will largely depend on the structural style and the rheology of the rocks at the time of deformation. Lithology and the state of lithification are factors influencing the fault geometries, such as their spatial distribution and the proportion of ductile drag contributing to the fault displacement. Below the resolvable limit of seismic reflectors (between 15 and 30 m depending on burial depth, processing parameters etc.), we may expect a wide range of geometries and scaling characteristics of faults as a result of lithological variations and various depths of faulting. Figure 16 shows how a curved seismic horizon can be accommodated by different arrangements of small fault arrays. The density of faults, which can be predicted from horizon curvature, will depend on the relative distribution of displacements in the fault array. In unconsolidated sediments, strain may be accommodated by ductile deformation (particulate flow) with less strain caused by fault deformation. Core studies have shown that faults exhibit a variety of geometries depending on original mineralogy, diagenesis, and cementation process along fault zones

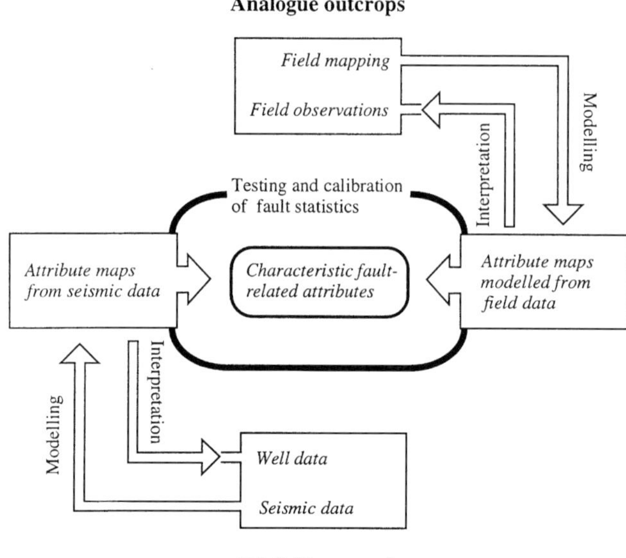

Fig. 17. A schematic presentation of the methodology for structural interpretation as proposed from this study. Attribute maps modelled from an outcrop analogue are interpreted on the basis of field observations. Geophysical attribute maps are interpreted by comparisons with modelled fault-related attributes constructed from the outcrops. Empirical relationships and methods used for fault predictions are calibrated by detailed structural data from available wells. See text for further discussion.

(Knipe 1992; Sverdrup & Bjørlykke 1997). Various folding mechanisms may occur in fault-related folds, such as flexural slip, shear folding, bending etc. Conjugate faults may be present in folded zones. It is therefore unrealistic to give exact estimations of fault density from seismic horizon curvature on the basis of empirical relationships. Horizontal well structural data seems very useful in assessing the folding mechanism that occurs below the seismic resolution (e.g. Fig. 15), and when combined with attribute maps can give quantitative inputs.

The methodology for the analysis of seismic attribute maps is depicted in Fig. 17. The main advantage with this method is the potential to integrate two independent data sources, which serve as both qualitative and quantitative assessments of seismic attribute map interpretation. Outcrop observations allow characterization of typical structural styles at the subseismic scale, e.g. relay ramp geometries, drag folds, roll-over folds, domino-style fault blocks. Integrating the knowledge of these styles into the description of faults from seismic attribute maps, means that the interpretations can be made with greater confidence. However, true quantitative predictions on the basis of outcrop data seem to be less viable because of the variability of structural styles observed at the subseismic scale. Horizontal wells gives new perspectives on how well data can be integrated with seismic data, as envisioned on horizon maps. Horizontal well structural data are the source for quantifying lateral variations of structural style, bedding orientation and fault frequency. Integration of such information with seismic data will improve the utilization of outcrop data in subsurface analyses.

Conclusions

(1) The change in horizon shape as seen from seismic data, can often be attributed to subseismic faults. The mechanisms of folding below the scale of resolution need to be known to predict the spatial distribution of small faults from seismic horizon geometry. In fault-related folds, information about the master fault geometry is crucial to predict the distribution of small faults.
(2) Automatic tracking combined with dip and azimuth modelling of high quality seismic data can contribute to detailed structural interpretation. Synthetic attribute maps constructed from Kilve outcrops indicate that small faults are most likely to be displayed

by the dip map. Azimuth seems efficient for the display of folds and flexures, but can display more subtle features which dip in directions other than that of the faulted surface. The use of dip and azimuth in detailed structural characterization will be improved by drawing attention to the style of folds and faults in the area. Combining dip and azimuth brings 3D geometries to the interpreters attention and increases the usefulness of both maps.

(3) Detailed field observations and modelling of synthetic attribute maps from Kilve in Somerset indicate that fault density increases as a logarithmic function with the maximum directional curvature. This relationship is not well established, since software programs only give an approximation of the true maximum directional curvature.

(4) Dip change modelling can be used to highlight areas of probable subseismic fault occurence. Horizontal well structural data from the Snorre Field in the North Sea show that the highest fault frequencies occur in regions with intermediate and high dip change.

(5) Subseismic fault predictions must rely on assumptions of fault geometry and deformation mechanisms below the scale of seismic resolution. We advise against inferences of the main deformation mechanism at subseismic scales on the basis of analogue outcrops or synthetic models. More realistic predictions of subseismic fault density can be obtained by detailed observations from well data and determination of subseismic fault distribution, real bedding dip and apparent dip of seismic horizons.

The research was funded by Saga Petroleum ASA and by the Norwegian Research Council. The authors thank Saga Petroleum ASA for giving us permission to publish the results in this paper. We also thank A. Edvardsen for providing the seismic sections of which Fig. 15a was scanned. The comments and suggestions from G. Jones, C. Townsend and an anonymous reviewer are greatly appreciated.

References

ANTONELLINI, A. & AYDIN, A. 1995. Effect of faulting on fluid flow in porous sandstones: geometry and spatial distribution. *The American Association of Petroleum Geologists Bulletin*, **79**, 642–671.

BADLEY, M. E., PRICE, J. D., RAMBECH DAHL, C. & AGDESTEIN, T. 1988. The structural evolution of the nortern Viking Graben and its bearing upon extensional modes of basin formation. *Journal of Geological Society of London Bulletin*, **145**, 455–472.

BOUVIER, J. D., KAARS-SIJPESTEIJN, C. H., KLUESNER, D. F., ONYEJEKWE, C. C. & VAN DER PAL, R. C. 1989. Three-dimensional seismic interpretation and fault sealing investigations, Nun River Field, Nigeria. *The American Association of Petroleum Geologists Bulletin*, **73**, 1397–1414.

CLOOS, E. 1968. Experimental analysis of Gulf coast fracture patterns. *The American Association of Petroleum Geologists Bulletin*, **52**, 420–444.

COWIE, P. A., KNIPE, R. J. & MAIN, I. G. 1996. Introduction to the Special Issue. *In*: COWIE, P. A., KNIPE, R. J. & MAIN, I. G. (eds) *Scaling laws for fault and fracture populations – analyses and applications. Journal of Structural Geology*, **18**, v–xi.

DALLEY, R. M., GEVERS, E. C. A., STAMPFLI, G. M., DAVIES, D. J., GASTALDI, C. N., RUIJTENBERG, P. A. & VERMEER, G. J. O. 1989. Dip and azimuth displays for 3D seismic interpretation. *First Break*, **7**, 86–95.

DART, C. J., MCCLAY, K. & HOLLINGS, P. N. 1995. 3D analysis of inverted extensional fault systems, southern Bristol Channel Basin, U.K. *In*: BUCHANAN, J. G. & BUCHANAN, P. G. (eds) *Basin Inversion*, Geological Society, London, Special Publication, **88**, 393–413.

DULA, W. F. 1991. Geometric models of listric normal faults and roll-over folds. *The American Association of Petroleum Geologists Bulletin*, **75**, 1609–1625.

GAUTHIER, B. D. & LAKE, S. D. 1993. Probalistic modelling of faults below the limit of seismic resolution in Pelican Field, North Sea, offshore United Kingdom. *The American Association of Petroleum Geologists Bulletin*, **77**, 761–777.

GIBBS, A. D. 1983. Balanced cross-section construction from seismic sections in areas of extensional tectonics. *Journal of Structural Geology*, **5**, 153–160.

HESTHAMMER, J. & FOSSEN, H. 1997a. Seismic attribute analysis in structural interpretation of the Gullfaks Field, northern North Sea. *Petroleum Geoscience*, **3**, 13–26.

—— & —— 1997b. The influence of seismic noise in structural interpretation of seismic attribute maps. *First Break*, **15**, 209–219.

HIGGS, W. G., WILLIAMS, G. D. & POWELL, C. M. 1991. Evidence of flexural shear folding associated with extensional faults. *Geological Society of America Bulletin*, **103**, 710–717.

HOETZ, H. L. J. G. & WATTERS, D. G. 1992. Seismic horizon attribute mapping for the Annerveen Gasfield, The Netherlands. *First Break*, **10**, 41–51.

JAMISON, W. R. & STEARNS, D. W. 1982. Tectonic deformation of Wingate Sandstone, Colorado National Monument. *The American Association of Petroleum Geologists Bulletin*, **66**, 2584–2608.

JONES, G. & KNIPE, R. J. 1996. Seismic attributes; application to structural interpretation and fault seal analysis in the North Sea Basin. *First Break*, **14**, 449–461.

KERR, H. & WHITE, N. 1996. Application of an inverse method for calculating three-dimensional fault geometries and slip vectors, Nun River Field, Nigeria. *The American Association of Petroleum Geologists Bulletin*, **80**, 432–444.

KNIPE, R. J. 1992. Faulting processes and fault seal. *In*: LARSEN, R. M., BREKKE, H., LARSEN, B. T. & TALLERAAS, E. (eds). *Structural and Tectonic Modelling and its Application to Petroleum Geology*. NPF Special Publication, **1**, Elsevier, New York, 325–342.

KNOTT, S., BEACH, A., BROCKBANK, P. J., BROWN, J. L., MCCALLUM, J. E. & WELBON, A. I. 1996. Spatial and mechanical controls on normal fault populations. *Journal of Structural Geology*, **18**, 359–372.

LISLE, R. J. & ROBINSON, J. M. 1995. The Mohr circle for curvature and its application to fold description. *Journal of Structural Geology*, **17**, 739–750.

LITTLE, T. 1996. Faulting-related displacement gradients and strain adjacent to the Awatere strike-slip fault in New Zealand. *Journal of Structural Geology*, **18**, 321–342.

LOWELL, J. D. 1985. Structural styles in Petroleum Exploration. OGCI Publications, Tulsa.

MONDT, J. C. 1993. Use of dip and azimuth horizon attributes in 3D seismic interpretation. SPE Formation Evaluation, **8**, 253–257.

MORLEY, C. K., NELSON, R. A., PATTON, T. L. & MUNN, S. G. 1990. Transfer zones in the East African Rift system and their relevance to hydrocarbon exploration in rifts. *The American Association of Petroleum Geologists Bulletin*, **74**, 1234–1253.

NARR, W. 1991. Fracture density in the deep subsurface: Techniques with application to Point Arguello Oil Field. *The American Association of Petroleum Geologists Bulletin*, **75**, 1300–1323.

PEACOCK, D. 1996. Field examples of variations in fault patterns at different scales. *Terra Nova*, **8**, 361–371.

—— & SANDERSON, D. 1994. Geometry and development of relay ramps in normal fault systems. *The American Association of Petroleum Geologists Bulletin*, **78**, 147–165.

RAMSAY, J. G. & HUBER, M. I. 1983. *The Techniques of Modern Geology*. Volume 1: Strain Analysis. Academic Press, London.

SVERDRUP, E. & BJØRLYKKE, K. 1997. Fault properties and development of cemented fault zones in sedimentary basins: Field examples and predictive models. *In*: MØLLER-PEDERSEN, P. & KOESTLER, A. G. (eds) *Hydrocarbon Seals: Importance for Exploration and Production*. NPF Special Publication, **7**, Elsevier, New York, 91–106.

TOWNSEND, C., FIRTH, I. R., WESTERMAN, R., KIRKEVOLLEN, L., HÅRDE, M. & ANDERSON, T. 1998. Small seismic-scale fault identification and mapping. *In*: JONES, G., FISHER, Q. & KNIPE, R. (eds) *Faulting, Fault Seal and Fluid Flow in Hydrocarbon Reservoirs*. Geological Society, London, Special Publications (this volume).

TRUDGILL, B. & CARTWRIGHT, J. 1994. Relay-ramp forms and normal-fault linkages, Canyonlands National Park, Utah. *Geological Society of America Bulletin*, **106**, 1143–1157.

UNDERHILL, J. R. 1988. Triassic evaporites and Plio–Quaternary diapirism in western Greece. *Journal of Geological Society*, London, **145**, 269–282.

WALSH, J. J. & WATTERSON, J. 1991. Geometric and kinematic coherence and scale effects in normal fault systems. *In*: ROBERTS, A. M., YIELDING, G. and FREEMAN, B. (eds) *The Geometry of Normal Faults*, Geological Society, London, Special Publication, **56**, 193–203.

WALSH, J. J., WATTERSON, J., CHILDS, C. & NICOL, A. 1996. Ductile strain effects in the analysis of seismic interpretations of normal fault systems. *In*: BUCHANAN, P. G. & NIEUWLAND, D. A. (eds.) *Modern Developments in Structural Interpretation, Validation and Modelling*. Geological Society, London, Special Publications, **99**, 193–203.

WHITE, N. 1992. A method for automatically determining normal fault geometry at depth. *Journal of Geophysical Research*, **97**, 1715–1733.

WITHJACK, M. O., OLSON, J. & PETERSON, E. 1990. Experimental models of extensional forced folds. *The American Association of Petroleum Geologists Bulletin*, **74**, 1038–1054.

YIELDING, G., WALSH, J. J. & WATTERSON, J. 1992. The prediction of small-scale faulting in reservoirs. *First Break*, **10**, 449–460.

Space and time propagation processes of normal faults

D. MARCHAL[1], M. GUIRAUD[2], T. RIVES[3] & J. VAN DEN DRIESSCHE[3*]

[1] *GES, University Nancy I, BP239, 54506 Vandoeuvre-Lès-Nancy cedex, France*
[2] *UMR–CNRS No. 5561, Centre des Sciences de la Terre, 6 bvd Gabriel, 21000 Dijon, France.*
[3] *Elf-Aquitaine (Production), CSTJF, Av. Larribau, 64018 Pau cedex, France*
[*] *Present address: Geosciences Rennes, University Rennes I, 35042 Rennes cedex, France*

Abstract: Studies of normal fault development in space and time by X-ray tomography imaged analogue modelling suggest that different processes occur during the propagation of these faults. The conceptual model proposed distinguishes two types of propagation: horizontal and vertical. Examination of horizontal propagation reveals two separate levels of organization. Firstly, an isolated fault segment develops, alternately over time, by radial propagation and by linkage with secondary fault segments (tip faults) arising at the end of the main fault segment (parent fault). These are arranged either in an en echelon or stepwise pattern. Secondly, isolated faults evolve by linkage. Two isolated faults connect up via a relay fault area which arises in the relay zone of the two isolated faults. The relay faults develop by radial propagation and link with the two major faults. Undulating fault traces result from propagation by tip-to-parent fault segment linkage or isolated-to-isolated fault linkage. Upward and downward vertical propagation are identified and exhibit equivalent propagation processes to those involved in horizontal propagation. Space–time sequences combining horizontal and vertical propagations over time suggest that the evolution of normal faults observed in space also reflects their evolution over time.

Faults are essential features in the formation of hydrocarbon traps. Evaluating the quantities of hydrocarbons trapped, especially at structural boundaries, demands thorough understanding of the 3D geometry of faults and their terminations. Tip propagation is one important mechanism of fault growth (e.g. Cowie & Scholz 1992*a*; Reches & Lockner 1994). The 3D geometry of fault tips is therefore directly related to the propagation process. Understanding the ultimate 3D geometry involves understanding the mechanisms of growth.

Fault propagation has been actively studied in the last ten years. Different approaches have been used to investigate fault growth mechanisms. Several workers have analysed normal fault geometry, displacement or the strain and stress fields at the fault tips from field studies (Hildebrand-Mittlefehldt 1980; Granier 1985; Walsh & Watterson 1989; Peacock & Sanderson 1991; Dawers *et al.* 1993; Trudgill & Cartwright 1994; Dawers & Anders 1995; McGrath & Davison 1995), seismic datasets (Clausen & Korstgard 1994; Mansfield & Cartwright 1996), laboratory crack experiments (Cox & Scholz 1988; Petit & Barquins 1988), analogue modelling (Vendeville 1987; McClay & Ellis 1987; McClay 1990; Childs *et al.* 1993), mathematical simulations (Pollard & Segall 1987; Cowie *et al.* 1993) or combined approaches (Willemse *et al.* 1996).

Two basic models can be identified (Cartwright *et al.* 1995; Cladouhos & Marret 1996): the radial propagation/fault growth model (Watterson 1986; Walsh & Watterson 1988; Cowie & Scholz 1992*a*,*b*) and the segment linkage model (Segall & Pollard 1980; Etchecopar *et al.* 1986; Peacock & Sanderson 1991; Anders & Schlische 1994; Cartwright *et al.* 1995). The first type of models considers growth of individual faults by a process of tip propagation in which the rupture dimensions for each slip event depends on the size of the earthquake and the length of the fault (Watterson 1986; Cowie & Scholz 1992*b*). A fault growing by radial propagation follows a growth path that is represented by a scaling law (Watterson 1986; Walsh & Watterson 1988; Cowie & Scholz 1992*a*). The second type of models incorporate fault linkage, consisting of propagation (initial radial propagation), interaction and linkage of segments (Peacock & Sanderson 1991; Anders & Schlische 1994; Cartwright *et al.* 1996). A fault growing by segment linkage follows a step-like growth path not described by the scaling relationship cited above (Cartwright *et al.* 1995). These two basic models are often contrasted because the two mechanisms are relevant to the overall displacement–length scaling of fault populations which is still very much debated, particularly in view of the importance of this scaling law for predictive fault analysis (Cartwright *et al.* 1996).

Marchal, D., Guiraud, M., Rives, T. & van den Driessche, J. 1998. Space and time propagation processes of normal faults. *In*: Jones, G., Fisher, Q. J. & Knipe, R. J. (eds) *Faulting, Fault Sealing and Fluid Flow in Hydrocarbon Reservoirs*. Geological Society, London, Special Publications, **147**, 51–70.

The type of approach used limits the spatio–temporal analysis of fault propagation. For example, analysis of field and seismic data shows 3D structures in a finite state; while analogue modelling, laboratory rock-mechanics experiments and mathematical simulations are usually limited to 2D views over time or to a final 3D geometry of microcracks or faults (i.e. not truely in space and time) because of the experimental design or data acquisition methods. To our knowledge, only recent experiments that utilize acoustic emission events to monitor faulting processes in rock samples allow space–time analysis of fault growth (Reches & Lockner 1994) but without any direct and precise observations of the evolving geometry of a fault.

Analogue modelling with X-ray tomography overcomes this problem by allowing the analysis of fault geometry in 3D over time (i.e. fault propagation in 4D). The other advantage of analogue modelling is that finite displacements can be achieved. This is hard to do in numerical models because re-gridding is required, which is computationally expensive and hard to implement. To study mechanisms specific to normal fault propagation while preventing any interference from other tectonic phenomena, we have deliberately opted for an extremely simplified tectonic setting. The study of neoformed faults (e.g. newly formed, without reactivation of an older tectonic structure) in the gravitational context seems appropriate for analysis of the process of fault propagation.

The purpose of this study is to analyse the evolution of neoformed normal fault propagation in space and time by analogue modelling with X-ray computerized tomography. The results are used to propose a model for propagation of these faults with regard to the space–time dimension of the phenomenon. In order to validate the theoretical model constructed from these experiments, the 3D geometry of the faults developed in the analogue model is compared with that of natural faults from 3D seismic datasets and with that of natural fault arrays from the literature. A conceptual model is then discussed and compared with other existing conceptual models.

Analogue modelling with X-ray computerized tomography

Ideally, the study of normal fault propagation involves analysing the development of natural faults in both space and time. While the spatial approach is resolved in part by the analysis of 3D seismic datasets, the time dimension remains difficult to cope with. Analogue modelling with X-ray computerized tomography can be used to overcome this problem in part. Many works based on analogue models have already investigated the evolution of extensional tectonic structures and their relationships with rheology of materials, sedimentation, etc. (e.g. McClay & Ellis 1987; Vendeville 1987; Childs *et al.* 1993; Gaullier *et al.* 1993). The approach set out here is original in that it analyses normal fault propagation from several 3D tomographic blocks formed at different stages of deformation over time.

Experimental procedure

The brittle–ductile two-layer analogue models used in this study are composed of a layer of silicone putty (PDMS-SCM36, $d = 0.97$) overlain by alternating layers of sand ($\varnothing = 50$ to $100\,\mu$m) and of Pyrex powder ($\varnothing = 100\,\mu$m) (Fig. 1). The silicone putty forms a ductile layer that behaves like plastic strata (evaporites, subcompacted clays, etc.). The sand-Pyrex unit forms a brittle layer that behaves in a similar way to indurated sediments. All the models were constructed according to the same principle. A 3 cm thick layer of silicone was spread over the bottom of a wooden box ($42 \times 46 \times 10$ cm) covered beforehand with Rhodoid film to provide a perfectly smooth surface. A moving wall and removable cards were placed at the free edge to block the model while it was constructed. Four layers of sand and three layers of Pyrex powder were then deposited uniformly at the surface of the model forming a 6.1 cm thick brittle layer. Before this, a thin layer of silicone putty was spread on each side wall to reduce the edge effects.

In this type of modelling, normal faults form in response to extension caused by the creep of silicone under gravity. The deformation process is enhanced by a 1° slope of the base (Fig. 1a). Only the early stages of deformation are analysed here. In all instances, apart from the two half-grabens reflecting the edge effects of the moving and fixed walls, a graben appeared and developed in the centre of the model (F1 and F2, Fig. 1b) at a mean extensional speed of $0.55\,\mathrm{cm\,h^{-1}}$. In analogue models, the deformation affecting the brittle part are narrow shear zones representing faults (McClay 1990; Vendeville & Jackson 1992). In granular materials (sand and Pyrex powder), a fault plane corresponds to a dilatancy zone along which the normal grain arrangement is disturbed. The fault zones have lower densities and exhibit a lower radiological density than the surrounding

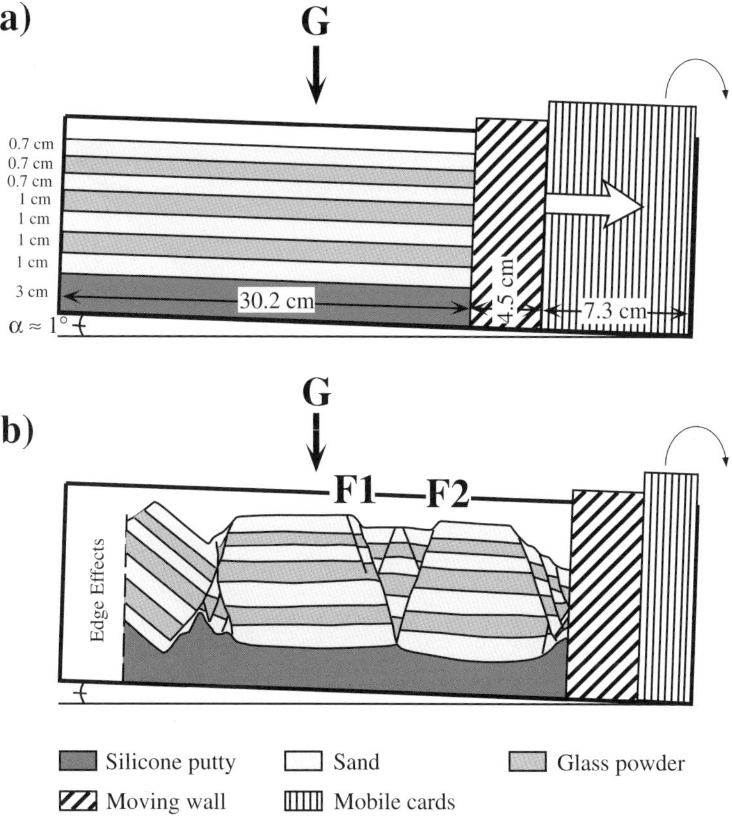

Fig. 1. Experimental apparatus. (**a**) Before deformation. The white arrow indicates the direction of extension. (**b**) At the end of the experiment. The line drawing of the central graben and the two edge-induced half-grabens is from an X-ray tomographic cross-section. The central graben exhibits a synthetic fault (F1) and an antithetic fault (F2).

material (Colletta *et al.* 1991). The faults therefore appear darker than the sand or Pyrex on the tomographic images.

Two types of acquisition were carried out using the non-destructive X-ray tomography technique. The first was sectional and surface views of the development of faults during deformation. Three vertical cross-sections (termed time cross-sections) uniformly spaced parallel to the direction of extension, were made at different points in the model at six readings per hour on average. The second type of acquisition involved stopping deformation to

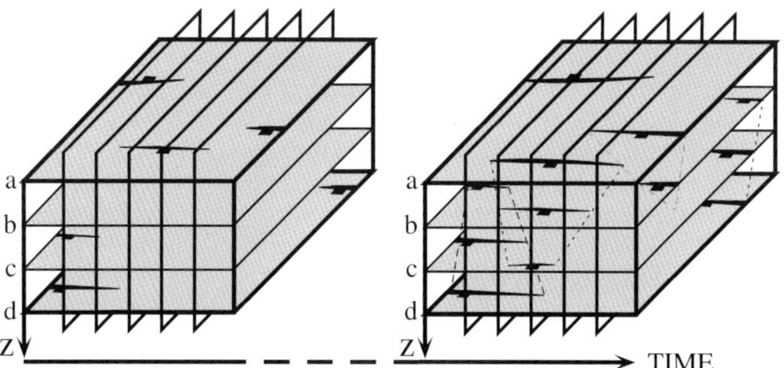

Fig. 2. The X-ray tomographic approach. The X-ray tomograph made serial cross-sections which were used to build a 3D block. The 3D block was made up from horizontal depth-slices (e.g. planes a, b, c, d). Three 3D blocks were constructed over time to analyse the evolution of 3D fault geometry.

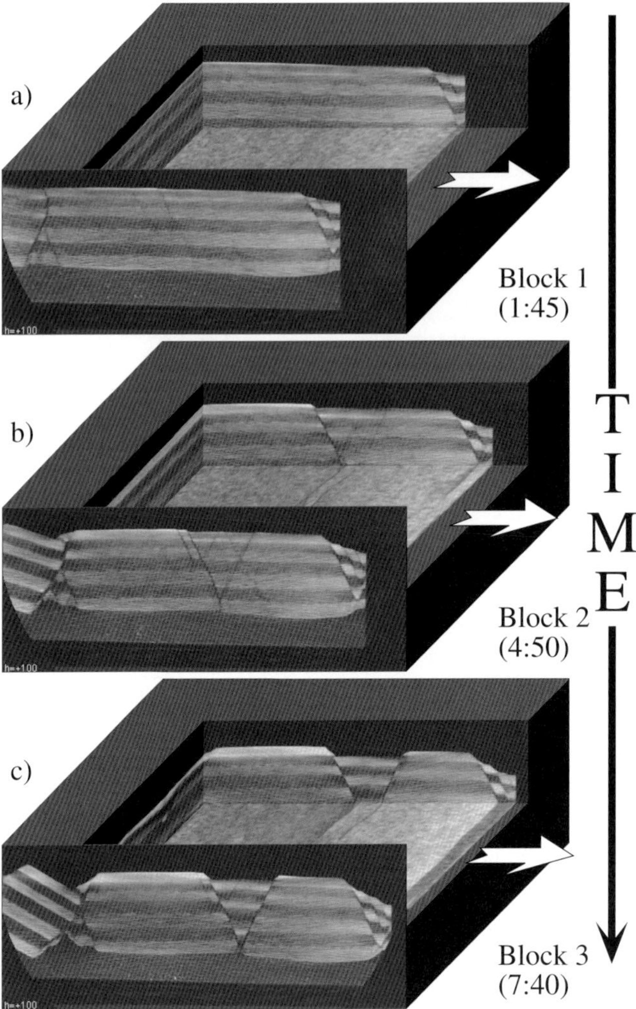

Fig. 3. The three 3D tomographic blocks of the experiment. (**a**) Block 1 reveals the potential faults of the central graben. (**b**) Block 2 displays the growing F1 and F2 fault systems. (**c**) Block 3 shows the well-developed central graben at the end of the experiment. The white arrows indicate the direction of extension.

make a set of serial vertical sections (termed 3D cross-sections) from which to reconstruct a 3D block of the model at a given stage of deformation (Fig. 2). The momentary stoppage of extension caused by jamming the piston against the card sheets was verified before and after 3D acquisition by means of vertical cross-sections and surface views. One 3D acquisition has an average duration of two hours. In addition, a control experiment without stoppage of deformation showed that temporarily blocking the model did not entail any major change in the sequence of fault development. Three 3D tomographical blocks were made to analyse the changing 3D geometry of normal faults over time (Fig. 3). Two types of dilatancy zone can be distinguished: potential faults with no visible throw (in the centre of the model, Fig. 3a) and faults in the narrow sense, termed here true faults, with throw (central graben, Fig. 3c).

For each 3D block we present a series of four evenly spaced depth-slices (Figs 4, 5 & 6a–d) at different depths (see Fig. 2). The fault network of the central graben is shown on each depth-slice (in cross-section in Figs 1 & 3). The fault pattern is controlled by observing serial 3D cross-sections making up the 3D block (Fig. 3). For descriptive purposes, the faults are oriented in the same way as the depth-slices (Figs 4a, 5a & 6a); i.e. the right-hand edge of a fault is at the right-hand side of a depth-slice.

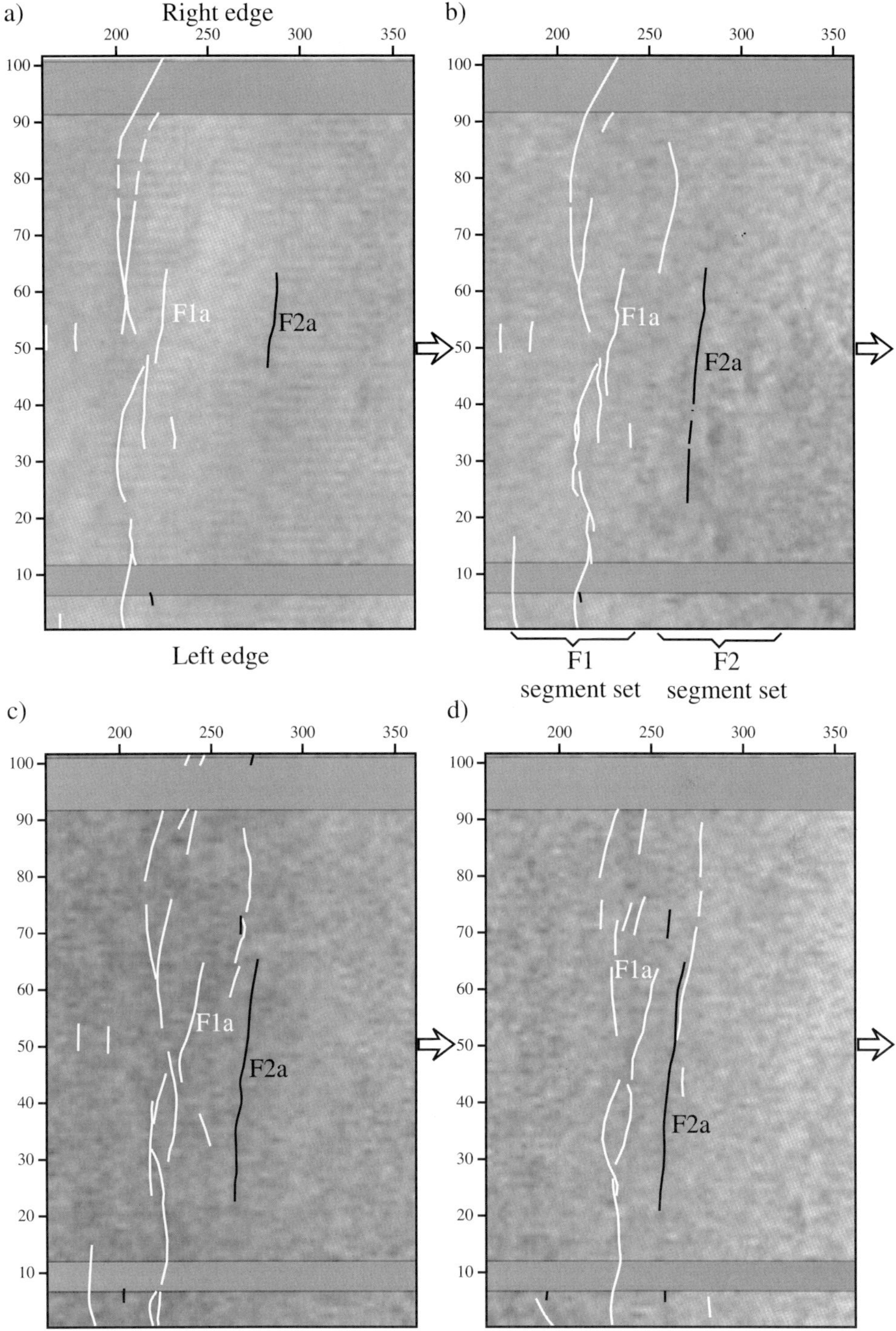

Fig. 4. Four evenly spaced depth-slices of block 1 showing segment sets F1 and F2. Line drawings of segments were made to highlight the fault network: synthetic faults are white and antithetic faults are black. (**a**) Upper level (S.035). (**b**) Upper-middle level (S.050). (**c**) Lower-middle level (S.065). (**d**) Lower level (S.080). White arrows indicate the direction of extension. The grey bands mark gaps in acquisition. See main text for details.

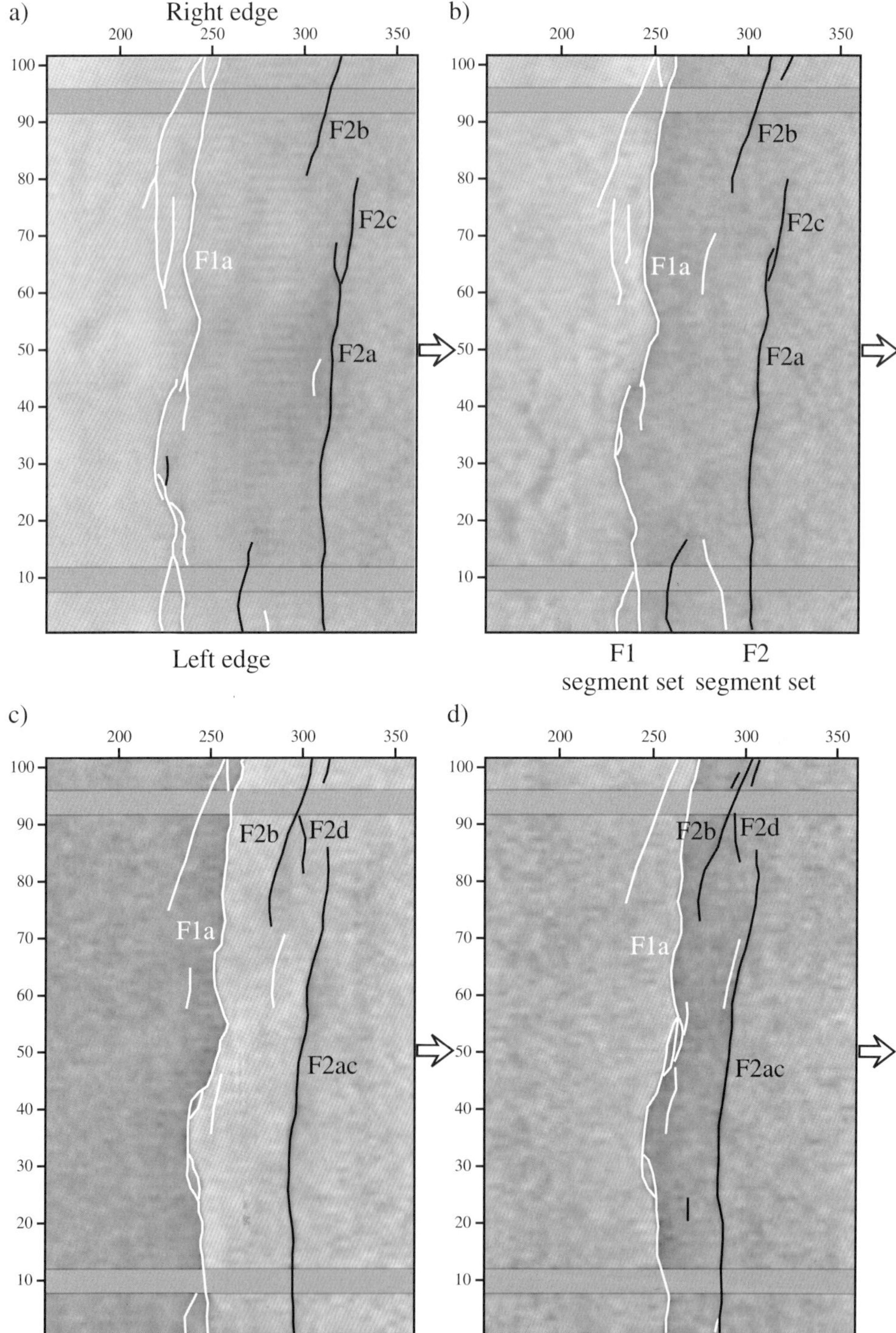

Fig. 5. Four evenly spaced depth-slices of block 2 (at the same depth as block 1 in a, b, c, d). Notation and remarks as for Fig. 4. See main text for details.

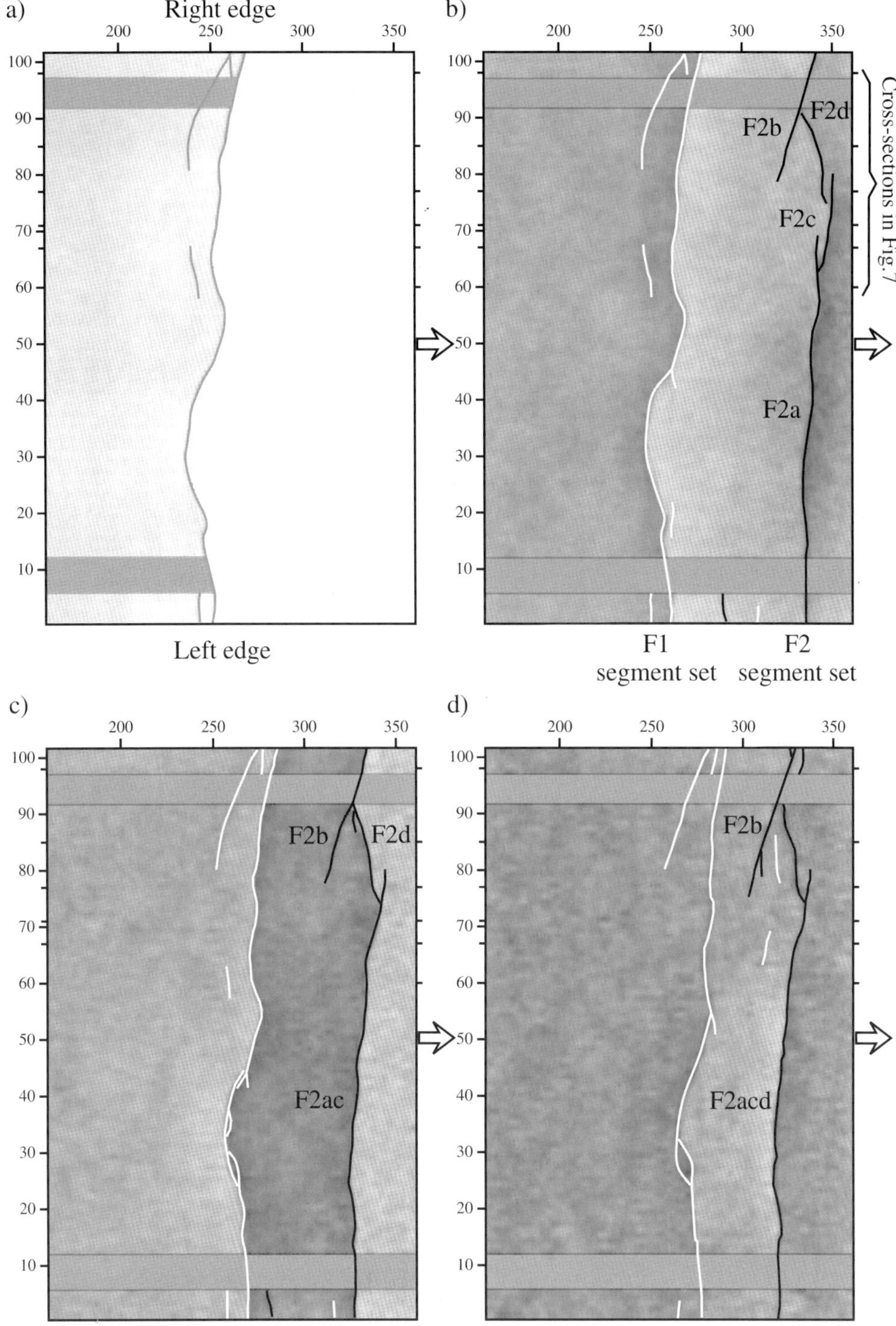

Fig. 6. Four evenly spaced depth-slices of block 3 (at the same depth as block 1 in a, b, c, d). Notation and remarks as for Fig. 4. To see faults set on (**a**), synthetic faults have been displayed in grey (note that the hanging-wall of F1 fault system is under the slice). See main text for details.

Experimental results

Figures 4, 5 & 6 are focused on the central graben. The two half-grabens at the walls are not shown. The first tomographic block (Block 1, Figs 3a & 4) was produced after the appearance on the time cross-sections of two potential faults in the centre of the model, one synthetic and one antithetic to the direction of extension (F1 and F2, respectively, Fig. 3a). The two dilatancy zones on the vertical sections of block 1 that give rise to the central graben display no detectable throw. The depth-slices (Fig. 4) show that segment set F1 is made up of numerous relay segments and runs the entire width of the model. Segment set F2 (Fig. 4) develops slightly later than F1 and therefore appears less advanced. Segment F2a (first segment of segment set F2) forms a comparatively straight line which lengthens with depth (Figs 4a–d).

At a more advanced stage of deformation (Block 2, Figs 3b & 5) certain potential faults become true faults with measurable offsets (e.g. F1 and F2, Block 2, Fig. 3b) while others vanish (e.g. the segment in the lower left corner of Figs 4a–d disappears in Figs 5a–d as does the segment to the right of F2a). It should be noticed that true fault terminations (with throw) look like potential faults (without throw; e.g. F2b, Fig.7c). The central segment of F1 (F1a) developed near the right-hand edge and displays several large undulations (Fig. 5). The different segments on the left of system F1 branch at depth (from Fig. 5a–d). The numerous relay zones exhibit a lens-shaped geometry (e.g. Fig. 5c). At the left-hand edge of the depth-slices (Fig. 5a & b) two small faults affect the interior of the central graben. New segments appear in segment set F2. Segment F2b is located at the right-hand edge near the centre of the model and is at a slight angle compared with F2a. The left-hand tip of F2a runs out of the scanner acquisition zone. The right-hand tip of F2a is characterized by the appearance of a new segment (F2c) which is smaller and runs in the same direction as F2a. The co-linear faults F2a and F2c form a right-stepping pattern. Downwards (from Fig. 5a–d) the spatial relationships between faults F2a and F2c can be seen to change:

(i) In depth-slice Fig. 5a segments F2a and F2c are in a relay arrangement. The left-hand tip of F2c links up with F2a.
(ii) The next depth-slice (Fig. 5b) shows the same arrangement but the right-hand tip of F2a curves round to connect with F2c. The F2a–F2c relay maps as a lense-shape.
(iii) Deeper in the model (Figs 5c & d) a single wavy line (F2ac) can be identified at the location of faults F2a and F2c on the previous depth-slice (Fig. 5b). The fault line F2ac is longer in Figs 5c & d than the sum of the lines F2a and F2c in Figs 5a & b. Both depth-slices (Figs 5c & d) show a segment (F2d) connected to F2b and forming a relay with F2ac.

The third and final 3D block (Block 3, Fig. 3c & 6) depicts the last stage of deformation at the end of the experiment. The upper-depth slice (Fig. 6a) exhibits a segment set F1 with only a single very undulating line running the full width of block 3. Segment F2d, which cannot be perceived at depth 'b' on block 2 (Fig. 5b), is henceforth visible at the same depth in block 3 (Fig. 6b). On the same depth-slice, the right-hand tip of F2d is connected to F2b and its left tip forms a relay with F2c. At depth (Fig. 6c & d), F2d connects up with F2ac. Downwards, F2a and F2c exhibit similar relationships to those described for block 2. In Fig. 6b the left-hand tip of F2c forms a relay with F2a and connects up with it. At depth (Fig. 6c & d), F2a and F2c merge to form a single line: F2ac. In Fig. 6c the throws of segment sets F1 and F2 are displayed by the colour contrast between the central graben (dark – Pyrex) and the two respective footwalls (light – sand). Analysis of throw along segment set F2 shows that the line can be defined representing the major fault, from right to left on the model, as follows: F2b as far as the connection with F2d, F2d as far as the connection with F2ac, then F2ac. At depths (Figs 6c & d) segment set F2 is represented by a single undulating line (F2 acdb) which has retained the lesser throw branches of F2b and F2ac. The line F2acdb exhibits on the right-hand half three major undulations.

Interpretation

Individual analysis of each 3D block specifies the 3D geometry of the faults. By comparing the 3D blocks, the changes in geometry in space and time can be analysed. The description of the first 3D tomographic blocks (Block 1, Fig. 4) shows that the faults first appear as potential faults that are clearly segmented (cf. F1). The strike dimension of the segments can be small (cf. F1a) or large (cf. F2a). Segment F2a propagates from the decollement surface (sand/silicone interface) upwards and shortens towards the surface. The transition from block 1 to block 2 shows that some potential faults disappear, by particle rearrangement, leaving faults to develop from potential to true faults. Over time deformation is concentrated on certain faults allowing them to evolve (in agreement with Cowie *et al.* 1993).

In the second block (block 2, Fig. 5) the right-hand tip of F1a has migrated to the right-hand edge and leaves the acquisition zone. The neo-formed part of F1a exhibits several major undulations (Fig. 5) as probable evidence of a series of connections between the different segments (see evolution of segment set F2 below). Segment F2b propagates from the right-hand edge towards the centre of the model and is slightly oblique, a feature which is ascribed to edge effects, relative to the axis of the central graben. The left-hand tip of F2a has developed towards the left-hand edge of the box and extends out of the scanner acquisition zone. Analysis of the relationships between F2a and F2c shows the hierarchy between the main fault (F2a) and the secondary fault (F2c) located at the footwall of the termination of F2a. Due to the obvious genetic link between the primary and secondary segments, these are termed the parent fault and the tip fault, respectively. The tip fault (F2c) and the parent fault (F2a) join up by linkage of the two fault planes involved in the relay zone (Figs 5b–c), producing a new wavy parent fault designated F2ac. The right-stepping pattern of the tip fault produces the undulating line resulting from tip-to-parent fault connection (F2ac; Figs 5c & d). The linkage is made at the same time as, or subsequently to the growth of the tip-fault by radial propagation (right tip of F2ac; Fig. 5c & d). Fault F2d is a relay fault (connecting fault, cf. Peacock & Sanderson 1994) which allows isolated faults F2ac and F2b to connect up. The spatial arrangement of F2d depends on the respective positions of F2ac and F2b. The transition from block 2 to block 3 shows that fault F2d propagated radially both horizontally (Figs 5c to 6c) and vertically (Figs 5b to 6b).

The final stage (block 3, Fig. 6) shows the overlapping of relay fault F2d and F2c on the shallower depth-slice (Fig. 6b). At depth, the connection of the relay fault F2d with the parent fault F2ac allows for interaction between the two parent faults/isolated faults (F2ac and F2b, Figs 6c & d). The isolated-relay-isolated faults create a single active fault plane. The branches of the isolated faults overlapping with the relay fault are inactive (Fig. 7). The resulting (active) fault is therefore very wavy. The right-hand half of segment set F2 exhibits three main waves: one from the connection between F2a and F2c and two from the connection of F2d with F2ac and with F2b. The propagation processes described for segment set F2 probably account for the single and wavy line of segment set F1 on the different depth-slices of block 3.

To resume, two major types of propagation are observed: firstly, processes related to propagation of isolated faults and secondly, processes associated with the evolution of sets of isolated faults. Isolated fault propagation is controlled by the tip–parent fault system. The lengthening of the isolated fault occurs in three stages:

(1) an overlapping co-linear secondary fault (tip fault) appears at the tip of the main fault (parent fault);
(2) both tip and parent faults lengthen by radial propagation;
(3) tip and parent faults connect up to form a new, longer, undulating fault.

Isolated fault set propagation, which is governed by interaction and linkage of the two isolated faults via a relay fault, occurs in three steps:

(1) the lengthening of isolated faults runs into the creation of an overlapping zone between the two isolated faults;
(2) a secondary fault (relay fault) appears in the relay zone to break the relay ramp;
(3) the active fault resulting from the isolated fault linkage is composed by the major parts of the two isolated faults and by the secondary relay fault. Both parts of the isolated faults overlapping the relay fault become inactive.

The undulating pattern of the faults results from the linkage processes: linkage of the tip and parent faults to the isolated fault system and linkage of two isolated faults by a relay fault.

Model of normal fault propagation in space and time

The experiments analysed earlier suggest a model integrating the different processes of normal fault propagation. Two types of propagation are distinguished conceptually: horizontal propagation (Fig. 8) and vertical propagation (Fig. 9). This idealized separation is induced by observation planes. Horizontal propagation occurs in the strike direction of the fault plane. This lateral growth is better seen in plan-view. Vertical propagation arises in the dip direction of the fault plane and is studied in cross-section.

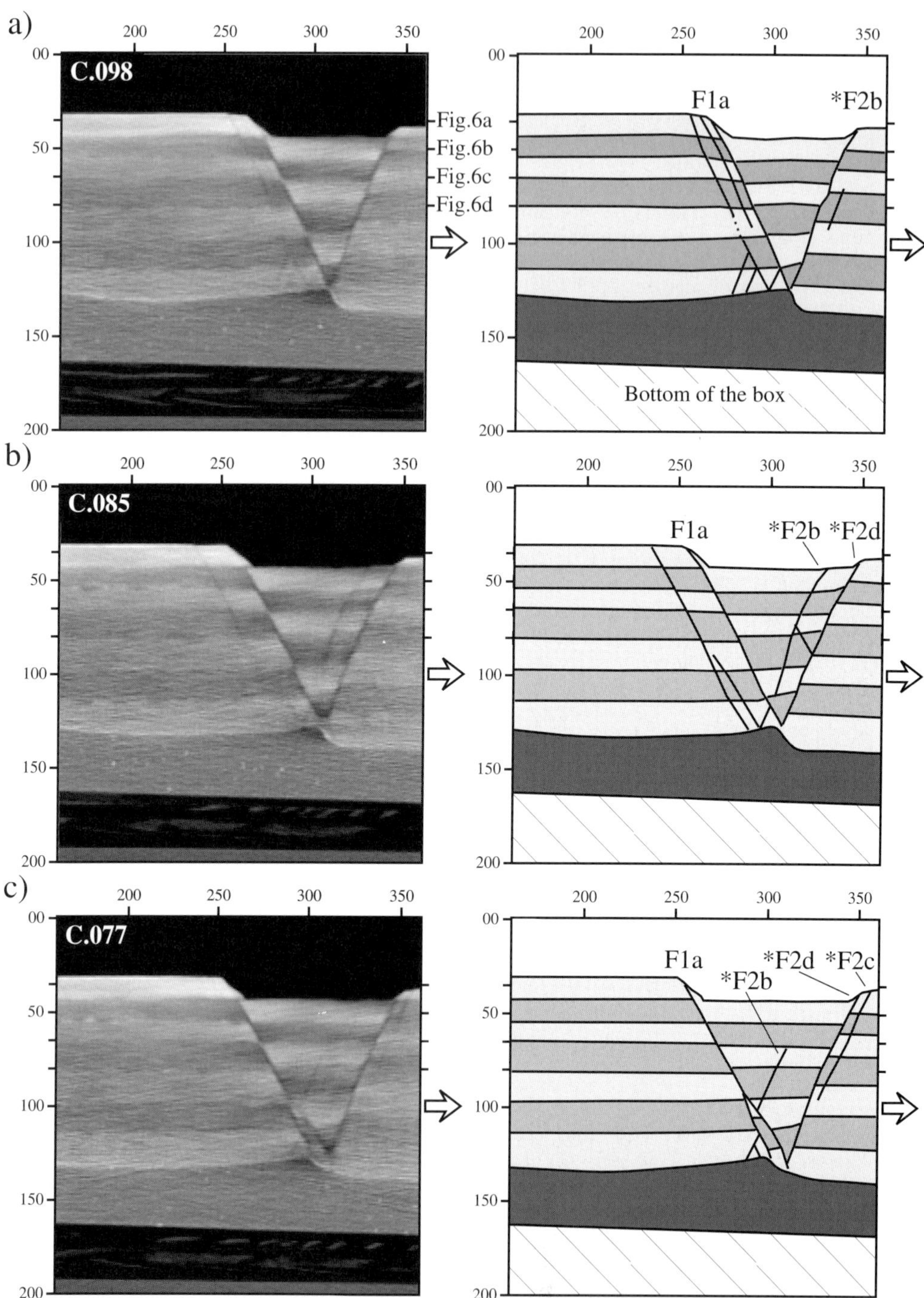

Fig. 7. 3D cross-sections of block 3 (location on Fig. 6). The line drawing (see Fig. 1 for key) of cross-sections (**a**) to (**f**) shows the active part (the initial segments are marked by asterisks) of fault F2abcd: *F2b on (**a**), *F2d on (**b**) and (**c**) and *F2a–*F2c on (**d**) to (**f**). The inactive parts of *F2b and *F2ac display very little displacement on (**b**) and (**c**). On (**c**), the termination of *F2b within the graben takes the form of a potential fault.

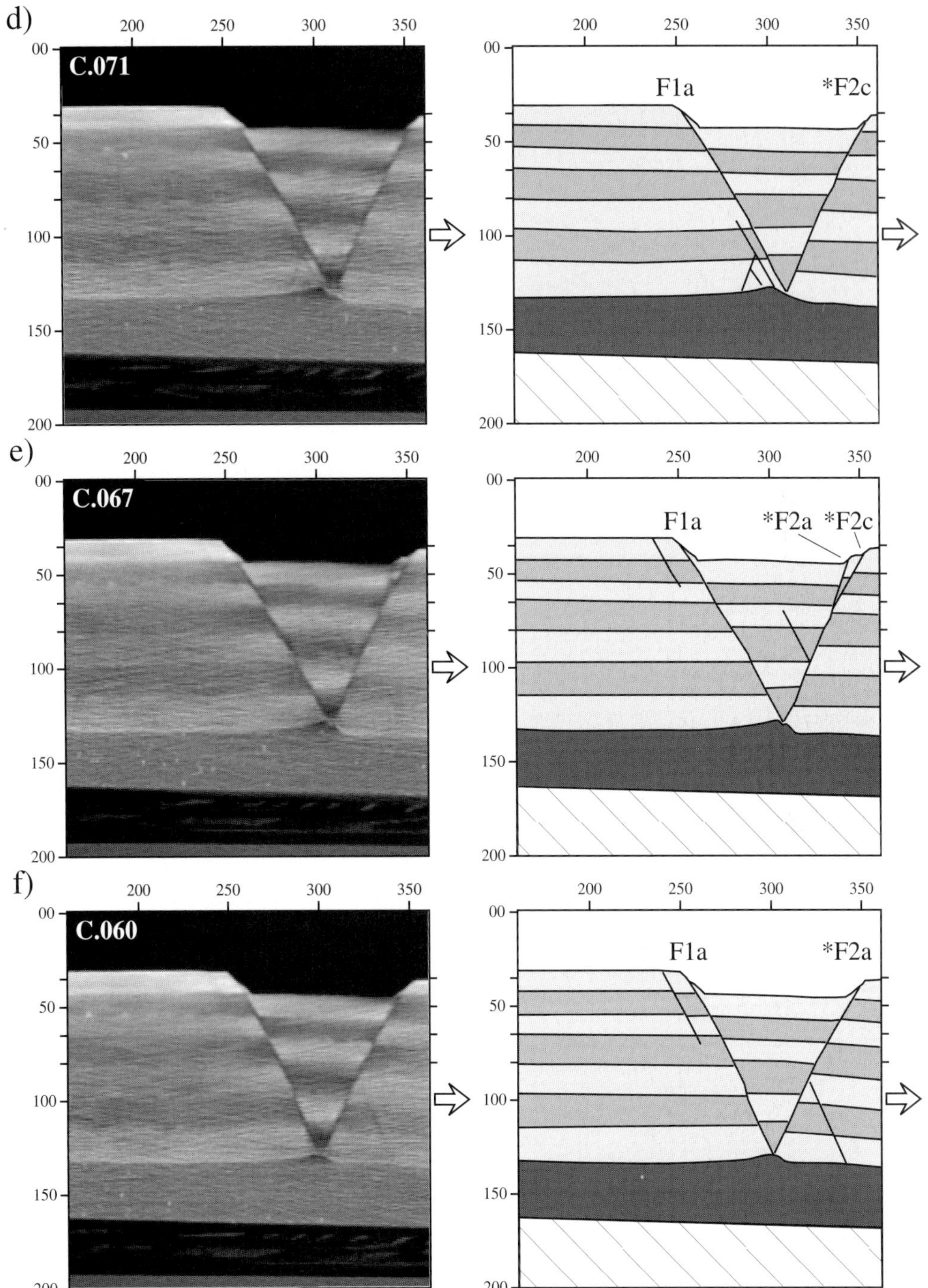
d)
C.071
F1a
*F2c
e)
C.067
F1a
*F2a
*F2c
f)
C.060
F1a
*F2a

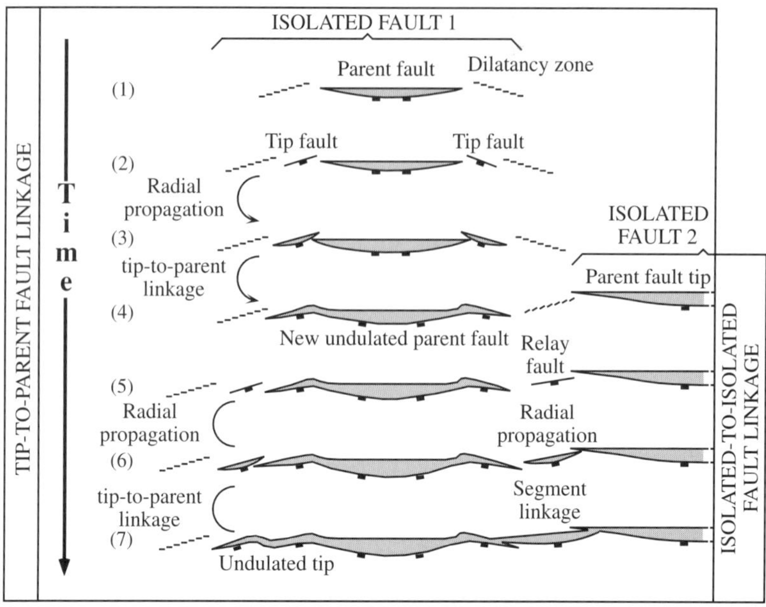

Fig. 8. General model of normal fault propagation. Seven stages of propagation of a growing fault set are shown. Two ranks of organisation are displayed: fault set with two (or more) isolated faults and isolated fault system with tip faults and parent fault. An isolated fault system evolves by tip-to-parent linkage, the tip fault appearing and evolving by radial propagation at the termination of the parent fault (1, 2, 3, 4). Isolated fault sets evolve by linkage of isolated faults via a fault relay which appears in the relay zone of the two isolated faults (5). This relay fault develops by radial propagation (6) and connects up with the two major faults (7). The final fault trace is wavy (7). The undulations result from two propagation by linkage processes: either tip-to-parent fault linkage (7 – left tip) and/or isolated-to-isolated fault linkage (7 – right).

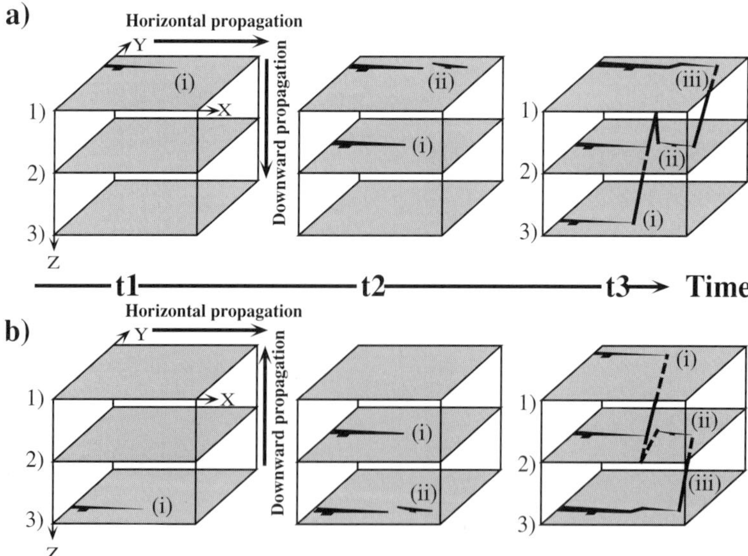

Fig. 9. Schematic spatio–temporal propagation sequences. (**a**) Downward sequence. (**b**) Upward sequence. These sequences can be viewed in 3D (1 to 3) and over time (t1 to t3). For the same horizontal propagation process (i to iii), the fault terminations on the final two blocks (at t3) look different depending on the direction of vertical propagation. The 4D sequence shows that the sequencing of the processes (i to iii) described for the same surface over time (a-1: t1 to t3 and b-3: t1 to t3) is the same as for the different surfaces at any given moment (a-t3: 3 to 1 and b-t3 1 to 3).

Horizontal propagation processes

The description of horizontal propagation processes involves the separation between propagation linking of tip fault to parent fault (isolated fault) with the linking between fault arrays (set of isolated faults).

Isolated faults propagate radially and by segment connection (Fig. 8). Radial propagation results in lengthening of fault traces by tip migration. The mechanism of radial propagation is discussed in detail in Cowie & Scholz (1992*a*, *b*). Segment connection with isolated faults results from the interaction between a parent fault and tip fault. This hierarchical arrangement of different segments was suggested by Wu & Bruhn (1994). A secondary segment (tip fault) appears and overlaps the end of the major segment (parent fault). The tip fault is either en echelon or co-linear in a stepwise pattern (right- and left-stepping pattern respectively, on the left-hand and right-hand ends looking towards the hanging-wall of the fault plane). Examples of this configuration can be seen in Fig. 11b and in Wu & Bruhn (1994), Jackson & Leeder (1994), Trudgill & Cartwright (1994) and Childs *et al.* (1995). Tip faults and parent faults are genetically related and the tip fault is subordinate to the parent fault. The tip fault connects up to the parent fault producing a single fault plane. The two processes described earlier (radial propagation and tip-to-parent fault linkage are separate on a given scale) can be reproduced if the fault is still isolated. Fault undulation results, therefore, from the relative positions of the parent fault and tip fault and the tip-to-parent fault linkage process.

The evolution of a set of isolated faults results in the ability to link to adjacent isolated fault arrays. The two isolated faults are linked by the appearance and connection of a relay fault (connecting fault, cf. Peacock & Sanderson 1994) with two isolated faults in a relay arrangement (Fig. 8). The relay fault develops by radial propagation. The fault trace geometry at the connection zones (relay zones) depends on the respective positions of the two isolated faults. The linkage between two isolated faults also generates a very wavy final trace. It should be noticed that connection of isolated faults is very well documented in Cartwright *et al.* (1995). At one stage of advanced deformation, the isolated-relay-isolated fault arrangement forms a single active fault plane (extending the entire width of the model), while the branches of the isolated faults overlapping with the relay fault are inactive.

We suggest that the fundamental mode of propagation of an isolated normal fault results alternately from radial propagation and tip-to-parent fault connection. Generally, fault undulation results from a propagation process by linkage: either tip-to-parent fault linkage and/or linkage from a relay to an adjacent established isolated fault.

Vertical propagation processes

Two fundamental types of vertical propagation have been differentiated on the basis of 3D tomographic blocks and kinematic sections (Fig. 9). Upward propagation is most often observed and occurs from the decollement surface (sand/silicone interface) towards the surface of the model. Downward propagation from the model surface toward the decollement surface is less common. Three-dimensional block analysis (3D cross-sections and depth-slices) indicates equivalent propagation processes to those of horizontal propagation: i.e. radial propagation and segment connection (see also Mansfield & Cartwright 1996).

It should be noted that, on time cross-sections, some small faults seem to be initiated at the sand/Pyrex powder interfaces, generally at the intersection between these interfaces with the graben building fault systems (F1 and F2). Some faults appear initially as a dilatancy zone which has the same dimensions as the entire plane of the future fault. Thus, the direction of propagation should be determined by the analysis of the evolution of the measurable offsets.

Space–time sequences of normal fault propagation

The combination of horizontal and vertical propagation processes over time provides examples of 4D sequences (space and time) of isolated normal fault tip evolution (Fig. 9). To simplify the sequence, only radial vertical propagation is considered at the fault tip. From a single horizontal succession (Fig. 8) two spatio-temporal sequences are constructed, depending on the direction of vertical propagation of the fault as described above (Figs 9a & b). The difference between the two 3D geometries obtained (Figs 9a-t3 & 9b-t3) depends entirely on the direction of vertical propagation of the fault.

The 4D sequences show that the processes (i–iii) described for the same surface over time (Fig. 9a-1-t1 to 9a-1-t3 and Fig. 9b-3-t1 to 9b-t3–3) are concatenated in the same way as for different surfaces at a given time (Fig. 9a-t3–3 to 9a-t3–1 and Fig. 9b-t3–1 to 9b-t3–3). This suggests that the evolution observed in space also reflects evolution over time. Natural examples displaying final 3D geometry similar to that described in the final block of each 4D sequence are presented later.

Natural examples

Niger Delta

The model presented above was tested in the Niger Delta. The neoformed normal faults developed during Tertiary times under gravitational sliding. Growth faults are the main structures observed. Rollovers of collapsed compartments are the result of the evolutionary geometry of listric faults and the weight of delta sediments overlying the ductile clays acting as the decollement surface (Weber 1987; Doust & Omatsola 1990). The example shown (Fig. 10) is located in the onshore part of the Delta. The field is bounded by WNW–ESE gravitational faults, dipping to the SSW. These faults exhibit large throws (F1, Fig. 10a) and affect a 1500–2000 m thick clay-sandstone series. The overall geometry is that of a graben on the crest of a rollover controlled by a major listric fault (F1, Fig. 10a). The other faults (F2–F14) form a synthetic and antithetic system relative to fault F1. These faults with smaller throws divide the field into compartments. Interpretation of the 3D seismic block (composed of 335 X lines and 215 Y lines) is used to visualize the 3D structure of the area. The different vertical cross-sections were correlated using time-slices and maps of seismic signal amplitude recorded with different reflectors (horizons A–E). The structural maps produced in this way displayed the organization of the fault network affecting the marker horizons (Fig. 10b).

Analysis of the structural maps shows that the length and throw of the faults affecting the rollover generally increase from the E horizon to the A horizon. Analysis of several maps at

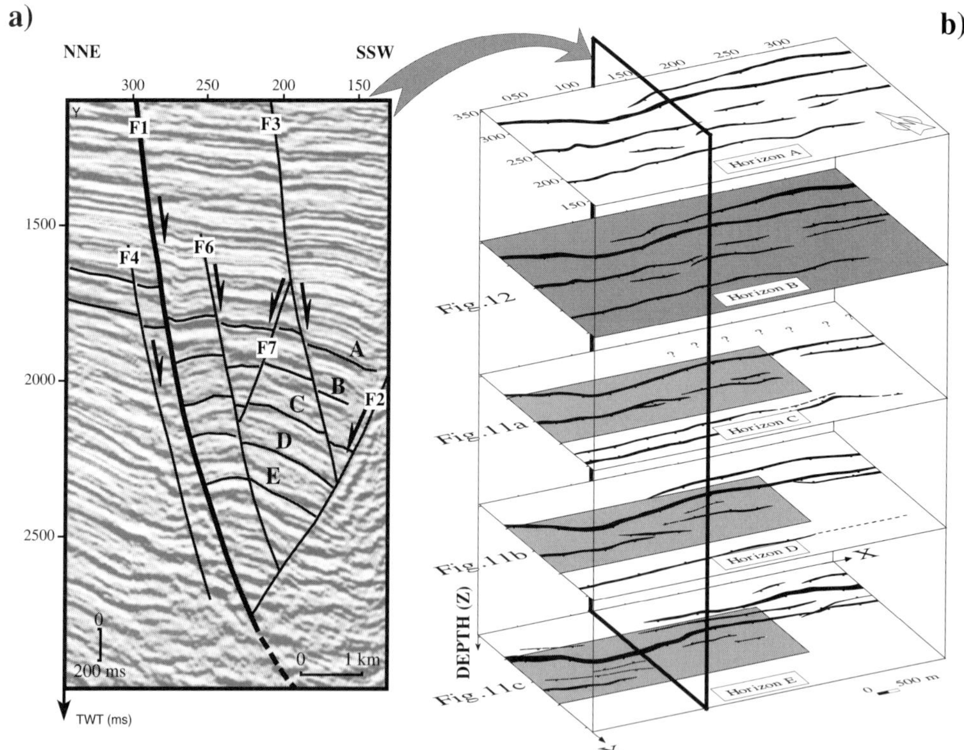

Fig. 10. 3D seismic data set of the Niger Delta. (**a**) Cross-section showing crestal synthetic and antithetic normal faults in a rollover structure which developed in the hanging wall of a major growth fault (F1). (**b**) The fault network is built from several cross-sections and several horizons (A to E). The shaded boxes outline the areas shown in the indicated figures.

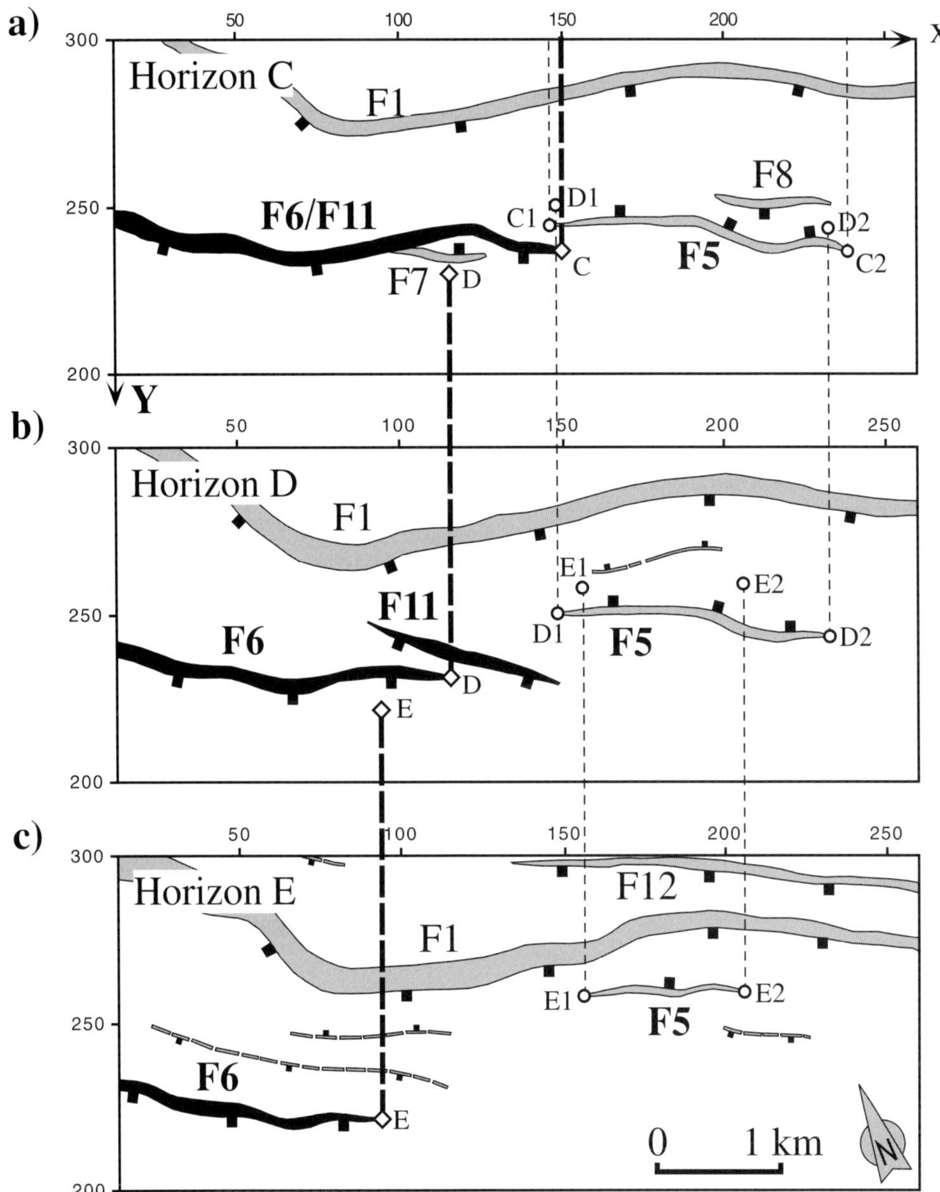

Fig. 11. Parent fault F6 termination traces on three horizon maps. (**a**) Upper horizon C shows the long undulating termination of the parent fault 'F6–F11'. (**b**) At the middle horizon D, the en echelon overlapping tip fault F11 is distinct from the parent fault F6. (**c**) The lower horizon E displays a shorter, simple undulating termination of fault F6 than horizon D. Rhombs and circles are located at the tip of faults F6 and F5, respectively for each horizon and are transposed to the horizon above.

varying depths provides detailed 3D geometry of the fault tips. A close up of the tip of one such fault (F6, Fig. 11) shows how fault length varies with depth. F6 and F5 lengthen upwards. At horizon E (Fig. 11c) the tip of fault F6 forms a single wavy trace. The transition to the overlying horizon D shows that the trace of fault F6 is slightly longer and that a small fault appears (F11). The en echelon fault F11 is oriented N130E and overlaps fault F6. The transition to horizon C is marked by the linkage between F6 and F11 thus producing an undulating trace. The point of inflection of the trace 'F6–F11' is located at the former relay zone.

This geometry reported suggests that the wavy trace of neoformed normal faults observed across the map on the scale of an oil field (e.g. Fig. 12) result from the successive linkage of secondary segments (tip faults) at the ends of main faults (parent faults) and the oblique alignment of the connected segments. The similarity between the 3D geometry of fault F6 (seismic, Fig. 11) and the theoretical fault of the advanced stage of the downward propagation sequence (Fig. 9a-t3) will also be noticed. We suggest that the evolution of a fault line on a map in space can also reflect evolution over time. Thus fault F6, located on the crest of a rollover, is thought to have developed from top to bottom. In real settings, the waviness of isolated normal faults reflects their mode of propagation. Each major undulation is thought to be a propagational increment reflecting a former stage of extension by linkage. Consequently, the point of inflection of the undulation corresponds to a former relay zone.

Bibliographic example

A further example of a network of neoformed normal faults in the Niger Delta is analysed in Bouvier *et al.* (1989; Fig. 13). The 3D geometry of fault Fx viewed on maps D5, E2 and F2 constructed from a 3D seismic block displays a horizontal propagation process similar to the one described above. However, the succession of the different stages of propagation through space (F2 to D5) is reversed compared with that of F6 of the seismic map above. This spatial sequence is thought to correspond to an upward propagation sequence (see Fig. 9b) equivalent to that presented in the analogue models.

A field example is provided in the Canyonlands Graben region (Utah, USA) which displays a complex zone of normal faulting (Trudgill & Cartwright 1994). These faults result from gravitational sliding of a 450–500 m thick clay–sandstone sediment stack (Pennsylvanian–Permian period) over a decollement surface formed by evaporite strata (Pennsylvanian system). The structure is still extending, indicating its relatively young age (around 0.5 Ma) which explains its excellent state of conservation. The undulating patterns of the normal fault lines on aerial photographs (Trudgill & Cartwright

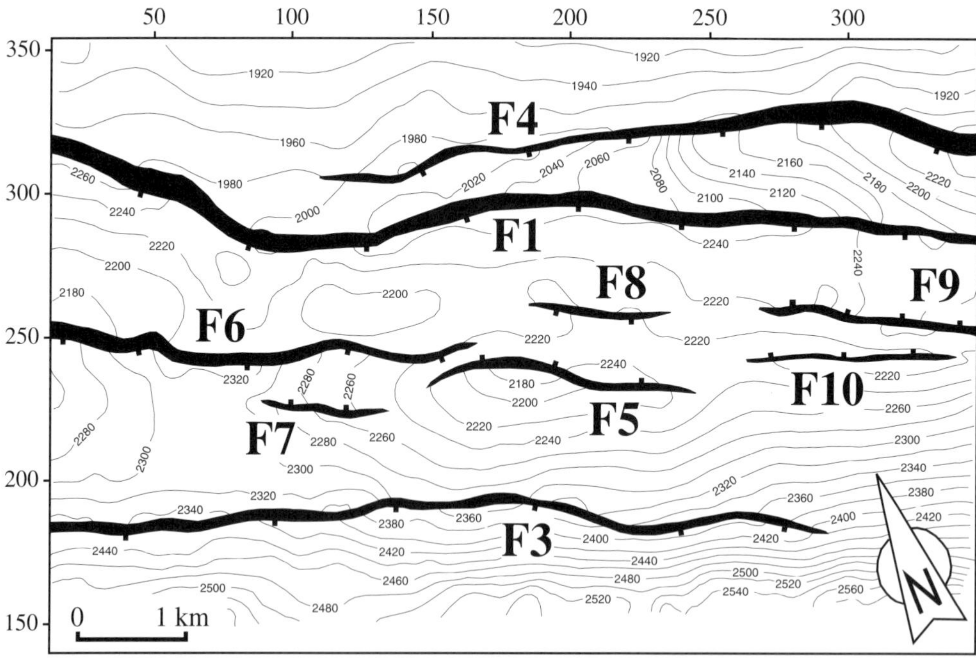

Fig. 12. Horizon B. The fault network contains faults of different lengths, some long, some short. Detailed mapping shows that every single one of the fault traces undulates along its entire length.

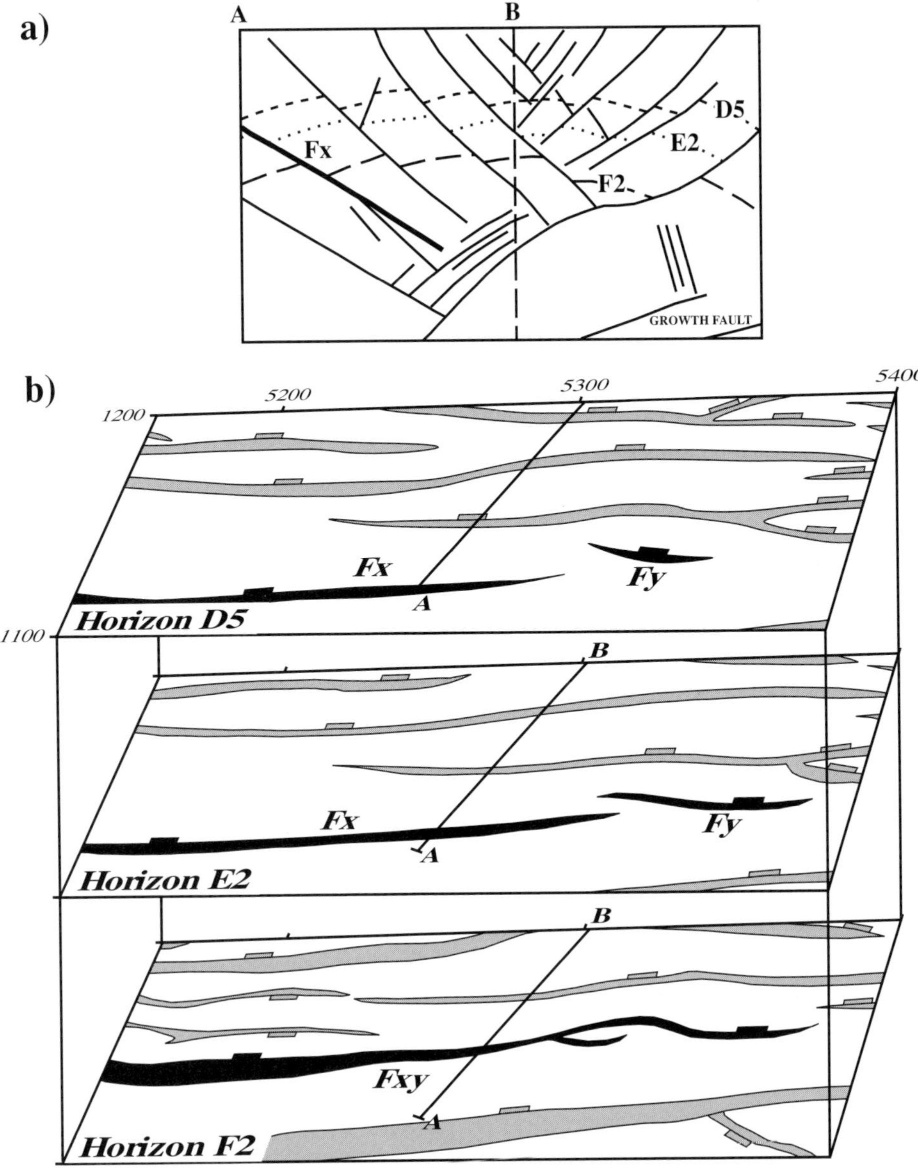

Fig. 13. Example of tip-to-parent fault linkage on three horizon maps (Nun River Field, Niger Delta) illustrating a probable upward propagation geometry (see Fig. 9b-t3). From Bouvier *et al.* (1989).

1994, figs 2 and 7) are similar to those described in the Niger Delta. Such wavy fault lines suggest propagation by segment connection, as demonstrated by Trudgill & Cartwright (1994).

Discussion and conclusions

Detailed analysis of the three 3D tomographic blocks shows how gravitational structures very similar to those observed in naturally deformed rocks may be reproduced in the laboratory using scaled-down analogue models. Analogue modelling using X-ray tomography is an effective technique for investigating fault propagation in space and time. The systematic study of spatio–temporal propagation of the different types of fault (normal faults, reverse faults, strike-slip faults, etc.) in varied tectonic environments would provide insight into the propagation process and would be an excellent test for confirming or refuting the different models of fault growth. This type of study is clearly valuable in petroleum exploration. The precise determination of the 3D geometry of fault tips may resolve a number of questions about possible points of leakage from traps and thus better define prospect boundaries. Fault propagation models can be used to explain the undulations in the cartographic trace of a fault and the associated geometric structures over time.

The different characteristics of fault propagation created in analogue models are synthesized in an overall conceptual model (Fig. 8). The process described by the model implies separating two major types of propagation, at a given scale of observation. The first family of processes relates to the propagation of isolated faults. The second family concerns the evolution of sets of isolated faults. Isolated fault propagation is governed by the tip-to-parent fault system, in which lengthening (i.e. propagation) of the fault occurs in three stages:

(1) a tip fault (secondary fault), either en echelon or co-linear, appears near the end of the parent fault (main fault) and overlaps the latter;
(2) both tip and parent faults propagate radially;
(3) the tip and parent faults connect up creating a new, longer, undulating parent fault.

The process described may be repeated later. Isolated fault set propagation is reflected by interaction and linkage of isolated faults in a relay arrangement. Two isolated faults are connected via a secondary fault (relay fault) appearing at the top of the relay ramp in the relay zone. This connection generates major undulations on the graben-bounding 'mega' fault. The experiments show that the undulating pattern of normal faults results mainly from the linkage process either of one tip fault and one parent fault or of two isolated faults via a relay fault. Any undulation (fault bend in Childs *et al.* 1995) is evidence of an earlier connection zone.

Two main models of fault propagation are contrasted in the literature. The first model (radial propagation, Cowie & Scholz 1992*a*) considers growth of individual faults in their own plane by tip rupture during slip events (earthquakes) in which each propagation increment depends on the size of the slip event and the length of the fault. The second model (segment linkage model, cf. Cartwright *et al.* 1995) suggests that faults propagate first by radial propagation and then by segment connection (isolated faults). Our model evokes, on a given scale, first radial propagation and tip-to-parent fault linkage processes, alternating in time and space and then an isolated-to-isolated fault linkage process (segment linkage, cf. Cartwright *et al.* 1995). Thus the apparent opposition of the two models currently developed in the literature appears unfounded. The process involved in these models is thought to correspond to two different stages in the overall process of fault and/or fault set propagation.

Analysis of 4D propagation (time and space) characterizes the horizontal and vertical propagation process. By combining these two types of process, it is possible to construct evolutionary sequences in space and time. This spatio–temporal pattern of propagation suggests that the evolution observed in space also reflects evolution over time. These 4D propagation sequences contribute to our understanding of fault propagation. This type of analysis is an excellent guide to interpreting 3D seismic data.

This research was funded by Elf Aquitaine (Production) (contract RADE.93.04). We are grateful to the company for providing the 3D seismic data and for permission to publish this paper. We thank the Institut Français du Pétrole, especially B. Colletta, J.-M. Mengus, M.-T. Bieber and C. Schlitter, for the analogue model design and the use of the X-ray tomograph. We would also like to express our gratitude to P. Cowie, R. J. Knipe and the editors of this volume for their constructive comments and helpful suggestions to an earlier version of this paper.

References

Anders, M. H. & Schlische, R. W. 1994. Overlapping faults, intrabasin highs, and the growth of normal faults. *Journal of Geology*, **102**, 165–180.

Bouvier, J. D., Kaars-Sijpesteijn, C. H., Kluesner, D. J., Onyejerkwe, C. C. & Van Der Pal, R. C. 1989. Three-Dimensional seismic interpretation and fault sealing investigations, Nun River field, Nigeria. *American Association of Petroleum Geologists Bulletin*, **73**, 1397–1414.

Cartwright, J. A., Mansfield, C. S. & Trudgill, B. D. 1996. The growth of normal faults by segment linkage. *In*: Buchanan, P. G. & Nieuwland, D. A. (eds.) *Modern Developments in Structural Interpretation, Validation and Modelling*. Geological Society, London, Special Publications, **99**, 163–177.

——, Trudgill, B. D. & Mansfield, C. S. 1995. Fault growth by segment linkage: an explanation for scatter in maximum displacement and trace length data from the Canyonlands Grabens of SE Utah. *Journal of Structural Geology*, **17**, 1319–1326.

CHILDS, C., EASTON, S. J., VENDEVILLE, B. C., JACKSON, M. P. A., LIN, S. T., WALSH, J. J. & WATTERSON, J. 1993. Kinematic analysis of faults in a physical model of growth faulting above a viscous salt analogue. *Tectonophysics*, **228**, 313–329.

——, WATTERSON, J. & WALSH, J. J. 1995. Fault overlap zones within developing normal fault systems. *Journal of the Geological Society, London*, **152**, 535–549.

CLADOUHOS, T. T. & MARRETT, R. 1996. Are fault growth and linkage models consistent with power-law distributions of fault lengths? *Journal of Structural Geology*, **18**, 281–293.

CLAUSEN, O. R. & KÖRSTGARD, J. A. 1994. Displacement geometries along graben bounding faults in the Horn Graben, Offshore Denmark. *First Break*, **12**, 305–315.

COLLETTA, B., LETOUZEY, J., PINEDO, R., BALLARD, J. F. & BALÉ, P. 1991. Computerized X-ray tomography analysis of sandbox models: Examples of thin-skinned thrust systems. *Geology*, **19**, 1063–1067.

COWIE, P. A. & SCHOLZ, C. H. 1992*a*. Physical explanation for the displacement – length relationship of faults using a post-yield fracture mechanics model. *Journal of Structural Geology*, **14**, 1133–1148.

—— & —— 1992*b*. Growth of Faults by Accumulation of Seismic Slip. *Journal of Geophysical Research*, **97**, B7, 11085–11095.

——, VANNESTE, C. & SORNETTE, D. 1993. Statistical Physics Model for the Spatiotemporal Evolution of Faults. *Journal of Geophysical Research*, **98**, B12, 21809–21821.

COX, S. J. D. & SCHOLZ, C. H. 1988. On the formation and growth of faults: an experimental study. *Journal of Structural Geology*, **10**, 413–430.

DAWERS, N. H. & ANDERS, M. H. 1995. Displacement – length scaling and fault linkage. *Journal of Structural Geology*, **17**, 607–614.

——, —— & SCHOLZ, C. H. 1993. Growth of normal faults: Displacement – length scaling. *Geology*, **21**, 1107–1110.

DOUST, H. & OMATSOLA, E. 1990. Niger Delta. *In*: EDWARDS, J. D. & SANTOGROSSI, P. A. (eds) *Divergent/Passive Margin Basins*. AAPG memoir, **48**, 201–238.

ETCHECOPAR, A., GRANIER, T. & LARROQUE, J. M. 1986. Origine des fentes en échelon: propagation des failles. *Comptes rendus de l'Académie des Sciences, Paris*, **302**, 479–484.

GAULLIER, V., BRUN, J. P., GUERIN, G. & LECANU, H. 1993. Raft tectonics: the effects of residual topography below a salt décollement. *Tectonophysics*, **228**, 363–381.

GRANIER, T. 1985. Origin, damping, and pattern of development of faults in granite. *Tectonics*, **4**, 721–737.

HILDEBRAND-MITTLEFEHLDT, N. 1980. Deformation near a fault termination, part II: a normal fault in shales. *Tectonophysics*, **64**, 211–234.

JACKSON, J. & LEEDER, M. 1994. Drainage systems and the development of normal faults: an example from Pleasant Valley, Nevada. *Journal of Structural Geology*, **16**, 1041–1059.

MCCLAY, K. R. 1990. Deformation mechanics in analogue models of extensional fault systems. *In*: KNIPE, R. J. & RUTTER, E. H. (eds) *Deformation Mechanisms, Rheology and Tectonics*. Geological Society, London, Special Publications, **54**, 445–453.

—— & ELLIS, P. G. 1987. Analogue models of extensional fault geometries. *In*: COWARD, M. P., DEWEY, J. F. & HANCOCK, P. L. (eds) *Continental Extensional Tectonics*. Geological Society, London, Special Publications, **28**, 109–125.

MCGRATH, A. G. & DAVISON, I. 1995. Damage zone geometry around fault tips. *Journal of Structural Geology*, **17**, 1011–1024.

MANSFIELD, C. S. & CARTWRIGHT, J. A. 1996. High resolution fault displacement mapping from three-dimensional seismic data: evidence for dip linkage during fault growth. *Journal of Structural Geology*, **18**, 249–263.

PEACOCK, D. C. P. & SANDERSON, D. J. 1991. Displacements, segment linkage and relay ramps in normal fault zones. *Journal of Structural Geology*, **13**, 721–733.

—— & —— 1994. Geometry and development of relay ramps in normal fault systems. *American Association of Petroleum Geologists Bulletin*, **78**, 147–165.

PETIT, J. P. & BARQUINS, M. 1988. Can natural faults propagate under mode II conditions? *Tectonics*, **7**, 1243–1256.

POLLARD, D. D. & SEGALL, P. 1987. Theoretical displacements and stresses near fractures in rock: with applications to faults, joints, veins, dikes, and solution surfaces. *In*: ATKINSON, B. K. (ed.) *Fracture Mechanics of Rock*. Academic Press, 277–350.

RECHES, Z. & LOCKNER, D. A. 1994. Nucleation and growth of faults in brittle rocks. *Journal of Geophysical Research*, **99**, B9, 18,159–18,173.

SEGALL, P. & POLLARD, D. D. 1980. Mechanics of discontinuous faults. *Journal of Geophysical Research*, **85**, B8, 4337–4350.

TRUDGILL, B. & CARTWRIGHT, J. 1994. Relay-ramp forms and normal-fault linkages, Canyonlands National Park, Utah. *Geological Society of America Bulletin*, **106**, 1143–1157.

VENDEVILLE, B. C. 1987. Champs de failles et tectonique en extension : modélisation expérimentale. *Mémoires et Documents du Centre Armoricain d'Etude Structurale des Socles*, **15**, 1–392.

—— & JACKSON, P. M. A. 1992. The rise of diapirs during thin-skinned extension. *Marine and Petroleum Geology*, **9**, 331–353.

WALSH, J. J. & WATTERSON, J. 1988. Analysis of the relationship between displacements and dimensions of faults. *Journal of Structural Geology*, **10**, 239–247.

—— & —— 1989. Displacement gradients on fault surfaces. *Journal of Structural Geology*, **11**, 307–316.

WATTERSON, J. 1986. Fault dimensions, displacements and growth. *Pure & Applied Geophysics*, **124**, 365–373.

WEBER, K. J. 1987. Hydrocarbon distribution patterns in Nigerian growth fault structures controlled by structural style and stratigraphy. *Journal of Petroleum Science and Engineering*, **1**, 91–104.

WILLEMSE, E. J. M., POLLARD, D. D. & AYDIN, A. 1996. Three-dimensional analyses of slip distributions on normal fault arrays with consequences for fault scaling. *Journal of Structural Geology*, **18**, 295–309.

WU, D. & BRUHN, R. L. 1994. Geometry and kinematics of active normal faults, South Oquirrh Mountains, Utah: implication for fault growth. *Journal of Structural Geology*, **16**, 1061–1075.

The appearance of potential sealing faults on borehole images

J. T. ADAMS[1] & C. DART[2*]

[1] *Z&S Geology Ltd, Kettock Lodge, Balgownie Drive, Bridge of Don, Aberdeen AB22 8GU, UK*

[2] *Z&S Geologi a/s, Sverdrupsgate 23, N-4007 Stavanger, Norway*

[*] *Present address: Norsk Hydro, Research Centre, N-5020 Bergen, Norway*

Abstract: Electrical and acoustic borehole images provide data on the structure, width, spacing and orientation of potential sealing faults in the subsurface. These data bridge the 'scale of observation gap' between the resolution limits of more conventional data such as seismic data and core, and provide continuous records often up to several thousand feet long. Potential fault seals can be measured from vertical, deviated or horizontal wells and from either clastic or carbonate rocks. Five categories of potential fault seals can be described from images: *juxtaposition fault seals* place reservoir against seal lithologies and appear as sharp image contrasts; *clay lined fault seals* commonly appear as shale rich bands on electrical images, which may be either resistive or conductive depending on the resistivity of the host lithology, and low amplitude bands on acoustic images; *grain size reduction fault seals* appear as resistive or high amplitude features; *cemented fault seals* appear as highly resistive or very high amplitude features, and *open/vuggy faults* appear as conductive or very low amplitude features. An integrated approach to image interpretation incorporates not only borehole image data but also core, conventional wireline logs, pressure data and production information.

Sealing faults are responsible for reservoir compartmentalization and permeability anisotropy at all scales. Faults are complex structures, composed of concentrations of large and small-scale faults and fractures known as damage zones (Engelder 1974). At reservoir scale they comprise an inner zone of highly deformed fault rock, and an outer zone of smaller faults and fractures (Jones & Knipe 1996).

Sealing potential of a fault is influenced by the overall fault zone or fault damage zone structure and by the fault rock types present. These are in turn controlled by deformation conditions, deformation history, fault displacement and host rock lithology (Knipe *et al.* 1997). Fault sealing is generally due to porosity reduction by disaggregation, cataclasis, pressure solution, clay smear and cementation (Fisher & Knipe this volume), and may indeed be a combination of these processes. Fault rock properties are likely to be different at different locations on the fault surface, resulting in spatial variations in sealing capacity.

While damage zones may be large, extending for up to 100 m away from faults with displacements of up to 100 m, internal fault damage zone structure is rarely resolved on 3D seismic. The sealing potential of these smaller-scale faults can be assessed from core, although core through fault zones is generally difficult to obtain, due to mechanical coring problems, and difficult to interpret due to induced fracture overprints. Borehole images provide continuous data cover across fault zones, and provide a useful means of estimating the sealing potential of faults penetrated by a borehole.

The objective of this paper is to present a range of examples of the use of borehole images in fault seal characterization. These examples show that systematic description of images by an experienced interpreter can provide qualitative information on fault sealing potential. This observational approach is greatly enhanced by integration with dynamic reservoir data and examples of such integration are also presented here. However, it should be stressed at the outset that borehole images cannot provide quantitative indications of fault rock permeability or fault transmissibility. This paper is primarily concerned with faults (i.e. discontinuities within a rock with a dominant shear component). Obviously, joint systems are significant to fluid flow within reservoir rocks (Nelson 1985), but these are not considered here.

Borehole images

Two types of borehole image tool are commonly used in the petroleum industry (Table 1).

Advantages and limitations of borehole images

Borehole image logs are commonly run over a long section of borehole, with a much higher

ADAMS, J. T. & DART, C. 1998. The appearance of potential sealing faults on borehole images. *In*: JONES, G., FISHER, Q. J. & KNIPE, R. J. (eds) *Faulting, Fault Sealing and Fluid Flow in Hydrocarbon Reservoirs*. Geological Society, London, Special Publications, **147**, 71–86.

Table 1. *Borehole image tool types*

Tool type	Electrical	Acoustic
Examples	Schlumberger FMS/FMI Halliburton EMI Atlas Wireline STAR (resistivity) (Anadrill RAB)[1]	Atlas Wireline CBIL/STAR (acoustic) Schlumberger UBI Halliburton CAST BPB AST
Tool technology	Pad-mounted arrays of imaging electrodes. Number of pads varies with tool type: • FMS = 4 • EMI/STAR = 6 • FMI = 8	Rotating ultrasonic transducer fires a series of sonic pulses and records their echo from the borehole wall. Both transit time (time taken to travel from the tool-formation-tool) and amplitude (amplitude of the returning echo) are recorded.
Tool response	Resistivity measurements respond to variations in rock texture, fluid type and saturation and mineralogy. Wide dynamic range.	Transit time represents hole size/shape variations. Amplitude responds to acoustic impedance contrasts, related closely to density and therefore porosity. Generally narrow dynamic range.
Bed & feature resolution	Generally *c.* 1 cm feature resolution, though smaller features can be resolved (Williams *et al.* 1995).	Vertical resolution depends on transducer type, ranges from 0.2–0.4 inches, with bed resolution *c.* 2.5–5 cm. Resolution is also dependent on logging speed.
Limitations	Require a conductive (water-based) mud system. Pad devices affected by poor hole conditions.	Work in both water-based and oil-based mud. Affected by mud type (solids content, mud weight, emulsions etc.). Hole shape and tool eccentricity are also critical factors.
References	Ekstrom *et al.* (1987) Halliburton (1995) Schlumberger (1992) Western Atlas (1996)	Faraguna *et al.* (1989) Schlumberger (1993) Atlas Wireline Services (1992)

[1] The Anadrill RAB is a Logging While Drilling measurement, on a much coarser scale than the wireline-conveyed electrical or acoustic images.

vertical and horizontal sampling rate than conventional wireline logs. This high sampling rate means that high-resolution false-colour electrical or acoustic images of the borehole can be generated and these permit bedding and fractures to be identified and oriented by visual interpretation. A wide spectrum of structural features may be confidently identified from borehole images, ranging from single deformation bands to major fault zones, with features observed over at least 5 orders of magnitude (<1 cm–100 m) bridging the scale gap between 3D seismic and core (Fig. 1).

Many of the fault image examples presented in this paper are from horizontal wells. Such wells are a key tool in understanding faults, as they provide a cross-sectional perspective of faults and their damage zones, allowing both fault rock characteristics and spatial distribution to be described. By convention, images from horizontal wells are presented with a horizontal borehole axis, as shown later in this paper. On such representations the low side is at the centre and the high side at the periphery of the image.

Borehole images are not photographs, but electrical or acoustic representations of the borehole wall and need to be interpreted as such. For example, if no resistivity or acoustic contrast exists between a fracture and the host rock, then the tool will not be able to image the fracture. Images have a lower resolution limit (*c.* 1 cm bed resolution, although most experience suggests that deformation bands only a few mm wide can be resolved), and are therefore unable to resolve every fracture present in a densely fractured rock. Borehole images also have a limited

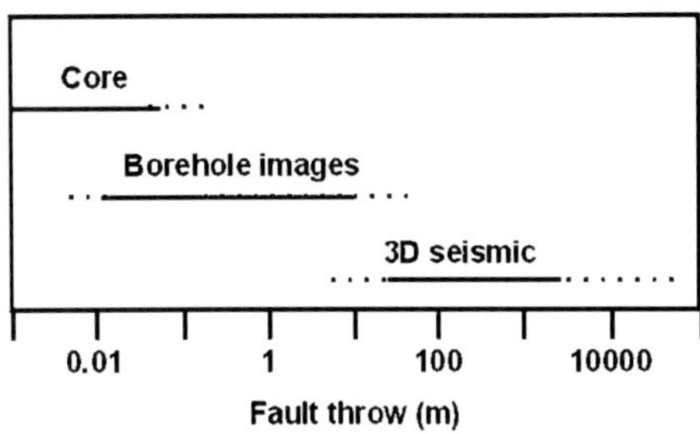

Fig. 1. Investigation scale of borehole images compared with core and 3D seismic. Borehole images can be used to identify a wide range of fracture scales, ranging from individual fractures down to <1 mm width and 1 cm displacement, to large fault zones with many metres of throw.

depth of investigation: acoustic images measure a reflection of the borehole wall, whereas electrical images are focussed at around 1.5–3 cm into the formation. They cannot therefore directly measure the characteristics of the rock fabric away from the zone of influence of the borehole. Finally, the interpretation possibilities of the images are limited by the experience of the interpreter.

Interpretation approach

The approach to interpreting borehole images is listed in Table 2, and involves an initial descriptive phase, followed by interpretation and integration with other data.

Integration of other measurements

Images alone can only provide a static description of the faults encountered in a well. Integration with other data is essential in order to establish the extent of fault sealing. Integration is required with the following types of data:

Direct observation in core. Integration of core with borehole image data contributes considerably to an understanding of fault zones and the contribution of core should never be underestimated. The key advantages of core integration in understanding fault sealing properties are:

(1) Direct observation on core of a range of fracture types present in the reservoir to allocate fault seal types and estimate their distribution to calibrate borehole images. Core observations permit quantification of image confidence and resolution limits, particularly where core and image logs are gathered over the same interval of reservoir.
(2) Direct orientation of core features to match with those seen on images. This allows the same features to be identified and matched on both datasets.
(3) Ability to carry out other measurements to investigate flow properties (e.g. whole-core permeability, plug permeability vs orientation, plug permeability vs fracture density, probe permeameter imaging).

The main limitations of core to image comparison in fault zones are the difficulty of obtaining intact faulted core (particularly where the fracture density is high), and short cored intervals may not necessarily be representative of the fracture density within major fault zones as they only sample a small part of much larger features.

In situ *stress field.* The *in situ* stress field has been shown to have an effect on the sealing capability of faults (Barton & Zoback 1992, 1994; Knott 1993; Heffer & Koutsabeloulis 1995; Hillis 1997). The key parameters influencing sealing capability are the orientation of the fault relative

Table 2. *Fault seal interpretation procedure*

Order	Interpretation phase	Key actions
1	Quality control	Log quality control including recognition of artifacts (Bourke 1989)
2	Feature recognition	Visual recognition of fractures from images
3	Parameter recording	Specific recording of fracture position (depth), orientation, fracture style, width, apparent width, conductivity or acoustic properties
4	Interpretation	Interpretation of large-scale fault properties from fracture density and orientation data
5	Integration	Integration with other well data (core, dynamic measurements, *in situ* stress)

to the maximum horizontal stress (Sh_{max}) direction, the magnitudes of the principal stresses and the pore pressure. In general, a fault will tend to be sealing if the normal stress acting on it is greater than the pore pressure. Faults striking perpendicular to Sh_{max} will have a tendency to be shut-in (i.e. be sealing) while those striking parallel will tend to be open. Information on the in situ stress field may be obtained directly from borehole images by observation of the orientation of borehole breakout and drilling-induced tension fractures (Plumb & Hickman 1985; Zoback *et al.* 1985, 1995; Plumb & Cox 1987; Bell 1990).

Full-waveform acoustic logs. Full-waveform sonic logs can be used to identify potential open fractures. The Stoneley tube wave will be reflected from open fracture surfaces due to the high acoustic contrast, whereas the wave will propagate across any closed fractures. Zones of open fractures can therefore be identified by the presence of chevron patterns or 'criss-cross' energy. Such patterns can also be created by ledges on the borehole wall and by changes in lithology, especially where there is a strong change in acoustic properties. Calibration of the acoustic waveforms with calipers and other openhole logs is therefore essential.

Dynamic measurements

(1) Mud-losses: The shape of the mud loss curve can be used to differentiate open natural fracture zones from drilling-induced fractures (Dyke *et al.* 1995).
(2) Production logs: Production logs help identify which fracture zones within a well contribute to flow (i.e. are open and vuggy)

Table 3. *Summary of fracture sealing potential from borehole images*

Sealing type	Definition	Effect on fluid flow	Image character	Interpretation pitfalls
Juxtapositon seal	Reservoir rocks placed directly against sealing lithologies across a fault plane/zone.	High sealing potential if sealing lithology is impermeable. Sealing potential depends on bed thickness, fault throw, stratigraphical template.	Sharp changes in image character across fault planes reflecting the change in lithology recorded on the openhole logs.	Recognition of lithology of hanging wall and footwall Fault plane recognition if close to bedding orientation
Clay-lined seal	Clay-lined fracture formed by shale smearing, mineral growth, alteration, washed-in clays and pressure solution.	Moderate to high, depending on the thickness/content of the shale smear and areal extent on the fault surface.	Discordant features, showing shale resistivity or acoustic character, with corresponding openhole log response.	Conflict with open fractures, which also appear conductive/low amplitude.
Grain size reduction seal	Common in clean high porosity sandstones, with dominant grain rotation and cataclastic deformation mechanisms.	Moderate sealing potential, depending on fracture density and extent of deformation.	Commonly more resistive than surrounding rock on electrical images, high amplitude on acoustic images.	Differentiation from cementation seals based on strength of image response, and openhole log character.
Cementation seal	Seal formed by cementation along the fault plane.	High sealing potential.	Highly resistive on electrical images, very high amplitude on acoustic images, accompanied by strong response on openhole logs.	Differentiation from grain size reduction seals based on strength of image response, and openhole log character.
Open/vuggy fractures	Flow may occur along a fracture or between fracture and matrix. Fracture may allow fluid flow across the fracture.	Not sealing, allow fluid flow along and across fracture.	Filled with drilling mud, therefore should appear conductive on electrical images. Low amplitude features on acoustic images.	Conflict with clay-lined fractures, which have a similar image appearance.

when displayed alongside a categorised fracture density curve generated from the borehole image interpretation.

(3) Formation pressure measurements: Different pressures in fault-bound reservoir compartments may be due either to virgin pressure differences or to pressure changes from production or injection. Multiprobe formation pressure sensors allow individual faults to be investigated on a small scale and are a valuable contributor to understanding local fault sealing properties.

(4) Well-tests: Well-test intervals are often too long to identify specific zones of production, but can be used to define broad flow zones. Drawdown history also allows local barriers to be identified.

Fault seal categories and appearance on borehole images

Five main categories of potential fault seals are recognized on borehole images:

(i) juxtaposition fault seals;
(ii) clay lined fault seals;
(iii) grain size reduction fault seals;
(iv) cementation fault seals;
(v) open/vuggy faults.

The fault seal types are summarized in Table 3, and their relative sealing capability is shown in Fig. 2.

Juxtaposition fault seals

Juxtaposition fault seals occur where reservoir rocks are placed against sealing lithologies across a fault plane. Where the sealing lithology is highly impermeable, juxtaposition seals have a strong negative impact on fluid flow. The sealing potential is dependent on net/gross ratio, thickness, distribution and connectivity of permeable and impermeable beds on the hanging wall and footwall (Knipe 1992; Knott 1993).

Juxtaposition fault seals are characterized by sharp changes in image character across fault planes reflecting the change in lithology recorded on the openhole logs. The example presented in Fig. 3 illustrates a juxtaposition fault zone cutting a horizontal well in a chalk reservoir. The left-most side of the image is in well-bedded mudstone and is separated from the right side of the image by a steeply-dipping fault zone. The sealing potential of this part of the fault is high. Sealing potential away from this location is related to the reservoir connectivity which is a function of displacement and reservoir stratigraphy. Biostratigraphical data can be used to allow the modelling of the throw of each major fault and therefore to constrain the likely juxtapositioning along the fault away from the borehole. Figure 4 shows a similar juxtaposition fault seal on an acoustic imaging tool. The sandstone in the hanging wall (dark coloured image) is juxtaposed against the lighter-coloured mudstone in the footwall. The local sealing potential of this fault is high. Both faults are sharply defined on the images, ranging from a single plane below the resolution of the tool (Fig. 4) to a fault zone 30 cm (1 foot) wide (Fig. 3).

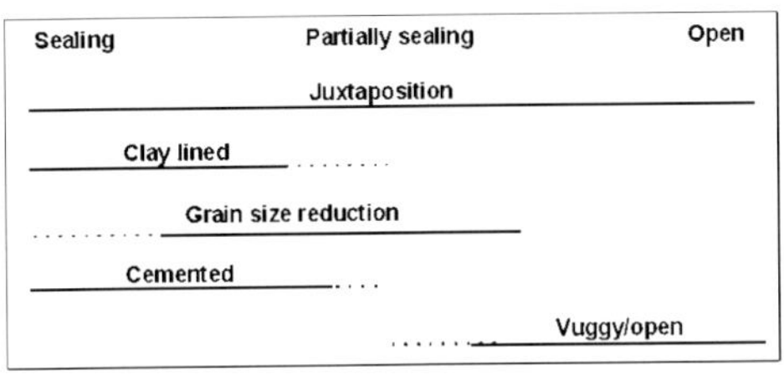

Fig. 2. Relative effectiveness of different fault seal types. The boundaries between the fault seal types are gradational.

Clay lined fault seals

Clay lining is an effective seal and can be formed by a number of mechanisms, the most common of which are shale smearing (especially in interbedded sands and shales), mineral growth, alteration and pressure solution.

Shale smear fault seals form by abrasion, shear or injection in an interbedded sandstone–shale sequence (Lindsay *et al.* 1990). The thickness and extent of the shale smear will be related to the proportion of shale beds contributing to the seal. Sealing potential therefore varies from moderate to high.

Shale smear fault seals appear as discordant features on electrical images with conductivity determined by the host rock. If the host rock is conductive brine-filled sandstone, the shale smear will appear more resistive; where the host rock is hydrocarbon-saturated sandstone, the shale smear will appear more conductive. An independent calibration of lithology from openhole logs or core is therefore a critical interpretation step. Shale smear features commonly appear as relatively low amplitude discordant features on acoustic images. The sealing potential of shale smeared fault zones may be ranked according to the fault zone width and the percentage of

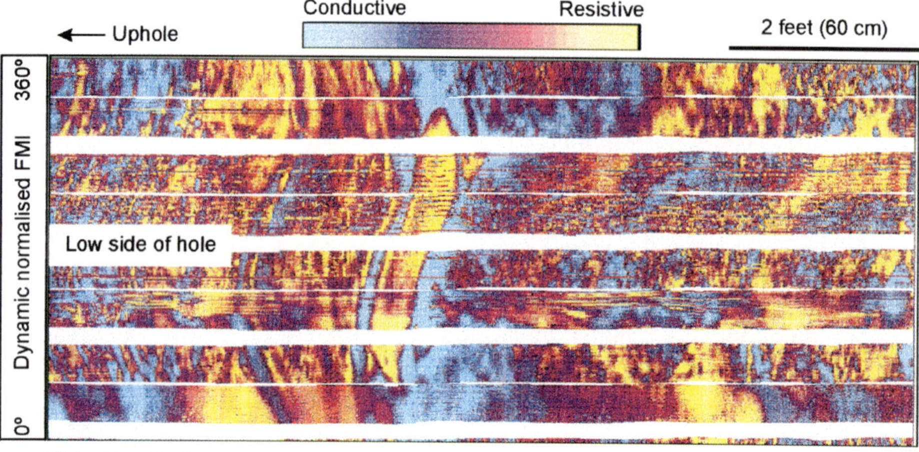

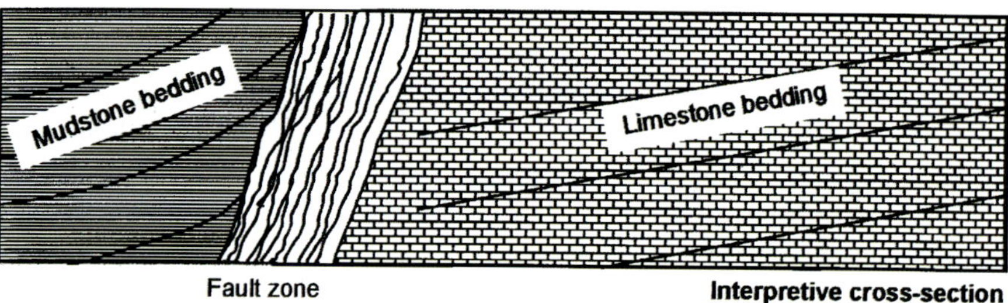

Fig. 3. Juxtaposition fault seal in chalk, horizontal well, FMI image. This fault juxtaposes impermeable mudrocks with permeable limestones, and therefore is of high sealing potential at this location. The fault zone is approximately 60 cm (2 feet) wide, formed of anastomosing slip zones. Throw on this fault could be further constrained following integration of biostratigraphical information, which may allow further evaluation of the seal potential.

potentially sealing material present within the fault zone.

Figure 5 shows 40 m (*c.* 120 ft) of an electrical image in a horizontal well through a sandstone reservoir (Follows 1997). A fault zone is present in the centre of the image. Bedding on the uphole side of the fault dips towards the SW, bedding on the downhole side of the fault dips to the NE, confirming significant displacement across the fault. The fault itself can be subdivided into a central conductive (dark coloured) fault zone and more resistive (light coloured) damage zones on each side. The central fault zone is 1.7 m (5 ft) wide. The conductive nature of the fault zone, and the strong shale response on the openhole logs, indicates that the fault zone is shale smeared and is therefore probably a significant seal. The adjacent damage zone comprises a series of discrete deformation bands and zones of deformation bands (see cementation and grain size reduction fault seals), which also act as baffles to fluid flow. The overall sealing potential of this fault zone is high. In contrast, Fig. 6 shows similar features in a horizontal well through chalk. This image features two conductive (light blue) faults. The fault seal is interpreted to be carbonaceous material related to increased compaction and pressure-solution along the fault zone. This fault zone has a high sealing potential. Note the bedding dip change across the fault at the right-hand side of the image. Clay smear fault seals are uncommon in carbonates due to the lack of terrigenous clay material in the system.

Grain size reduction fault seals

Grain size reduction fault seals are a common product of faulting within high porosity sandstones and carbonates, with deformation principally associated with grain rotation and

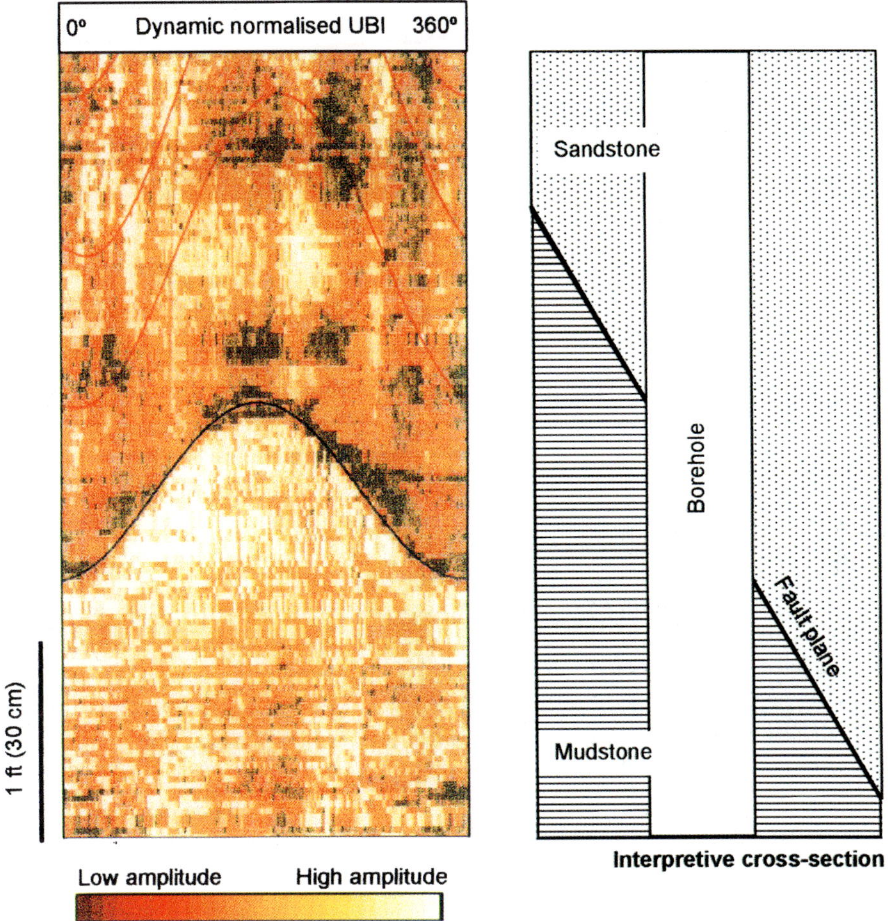

Fig. 4. Juxtaposition fault seal in clastics, UBI image. This fault juxtaposes mudstone in the footwall with sandstone in the hanging wall, resulting in an effective seal. The fault in this case is a single plane, and fault width is not resolvable from this image.

cataclasis (Aydin 1978; Underhill & Woodcock 1987; Antonellini & Aydin 1994). Sealing potential is dependent on the host lithology (grain size, clay content) and the extent of grain size reduction and post faulting lithification (Fisher & Knipe this volume). Sealing potential is moderate to low. Commonly quoted figures are a permeability reduction of 2–6 orders of magnitude. Fisher & Knipe (this volume) however indicate fault rock permeabilities of 0.001–0.1 mD.

Grain size reduction fault seals appear as resistive features on electrical images and high amplitude reflections on acoustic images. A typical grain size reduction fault seal in clastic rocks is illustrated in Fig. 7 from an electrical image. The fault is a zone of deformation bands (Antonellini & Aydin 1994), and the internal fabric may be observed in the resistive zone. A single deformation band is offset by later movement across the fault zone. The resistive nature of the fault zone is due to the decreased permeability caused by cataclastic deformation processes. This image forms part of a major fault zone highlighted in the well overview display (Fig. 8). Deformation in the hanging wall of this fault zone is accommodated by ductile bedding rotation (folding) as well as minor faulting; deformation in the footwall is primarily associated with minor faulting. The width of the damage zone is indicated by a fracture density curve (number of fractures per foot along hole). A significant drop in formation pressure (as recorded by formation pressure data) occurs across the fault zone. This indicates that the fault has acted as a baffle to fluid flow during the production life

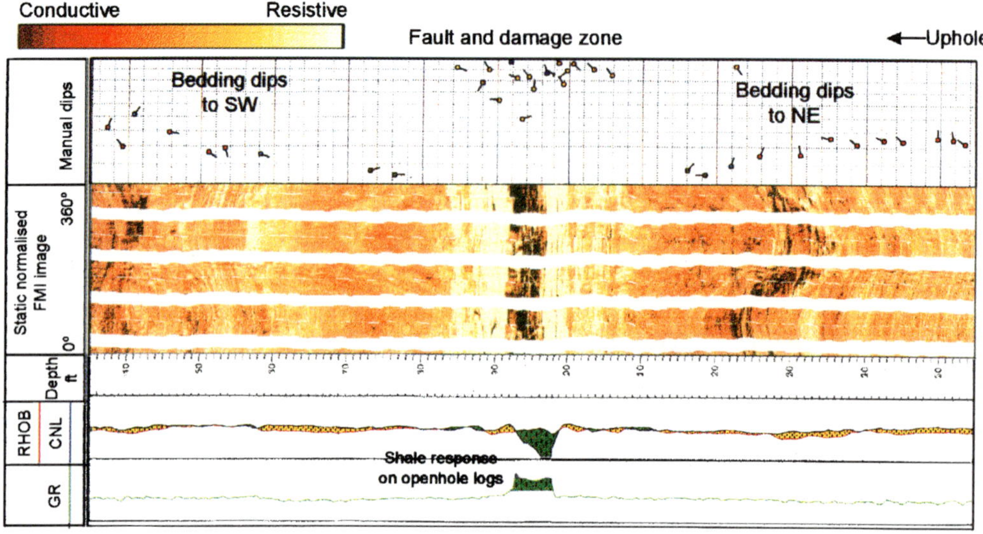

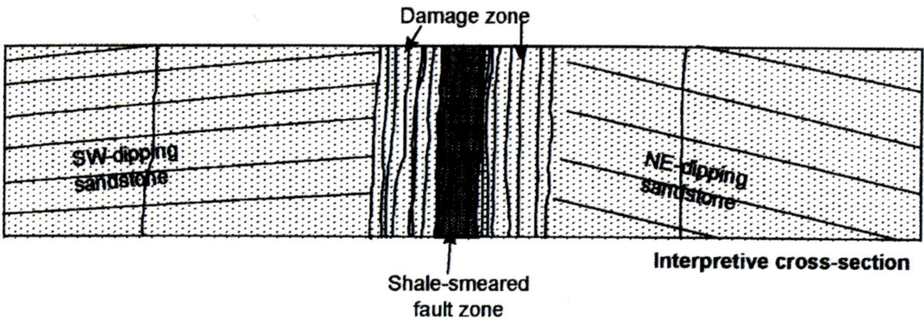

Fig. 5. Shale smear fault zone in sandstone, horizontal well, FMI image. This image is from an aeolian sandstone. The central fault zone shows a 1.2 m (4 ft) wide central fault zone, interpreted as clay-smeared on the basis of the high conductivity (dark colour) and openhole log response. The damage zone on either side of the fault is more resistive than the surrounding sandstone, with elevated density log readings, and is interpreted to comprise a series of deformation bands, with possible associated cementation. Such features were seen in cored intervals of nearby wells.

of this field, and thereby effectively compartmentalized this part of the reservoir.

Cementation fault seals

Cemented fault seals are where cementation is the main porosity reduction mechanism. Cements may have developed from fluids flowing along the fault zone. Alternatively, cements may be generated locally by processes such as pressure solution, and preferentially precipitate within faults due to their greater number of crystal growth sites (Fisher & Knipe this volume). Cementation leads to a dramatic reduction in cross-fault permeability compared with uncemented faults. Fisher & Knipe (this volume) show that quartz-cemented cataclasites often have permeabilities below 0.001 mD compared with values of above 0.01 mD for uncemented cataclasites. Cemented faults are therefore excellent seals.

Cemented faults appear as very resistive features on electrical images and very high amplitude reflections on acoustic images. A resistivity image across a cemented fault zone is shown in Fig. 9 (Harker *et al.* 1990). The hanging wall appears to be more conductive than the footwall, and as the lithology in both blocks is similar (Claymore Sandstone Member of the Kimmeridge Clay Formation), this suggests that the hanging wall has a lower oil saturation than the footwall. In addition, RFT pressure measurements showed a 1000 psi difference between

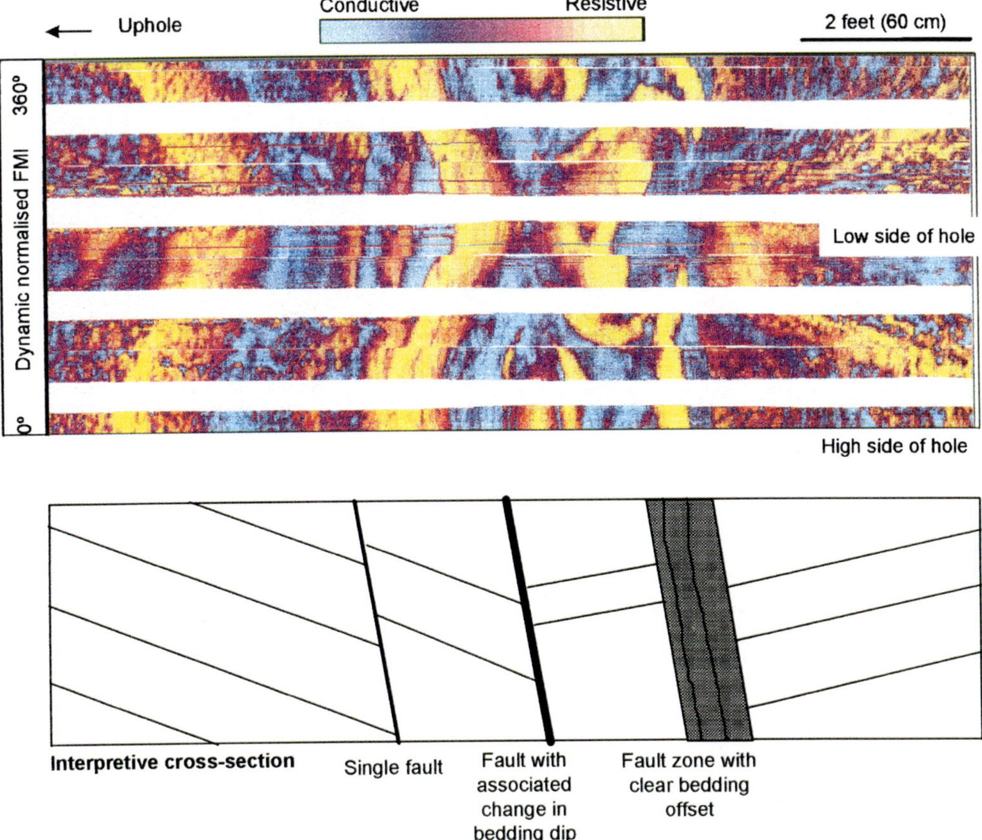

Fig. 6. Clay-lined faults in a horizontal well through chalk (FMI images). These faults are interpreted as clay lined based on their high conductivity (light blue) response. The main fault is located in the middle of the image, and has an associated bedding dip change.

the hanging wall and footwall, indicating that the fault is acting as a barrier during the production phase of this field. The fault zone is highly resistive, due to cementation (confirmed by the strong response from the density–neutron log) and appears to be similar in structure to the zone of deformation bands observed in Fig. 7. This feature could easily be mistaken for a cemented nodule if the images had not been available to indicate the planar nature and the internal fabric. The same fault zone, with a similar pressure drop, was also observed in two neighbouring wells in the same field (Harker *et al.* 1990).

Open/vuggy faults

Fault zones and joint systems within petroleum reservoirs which have not been entirely occluded by cementation or the products of deformation processes often act as conduits for fluid flow. Flow may occur along the fault or the fault may enhance communication between matrix porosity and the wellbore. This category includes features ranging from open faults and joints, to partially cemented structures which have a connected vuggy fabric. In some reservoirs, fractures represent the main source of permeability and therefore determine the viability of the prospect (Dart & Priisholm 1995; Ericsson *et al.* this volume). Flow along the fault planes may be variable, depending on the amount of channelling, but cross-fault sealing potential is dependent on the development of adjacent shale smears, deformation bands and cementation halos. Cross-fault sealing potential may be ranked by considering the number and width of open fault planes.

Open faults are filled with drilling mud, and therefore generally appear as conductive features on electrical images with an overall appearance similar to conductive shale smear type fault zones. On acoustic images, they appear as low

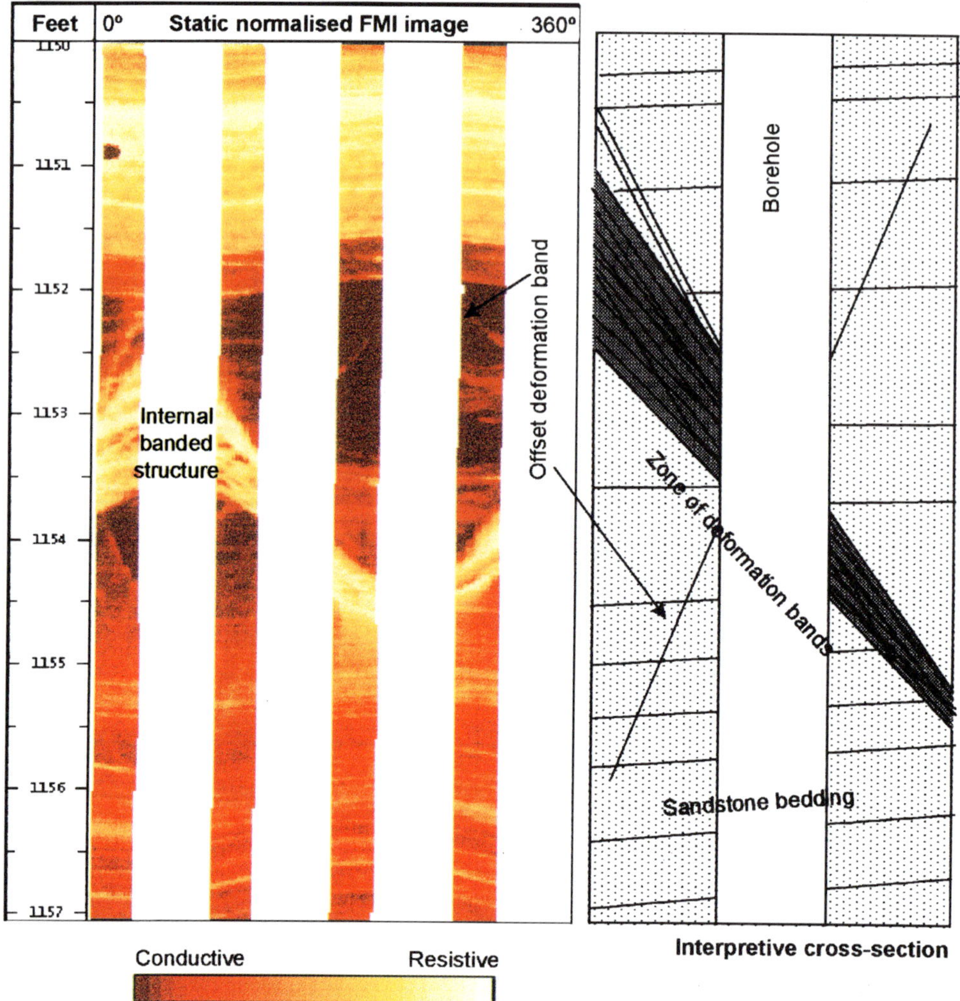

Fig. 7. Grain-size reduction fault seal in clastic rocks (FMS image). This image forms part of a major fault zone shown in Fig. 8. The individual fault shown here comprises a zone of deformation bands, formed by repeated strain causing widening of the deformation zone by strain hardening. The internal banded structure of the zone of deformation bands is a typical characteristic. Cement may be present, but the openhole logs do not show much change, suggesting that any decrease in porosity is primarily due to grain-size reduction.

amplitude features and have high transit times (i.e. they have an expression on the borehole wall). Figure 10 shows a major open fault in well-bedded shales. The fault plane on the images is dark (conductive) and irregular and is associated with both poor hole conditions and with heavy mud losses during drilling. In general terms, the width of the conductive fault planes represents the width of the open 'hole' in the borehole wall. However, this parameter will have been enhanced by spalling and abrasion during drilling and so may not represent the true width of the open fault. Such fractures are often enhanced by preferential opening near the borehole in the minimum horizontal stress direction (i.e. perpendicular to the maximum horizontal stress direction). Figure 11 shows a vuggy fault in chalk. This fault has a conductive (blue) core, with a resistive halo (yellow) suggesting partial cementation.

Fracture aperture measurements. Fracture aperture measurements are often performed on electrical images which show conductive fractures. Most methods are based on the 'excessive conductivity' approach published by Luthi & Souhaité (1990). This is a quantitative approach

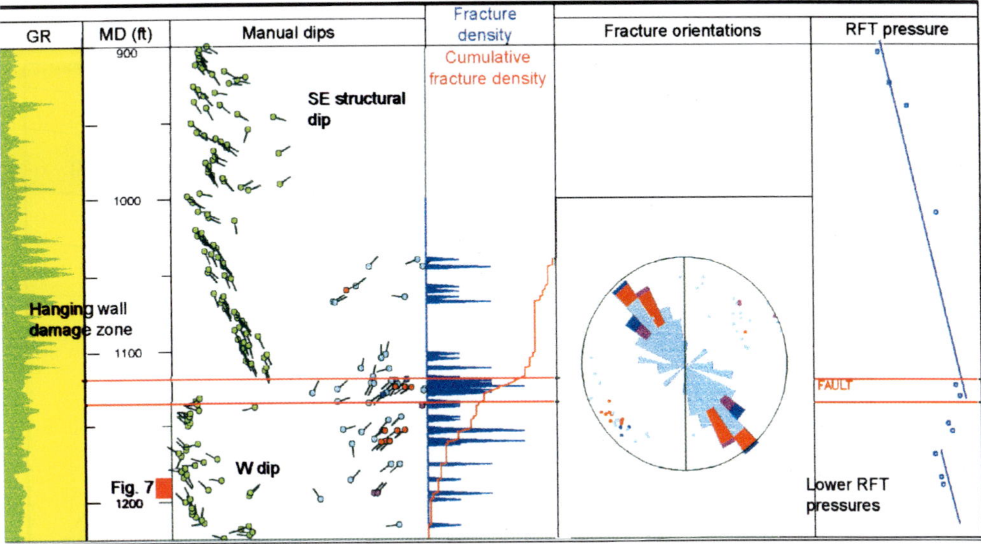

Fig. 8. Grain-size reduction fault seals in clastic rock. Integration of the fault zone shown in Fig. 7 with additional data, including formation pressure (RFT) data.

to estimating fracture width based on the relative contrast of the fracture fill and the host rock. It logically leads to estimates of fracture porosity and storage volume for reservoir modelling work. Recent work (Evans *et al.* 1996) has shown a good agreement between fracture widths modelled from electrical images and SEM measurements. However, we believe that the resistivity based fracture aperture methodology should be treated with caution, particularly in the following areas:

(1) An image-based aperture measurement is carried out at a very shallow depth of investigation, on rock that has been subjected to major stresses related to the drilling process. The apertures calculated from the borehole wall may therefore not reflect true apertures in the virgin reservoir.
(2) The application of this methodology needs to make a distinction between natural fractures and drilling-induced fractures, as they have significantly different distribution and flow characteristics.
(3) Aperture measurements generally do not correlate directly with fluid flow (Barton & Zoback 1992). This is because the key factor is not fracture width, but fracture connectivity to a wider open fracture network. This can only be predicted by building a model of the 3D fracture network geometry, incorporating stress field information, and tested during welltests or measurement of inflow performance.
(4) A conductive reading on a borehole image may be caused by the fracture being invaded by drilling mud, thus indicating that the fracture is open (at least locally, at the borehole wall). However, conductive readings can also occur where there is any other conductive fracture fill (e.g. clay smear, pyrite cement). Fracture aperture calculations carried out using an undifferentiated conductive reading may therefore be inaccurate. For example, a single conductive spike caused by pyrite may create a large apparent aperture using the Luthi & Souhaité method.

In conclusion, we believe that fracture aperture measurements made from electrical borehole images should be treated with caution, and only used quantitatively where verified by dynamic reservoir measurements. Figure 12 summarizes the thought processes which led to interpretations of fault seal properties from images and other relevant data.

Discussion

Applications

Borehole image based evaluation of fault seal types has a number of key applications in reservoir characterization. Images can assist in planning of completion strategies, including the setting of packers for testing specific fault

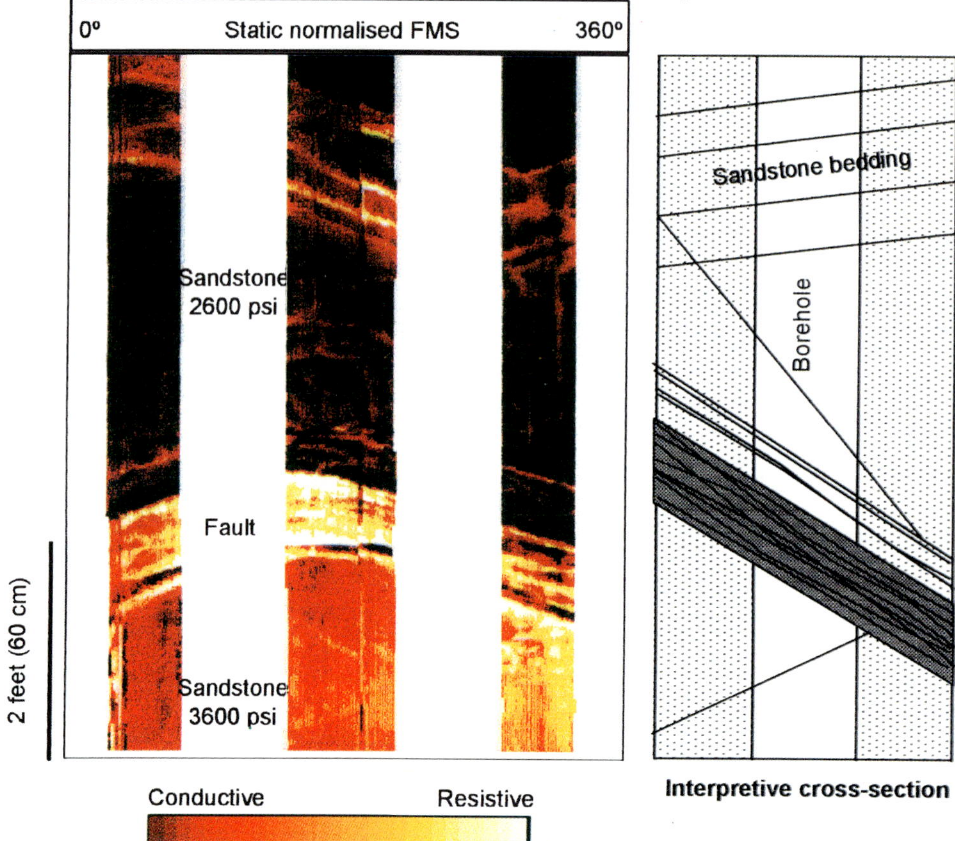

Fig. 9. Electrical (FMS) image across a cemented fault zone. This fault occurs in the moderate porosity sandstones of the Claymore Sandstone Member of the Kimmeridge Clay Formation, Claymore Field, Central North Sea (Harker *et al.* 1990). The openhole logs show a very strong response (high density, low neutron), indicating that the fault is cemented, and this is verified by the very strong resistivity of the fault zone. RFT formation pressure measurements indicate that this fault is responsible for a 1000 psi pressure difference between the hanging wall and footwall. The darker colour of the hanging wall suggests a higher water saturation than the footwall.

blocks or open fracture zones. After core calibration, confident image interpretation can be performed outside the cored interval, thus providing greater information on the nature and size of fault zones. Image interpretations can be integrated with seismic data, to allow near sub-seismic faults to be confidently interpreted. Fault rock data gathered from images can be directly input into fractured reservoir modelling (Rawnsley *et al.* 1997, Swaby & Rawnsley 1997).

Advantages and limitations of an image-based technique

An image-based technique can provide valuable information on the fault zone structure and nature of the related deformation. Images provide continuity within a borehole, which may not be achieved using core. Images are often run over long sections of borehole (commonly 10–100 times longer than the cored sections), allowing the distribution and structure of more features to be evaluated. However, borehole image interpretation is only one aspect of overall fault rock characterization. Images may be affected by borehole conditions and acquisition practice, which will adversely affect the ability to interpret fractures from the images. Resistivity or acoustic properties are not direct indicators of lithology and require calibration to openhole logs. Most importantly, borehole images cannot provide quantitative indications of fault rock permeability or fault transmissibility,

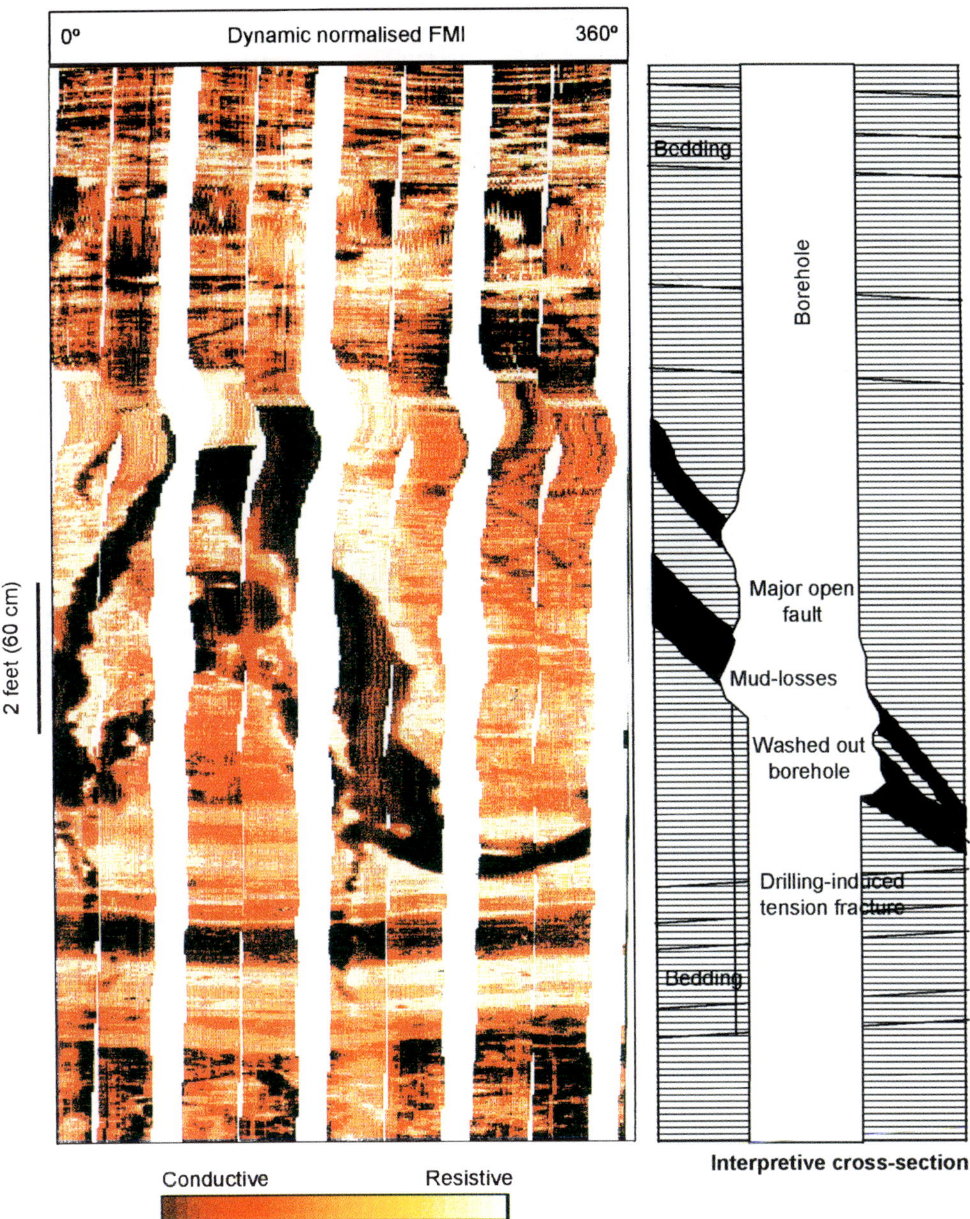

Fig. 10. Major open fault in well-bedded shale (FMI image). This fault was classified as being open due to associated heavy mud losses during drilling. The other lower confidence indicators of an open fault existing here are the lower resistivity (darker colour) and the associated borehole washout. However, these could also indicate a clay-smeared fault zone (compare with Fig. 5).

and integration with dynamic measurements is therefore required. In addition, images are of limited lateral extent, being confined to the borehole. Fault rock properties are likely to be different at different locations on the fault surface, resulting in spatial variations in sealing properties. This is obviously an important consideration when attempting to extrapolate

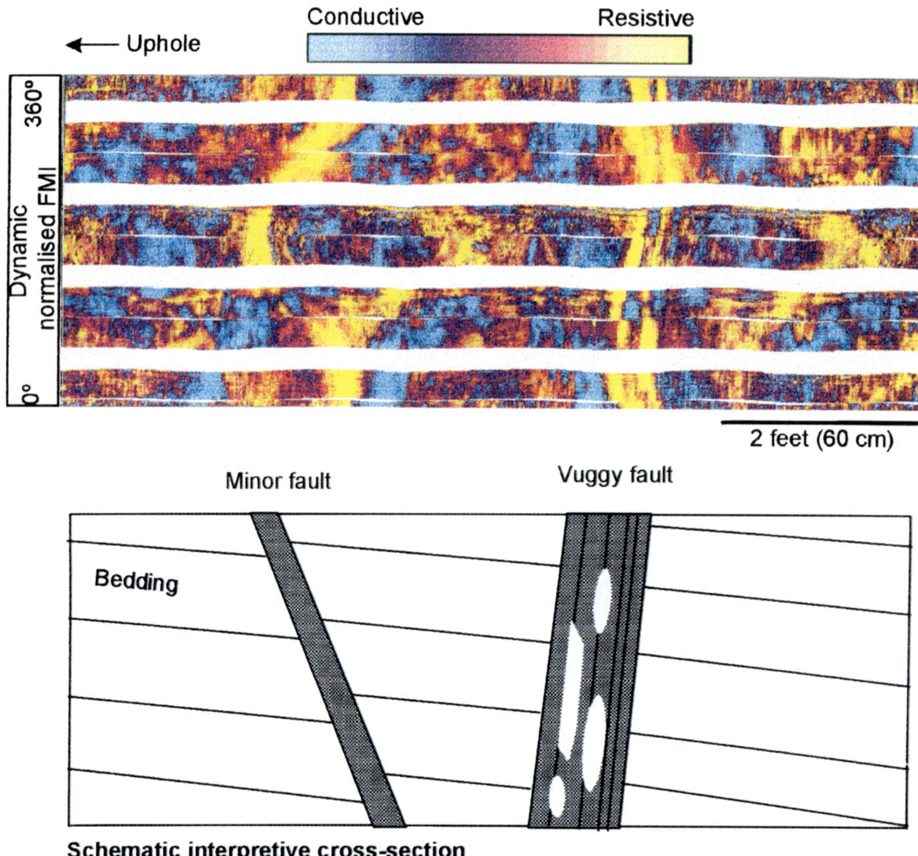

Fig. 11. Vuggy fault in chalk (FMI image). This fault has a conductive (blue) vuggy core, with a resistive halo (yellow) suggesting partial cementation.

single-point borehole measurements to the entire fault surface.

Conclusions

(1) Borehole images allow fault zones encountered within wellbores to be characterized in terms of fault rock structure, geometry and sealing potential. Five categories of fault seals can be described from borehole images:

  (i) *juxtaposition fault seals*: the reservoir is placed against seal lithologies and appear as sharp image contrasts;
  (ii) *clay lined fault seals* within fault zones appear as contrasting resistivity bands on electrical images (depending on the resistivity of the host rock), and relatively low amplitude bands on acoustic images;
  (iii) *grain size reduction fault seals* appear as resistive or high amplitude features;
  (iv) *cemented fault seals* appear as very resistive or very high amplitude features;
  (v) *open/vuggy faults* appear as conductive features on electrical images, and as low amplitude and high transit time features on acoustic images.

(2) Borehole images allow qualitative characterization of potential fault seals by permitting detailed fault rock structure to be described.
(3) Borehole images only provide assessment of sealing potential in the immediate vicinity of the wellbore. Along-fault seal properties may change.
(4) Caution must be exercised in the use of quantitative aperture measurements obtained from electrical borehole images.
(5) Borehole images only provide a qualitative estimate of sealing potential and must be integrated with dynamic measurements such

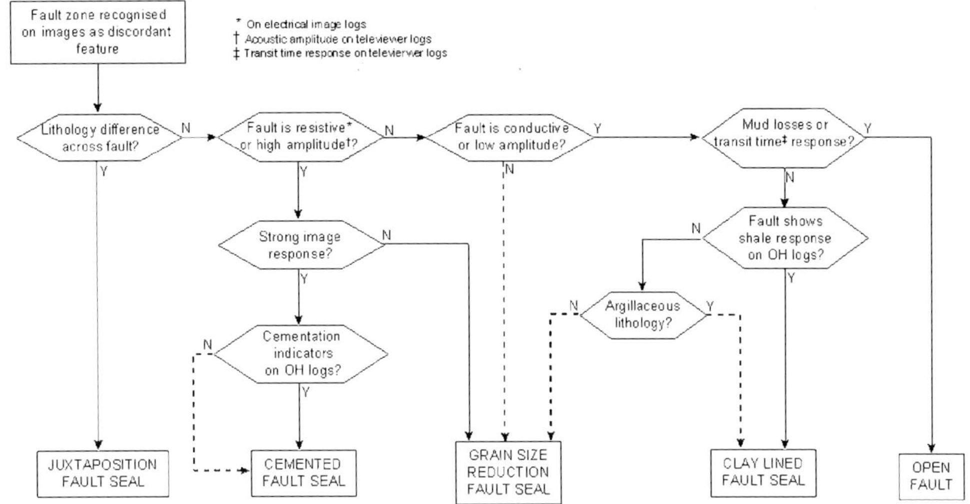

Fig. 12. Simplified image interpretation process map for fault seal interpretation. This illustrates possible thought processes towards determining fault sealing properties from borehole image logs. Such an interpretation is never an isolated process, and integration with other data (core, logs, dynamic measurements) is essential. Dashed lines indicate lower confidence interpretation.

as formation pressure, flowmeter data or well tests to fully determine sealing potential.

We acknowledge the permission of Z&S consultancy clients to use the images presented in this paper. Particular contributions were made by C. Glass, I. Tribe, C. Ottesen, S. Sadler and R. McGarva. We are grateful to E. Follows, G. Jones and M. Lovell for constructive and challenging reviews. The corresponding author can be contacted by email on john.adams@zands.com.

References

ANTONELLINI, M. & AYDIN, A. 1994. Effect of faulting on fluid flow in porous sandstones: petrophysical properties. *American Association of Petroleum Geologists Bulletin*, **78**, 355–377.

ATLAS WIRELINE SERVICES 1992. *Digital Circumferential Borehole Imaging Log (CBIL)*, Brochure, AT92–082.

AYDIN, A. 1978. Small faults formed as deformation bands in sandstone. *Pure and Applied Geophysics* **116**, 913–930.

BARTON, C. A. & ZOBACK, M. D. 1992. Self-similar distribution and properties of macroscopic fractures at depth in crystalline rock in the Cajon Pass scientific drillhole. *Journal of Geophysical Research*, **97**, B4.

—— & —— 1994. Stress perturbations associated with active faults penetrated by boreholes: Possible evidence for near-complete stress drop and a new technique for stress magnitude measurement. *Journal of Geophysical Research*, **99**, 9373–9390.

BELL, J. S. 1990. Investigating stress regimes in sedimentary basins using information from oil industry wireline logs and drilling records. *In*: HURST, A., LOVELL, M. A. & MORTON, A. (eds), *Geological Applications of Wireline Logs*, Geological Society, London, Special Publications, **48**, 305–325.

BOURKE, L. T. 1989. Recognising artifact images of the Formation MicroScanner. *Proceedings of Society of Professional Well Log Analysts 13th Annual Logging Symposium*, Denver, June, Paper WW.

DART, C. J. & PRIISHOLM, S. 1995. 3D characterisation of fracture networks in chalk using electrical borehole images, Danish Central Graben, North Sea. Abstract in *American Association of Petroleum Geologists Bulletin*, **79**, 1206.

DYKE, C. G., WU, B. & MILTON-TAYLER, D. 1995. Advances in characterising natural-fracture permeability from mud-log data. *SPE Formation Evaluation*, **10**, 160–166.

EKSTROM, M. P., DAHAN, C. A., CHEN, M. Y., LLOYD, P. & ROSSI, D. J. 1987, Formation imaging with microelectrical scanning arrays. *The Log Analyst*, **28**, 294–306.

ENGELDER, J. T. 1974. Cataclasis and the generation of fault gouge. *Bulletin of the Geological Society of America*, **85**, 1515–1522.

ERICSSON, J. B., MCKEAN, H. C. & HOOPER, R. J. 1998. Facies and curvature controlled 3D fracture models in a Cretaceous reservoir, Arabian Gulf. *This volume*.

EVANS, L. W., THORN, D. & DUNN, T. L. 1996. Formation MicroImager, MicroScanner, and core characterisation of natural fractures in a horizontal well in the Upper Almond Bar Sand, Echo Springs Field, Wyoming. *Gulf Coast Section of Society of Economic Palaeontologists and Mineralogists (SSEPM) Foundation 17th Annual Research Conference, Stratigraphic Analysis*, December 8–11, 1996.

FARAGUNA, J. K., CHACE, D. M. & SCHMIDT, M. G. 1989. An improved borehole televiewer system: image acquisition, analysis and integration, *Proceedings of Society of Professional Well Log Analysts Thirtieth Annual Logging Symposium*, June 11–14, Denver.

FISHER, Q. J. & KNIPE, R. J. 1998. Microstructural controls on the petrophysical properties of deformation features. *This volume*.

FOLLOWS, E. 1997. Integration of inclined pilot hole core with horizontal image logs to appraise an aeolian reservoir, Auk Field, Central North Sea. *Petroleum Geoscience*, **3**, 43–55.

HALLIBURTON ENERGY SERVICES LTD. 1995. *Electrical Micro Imaging Service*, Brochure, EL1076.

HARKER, S. D., MCGANN, G. J., BOURKE, L. T. & ADAMS, J. T. 1990. Methodology of Formation MicroScanner Tool image interpretation in Claymore and Scapa Fields (North Sea). *In*: HURST, A., LOVELL, M. A. & MORTON, A. C. (eds), *Geological Applications of Wireline Logs*, Geological Society Special Publication No. **48**, 81–88.

HEFFER, K. & KOUTSABELOULIS, N. C. 1995, Stress effects on reservoir flow: – Numerical modelling used to reproduce field data. *In*: DE HAAN, H. J. (ed.) *New Developments in Improved Oil Recovery*, Geological Society, London, Special Publications, **84**, 81–88.

HILLIS, R. R. 1997. Does the in situ stress field control the orientation of open natural fractures in subsurface reservoirs? *Exploration Geophysics*, **28**, 80–87.

JONES, G. & KNIPE, R. J. 1996. Seismic attribute maps; application to structural interpretation and fault seal analysis in the North Sea Basin. *First Break*, **14**, 449–461.

KNIPE, R. J. 1992. Faulting processes and fault seal. *In*: LARSEN, R. M., BREKKE, H., LARSEN, B. T. & TALLERAS, E. (eds) *Structural and tectonic modelling and its application to petroleum geology*, Norwegian Petroleum Society, Special Publication, **1**, 325–342.

——, FISHER, Q. J., JONES, G. CLENNELL, M. R., FARMER, A. B., HARRISON, A., KIDD, B., MCALLISTER, E., PORTER, J. R. & WHITE, E. A. 1997. Fault seal analysis: successful methodologies, application and future directions. *In*: MØLLER-PEDERSEN, P. & KOESTLER, A. G. (eds) *Hydrocarbon Seals: Importance for Exploration and Production*. Norwegian Petroleum Society (NPF) Special Publication, **7**, 1–24.

KNOTT, S. D., 1993. Fault seal analysis in the North Sea. *American Association of Petroleum Geologists Bulletin*, **77**, 778–792.

LINDSAY, N. G., MURPHY, F. C., WALSH, J. J. & WATTERSON, J., 1990. Outcrop studies of shale smears on fault surfaces. *In*: FLINT, S. S. & BRYANT, I. D. (eds), *The Geological Modelling of Hydrocarbon Reservoirs and Outcrop Analogues*. Special Publication of the International Association of Sedimentologists, **15**, 113–124.

LUTHI, S. M. & SOUHAITE, P. 1990, Fracture apertures from electrical borehole scans, *Geophysics*, **55**, 821–833.

NELSON, R. A. 1985. *Geological Analysis of Naturally Fractured Reservoirs*. Gulf Publishing Company, Houston.

PLUMB, R. A. & HICKMAN, S. H. 1985. Stress-induced borehole elongation: a comparison between the four-arm dipmeter and the borehole televiewer in the Auburn Geothermal Well. *Journal of Geophysical Research*, **90**, 5513–5521.

—— & COX, J. W. 1987, Stress directions in eastern North America determined to 4.5 km from borehole elongation measurements. *Journal of Geophysical Research*, **92**, 4805–4816.

RAWNSLEY, K., AUZIAS, V., PETIT, J. P. & RIVES, T. 1997. Extrapolating fracture orientations from horizontal wells using stress trajectory models. *Petroleum Geoscience*, **3**, 145–152.

SCHLUMBERGER 1992. *FMI Fullbore Formation MicroImager*, Brochure, SMP 9210.

—— 1993. *Ultrasonic Imaging. USI UltraSonic Imager, UBI Ultrasonic Borehole Imager*, Brochure, SMP-9230.

SWABY, P. A. & RAWNSLEY, K. D. 1997. An interactive 3D fracture-modeling environment. *Society of Petroleum Engineers Computer Applications*, June 1997, 82–87.

UNDERHILL, J. R. & WOODCOCK, N. H. 1987. Faulting mechanisms in high porosity sandstones; New Red Sandstone, Arran, Scotland. *In*: JONES, M. E. & PRESTON, R. M. F. (eds) *Deformation of Sediments and Sedimentary Rocks*, Geological Society, London, Special Publications, **29**, 91–105.

WESTERN ATLAS INTERNATIONAL INC. 1996. *STAR Imager* Brochure, L96–037.

WILLIAMS, C. G., JACKSON, P., LOVELL, M. A., HARVEY, P. K. & REECE, G. 1995. Numerical simulation of downhole electrical conductance imaging. *Proceedings of Society of Professional Well Log Analysts 16th European Formation Evaluation Symposium*, Aberdeen, Paper O.

ZOBACK, M. D., BARTON, C. A., BRUDSY, M., CHANG, C., MOOS, D., PESKA, P. & VERNIK, V. 1995. A review of some methods for determining the in situ stress state from observations of borehole failure with applications to borehole stability and enhanced production in the North Sea, *Workshop on Rock Stresses in the North Sea*, Trondheim, Norway, Feb 13–14, 1995.

——, MOOS, D. & MASTIN, L. 1985. Well bore breakouts and in situ stress. *Journal of Geophysical Research*, **90**, 5523–5531.

Structure and content of the Moab Fault Zone, Utah, USA, and its implications for fault seal prediction

K. A. FOXFORD[1], J. J. WALSH[1], J. WATTERSON[1], I. R. GARDEN[2*], S. C. GUSCOTT[3†] & S. D. BURLEY[3*‡]

[1] *Fault Analysis Group, Department of Earth Sciences, University of Liverpool, Liverpool L69 3BX, UK*

[2] *Reservoir Description Research Group, Department of Petroleum Engineering, Heriot–Watt University, Riccarton, Edinburgh EH14 4AS, UK*

[3] *Diagenesis Research Group, Department of Earth Sciences, University of Manchester, Manchester M13 9PL, UK*

[*] *Present address: BG Technology, Ashby Road, Loughborough, Leicestershire LE11 3GR, UK*

[†] *Present address: Amerada Hess, Scott House, Altness, Aberdeen AB1 4LE, UK*

[‡] *Present address: Basin Dynamics Research Centre, Department of Earth Sciences, University of Keele, Staffordshire ST 5BG, UK*

Abstract: The structure and content of the Moab Fault zone are described for 37 transects across the fault zone where throws range from less than 100 m to *c.* 960 m. The 45 km long fault trace intersects a sedimentary sequence containing a high proportion of sandstones with good reservoir properties, interspersed with numerous mudstone layers. Typically, the fault zone is bounded by two external slip zones with the fault zone components separated by up to nine internal slip zones. Fault zone components are tabular lenses of variably deformed sandstones and sandstone cataclasites and breccia, with a wide size range, usually enclosed in a matrix of shaley fault gouge containing mm to m scale entrained sandstone fragments. Neither fault zone structure nor content can be predicted by extrapolation over distances as little as 10 m. Although variable in thickness, shaley gouge is always present except where the mudstone is <*c.* 20% of the faulted sequence. The distribution of shaley gouge conforms with exis•ng algorithms for predicting the presence or absence of shaley gouge in subsurface fault zones. The fault zone heterogeneity is attributed to tip-line and asperity bifurcation processes.

Assessment of the sealing potential or capacity of fault zones in the sub-surface is an important aspect of both the search for hydrocarbon reservoirs and their exploitation. The sealing potential of a fault is a function of the sequence juxtapositions across the fault and of the presence or absence of membrane seals resulting from the structure and content of the fault zone (Weber *et al.* 1978; Downey 1984; Allan 1989; Nybakken 1991; Knipe 1992; Gibson 1994; Berg & Avery 1995; Fristad *et al.* 1997). There are two possible, end-member, approaches to the prediction of membrane seals, one largely empirical and the other deterministic. The empirical approach takes no account of the possible structure of the fault zone at a sub-seismic scale, which is considered to be effectively unpredictable, but relies on a variety of algorithms to predict fault seal potential (e.g. due to shale smear/gouge formation) by analysis of the sequence and sequence geometry adjacent to a fault (e.g. Gibson 1994; Fristad *et al.* 1997). An essential element of this approach is the local calibration of results, based on comparison with data from faults for which the across-fault pressure histories are known. The deterministic approach places emphasis on the prediction of the presence and properties of membrane seals by extrapolating data from wells intersecting either the target fault or faults believed to be comparable with it (Fisher & Knipe 1998, this volume). Our earlier conclusion that the structure and content of fault zones are largely unpredictable, based on studies of relatively small faults in the UK (Childs *et al.* 1997*a*,*b*), are reinforced by the observations reported here on the Moab Fault in SW Utah. The highly irregular nature of this fault zone points strongly towards empirical methods for prediction of fault seal potential. Data are presented on the structure and content of the Moab Fault zone, obtained from 35 transects across the fault, supplemented by transects across two smaller faults at the crest of a faulted anticline in the hanging wall of the Moab Fault. These transects have been observed either in natural exposures or in trenches excavated for the purpose. Conclusions are drawn concerning the geometry and evolution of the fault zone

Foxford, K. A., Walsh, J. J., Watterson, J. *et al.* 1998. Structure and content of the Moab Fault Zone, Utah, USA, and its implications for fault seal prediction. *In*: Jones, G., Fisher, Q. J. & Knipe, R. J. (eds) *Faulting, Fault Sealing and Fluid Flow in Hydrocarbon Reservoirs*. Geological Society, London, Special Publications, **147**, 87–103.

and the prediction of fault zone hydraulic properties in the subsurface.

Regional context

The Moab Fault is a salt-related normal fault (Fig. 1), with a 45 km long surface trace and maximum throw at surface of *c*. 960 m, offsetting a Pennsylvanian to Cretaceous sequence (Figs 1 & 2; Doelling 1988; Foxford *et al.* 1996). The maximum surface throw is on the relatively simple southern segment of the fault trace and throw decreases northwards to zero at the end of a more complicated northern segment which has numerous splays (Fig. 3; Foxford *et al.* 1996). Along the southern segment, footwall bed dips define a structural high symmetrically disposed about the point of maximum throw. A prominent hanging wall feature of the southern segment is the Moab Anticline, with a crestal collapse graben accommodated by an array of normal faults. The Moab Fault was active from

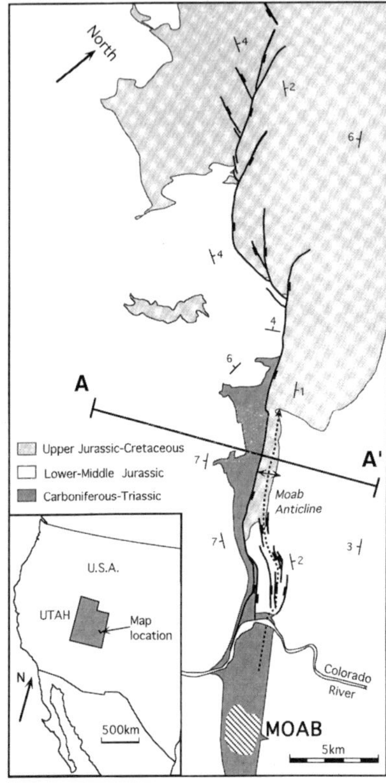

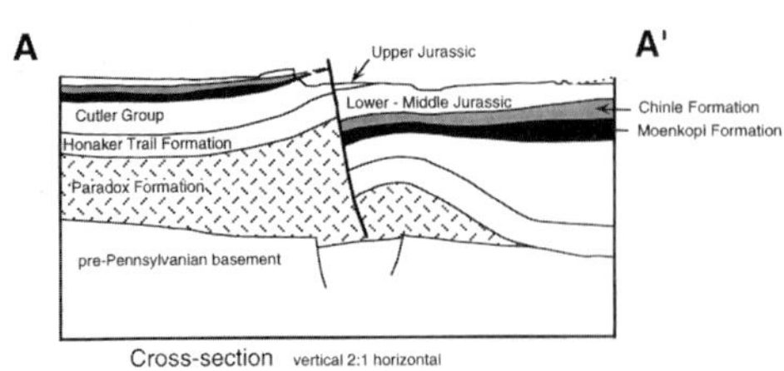

Fig. 1. (**a**) Geological map of the Moab area, showing the surface trace of the Moab Fault and representative attitudes of bedding. The inset map is a location map showing the western portion of the USA. (**b**) Cross-section of Moab Fault derived from outcrop mapping and interpreted seismic data (Foxford *et al.* 1996). Location of cross-section is shown in (**a**).

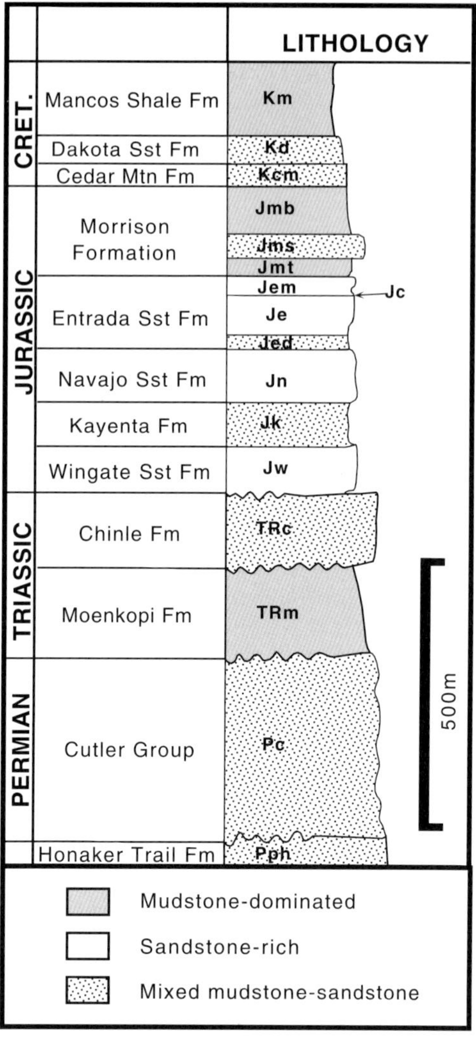

Fig. 2. Schematic stratigraphic section illustrating the nature of the sequence cut by the Moab Fault. Note the distribution of mudstone, sandstone-rich and mixed mudstone–sandstone sequences. The stratigraphic abbreviations are used in Figs 4, 5 & 6.

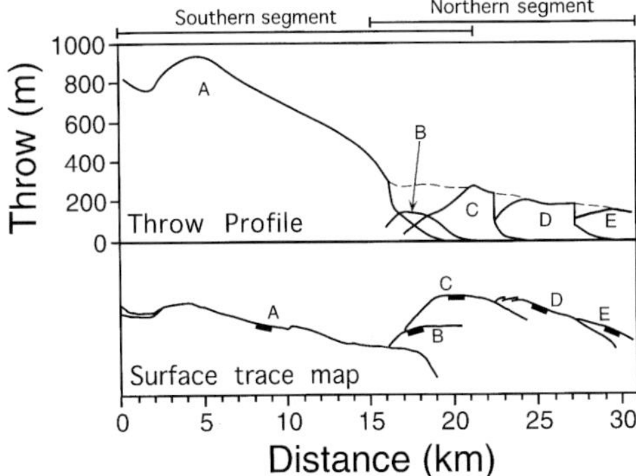

Fig. 3. Plot of throw versus distance along the surface trace of the Moab Fault. Separate throw profiles are shown for each of the major fault splays (solid lines) together with an aggregated profile for the fault system as a whole (broken line). The lateral extent of both the northern and the southern segments are shown (top of diagram). Throw values are calculated using locally determined stratigraphic offsets and include the component of normal drag. Throw transfer between the northern splay faults is consistent with their origin as breached relay zones. The relay zones are important because it is only along these portions of the surface trace that low throw values exist (e.g. where segments A, B & C overlap).

the Triassic until at least the Early Tertiary, but with a break from mid-Jurassic until at least mid-Cretaceous (Foxford *et al.* 1996). Details of the fault and country rocks, which are underlain by either a salt anticline or salt roller (Jackson & Talbot 1994), are given in Foxford *et al.* (1996). None of the fault zone features described is due specifically to salt tectonics.

Those parts of the sequence which either abut the fault at outcrop or which have been displaced past the present outcrop level, and which may have contributed material to the outcropping fault zone, are shown in the lithostratigraphic tables in Fig. 2. The faulted interval can be separated into three lithological groups, i.e. mudstone-dominated, mixed mudstone–sandstone and sandstone-rich sequences. Mudstone-dominated sequences comprise the Moenkopi Formation (91–158 m thick) together with the Tidwell (10–15m) and Brushy Basin (80–90 m) members of the Morrison Formation. Mixed mudstone-sandstone intervals, which comprise interbedded fluvial and aeolian sandstones and floodplain or lacustrine mudstones and siltstones, include the Cutler Group, (10–400 m), the Chinle (98–122 m), Kayenta (67–104 m), Curtis (0–20 m) and Cedar Mountain (30–60 m) Formations, the Dewey Bridge Member of the Entrada Sandstone (40–60 m), and the Salt Wash Member of the Morrison Formation (*c.* 40 m). Sandstone-rich units are all predominantly of aeolian origin and comprise Wingate and Navajo sandstones, and the Slick-Rock and Moab Tongue members of the Entrada Sandstone. The sandstone-rich units together have an aggregate thickness of between 165m and 290 m. The aeolian sandstones are well-connected reservoir rocks with porosities of 15–30% and permeabilities of 100–4000 mD.

Main characteristics of the fault zone

Transects across the fault zone are exposed in several canyons (e.g. Fig. 4) which cross the fault trace at a high angle, draining the topographic highs of the Moab Fault footwall and of the Moab Anticline. Elsewhere, shallow trenches were excavated across the fault. These exposures provide good data on lateral variation in the fault zone on scales of 10s to 100s of metres, but vertical variations can be determined over intervals of 1 to 20 m. The locations of the transects are shown on Fig. 5.

The Moab Fault zone is, with few exceptions, a sharply defined brittle shear zone, 1–10 m wide, externally bounded by major slip zones which separate fault rocks from relatively undeformed wallrock. External to the fault zone proper but closely associated with it, are concentrations of cataclastic, and less commonly non-cataclastic, slip bands (or granulation seams; Aydin 1978; Aydin & Johnson 1978, 1983; Antonellini *et al.* 1994; Fowles & Burley 1994), which occur only in massive aeolian sandstones within *c.* 40 m of the fault zone and accommodate insignificant displacement (usually <*c.* 10 m). Concentrations of slip bands constitute a damage zone in the sense of Knipe *et al.* (1997) and are often accompanied by polished and striated slip surfaces where they are within a few metres of the fault zone. Normal drag outside the fault zone accommodates a small but significant proportion of the total displacement, particularly in the hanging wall. The throws at fault zone transect localities are derived from locally determined stratigraphic offsets and neither include the normal drag component nor take account of Triassic–Lower Jurassic stratigraphic growth across the fault; the latter simplification is not significant to this article.

At outcrop the fault zone dip is variable, averaging 51°NE along the southern segment and 71°NE along the northern segment, with sub-vertical dips restricted to E–W striking splays in the vicinity of branch-points (Foxford *et al.* 1996). A variety of slip indicators, e.g. striations, corrugations, tool tracks, pluck marks, gutters and breccia pods (Hancock & Barka 1987), indicates that dip-parallel slip dominates. There

Fig. 4. Outcrop photographs showing rapid variations in structure and content of the Moab Fault at R191 Canyon (for location see Fig. 5). The fault is well exposed in both the NW (top) and SE (bottom) sides of the canyon. Note the disparity in fault zone structure, i.e. numbers of slip zones (white lines), sheets of shaley gouge (S) and breccia, between both sides of the canyon, in particular note the absence of the sandstone lens (Jem) in the NW side. Throw at this locality is *c.* 960 m. Stratigraphic abbreviations used are shown on Fig. 2. For more details see text and Fig. 5(d).

are two situations in which oblique-slip occurs. The first of these is along prominent fault bends e.g. opposite the Visitors Centre at the entrance to Arches National Park, where lineations pitch 62°E, and along the footwall splay at the Railway Tunnel location (Fig. 5), where lineations pitch 52°N. The second exception occurs on minor slip surfaces within the fault zone where it contains metre-scale blocks of massive sandstone. The slip surfaces dissect the massive sandstone blocks and show several sets of striations, many of which are sub-horizontal. These apparently anomalous slip directions are attributed to rotations which locally accommodated relative movement between sandstone blocks in the fault zone during faulting.

Fault zone components

The principal fault zone components are either undeformed or cataclastic derivatives of host sequence lithologies, and comprise:

(i) sandstone lenses or blocks;
(ii) sandstone cataclasites and breccias;
(iii) shaley fault gouge;
(iv) slip zones; and
(v) veins.

Sandstone lenses and blocks

Sandstones occur in the fault zone either as fault-parallel lenses (Fig. 4), which vary internally from undeformed to strongly faulted, or as variably deformed blocks enclosed by sheets of shaley fault gouge. Sandstone lenses are always isolated from the wallrock sequence by intervening slip zones and occupy a continuous range from a few cm to several 100's of metres in length, and from cm to *c.* 50 m in thickness. Sandstone blocks are often angular and equidimensional and are less than *c.* 10 m in size.

Sandstone cataclasites and breccias

Sandstone cataclasites and cohesive fault breccias (Sibson 1977) are a widespread and distinctive component of the fault zone (see Figs 4 & 5), occurring within 1cm to 1m thick sheets which in some cases can be traced to their source layer. Cataclasites and breccias contain lens-shaped and variably deformed, mm to cm scale clasts of coarse-grained host- and fault-rocks surrounded by their foliated and intensely cataclased derivatives. Many clasts have been reworked within the fault zone to form cataclasite porphyroclasts.

Shaley fault gouge

Mudstones incorporated in the fault zone are present as fault-parallel sheets of shaley fault gouge, of widely varying thickness (cm to *c.* 10 m thick), which in some cases are not isolated from their source layer. One or more layers of shaley gouge are present in all of the figured transects at which the throw is $>c.$ 100 m. The number of shaley gouge layers in a fault zone increases with the number of slip zones (see below) and the shaley gouge content of the fault zone varies from 100% to zero (Figs 5 & 6). The term shaley fault gouge is used for fine grained rocks which cannot be more precisely classified on the basis of outcrop characteristics. Petrographic examination shows that shaley gouge derived from the post-Wingate Sandstone sequence, i.e. Jurassic or younger units, is dominated by phyllosilicates, whereas gouge derived from older sequences contains a higher proportion of cataclased siltstone and sandstone: the latter type is partly due to comminution of micaceous fluvial sandstones which are very common in the Permo-Triassic sequence. Layers of shaley fault gouge have a pronounced shear fabric and contain 10 mm–2 m blocks of sandstone which are often little deformed and are frequently angular. Rarely, injection or flowage of shaley fault gouge into fractures within sandstone lenses or blocks has been observed.

Slip zones

Slip zones are strongly foliated zones which accommodate most of the fault displacement (Figs 4, 5 & 6). They are rich in shaley gouge, sometimes containing sandstone cataclasites with polished slip surfaces, and range in thickness from 1mm to 1m. The minimum number of slip zones on any one transect of the fault zone is two, in which case they form the external bounding surfaces to the zone, separating fault rocks from relatively undeformed wallrocks. Internal slip zones are common and often separate lithologically distinct components of the fault zone. The internal slip zones, although broadly parallel to the margins of the fault zone, are often locally oblique to the margins and follow or define the boundaries of lensoid bodies of fault rock. It is likely that some internal slip zones extend from one external boundary of the fault zone to the other.

a Railway Tunnel

b Arches Entrance

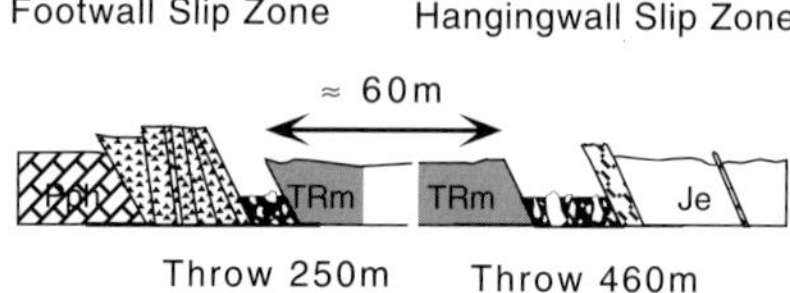

c Moab Canyon

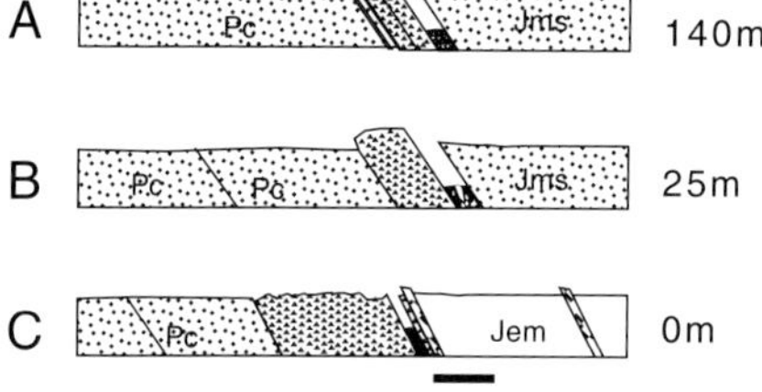

d R191 Canyon

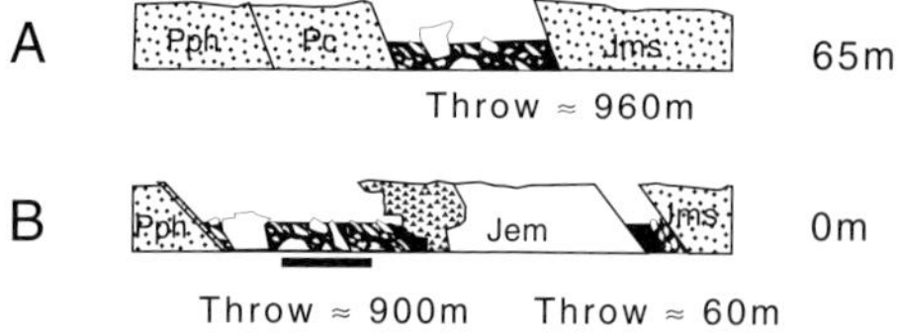

e Sevenmile Canyon

f Corral Canyon

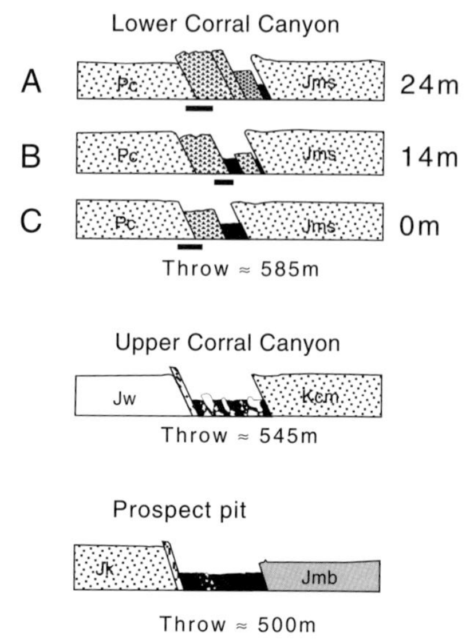

g Courthouse Mine

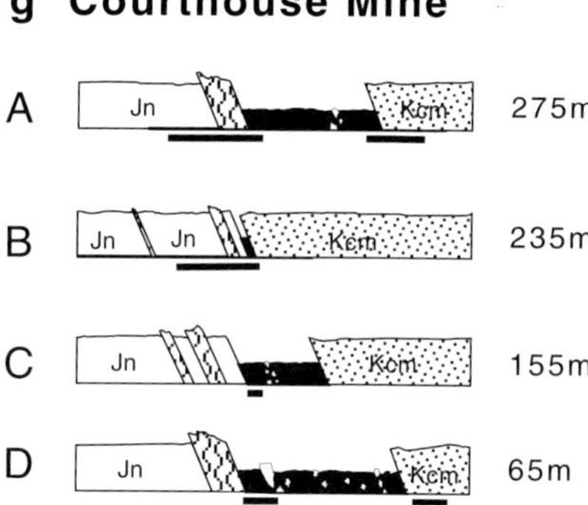

h Courthouse

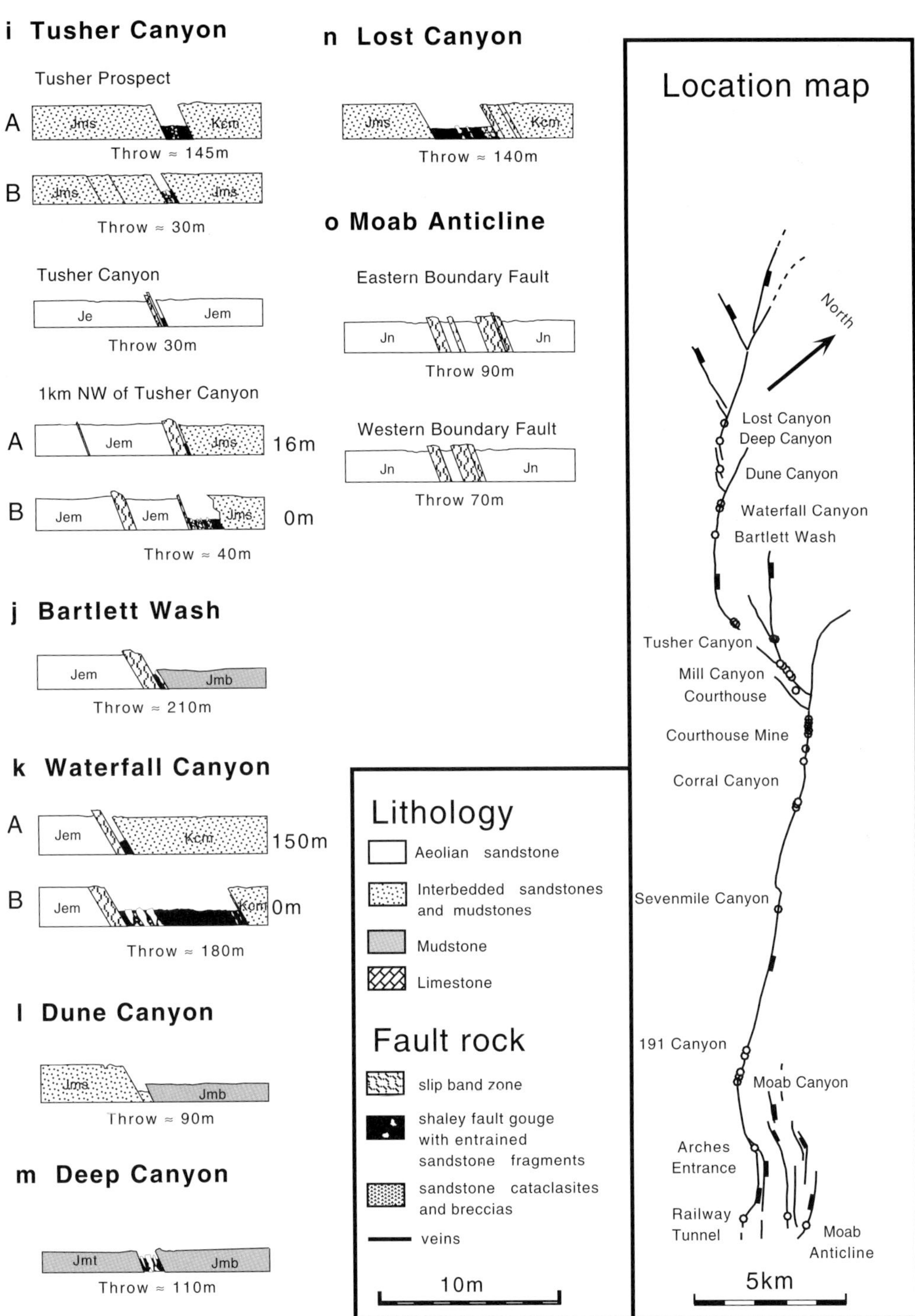

Fig. 5. Schematic transects for 32 logs across the surface trace of the Moab Fault. Insets show the key to the transects and a surface trace map outlining the locations of the transects. At localities with more than 1 transect, the transects are arranged in order with the most south-easterly transect at the bottom. Distances between transects are measured from the most south-easterly transect at each locality.

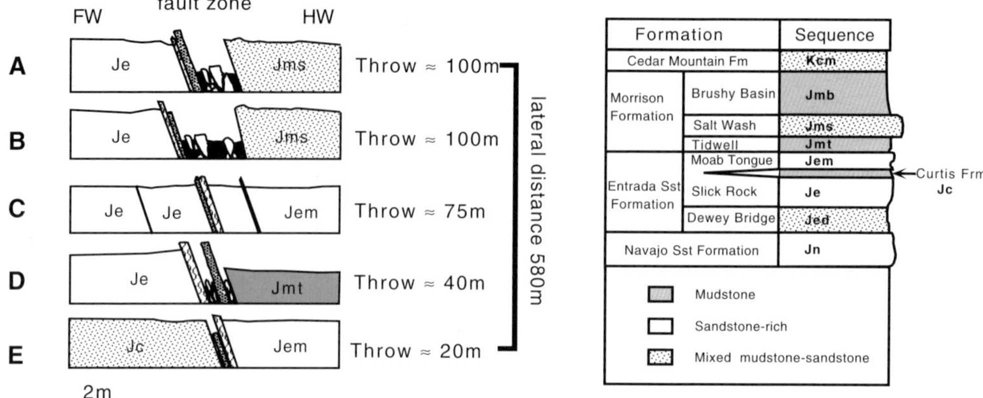

Fig. 6. Series of 5 transects along a single splay of the Moab Fault at Mill Canyon, illustrating along-strike variations in fault zone structure and content. Also shown is the detailed stratigraphy offset by the fault (see Fig. 5 for key and location map).

Veins

Veins are a subordinate component of the fault zone (Fig. 5). Along most of the fault zone these are carbonate-filled veins which include drusy, vuggy and fibrous calcite cement textures. These veins are little deformed. By contrast, chalcedonic veins which occur locally where the fault cuts early silicified Cedar Mountain Formation sandstones, are often intensely deformed within the fault zone.

Fault zone complexity

The internal geometry of the Moab fault zone (Figs 4, 5 & 6) is complex in terms of the numbers of slip zones, the partitioning of throw between them and the distribution of fault rocks, all of which vary over the fault surface. Rapid lateral changes in structure, content and thickness of the fault zone (on scales of tens to hundreds of metres) are usual and comparable vertical changes are likely, although at outcrop vertical variations can be determined only on scales of 1 to 20 metres. The complex structure of the fault zone is well illustrated by consideration of four localities along the fault trace: (i) R191 Canyon, (ii) Corral Canyon, (iii) Courthouse Mine and (iv) Mill Canyon (see Figs 5 & 6).

R191 Canyon

The two transects, *c.* 50 m apart and each accommodating a throw of almost 1 km, differ in both thickness and content of the fault zone (Figs 4 & 5d). In the SE transect (Fig. 5d B), fault zone rocks are derived from at least two stratigraphic intervals. Of the four main slip zones in the SE transect, the lowermost two slip zones bound a sheet of foliated shaley fault gouge with mm–m scale entrained sandstone blocks derived from the pre-Wingate Sandstone sequence. These slip zones separate the Honaker Trail Formation in the footwall from a central zone consisting of an 8–10 m thick sheet of variably deformed Moab Tongue sandstone, the footwall margin of which is brecciated where it is in contact with the shaley gouge. The lowermost slip zone accommodates relatively little throw. Most of the throw, *c.* 900 m, is accommodated along the second slip zone which contains several prominent slip surfaces. The uppermost two slip zones separate the sheet of Moab Tongue sandstone from sandstones of the Salt Wash Member which form the hanging wall. The slip zones consist of two slip surfaces bounding a layer of shaley fault gouge, derived from Morrison Formation mudstones, and together accommodate *c.* 60 m throw. Sandstone clasts within shaley gouge have undergone iron oxide reduction, as have the upper *c.* 0.3m of the lowermost shaley gouge and the sheet of Moab Tongue sandstone.

The NE transect (Fig. 5d A) contains only three slip zones. Like the SE transect, the lowermost slip zone again accommodates little throw and juxtaposes the Honaker Trail Formation in the footwall against a little deformed fault-parallel sheet of the immediately overlying Cutler Formation. Above this sheet there is a single thick sheet of shaley fault gouge derived exclusively from the pre-Wingate Sandstone sequence, with abundant mm–m scale sandstone blocks, most of which are only weakly deformed and are derived from the Cutler Formation. Some of these sandstone blocks are injected by cm thick veins of shaley fault gouge. Most of the throw is probably accommodated on the uppermost of the two slip zones which bound the thick layer of shaley fault gouge. No shaley gouge derived from the post-Wingate Sandstone sequence occurs in the NW transect, although mudstones in the Salt Wash Member are thinned adjacent to the fault. The shaley gouge in the NW transect can be correlated with the lowermost shaley gouge in the SE transect. However, what appears on the SE transect to be a layer of Moab Tongue sandstone is not seen on the NW transect; consequently the layer must have the form of a large lens (Fig. 4). The same must also be true of other fault zone components at the R191 Canyon locality which on individual outcrops appear to be sheets; e.g. the uppermost shaley gouge of the SE transect which does not extend into the NW transect, and the lowermost Cutler Formation slice in the NW transect which does not extend into the SE transect.

Corral Canyon

Three closely spaced transects along 24 m of the fault trace at lower Corral Canyon (Fig. 5f A–C) further illustrate the rapid changes in structure and content of the fault zone. Topmost Cutler Formation footwall sandstones are separated from Salt Wash Member hanging wall sandstones by a fault zone with three to five slip zones and with layers of shaley fault gouge and of foliated sandstone cataclasite and breccia. The sandstone cataclasite and breccia layers appear to be derived from the Triassic sequence, but the partitioning of throw between individual slip zones cannot be established. The uppermost shaley gouge layer on all three transects consists of strongly foliated, phyllosilicate-dominated, shaley gouge layers (20 mm–240 mm thick) derived from basal Morrison Formation mudstones. The two southern-most transects also contain a thick lower shaley gouge layer (*c.* 1m) derived from the pre-Wingate Sandstone portion of the sequence, principally the Triassic mudstone-dominated Chinle Formation, and with relatively high proportions of entrained wallrock fragments and cataclased siltstone and micaceous sandstone. The interleaving of the fault zone components indicates that the forms of both the shaley gouge and the cataclasite and breccia layers are lenticular. Undeformed fault-parallel calcite veins occur in the immediate footwall of the fault and in the lowermost lens of cataclasite and breccia.

Courthouse Mine

To the north of Corral Canyon (Fig. 5h–i) the fault rocks are derived from the post Wingate sequence, and the fault zone contains a shaley gouge derived predominantly from Brushy Basin Member mudstones. The thickness of the shaley gouge is variable even when juxtaposition and throw are essentially constant. For example, along the Moab Fault at Courthouse mine (Fig. 5g) five localities along 275m of the fault trace just to the south of the first major branch-point, show transects of the fault zone where it accommodates *c.* 370 m throw. In each transect the footwall is composed of aeolian Navajo Sandstone and the hanging wall is composed of fluvial Cedar Mountain Formation sandstone, the latter showing pronounced normal drag. The transects are similar in that only shaley gouge, with sandstone blocks, and concentrations of slip bands are present. In all cases, only two major slip zones occur and these bound the layers of shaley fault gouge. The most distinctive variations are in the sandstone block content of the gouge and, more importantly, the thickness of the gouge layer, which varies between 0.2 m and 5.2 m. Variations in sandstone block content suggest that the shaley gouge layer may have a layered or anastomozing internal structure. The shaley gouge and sandstone blocks are derived principally from the Brushy Basin Member. Although the thickness of the shaley fault gouge varies by a factor of 26, a continuous layer of shaley gouge is always present. Concentrations of slip bands are restricted to the footwall aeolian sandstones, either immediately adjacent to the fault zone or as well defined zones further into the footwall. Exceptionally, many minor slip band zones occur in unsilicified portions of hanging wall Cedar Mountain Formation sandstones. Undeformed calcite veins are present in each transect, occurring in shaley gouge and in both the footwall and the hanging wall sandstones.

Mill Canyon

Changes in fault zone content, in particular the amount of shaley gouge, vary with the juxtaposition geometry of the faulted sequence. Along the splay of the Moab Fault at Mill Canyon (Fig. 6) a shaley gouge is not present over the entire fault surface. The five transects along 580 m of the fault trace, have throws which decrease westwards, from 100 m to 20 m, towards Mill Canyon branch-point, and each has two to four slip zones. The low throws are apparently anomalous due to their location on a fault segment bounding an unbreached relay zone (see below). Thickness and content of the fault zone both vary unpredictably along the fault trace, but in all transects the fault zone contains one layer of sandstone cataclasite, derived from the aeolian sandstones. Associated concentrations of slip band zones occur in the footwall and hanging wall indiscriminately, but only where there is aeolian sandstone. The fault zone contains at least one layer of shaley gouge, except in transect C. Along Mill Canyon, the Slick Rock Member is separated from the Moab Tongue by Curtis Formation mudstones. The Curtis Formation is the source of the shaley gouge layers in transect E, where they can be traced back to their source, and of the lowermost gouge layer in transect D. Shaley gouge is otherwise sourced from the basal Morrison Formation units, including the Salt Wash Member. The absence of a shaley gouge in transect C is attributed to the large throw on this transect in relation to the thickness of potential shaley gouge source layers, i.e. Curtis Formation mudstones.

Other data

Numerous other transects across the fault zone have been examined; most of these show offsets of aeolian sandstones in which mudstones are either absent or form only a small proportion of the sequence. These sandstone dominated fault zone transects show slip band zones and associated polished slip surfaces and sheets of sandstone cataclasite, with shaley gouge present only in transects adjacent to thin (up to 1 m thick) mudstone layers in the country rock.

Relationship between fault throw and fault zone thickness

There is a weak correlation between fault throw and fault zone thickness (Fig. 7), reflecting the wide variation in the ratio of fault throw to fault zone thickness, even between points separated by distances which are only small fractions of the fault throw. This weak correlation reflects the laterally variable nature of the fault zone with its discontinuous lenses of fault rock. At Courthouse Mine, for example, the thickness of shaley gouge varies by a factor of 26 over a distance of 170 m, equivalent to less than half the

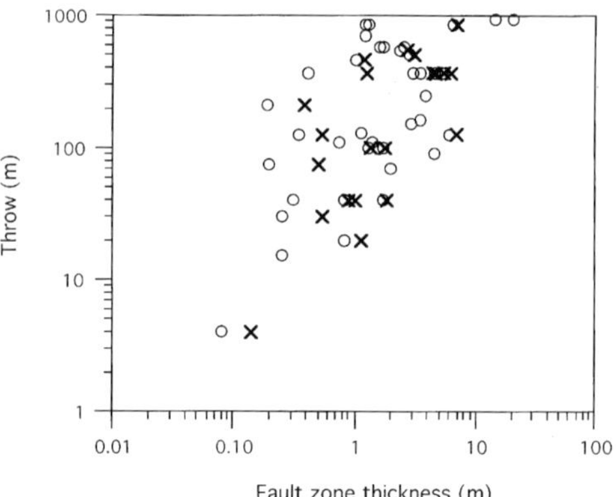

Fig. 7. Plot of fault zone thickness versus fault throw for the 37 logged fault transects shown in Fig. 5. Fault zone thickness either includes (×) or excludes (•) narrow zones (<1 m thick) with high densities (>50 m^{-1}) of slip bands and associated slip surfaces, but does not include the *c.* 40 m wide zone containing lower densities (<10 m^{-1}) of slip bands. Note the large variation in the ratio of fault zone thickness to fault throw.

fault throw at this locality (Fig. 5f). Over the same distance, the throw/thickness ratios vary from 70–900. Neither the thickness nor content of the fault zone at a point can be derived by extrapolating even a few metres from the point of observation. This apparently random variation in both structure and content of the fault zone is consistent with the lensoid form of the individual components of the fault zone and with the beaded form of the mapped fault zone trace. Dimensions of the lenses and wavelengths of the fault trace beads occupy a continuous range from a few cm to several 100's of metres. The beaded fault zone trace is on a mappable scale at several localities along the southern segment of the fault trace (Fig. 8). Observations of the limited vertical exposure available suggest that a beaded form also characterizes fault traces on vertical sections. It is not yet clear however whether the long dimensions of lenses and fault trace beads are preferentially oriented within the fault zone (i.e. down dip or along strike or otherwise).

Fault zone geometry and growth

The lensoid form of individual fault zone components and the beaded nature of fault zone traces, both of which occur on a wide range of scales, are attributed to two main processes:

(i) the progressive formation and occlusion of a range of sizes of relay zones (tip-line bifurcation of Childs *et al.* 1996); and

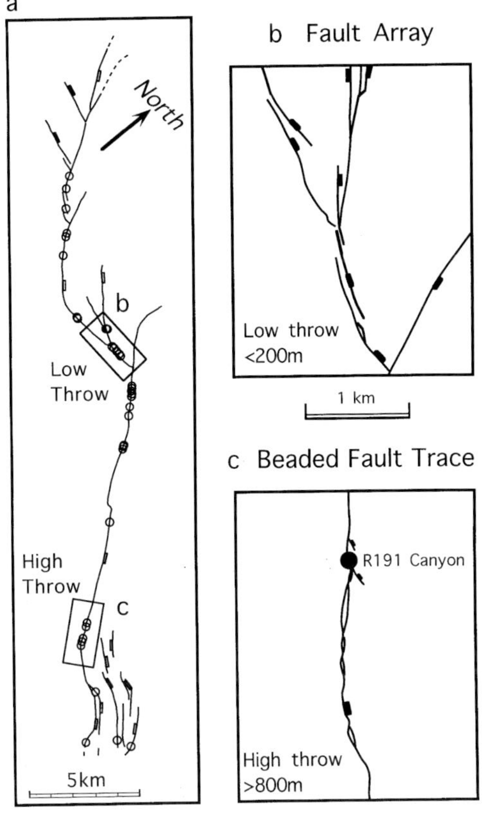

Fig. 8. Maps of the surface trace of the Moab Fault showing locations of fault transects and locations of (**b**) low and (**c**) high throw detail maps. In regions of low throw (<200 m), the surface trace contains a series of fault segments separated by relay zones. In areas of high throw (>800 m) the fault trace is beaded.

(ii) mechanical erosion and accretion by displacements on multiple slip zones within a fault zone (asperity bifurcation of Childs *et al.* 1996).

Along the lower displacement (<300 m) northern segment of the fault trace, several relay zones with separations (i.e. fault-normal distance between fault traces) in the range 10s–100s metres occur in the post-Navajo Sandstone sequence (Fig. 8). Along the higher displacement (>*c.* 300 m) southern segment, lenses of post-Navajo Sandstone sequence rocks, up to 50 m thick and *c.* 500 m long are attributed to occlusion (or breaching) of former relay zones (Peacock & Sanderson 1991; Childs *et al.* 1995), with sizes similar to those of existing relay zones along the northern segment. These occluded relay zones are responsible for the large-scale beaded outline of the fault zone trace (Fig. 8). Both existing and occluded relay zones are most common in the post-Navajo Sandstone sequence, i.e. Middle Jurassic and younger, which is affected only by the post-mid-Cretaceous phase of fault activity. When this phase of activity began, the already established fault propagated upwards into the overlying unfaulted sequence, providing circumstances which are conducive to tip-line bifurcation (Withjack *et al.* 1989; Walsh *et al.* 1996) and the formation of relay zones (Huggins *et al.* 1995; Childs *et al.* 1996). Small relay zones, with correspondingly small separations, inevitably become occluded at smaller displacements than larger ones but intact relays with aggregate displacements of up to 200 m, often with intense internal deformation, are preserved along the northern segment of the Moab Fault.

Rapid thickness changes of the fault zone are also due, at least in part, to two or more slightly divergent slip zones having been active at each point on the fault surface at some stage. Multiple slip zones either permit or arise from mechanical erosion and accretion processes (asperity bifurcation of Childs *et al.* 1996) by which fault zone thickness at a point can be reduced to almost zero, whilst nearby there is a complementary doubling in thickness. The operation of these processes throughout the evolution of the Moab Fault zone would also account for the broad range of throw/thickness ratios and for the weak positive correlation between fault throw and fault zone thickness.

Implications for fault seal prediction

The sealing properties of a fault can be attributed either to across-fault juxtaposition of reservoir and non-reservoir units or to the presence of fault rocks forming membrane seals or to a combination of both. Membrane seals are usually characterized either by shaley gouge or by cataclastic or cemented fault rock, or by both. Published work on fault sealing suggests that shaley gouge is the most effective fault rock in forming a positive seal as cataclased sandstone will generally support only a small hydrocarbon column on a geological time scale (Smith 1980; Nybakken 1991; Gibson 1994). Capillary entry pressure data (unpublished) for the cataclastic slip bands occurring within aeolian sandstones adjacent to the Moab Fault zone, suggest little contribution to the seal capacity of the fault, although data are not yet available for polished slip surfaces. Cementation of the Moab Fault zone and adjacent host rocks is patchy (Foxford *et al.* 1996; Garden *et al.* 1997) and does not provide a basis for widespread fault sealing. We therefore confine our discussion of membrane seals to the role of shaley fault gouge.

Allen diagrams and sequence juxtaposition

The existence of both intact and occluded relay zones and, more generally, the partitioning of throw between two or more slip surfaces within the Moab Fault zone raises a question concerning the interpretation of across-fault sequence juxtaposition or Allan diagrams (Allan 1989) and of shale smear calculations (Bouvier *et al.* 1989; Jev *et al.* 1993; Gibson 1994; Fristad *et al.* 1997). In both procedures a single slip surface is usually assumed, although there is abundant evidence from the Moab Fault and elsewhere (e.g. Childs *et al.* 1995, 1996, 1997*b*) for partitioning of throw onto at least two significant slip surfaces. Slip surfaces separated by a distance less than the lateral resolution of seismic data, often *c.* 50–100 m at typical reservoir depths, will appear on seismic data as a single fault (Childs *et al.* 1995; Jones & Knipe 1997). Calculations in which slip is partitioned between two or more surfaces will, in many cases, provide quite different results from those in which the aggregate throw is assumed to be accommodated on a single slip surface. A good example on the Moab Fault is on the Arches Entrance transect where the two fault strands, 60 m apart, would almost certainly be mapped seismically as a single fault surface. From a juxtaposition standpoint, the intervening 60 m of Moenkopi Formation mudstones renders the gross juxtaposition of footwall Honaker Trail against hanging wall Slick Rock Member invalid for seal calculation purposes. In shale smear calculations, the combined effect of two faults of 250 m and 460 m throw is quite different to that of a single 710 m throw fault. Although partitioning of throw on to two or more significant slip surfaces introduces some 'noise' into the empirical databases used in fault seal prediction studies, and a consequent increase in risk, the empirical methods retain a significant advantage over apparently deterministic methods. Prediction of multiple slip surfaces and throw partitioning, at a scale below the lateral resolution of seismic data and over an entire fault surface, from well data and from deterministic modelling is beyond present capabilities. A useful refinement of empirical methods, however, would be sensitivity studies of the likely impact of throw partitioning, given a range of plausible geometries based on outcrop analogues or well data.

Fault seal prediction – shale gouge/smear methods

The most significant feature of the Moab Fault zone from the standpoint of seal potential assessment is the almost constant presence of at least one shaley gouge layer, even though the thicknesses of shaley gouge layers are variable over short distances and are, in practice, unpredictable. Given the complex distribution and impersistence of other fault rocks within the fault zone, a prediction of the presence or absence of a shale layer is potentially more robust than is a prediction of either the content or the structure of the fault zone at any point on the fault surface.

Algorithms for predicting the presence or absence of a shale smear or gouge on fault surfaces (Bouvier *et al.* 1989; Jev *et al.* 1993; Lindsay *et al.* 1993; Gibson 1994; Fristad *et al.* 1997; Yielding *et al.* 1997) are based on either:

(i) the percentage of discrete shale, or mudstone, layers in the faulted sequence; or
(ii) the percentage shale of the sequence which was moved past a point on a fault surface; or
(iii) on the along-fault distance, in the slip direction, of a point on a fault surface from a potential shale source layer, or layers, and on the thickness of the layer.

These different types of algorithm often provide similar results (Yielding *et al.* 1997) and we have applied one of them to the Jurassic and Cretaceous sequences offset by the Moab Fault.

Shale gouge ratio is defined as the percentage shale (or mudstone) in the sequence which has moved past each point on the fault (Fristad *et al.* 1997; Yielding *et al.* 1997) and, for subsurface faults, is calculated from shale volume (vshale) curves derived from petrophysical logs. Shale gouge ratio therefore incorporates the contribution of all layers, including, for example, impure sandstones with phyllosilicate concentrations, to shaley gouge formation (Knipe 1997; Yielding *et al.* 1997; Fristad *et al.* 1997). Figure 9 shows a sequence/throw juxtaposition diagram (Bentley & Barry 1991; Knipe 1997; Childs *et al.* 1997*b*) for the Middle Jurassic to Lower Cretaceous sequence of the Moab area. Shale gouge ratios are shown for throw values which juxtapose footwall sandstone against hanging wall sandstone, i.e. potential connectivity, and are calculated on the basis of mudstones and sandstones representing 100% and 0% shale, respectively, a good approximation for this part of the faulted sequence. The sequence/throw juxtaposition diagram (Fig. 9) shows the positions of 21 points on the fault surface, most of which are represented by one of the transects shown

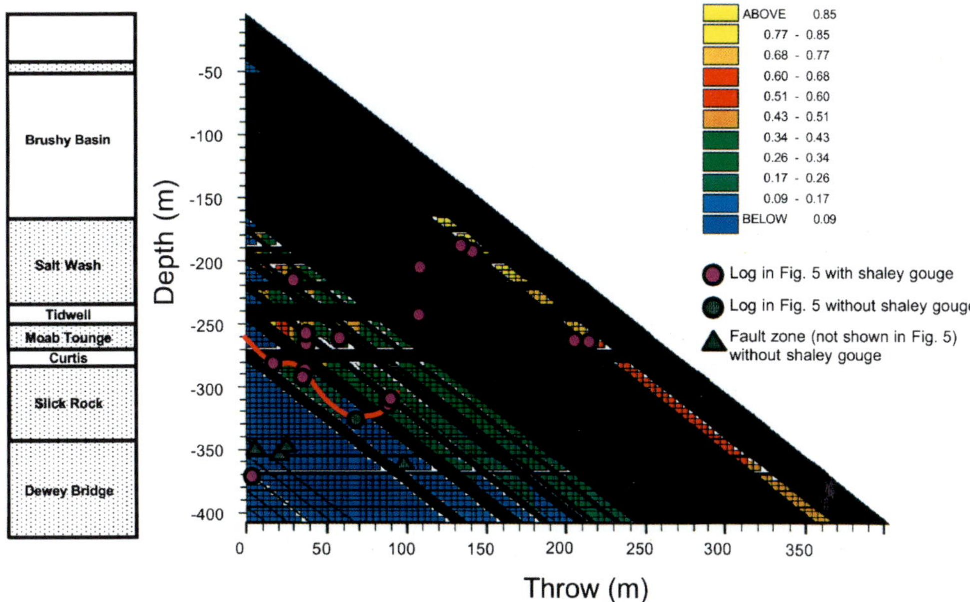

Fig. 9. Sequence/throw juxtaposition diagram (Bentley & Barry 1991; Knipe 1997; Childs *et al.* 1997*b*) for the Middle Jurassic to Lower Cretaceous portion of the sequence, which is contoured for the shale gouge ratio where footwall sandstones are juxtaposed against hanging wall sandstones. Mudstones on stratigraphic column and on sequence/throw juxtaposition diagram are shown black. The shale gouge ratio is expressed as the fraction of mudstone in the sequence moving past a point on the fault surface (e.g. 0.17 is equivalent to shale gouge ratio of 17%). Superimposed on the diagram is the surface trace of the Mill Canyon fault segment (red line), and the points plot the positions of 21 fault transects (those without shaley gouge are shown as grey points), most of which are shown on Figs 5 & 6.

in Figs 5 & 6, for which the presence or absence of shaley gouge is flagged by colour. Shaley gouge is absent only where fault displacement is $<c.$ 100 m and aeolian sandstone dominated sequences of Lower to Middle Jurassic age are juxtaposed. In both the Permo-Triassic (not shown on Fig. 9) and Upper Jurassic portions of the sequence, shaley gouge source layers are numerous, with maximum stratigraphic intervals between mudstone layers no greater than some tens of metres and commonly <10 m. The single Mill Canyon transect which does not contain shaley gouge provides a shale gouge ratio of $<20\%$, which is lower than that for the other Mill Canyon transects. The other data points indicate that a shale gouge ratio of less than *c.* 20% is characteristic of zones which do not contain shaley gouge. An apparent exception is the point at which the Dewey Bridge Member is juxtaposed against itself, but in this case the shale gouge ratios have been computed and are represented in Fig. 9 at an area scale which is too coarse to show that the shale gouge ratio at this point is actually greater than 20%.

The shale gouge ratio 'cutoff' value of 20% is similar to values reported from other studies. Fristad *et al.* (1997) showed that a shale gouge ratio of 18% defines the boundary between sealing and non-sealing faults of the Oseberg Field in the Northern North Sea. In a study of faults within Carboniferous fluvio–deltaic sandstones and shales, Lindsay *et al.* (1993) identified shale smears on all fault surfaces with shale gouge ratios $>c.$ 15%: this study used a measure of shale smear potential, the shale smear factor, which is equivalent to the reciprocal of the shale gouge ratio. Gibson (1994) applied the shale smear factor method to a faulted reservoir in Trinidad, and concluded that the faults seal at shale gouge ratios $>c.$ 25%. The Moab Fault data therefore provide observational support for using a shale gouge ratio of *c.* 20% to distinguish between the presence or absence of a shaley gouge and, implicitly, to distinguish those parts of faults which are sealing and non-sealing. However, although empirical data point to a relatively sharp 'cutoff' value in individual studies, it is anticipated that as more data become available for individual reservoirs or provinces, the 'cutoff' values determined will occupy increasingly broader ranges e.g. 20–30% (see Yielding *et al.* 1997).

Implications for sub-surface fault seal prediction

Although the Moab Fault zone data provide some justification for the use of existing shale smear prediction techniques, the application of these techniques to either exploration or production requires the use of high quality empirical databases, ideally including maps with across-fault pressure data (Fristad *et al.* 1997; Yielding *et al.* 1997). Such databases implicitly include the effects of sub-seismic throw partitioning which will, however, be reflected in a decreased confidence attached to fault seal predictions. If it is assumed that a database includes sufficient data from faults which are geologically comparable with a target fault, then the impact of multiple slip surfaces on fault sealing is incorporated implicitly in the risk assessment. Empirical seal prediction methods provide aggregate descriptions of fault zones which are appropriate to the problem. Given the observed internal complexity of fault zones, an empirical approach to fault seal prediction is more likely to be successful than methods which attempt disaggregated characterizations of fault zones through deterministic modelling or extrapolations from well data. The structure of the Moab Fault zone suggests that valid prediction of 3D fault zone structure and fault seal capacity from well data is not possible, and supports the view of Downey (1984) that

> 'the measured values from a random core sample, unfortunately, have little relevance to the problem of determining the weakest leak point of the seal. Just as little comfort can be taken from a guarantee that your parachute will (on the average) open, explorationists are not really interested in the average properties of an enclosing, sealing surface'.

Fault rock properties determined from core samples do, however, provide useful constraints for modelling fluid flow on the production time-scale (Knai & Knipe 1998, this volume), but estimates of the effective permeabilities of faults still require some knowledge of the spatial and frequency distributions of fault zone thicknesses and fault rock permeabilities over individual fault surfaces.

Much of the structure of the Moab Fault zone is attributed to brittle rather than soft-sediment deformation. It may yet be demonstrated that brittle deformation provides inherently less predictable fault zone structures than does soft-sediment deformation. The soft-sediment fault model given by James (1997), however, indicates that soft-sediment deformation also produces very complex and unpredictable fault zones (Childs *et al.* 1997) even in areas where shale gouge/smear techniques have been applied successfully, e.g. Niger Delta (Weber *et al.* 1978; Bouvier *et al.* 1989; Jev *et al.* 1993).

Fluid flow along the Moab Fault Zone

Although a shale layer in a fault zone may provide an effective seal to across-fault flow, along-fault flow may still be possible. Evidence for sub-vertical flow along fault zones is common, especially where faults are still active (*e.g.* Sibson *et al.* 1975; Sibson 1981, 1987). Localization of fluid flow along faults and fault localization of hydrothermal mineral concentrations have long been basic tenets of mineral exploration and mining practice (Newhouse 1942). In some cases, fault localization of mineralization may be due to faults having acted as barriers to flow, but in many other cases faults have provided conduits for mineralizing fluids. The Moab Fault is itself characterized by an abundance of mineral shows and workings (Foxford *et al.* 1996). Other indications that the Moab Fault zone may have been a conduit for fluid flow are:

(i) the presence of undeformed calcite veins either within or immediately adjacent to the fault zone;
(ii) the reduced state of shaley gouge adjacent to high permeability sandstones within the uppermost 30cm of the fault zone in two transects across the southern fault segment; and
(iii) locally accentuated calcite cementation and iron oxide reduction of aeolian sandstones adjacent to the fault (Garden *et al.* 1997).

Along-fault permeability may have been promoted by the highly anisotropic shaley gouge fabrics or by fault zone fractures. From an outcrop and geochemical analysis of carbonate cements and iron-oxide reduced sandstones on the Moab Anticline, Garden *et al.* (1997) suggest that subsequent to fault movement, the fault zone was the locus of vertical migration of over-pressured hydrocarbon and aqueous, carbonate-saturated, fluids. These observations, combined with the almost universal presence of shaley gouge, particularly at lower stratigraphic levels, point to the possibility that a fault can be a conduit for fluid flow at the same time as it supports significant across-fault pressure differences.

Conclusions

(1) The structure and content of the Moab Fault zone are highly variable laterally and neither can be predicted by extrapolation from data, even over distances of as little as 10 m. It is likely that vertical variation along the fault zone is equally unpredictable.

(2) The number of slip zones within the Moab Fault zone ranges from a minimum of two to a maximum of at least nine, and varies rapidly and unpredictably both laterally and vertically along the fault.

(3) Ratios of fault throw to fault zone thickness are in the range 50 to 900, and commonly vary by a factor of more than 12 over distances equivalent to only half of the local fault throw.

(4) Shale gouge is the most persistent fault rock within the fault zone. It is present where percentage mudstone in the sequence moving past a point on the fault surface is >*c*. 20%, although its aggregate thickness is highly variable and unpredictable. The presence or absence of shaley gouge appears to be the only predictive feature of fault zones and provides an observational rationale for empirical methods of predictive shale smear in the subsurface.

(5) Given the complexity of structure and content of fault zones, fault seal prediction methods should not place an over-reliance on extrapolation from well data and on deterministic modelling.

(6) Evidence from the Moab Fault supports the view that fault zones may be conduits for along-fault fluid flow and, at the same time, support significant across-fault pressure differences.

This work was carried out mostly during the course of the project 'Quantification of fault-related diagenetic variation of reservoir properties at outcrop' undertaken at Liverpool, Manchester and Heriot–Watt Universities, managed by the Petroleum Science and Technology Institute (now CMPT), and sponsored by Amerada Hess Ltd, Amoco UK, British Gas, Exxon Production Research Corporation, Japan National Oil Corporation, Mobil North Sea Ltd, Shell UK, and Total. Publication of this paper is with the permission of the sponsors. A part of the work concerning fault seal was undertaken during the course of a project funded by the EU Hydrocarbon Reservoir Research Programme (contract JOUF3-CT95–0006). M. Eeles is thanked for her help in draughting the diagrams.

References

ALLAN, U. S. 1989. Model for hydrocarbon migration and entrapment within faulted structures. *Bulletin of the American Association of Petroleum Geologists*, **73**, 803–811.

ANTONELLINI, M. A., AYDIN, A. & POLLARD, D. D. 1994. Microstructure of deformation bands in porous sandstones at Arches National Park, Utah. *Journal of Structural Geology*, **16**, 941–959.

AYDIN, A. 1978. Small faults formed as deformation bands in sandstone. *Pure and Applied Geophysics*, **116**, 913–930.

—— & JOHNSON, A. M. 1978. Development of faults as zones of deformation bands and as slip surfaces in sandstone. *Pure and Applied Geophysics*, **116**, 931–942.

—— & ——, A. M. 1983. Analysis of faulting in porous sandstones. *Journal of Structural Geology*, **5**, 19–31.

BENTLEY, M. R. & BARRY, J. J. 1991. Representation of fault sealing in a reservoir simulation: Cormorant block IV UK North Sea. *In*: *66th Annual Technical Conference & Exhibition of the Society of Petroleum Engineers*, Dallas Texas, 119–126.

BERG, R. R. & AVERY, A. H. 1995. Sealing properties of Tertiary growth faults, Texas Gulf Coast. *Bulletin of the American Association of Petroleum Geologists*, **79**, 375–393.

BOUVIER, J. D., KAARS-SIJPESTEIJN, C. H., KLUESNER, D. F., ONYEJEKWE, C. C. & VAN DER PAL, R. C. 1989. Three-dimensional seismic interpretation and fault sealing investigations, Nun River Field, Nigeria. *Bulletin of the American Association of Petroleum Geologists*, **73**, 1397–1414.

CHILDS, C., WATTERSON, J. & WALSH, J. J. 1995. Fault overlap zones within developing normal fault systems. *Journal of the Geological Society, London*, **152**, 535–549.

——, —— & —— 1996. A model for the structure and development of fault zones. *Journal of the Geological Society, London*, **153**, 337–340.

——, —— & —— 1997*a*. Discussion on a model for the structure and development of fault zones: reply. *Journal of the Geological Society, London*, **154**, 366–368.

——, WALSH, J. J. & WATTERSON, J. 1997*b*. Complexity in fault zones and its implications for fault seal prediction. *In*: *Hydrocarbon Seals–Importance for Exploration & Production*. Special Publication of the Norwegian Petroleum Society, Trondheim, Norway, **7**, 61–72.

DOELLING. H. H. 1988. Geology of the Salt Valley Anticline and Arches National Park, Grand County Utah. *In*: *Salt Deformation in the Paradox Region*, Utah Geological and Mineral Survey Bulletin **122**, 1–60.

DOWNEY, M. W. 1984. Evaluating seals for hydrocarbon accumulations. *Bulletin of the American Association of Petroleum Geologists*, **68**, 1752–1763.

FISHER, Q. J. & KNIPE. R. 1998. Fault sealing processes in siliciclastic sediments. *This volume*.

FOXFORD, K. A., GARDEN, I. R., GUSCOTT, S. C. BURLEY, S. D., LEWIS, J. J. M., WALSH, J. J. & WATTERSON, J. 1996. The field geology of the Moab Fault. *In*: HUFFMAN, A. C., LUND, W. R. & GODWIN, L. H. (eds) *Geology and Resources of the Paradox Basin*, Utah Geological Society Guidebook **25**, 265–283.

FOWLES, J. & BURLEY, S. 1994. Textural and permeability characteristics of faulted, high porosity sandstones. *Marine and Petroleum Geology*, **11**, 608–623.

FRISTAD, T., GROTH, A., YIELDING, G. & FREEMAN, B. 1997. Quantitative fault seal prediction – A case study from Oseberg Syd. *In*: *Hydrocarbon Seals – Importance for Exploration & Production*. Special Publication of the Norwegian Petroleum Society, Trondheim, Norway, **7**, 107–125.

GARDEN, R., GUSCOTT, S., FOXFORD, K. A., BURLEY, S., WALSH, J. J. & WATTERSON, J. 1997. An exhumed fill and spill hydrocarbon fairway in the Entrada sandstone of the Moab Anticline, Utah. *In:* HENDRY, J., CAREY, P., PARNELL, J, RUFFELL, A. & WORDEN, R. (eds) *Second International Conference on Fluid Evolution, Migration and Interaction in Sedimentary Basins and Orogenic Belts*, Belfast, Northern Ireland. 287–290.

GIBSON, R. G. 1994. Fault zone seals in siliciclastic strata of the Columbus Basin, offshore Trinidad. *Bulletin of the American Association of Petroleum Geologists*, **78**, 1372–1385.

HANCOCK, P. L. & BARKA, A. A. 1987. Kinematic indicators on active normal faults in western Turkey. *Journal of Structural Geology*, **9**, 573–584.

HUGGINS, P., WATTERSON, J., WALSH, J. J. & CHILDS, C. 1995. Relay zone geometry and displacement transfer between normal faults recorded in coalmine plans. *Journal of Structural Geology*, **17**, 1741–1755.

JACKSON, M. P. A. & TALBOT, C. J. 1994. Advances in salt tectonics. *In*: HANCOCK, P. L. (ed.) *Continental deformation*, Pergamon Press, Oxford, 159–179.

JAMES, D. M. D. 1997. Discussion on a model for the structure and development of fault zones. *Journal of the Geological Society, London*, **154**, 366.

JEV, B. I., KAARS-SIJPESTEIJN, C. H., PETERS, M. P. A. M., WATTS, N. L. & WILKIE, J. T. 1993. Akaso Field, Nigeria: use of integrated 3-D seismic, fault-slicing, clay smearing and RFT pressure data on fault trapping and dynamic leakage. *Bulletin of the American Association of Petroleum Geologists*, **77**, 1389–1404.

JONES, G. & KNIPE, R. J. 1997. Seismic attribute maps; application to structural interpretation and fault seal analysis in the North Sea Basin. *First Break* **14**, 449–461.

KNAI, T.-A. & KNIPE, R. J. 1998. The impact of faults on fluid flow in the Heidrun Field. *This volume*.

KNIPE, R. J. 1992. Faulting processes and fault seal. *In*: LARSEN, R. M., BREKKE, H., LARSEN, B. T. & TALLERAS, E. (eds) *Structural and Tectonic Modelling and its Application to Petroleum Geology*, Norwegian Petroleum Society Special Publication **1**, 325–342.

KNIPE, R. J. 1997. Juxtaposition and seal diagrams to help analyze fault seals in hydrocarbon reservoirs. *Bulletin of the American Association of Petroleum Geologists*, **81**, 187–195.

KNIPE, R. J., FISHER, Q. J., JONES, G., CLENNELL, M. B., FARMER, A. B., HARRISON, A., KIDD, B., MCALLISTER, E., PORTER, J. R. & WHITE, E. A. 1997. Fault seal analysis: successful methodologies, application and future directions. *In*: MØLLER-PEDERSON, P. & KOESTLER, A. G. (eds) *Hydrocarbon seals: Importance for Exploration and Production*, NPF Special Publication **7**, 1–24.

LINDSAY, N. G., MURPHY, F. C., WALSH, J. J. & WATTERSON, J. 1993. Outcrop studies of shale smears on fault surfaces. *In*: FLINT, S. & BRYANT, A. D. (eds) *The Geological Modelling of Hydrocarbon Reservoirs and Outcrop Analogues*, International Association of Sedimentology, **15**, 113–123.

NEWHOUSE, W. H. 1942. *Ore deposits as related to structural features*. Princetown University Press, Princeton.

NYBAKKEN, S. 1991. Sealing fault traps – an exploration concept in a mature petroleum province: Tampen Spur, northern North Sea. *First Break*, **9**, 209–222.

PEACOCK, D. C. P. & SANDERSON, D. J. 1991. Displacements, segment linkage and relay ramps in normal fault systems. *Journal of Structural Geology* 13, 721–733.

SIBSON, R. H. 1977. Fault rocks and fault mechanisms. *Journal of the Geological Society, London*, **133**, 191–213.

—— 1981. Fluid flow accompanying faulting: field evidence and models. *In*: SIMPSON, D. W. & RICHARDS, P. G. (eds) *Earthquake prediction: an international review*, American Geophysical Union Maurice Ewing Series for Geophysical Monographs, **4**, 593–603.

—— 1987. Earthquake rupturing as a mineralizing agent in hydrothermal systems. *Geology*, **15**, 701–704.

——, MOORE, J. M. & RANKIN, A. H. 1975. Seismic pumping–a hydrothermal fluid transport mechanism. *Journal of the Geological Society, London*, **131**, 653–659.

SMITH, D. A. 1980. Sealing and non-sealing faults in Louisiana Gulf Coast salt basin. *Bulletin of the American Association of Petroleum Geologists*, **64**, 145–172.

WALSH, J. J., WATTERSON, J., CHILDS, C. & NICOL, A. 1996. Ductile strain effects in the analysis of seismic interpretations of normal fault systems. *In*: BUCHANAN, P. G. & NIEUWLAND, D. A. (eds) *Modern Developments in Structural Interpretation, Validation and Modelling*, Geological Society, London, Special Publications, **99**, 27–40.

WEBER, K. J., MANDL, G., PILAAR, W. F., LEHNER, F. & PRECIOUS, R. G. 1978. The role of faults in hydrocarbon migration and trapping in Nigerian growth fault structures. *Offshore Technology Conference* **10**, paper OTC 3356, 2643–2653.

WITHJACK, M. O., MEISLING, K. E. & RUSSELL, L. R. 1989. Forced folding and basement-detached normal faulting in the Haltenbanken area, offshore Norway. *In*: TANKARD, A. J. & BALKWILL, H. R. (eds) *Extensional Tectonics and Stratigraphy of the North Atlantic Margins*, American Association of Petroleum Geologists Memoir, **46**, 567–575.

YIELDING, G., FREEMAN, B., & NEEDHAM, D. T. 1997. Quantitative fault seal prediction. *Bulletin of the American Association of Petroleum Geologists,* **81**, 897–917.

The relationship between faults and pressure solution seams in carbonate rocks and the implications for fluid flow

D. C. P. PEACOCK[1], Q. J. FISHER[1], E. J. M. WILLEMSE[2,3] & A. AYDIN[2]

[1] *Rock Deformation Research Group, School of Earth Sciences, University of Leeds, Leeds LS2 9JT, UK*

[2] *Rock Fracture Project, Department of Geological & Environmental Sciences, Stanford University, Stanford CA 94305-2115, USA*

[3] *Present address: Shell International, 2280 AB Rijswijk, The Netherlands*

Abstract: Pressure solution plays a major role in fault initiation and development in carbonate rocks. Strike-slip faults in the Triassic and Jurassic limestones of Somerset, UK, initiated as en echelon extension fractures, which became linked by pressure solution seams. Shear occurred along the pressure solution seams as the bridges between the veins rotated. The linked vein-pressure solution seam systems developed into pull-aparts and eventually into through-going faults. Normal fault planes in the Cretaceous Chalk at Flamborough Head, Yorkshire, UK, commonly have the pitted appearance of slickolites. Phyllosilicates are often concentrated along faults in Chalk, the thickness of the phyllosilicate gouge being proportional to fault displacement. The enrichment of phyllosilicates along the faults is due to pressure solution rather than simply to phyllosilicate smear. Pressure solution can be concentrated at the contractional quadrants of faults, particularly where there is a contractional overstep onto an adjacent fault. Metre-scale oversteps between strike-slip faults in Somerset often have pressure solution seams, while contractional oversteps and bends at Flamborough Head are accommodated by compaction of beds, with pressure solution apparently being important. The concentration of phyllosilicates by pressure solution can hinder fluid flow along and across faults in carbonate rocks. A high density of pressure solution seams between overstepping faults can effectively link these faults, increasing both the effective length of the barrier to fluid flow and the fault seal potential.

Carbonate reservoirs contain upwards of 60% of the Earth's recoverable hydrocarbons (Roehl & Choquette 1985). Faults, fractures and pressure solution seams within such reservoirs may control hydrocarbon flow both during migration and production. For example, it is well established that fractures may increase the permeability of carbonate reservoirs by many orders of magnitude (e.g. Price 1987). Pressure solution seams have also been suggested to play an important role in hydrocarbon migration (Sassen *et al.* 1987; Leythaeuser *et al.* 1995). Although less well studied, pressure solution seams within carbonate rocks may act as permeability barriers restricting hydrocarbon flow. An understanding of deformation processes within carbonate sediments is therefore of both economic and academic interest.

Recent work has shown that faults within carbonates rarely develop without modifying the stress field in the surrounding rock, particularly at fault tips where displacement gradients are high (e.g. Peacock & Sanderson 1995a; Petit & Mattauer 1995). A variety of deformation mechanisms may operate contemporaneously to accommodate this stress, depending upon factors such as temperature, strain rate, stress orientation and the material properties of the deforming sediment. This leads to the formation of a variety of associated deformation features. For example, faults, veins and pressure solution seams are often very closely related within carbonate sediments (Rispoli 1981; Gamond 1983, 1987; Peacock & Sanderson 1995a; Petit & Mattauer 1995; Willemse *et al.* 1997). In particular, pressure solution seams and veins occur within the contractional and extensional quadrants of faults, respectively (Fig. 1). Also, the position of pressure solution seams influences the nucleation and hence distribution of slip planes (Willemse *et al.* 1997).

The aim of this paper is to highlight the interaction of the various deformation mechanisms in carbonate rocks and to discuss their influence on fluid flow. Particular attention is paid to describing the possible ways in which pressure solution may accommodate strain and affect fluid flow in and around faults. Field and microstructural data are integrated into a model for fault propagation through carbonate rocks. The paper focuses on two main field areas, which taken together provide excellent examples of the

Peacock, D. C. P., Fisher, Q. J., Willemse, E. J. M. & Aydin, A. 1998. The relationship between faults and pressure solution seams in carbonate rocks and the implications for fluid flow. *In*: Jones, G., Fisher, Q. J. & Knipe, R. J. (eds) *Faulting, Fault Sealing and Fluid Flow in Hydrocarbon Reservoirs*. Geological Society, London, Special Publications, **147**, 105–115.

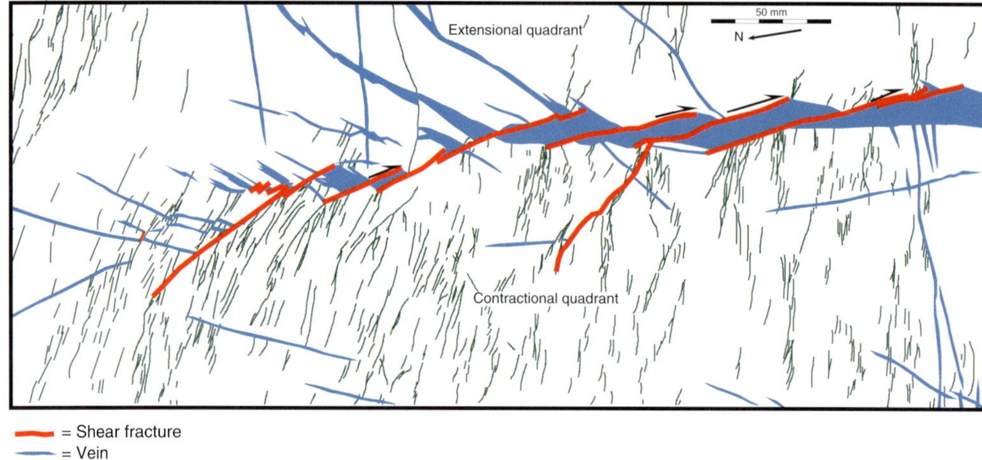

Fig. 1. Plan view of the tip of a dextral strike-slip fault in Jurassic limestone, at East Quantoxhead, Somerset. Pressure solution is concentrated in the contractional quadrant of the fault, and veining is concentrated in the extensional quadrant, allowing displacement to decrease rapidly at the fault tip. For similar examples from the Matelles exposure, Languedoc, France, see Rispoli (1981) and Petit & Mattauer (1995).

complex interplay between faulting, pressure solution and vein formation. Very well-exposed strike-slip faults from the Upper Triassic and Lower Jurassic limestones and shales of the Somerset coast, UK (Fig. 2; Bowyer & Kelly 1995; Peacock & Sanderson 1992, 1995*a*,*b*; Willemse *et al.* 1997) are used to demonstrate the role of pressure solution in fault propagation and vein formation. Normal and oblique-slip faults from the Cretaceous Chalk of Flamborough Head, Yorkshire, UK (Fig. 2; Peacock & Sanderson 1994) with displacements of up to several metres, are used to show how pressure solution can concentrate phyllosilicates on a fault surface. An example is also used from the Lower Jurassic limestones at Southerndown, South Glamorgan, Wales.

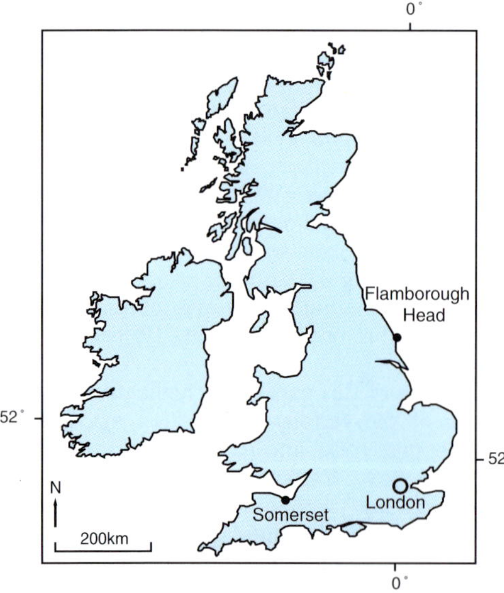

Fig. 2. Map of the British Isles, showing the locations of Flamborough Head and the Somerset coast field area.

Methods

Field work

Maps of faults with millimetre-scale displacements in Somerset were produced using photographs enlarged to scales of ×5 to ×0.5. These faults are well exposed on the smooth, upper surfaces of limestone beds, and maps were also made by tracing the faults directly onto acetate sheets taped to the rocks. The Chalk cliffs at Flamborough Head were logged, with the locations, orientations and displacements of 1339 faults over the 6 km section being measured (Peacock & Sanderson 1994). The geometries of fault planes were mapped onto photographs enlarged to scales of about ×5 (Peacock & Zhang 1994).

Microstructural examination

Microstructural examination was undertaken of calcite-cemented veins within an out-of-situ boulder collected on the beach at Lilstock. The deformation features are identical to the *in situ* features shown in Fig. 1. Polished blocks were prepared so that they could be examined using back-scattered electron (BSE) microscopy. The polished samples were made from $\approx 1-2$ cm^3 blocks, air dried and then impregnated with a low viscosity resin in a vacuum impregnation unit. Samples were examined with a CAMSCAN CS44 high performance scanning electron microscope. This is equipped with a high resolution solid state four quadrant back-scattered electron (BSE) detector, a secondary electron detector (SE), a cathode luminescence detector (CL) and an EDAX energy dispersive X-ray spectrometry (EDS) system. The EDS detector has an ultra thin window to allow the detection of low energy X-rays emitted from the light elements B, C, N, O and F.

Examples of combined pressure solution and faulting

Strike-slip faults at Lilstock, Somerset

The exposed Lower Jurassic limestone between East Quantoxhead (grid reference ST 122439) and Lilstock (ST 196462), Somerset, UK, is part of the Bristol Channel Basin which experienced N–S extension during the Mesozoic era (Chadwick 1986; Brooks *et al.* 1988) and N–S directed contraction during Late Cretaceous to mid Tertiary times (Peacock & Sanderson 1992). This latter period of deformation was responsible for the formation of strike-slip faults which initiated as conjugate arrays of veins, with individual vein segments striking approximately N–S (Peacock & Sanderson 1995*a*,*b*). Phyllosilicate-rich deformation features are present in the bridges between the vein segments in some arrays. These pressure solution seams have shear displacement in places where the bridges were rotated (Fig. 3; Willemse *et al.* 1997). Pull-aparts have developed in places where these sheared pressure solution seams have linked the veins. Through-going faults, with calcite along their planes, developed where the pull-aparts overlap. This evolutionary sequence (shown in Fig. 4) can be observed spatially at fault tips, and is inferred to have also occurred as a time series.

Microstructural work was carried out to determine the origin and sequence of various deformation structures. A BSE photomontage

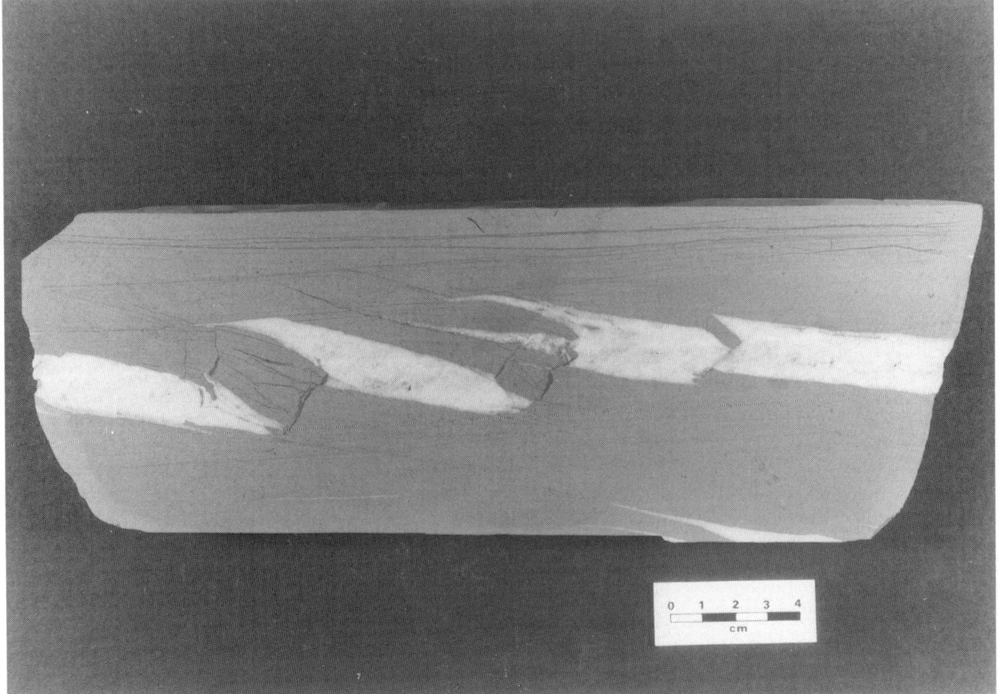

Fig. 3. Photograph of a cut slab of limestone containing overstepping veins which are linked by pressure solution seams to form incipient pull-aparts. Rotation of the bridges has allowed shear to occur on the pressure solution seams. This example is from the Upper Triassic limestones at Lilstock, Somerset, and was not *in situ*.

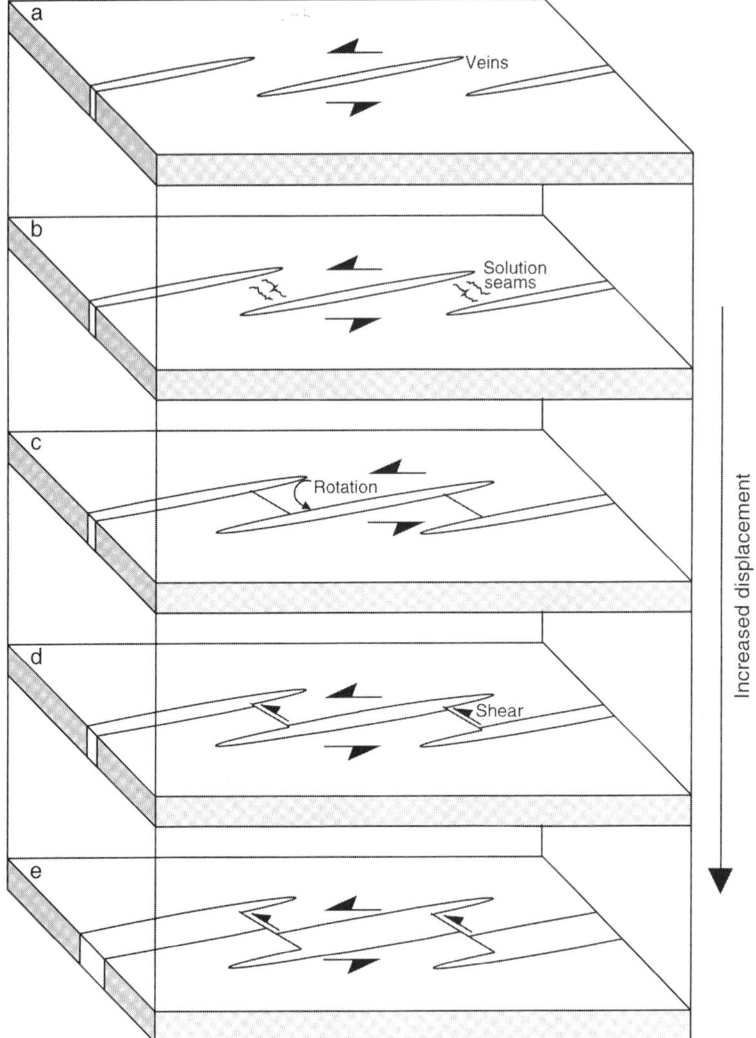

Fig. 4. Block diagram illustrating the possible sequence of development of a sinistral fault zone in a limestone bed (compare with Peacock & Sanderson 1995*a*, fig. 12). The levels (a–e) can represent a series of layers with displacement increasing downwards, i.e. spatial evolution. This sequence is often visible along the strike of strike-slip faults, from en echelon veins furthest from the fault, to linked pull-aparts which form a through-going fault. The levels (a–e) can also represent different stages in the evolution of the fault zone, i.e.temporal evolution. (**a**) Right-stepping en échelon veins develop. (**b**) Pressure solution starts to occur in the bridges between the veins. (**c**) Through-going pressure solution seams cut the bridges and link the overstepping veins. (**d**) Shear starts to occur on the pressure solution seams. (**e**) A fault is developed as a linked system of shear fractures and pull-aparts. For a full description of the model, see Willemse *et al.* (1997).

showing the relationship between the main structural elements is given in Fig. 5. This sample and photomontage forms the basis of the following analysis, although information is also drawn from the relationships established while examining other faults. The host sediment is a calcite mudrock containing small quantities of framboidal pyrite (Fig. 6a). The amount of detrital material varies significantly. For example, the host sediment on the right-hand side of the F1 fault in Fig. 5 contains only $\approx 10\%$ detrital material, whereas the sediment on the left-hand side of this fault contains upwards of 60% detrital material (Fig. 6b).

The most obvious deformation features within the sample are calcite-cemented veins up to 4 mm

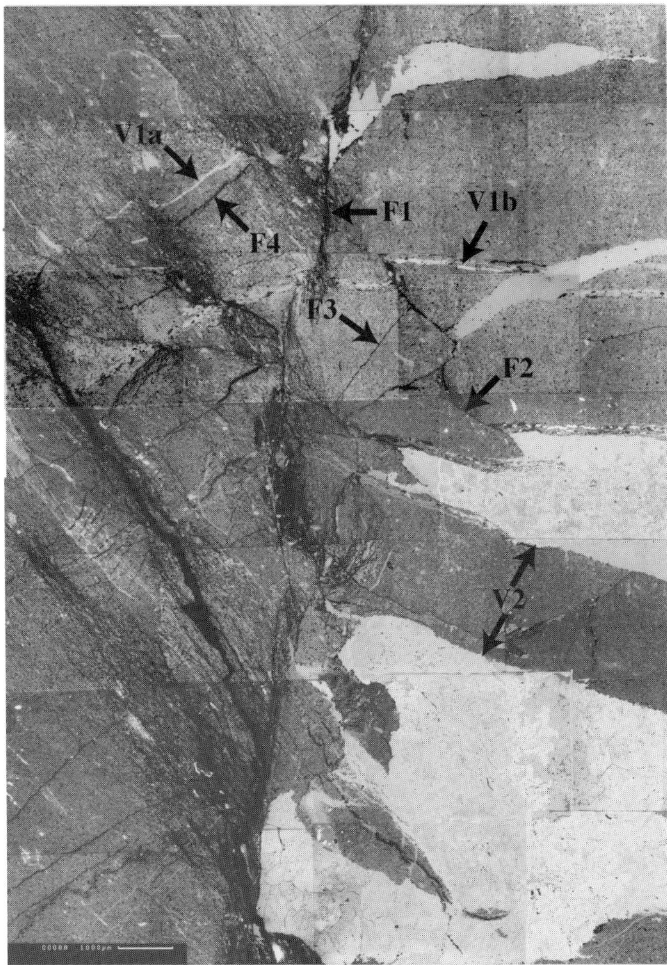

Fig. 5. BSE photomontage showing the relationships between the various deformation features present in one of the polished blocks examined. Note that 'F' denotes fault or fracture and 'V' denotes vein. The host sediment to the left of F1 has a strong pressure solution fabric.

wide (V2 in Fig. 5), which appear to terminate at a dextral fault (F1 in Fig. 5). The microstructure of both the host sediment and the deformation features are different on either side of F1. The overall structure of this sample is consistent with the calcite-cemented veins having formed as tail cracks near the tip of the dextral F1 fault. The host sediment to the left of F1 in Fig. 5 has a strong fabric orientated at $\approx 45°$ to F1. A series of calcite-cemented micro-veinlets (V1a in Fig. 5) are orientated perpendicular to this fabric, as are a series of open fractures (F4 in Fig. 5). The strong fabric formed due to pressure solution, as indicated by the enrichment of detrital material (Fig. 6b). High contrast BSE imagery shows that the micro-veinlets are composed of two distinct carbonate types. The first has a pure calcite composition, the second shows a minor enrichment in manganese. The manganese-rich calcite appears to have undergone partial dissolution in some of the veinlets. The deformation features to the right of F1 appear more complex (Fig. 5). As mentioned earlier, the most dominant features are the calcite-cemented veins (V2). There are, however, a variety of less obvious features, including:

(1) A set of micro-veinlets (V1b in Fig. 5) orientated sub-parallel to V2. These have the same microstructure as the V1a set of veins present to the left of F1.
(2) A set of phyllosilicate-rich strain features (F2 in Fig. 5) which link the V2 calcite-cemented veins, and which are probably sheared. They

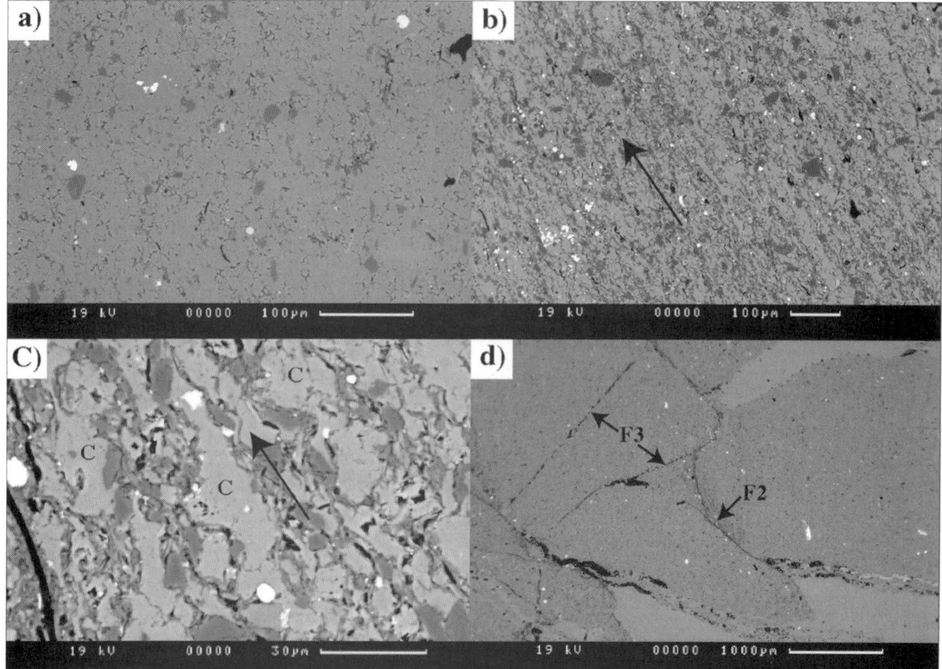

Fig. 6. BSE photomicrographs. (**a**) The host sediment to the right hand side of F1 (Fig. 5). Note that it is contains a very high concentration of calcite (the light coloured phase). (**b**) The host sediment to the left hand side of F1. Note that it has a strong fabric, orientated parallel to the arrow, formed due to pressure solution. Also note that it contains far more detrital material than the sediment shown in (**a**). (**c**) A high magnification view of the F2 deformation features. Note the strong fabric orientated parallel to the arrow. Note also that the calcite grains (C) have a corroded appearance following grain-to-grain dissolution (pressure solution). (**d**) The relationship between the F2 and F3 deformation features. F3 seems to have accommodated strain resulting from shear displacement along F2.

have a strong fabric orientated approximately perpendicular to V2 (Fig. 6c).

(3) A set of microfaults (F3 in Figs 5 and 6d) which appear to link F2 with V2. The F3 fractures seem to have formed in order to accommodate shear deformation along the tips of the F2 deformation features.

The V1 veins are displaced by the V2 veins, suggesting that they were the first features to form. The V1 veins are also displaced by the F2 sheared solution seams, suggesting that the strain continued to be accommodated on F2 structures following the formation of the V2 veins. The F2 pressure solution seams appear to have accommodated strain resulting from the formation of the V2 veins, suggesting that they formed at the same time. The last deformation event was the formation of the open fractures (F4). Uncemented dilation sites are also present along the dominant fault F1, suggesting that the F1 fault was reactivated after the supply of calcite cement had declined.

The deformation features present within this specimen are consistent with a simple model in which deformation occurred at two distinct phases. The first stage of deformation involved dextral movement along F1 and the formation of first V1 and then V2 and F2. As V1 continued to open, block rotation resulted in the rotation of the pressure solution seams (F2) which ultimately led to shear deformation along F2 and the formation of the minor F3 faults. A second deformation episode resulted in the reactivation of F1 and the formation of the F4 open fractures. Overall, the left and right side of F1 could be interpreted as its contractional and extensional quadrants, respectively (Rispoli 1981; Pollard & Segall 1987).

An aim of the microstructural work was to determine whether the deformation features (F2) which link the major calcite-cemented veins (V2), had deformed by mechanical (independent particulate flow or cataclasis) or chemical (pressure solution) processes. The corroded nature of the calcite grains is consistent with pressure solution. A similar conclusion is reached if one considers the source of the calcite cement. The individual calcite-cemented veins seem to be linked by pressure solution seams, suggesting that their linkage did not provide a conduit for fluid flow. A local source of cement was therefore probably needed for vein formation, the most obvious of which is localized grain-to-grain calcite dissolution (pressure solution). Indeed, this is entirely consistent with the results of isotopic studies conducted on similar veins within the Chalks of the North Sea, which suggests most carbonate is supplied locally (Egeberg & Saigal 1991).

These results are consistent with the model of Willemse *et al.* (1997) that the faults developed within a single stress regime without pre-existing fractures or mode II fracture propagation (Martel *et al.* 1988; Peacock & Sanderson 1995*a*).

Pressure solution along fault planes in Chalk, Flamborough Head

Faults in the Late Cretaceous Chalk are well-exposed in the cliffs at Flamborough Head. In total, 1339 normal and oblique-extensional faults were measured by Peacock & Sanderson (1994) along a 6 km section in the Chalk along the south coast, between Sewerby (grid reference TA 20166866) and High Stacks (TA 25787041). These have displacements of <6 m and have a wide range of orientations. Nearby, larger displacement faults strike approximately E–W. These include the Selwicks Bay Fault Zone (TA 255707), which has ≈ 23 m finite normal displacement (Rawson & Wright 1992). These larger faults show evidence of reverse-reactivation, but the faults with <6 m displacement show no evidence of this contractional event.

Fault planes in the Cretaceous Chalk at Flamborough Head are often pitted, with the appearance of oblique stylolites. The pitting is particularly well-developed at contractional bends along the fault planes (Fig. 7a). These fault surfaces could be termed *slickolites* (Ramsay & Huber 1987). Marl layers comprise only about 0.5% of the sequence (Rawson & Wright 1992, fig. 35), therefore phyllosilicate along the fault planes (Fig. 7b) probably did not simply result from smearing of phyllosilicate beds (Knipe 1992). Instead, it seems far more likely that phyllosilicates became concentrated due to the removal of carbonate by pressure solution. There are no data on the timing of the pressure solution on the fault surfaces, which may have occurred before, during and/or after faulting.

Pressure solution probably occurs because the fault planes are orientated at an angle to the maximum principal stress, hence a resolved normal stress (e.g. Price & Cosgrove 1990, fig. 1.4) exists on the fault plane. Brecciation, fluid flow and the locally high stresses that occur along fault planes could enhance pressure solution (Gratier & Gamond 1990). Sibson (1977) suggests that pressure solution may become important in fault zones when grain size is reduced to 10 μm. Koestler & Ehrmann (1991) show that phyllosilicate enrichment along faults in Chalk is proportional to fault displacement, with phyllosilicate layers up to 10 mm thick occurring on faults with metre-scale displacements. Corbett *et al.* (1987) show that smectite is concentrated along faults in the Austin chalk, Texas, and suggest that these phyllosilicates act as 'soft inclusions' which localize shear failure. The reason why phyllosilicates became concentrated in a plane at a high angle to bedding before faulting is not given by Corbett *et al.* (1987). An alternative explanation is that the phyllosilicates became concentrated along faults by enhanced pressure solution, during or after faulting.

Pressure solution at contractional oversteps between faults, Somerset and South Wales

Contractional oversteps in Somerset and at Southerndown, South Wales, occur between strike-slip faults which are well-exposed on the upper surface of Upper Triassic and Liassic limestones on the wave-cut platform. Pressure solution is often concentrated near the tips of faults (Fig. 1). High displacement gradients at fault tips are accommodated by contraction on one side of the fault (e.g. by solution seams) and extension on the other side (e.g. by veins; Fletcher & Pollard 1981; Pollard & Segall 1987). This effect is particularly pronounced at fault oversteps (Fig. 8a). Such *strike-slip relay ramps* are described by Peacock & Sanderson (1995*b*). Similarly, fault displacement at contractional oversteps along the normal faults at Flamborough Head (Peacock & Zhang 1994) appears to be accommodated by compaction and partly by pressure solution (Fig. 8b), with beds thinned by up to 90%.

Implications for fluid flow and fault seal

The examples cited earlier show that deformation of carbonate-rich rocks may result in the production of pressure solution seams and

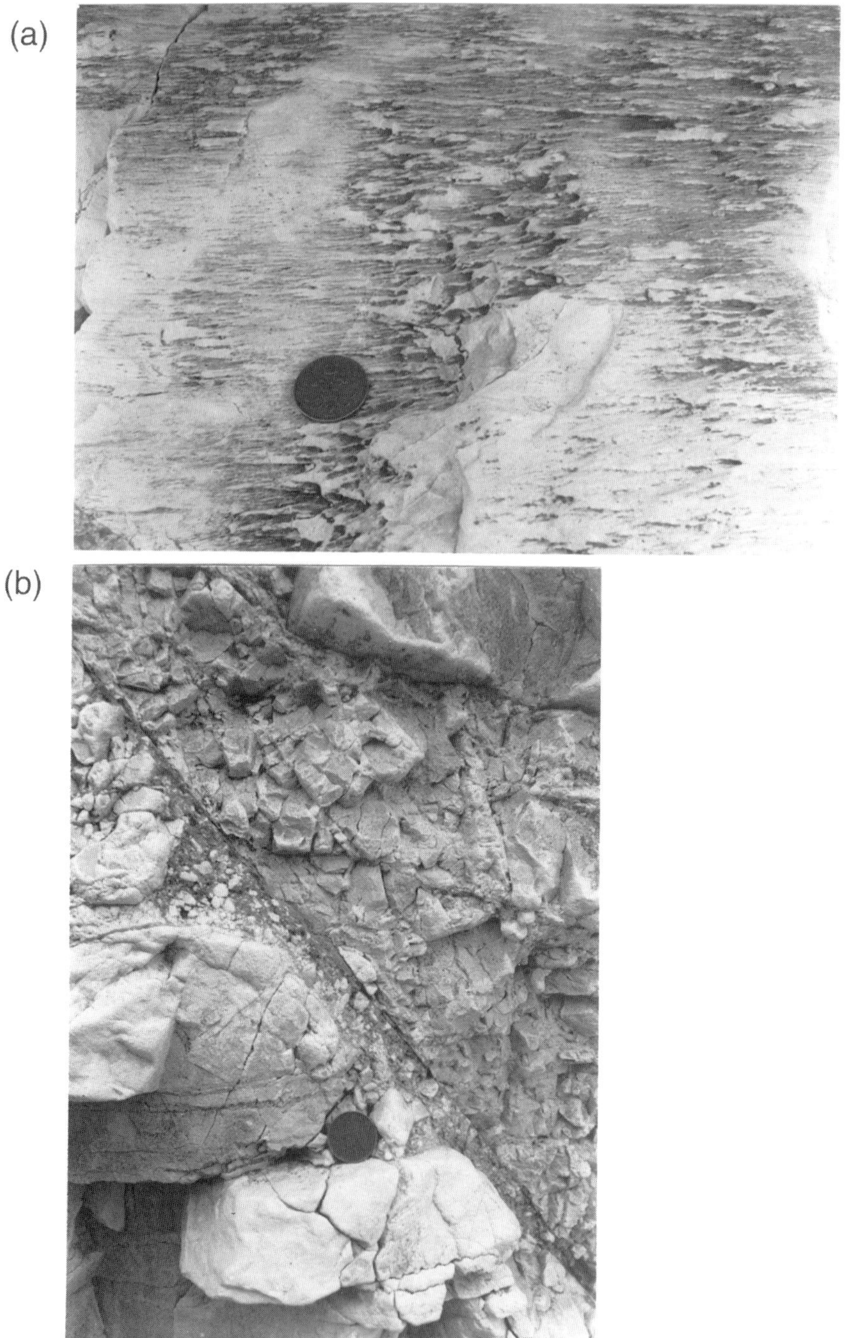

Fig. 7. Photographs of faults in the Chalk at Flamborough Head, Yorkshire. (**a**) Sinistral fault surface at Selwicks Bay (Starmer 1995) which shows the pitting which is typical of solution seams in the Chalk. The pitting is particularly pronounced at contractional bends on the fault surface. The view is approximately westwards. (**b**) Phyllosilicate along a fault plane with *c.* 6 m displacement. Such phyllosilicate layers have traditionally been attributed to smear of phyllosilicate-rich beds (e.g. Knipe 1992), but may be partly caused by pressure solution along the fault surface. The view is approximately northwards.

(a)

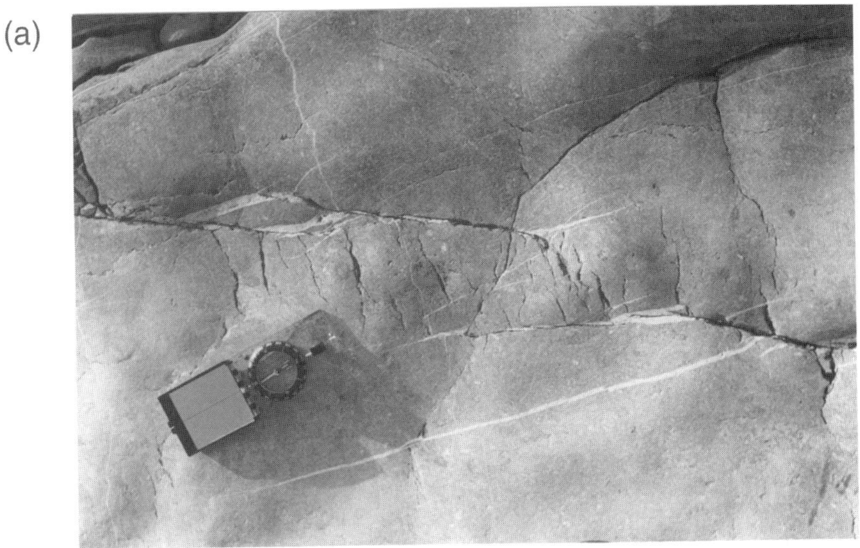

(b)

Fig. 8. Photographs of pressure solution at contractional oversteps along faults. (**a**) Overstep between strike-slip faults in Lower Jurassic limestones at Southerndown, South Wales (approximate grid reference SS 884732). (**b**) A contractional overstep in a marl-rich layer between overstepping normal faults in the Chalk at Flamborough Head. Compaction of the layer appears to be accommodated partly by pressure solution (Peacock & Zhang 1994).

phyllosilicate-rich shear fractures which link veins. Also, phyllosilicate-rich gouges can form along discrete fault planes, and the concentration of phyllosilicate by pressure solution may create barriers to fluid flow. The carbonate cemented veins are not sufficiently extensive by themselves to create spatially extensive barriers to fluid flow. Instead, they are likely to increase the tortuosity of flow paths. The linkage of these veins by pressure solution seams and phyllosilicate-rich shear fractures does, however, raise the possibility that a spatially extensive permeability barrier could form.

The features of brittle deformation of carbonate rocks by themselves are more likely to cause the formation of conduits for fluid flow rather than barriers. The results presented earlier suggest, however, that the concentration of phyllosilicates along the fault surface by pressure solution results in the formation of a phyllosilicate-rich gouge similar to the clay smear described by Lindsay *et al.* (1993), which are well known to act as barriers to fluid flow within siliciclastic hydrocarbon reservoirs (e.g. Jev *et al.* 1993; Knott 1993; Gibson 1994). Importantly, a phyllosilicate-rich gouge is likely to have the ability to deform in a ductile manner and therefore may not be breached by open fracture networks.

Conclusions

Detailed analysis of fault zones in carbonates shows that pressure solution can occur during fault propagation, and can accommodate slip along a fault:

(1) Field and SEM analyses suggest that pressure solution was important in the propagation of strike-slip faults in the Upper Triassic and Liassic limestones of Somerset. The pressure solution seams developed between en échelon veins, and sheared as the bridges between the veins rotated, causing the development of linked systems of veins and pressure solution seams. These systems developed into through-going faults as displacement increased.
(2) Pitting on the fault planes in the Chalk at Flamborough Head indicates that pressure solution has occurred. Phyllosilicate concentration along the faults may result from both pressure solution and clay smear.
(3) Pressure solution can be concentrated in the contractional quadrants near fault tips, particularly where there is a contractional overstep onto an adjacent fault. Metre-scale oversteps between strike-slip faults in Somerset often show a pressure solution-cleavage. Contractional oversteps and bends along normal faults at Flamborough Head are accommodated by thinning of beds, with pressure solution apparently being important.
(4) The concentration of phyllosilicates along faults by pressure solution can influence fluid flow along and across faults. Pressure solution seams between overstepping faults can effectively link faults, hindering fluid flow at the overstep.
(5) Pressure solution in carbonate sediments acts to concentrate phyllosilicates to form solution seams and phyllosilicate-rich fault gouges resembling clay-smears. These features have the potential to act as barriers to fluid flow.

DCPP is funded by a NERC ROPA award to R. Knipe. Careful reviews were given by E. Rutter and M. Safaricz.

References

Bowyer, M. O'N. & Kelly, P. G. 1995. Strain and scaling relationships of faults and veins at Kilve, Somerset. *Proceedings of the Ussher Society*, **8**, 411–415.

Brooks, M., Trayner, P. M. & Trimble, T. J. 1988. Mesozoic reactivation of Variscan thrusting in the Bristol Channel area, UK *Journal of the Geological Society of London*, **145**, 439–444.

Chadwick, R. A. 1986. Extensional tectonics in the Wessex Basin, southern England. *Journal of the Geological Society of London*, **143**, 444–465

Corbett, K., Friedman, M. & Spang, J. 1987. Fracture development and mechanical stratigraphy of Austin chalk, Texas. *American Association of Petroleum Geologists Bulletin*, **71**, 17–28.

Egeberg, P. K. & Saigal, G. C. 1991. North Sea Chalk diagenesis: cementation of Chalks and healing of fractures. *Chemical Geology*, **2**, 339–354.

Fletcher, R. C. & Pollard, D. D. 1981. Anti-crack model for pressure solution surfaces. *Geology*, **9**, 419–424.

Gamond, J. F. 1983. Displacement features associated with fault zones: a comparison between observed examples and experimental models. *Journal of Structural Geology*, **5**, 33–45.

—— 1987. Bridge structures as sense of displacement criteria in brittle fault zones. *Journal of Structural Geology*, **9**, 609–620.

Gibson, R. G. 1994. Fault-zone seals in siliclastic strata of the Columbus Basin, offshore Trinidad. *American Association of Petroleum Geologists Bulletin*, **78**, 1372–1385.

Gratier, J. P. & Gamond, J. F. 1990. Transition between seismic and aseismic deformation in the upper crust. *In*: Knipe, R. J. & Rutter, E. H. (eds) *Deformation mechanisms, rheology and tectonics*. Geological Society, London, Special Publication **54**, 461–473.

Jev, B. I., Kaars-Sijpesteijn, C. H., Peters, M. P. A. M., Watts, N. L. & Wilkie, J. T. 1993. Akaso field, Nigeria: Use of integrated 3-D seismic, fault slicing, clay smearing, and RFT pressure data on fault trapping and dynamic leakage. *American Association of Petroleum Geologists Bulletin*, **77**, 1389–1404.

Knipe, R. J. 1992. Faulting processes and fault seal. *In*: Larsen, R. M., Brekke, H., Larsen, B. T. & Talleraas, E. (eds) *Structural and tectonic modelling and its application to petroleum geology*. NPF Special Publications **1**, 325–342.

Knott, S. D. 1993. Fault seal analysis in the North Sea. *American Association of Petroleum Geologists Bulletin*, **77**, 778–792.

KOESTLER, A. G. & EHRMANN, W. U. 1991. Description of brittle extensional features in Chalk on the crest of a salt ridge (NW Germany). *In*: ROBERTS, A. M., YIELDING, G. & FREEMAN, B. (eds) *The geometry of normal faults*. Geological Society, London, Special Publications, **56**, 113–123.

LEYTHAEUSER, D., BORROMEO, O., MOSCA, F., DI PRIMIO, R., RADKE, M. & SCHAEFER, R. G. 1995. Pressure solution in carbonate source rocks and its control on petroleum generation and migration. *Marine and Petroleum Geology*, **12**, 717–733.

LINDSAY, N. G., MURPHY, F. C., WALSH, J. J. & WATTERSON, J. 1993. Outcrop studies of shale smears on fault surfaces. *Special Publication of the International Association of Sedimentologists*, **15**, 113–123.

MARTEL, S. J., POLLARD, D. D. & SEGALL, P. 1988. Development of simple strike-slip fault zones, Mount Abbot quadrangle, Sierra Nevada, California. *Geological Society of America Bulletin*, **100**, 1451–1465.

PEACOCK, D. C. P. & SANDERSON, D. J. 1992. Effects of layering and anisotropy on fault geometry. *Journal of the Geological Society of London*, **149**, 721–733.

—— & —— 1994. Strain and scaling of faults in the Chalk at Flamborough Head, UK. *Journal of Structural Geology*, **16**, 97–107.

—— & —— 1995a. Pull-aparts, shear fractures and pressure solution. *Tectonophysics*, **241**, 1–13.

—— & —— 1995b. Strike-slip relay ramps. *Journal of Structural Geology*, **17**, 1351–1360.

—— & ZHANG, X. 1994. Field examples and numerical modelling of oversteps and bends along normal faults in cross section. *Tectonophysics*, **234**, 147–167.

PETIT, J. P. & MATTAUER, M. 1995. Palaeostress superimposition deduced from mesoscale structures in limestone: the Matelles exposure, Languedoc, France. *Journal of Structural Geology*, **17**, 245–256.

POLLARD, D. D. & SEGALL, P. 1987. Theoretical displacements and stresses near fractures in rock: with applications to faults, joints, veins, dikes, and pressure solution surfaces. *In*: ATKINSON, B. K. (ed.) *Fracture mechanics of rock*. Academic Press, London, 277–349.

PRICE, M. 1987. Fluid flow in the Chalk of England. *In*: GOFF, J. C. & WILLIAMS, B. P. J. (eds) *Fluid flow in sedimentary basins and aquifers*. Geological Society, London, Special Publications, **34**, 141–157.

PRICE, N. J. & COSGROVE, J. W. 1990. *Analysis of geological structures*. Cambridge University Press, Cambridge.

RAMSAY, J. G. & HUBER, M. I. 1987. *The techniques of modern structural geology, volume 2: folds and fractures*. Academic Press, London.

RAWSON, P. F. & WRIGHT, J. K. 1992. *The Yorkshire Coast*. Geol. Assoc. Guide **34**.

RISPOLI, R. 1981. Stress fields about strike-slip faults inferred from stylolites and tension gashes. *Tectonophysics*, **75**, T29–T36.

ROEHL, P. O. & CHOQUETTE, P. W. 1985. Introduction. *In*: ROEHL, P. O. & CHOQUETTE, P. W. (eds) *Carbonate petroleum reservoirs*. Springer-Verlag.

SASSEN, R., MOORE, C. H. & MEENDSEN, F. C. 1987. Distribution of hydrocarbon source potential in the Jurassic Smackover Formation. *Organic Geochemistry*, **11**, 379–383.

SIBSON, R. H. 1977. Fault rocks and fault mechanisms. *Journal of the Geological Society of London*, **133**, 191–213.

STARMER, I. C. 1995. Deformation of the Upper Cretaceous Chalk at Selwicks Bay, Flamborough Head, Yorkshire: its significance in the structural evolution of north-east England and the North Sea Basin. *Proceedings of the Yorkshire Geological Society*, **50**, 213–228.

WILLEMSE, E. J. M., PEACOCK, D. C. P. & AYDIN, A. 1997. Nucleation and growth of strike-slip faults in limestone. *Journal of Structural Geology*, **19**, 1461–1477.

Fault sealing processes in siliciclastic sediments

Q. J. FISHER & R. J. KNIPE

Rock Deformation Research Group, Department of Earth Sciences, University of Leeds, Leeds LS2 9JT, UK

Abstract: The microstructure and petrophysical properties of fault rocks from siliciclastic hydrocarbon reservoirs of the North Sea are closely related to the effective stress, temperature and sediment composition at the time of deformation, as well as their post-deformation stress and temperature history. Low permeability fault rocks may develop due to a combination of processes including: the deformation induced mixing of heterogeneously distributed fine-grained material (principally clays) with framework grains, pressure solution, cataclasis, clay smear, and cementation. Fault rocks can be classified into various types (disaggregation zones, phyllosilicate-framework fault rocks, cataclasites, clay smears, and cemented faults/fractures) based upon their clay and cement content as well as the amount of cataclasis experienced. In the absence of extensive cementation, the distribution of fault rock types along a fault plane can often be predicted from a detailed knowledge of the reservoir sedimentology. The permeability of fault rocks can vary by over six orders of magnitude, depending on the extent to which the porosity reduction processes have operated. Utilizing the strong link between the petrophysical properties of fault rocks and their geohistory allows the risks associated with fault seal evaluation to be reduced.

Faults are a major heterogeneity in sedimentary sequences and play an important role in the entrapment and movement of hydrocarbons on both geological and production time-scales (Smith 1966, 1980; Berg 1975; Watts 1987). Barriers or restrictions to fluid flow may be formed by the juxtaposition of reservoir rocks against impermeable horizons (juxtaposition seal) or by the development of a fault rock with a high sealing capacity (fault seal *sensu stricto*). Juxtaposition seals are one of the most widely recognized hydrocarbon traps and are frequently targeted during exploration. The importance of intra-reservoir faults (which maintain reservoir/reservoir juxtapositions) is less well recognized and often ignored. An understanding of their sealing capacity is, however, important to both assess the likely distribution of hydrocarbons in a particular area and to plan an optimum production strategy. It is therefore surprising how little published work is available on the key factors which control the types and petrophysical properties (porosity, permeability, capillary pressure) of fault rocks within siliciclastic hydrocarbon reservoirs. Notable exceptions have discussed clay smears (Knipe 1992; Gibson 1994; Berg & Avery 1995; Yielding *et al.* 1997), clay gouges (Morrow *et al.* 1984) and cataclasites (Aydin 1978; Pittman 1981; Underhill & Woodcock 1987; Knipe 1992; Antonellini & Aydin 1994). Deformation features present within impure sandstones have only occasionally been studied (e.g. Sverdrup & Bjørlykke 1992).

The scarcity of published data in the petrophysical properties of fault rocks has led to the situation where fault sealing is often treated in an empirical manner. Beginning with the technical vocabulary, faults are usually either classified as sealing or non-sealing. The term 'sealing' is used to cover a range of faults from those that act as minor production baffles to those which have supported large hydrocarbon columns over geological time. This classification of fault sealing is over-simplistic, and neglects previous work which has shown that faults may act as membrane seals with finite sealing capacities (Watts 1987).

The lack of published data on the types and petrophysical properties of fault rocks limits the extent to which fluid distribution patterns and production data within established hydrocarbon fields may be interpreted. For example, a knowledge of the petrophysical properties of the various fault rock types is needed if production data are to be interpreted to assess which fault rock types control sealing capacity. Finally, an understanding of the fundamental processes which control the petrophysical properties of faults is necessary in order to extrapolate results gained from studying the sealing behaviour of a single fault, to other faults within a field and then to other fields.

The purpose of this paper is to describe the microstructural characteristics of fault rocks within siliciclastic sedimentary rocks and to outline the fundamental factors controlling their

FISHER, Q. J. & KNIPE, R. J. 1998. Fault sealing processes in siliciclastic sediments. *In*: JONES, G., FISHER, Q. J. & KNIPE, R. J. (eds) *Faulting, Fault Sealing and Fluid Flow in Hydrocarbon Reservoirs*. Geological Society, London, Special Publications, **147**, 117–134.

petrophysical properties. We show that a wide range of processes may reduce the permeability and increase the capillary pressure of fault rocks. The extent to which these operate is shown to vary significantly depending upon the sediment composition, as well as the stress and temperature histories experienced pre-, syn- and post-deformation. Understanding these processes and relationships will provide a valuable tool for the prediction of fault seal capacity.

The paper uses microstructural and physical property analyses obtained, over the last five years, from faults and fractures present within cores from 38 hydrocarbon reservoirs from the North Sea. To supplement this database, some results obtained during the study of an oil field in Australasia have also been included. The fields studied are diverse both in terms of their reservoir rocks and burial and deformation histories experienced. This paper integrates these observations with published data on sandstone diagenesis and rock deformation experiments.

Methods

Laboratory based analysis was aimed at characterizing the microstructural and petrophysical properties inside and outside of faults and fractures sampled, identifying the deformation mechanisms and sealing processes experienced, and constraining the timing of deformation relative to the diagenetic history.

Microstructural examination

The microstructure of samples were examined by secondary electron (SE), back-scattered electron (BSE) and cathode luminescence (CL) imagery using the same instrumentation as outlined in Peacock *et al.* (this volume). The modal composition of deformed and undeformed sediments was determined by image analysis of BSE and Cl micrographs as well as X-ray dot maps, as outlined in Fisher *et al.* (in press). Point counting was used to quantify mineral phases which could not be differentiated using image analysis techniques.

Pore aperture size measurement and threshold pressure determinations

The pore aperture size distributions of samples was measured using mercury porosimetry. The Micromeritics Autopore II 9220 porosimeter used injects mercury into evacuated samples in pressure increments up to 60 000 psi. After reaching equilibrium at each pressure increment, the volume of mercury intruded was recorded. The pressure was plotted against the mercury saturation to produce an injection curve from which the threshold pressure, and hence capillary pressure, was determined (Katz & Thompson 1986, 1987).

Ideally, separate plugs from inside and outside the fault zone should be used. Most faults examined were, however, too narrow to cut a separate sample. In such cases, a plug was cored with the fault zone cutting horizontally across its centre. The sample was sealed with epoxy resin on all sides except its base. The mercury injection curves from these sealed samples have two threshold pressure indications. The first threshold is usually characteristically low and represents the host sediment. The second threshold represents the pressure at which the fault is breached.

Water permeability

Traditional methods of gas permeametry, such as the Hassler sleeve, are unable to measure permeabilities below about 0.01 mD, and the results below 1 mD are probably not very accurate because of lateral sealing problems and uncertainties in the Klinkenberg correction. Many fault rocks have far lower permeabilities; hydraulic conductivity tests were therefore performed using a custom-designed flow-pump permeameter (Olson & Daniel 1981; Olsen *et al.* 1991). The computerized permeameter, which uses an accurate constant-rate-of-flow pump, and a highly linear differential pressure transducer, is capable of measuring permeabilities in the range of a few darcies to a few hundredths of a microdarcy.

Permeability measurements were made on 27 mm diameter plugs or $\sim$ 25 mm $\times$ 50 mm rectilinear blocks. The samples were cut so that flow was measured across the fault. The cylindrical cores were placed into tightly fitting silicone rubber sleeves. The rectilinear samples were confined in polymer sleeves individually moulded to fit the specimens. The sleeved samples were placed in a secondary rubber or neoprene membrane and confined in a cell at a pressure of around 800 kPa to prevent any leakage of permeant occurring down the sides of the sample.

The fault rocks examined tended to be thinner than the sample size used for testing. The permeability of the faults was therefore calculated by assuming that the permeability of the sample containing the fault represented the harmonic mean of the fault rock and the host sediment; the permeability of the latter was measured separately.

The water flow permeameter is accurate to <5% down to permeabilities of <1 μD. Inaccuracies in permeability analysis may, however, occur due to the measurement of fault rock thickness, temperature variation in the testing laboratory, the presence of microcracks formed during sample preparation or as a result of stress relaxation following coring. It seems likely, however, that the accuracy of these permeability measurements is far higher than the intrinsic permeability variation that exists within individual specimens. This is consistent with the excellent correlation obtained between permeability and mercury–air threshold pressures.

Results

The microstructure and petrophysical properties of most faults examined, as well as the sealing processes experienced, were found to be highly dependent on the composition of the sediment at the time of deformation. The main situation where this proved not to be the case was where cements precipitated from fault- and fracture-transported fluids. The fault rocks described later are therefore divided in terms of the composition of the host sediment at the time of deformation. Faults rocks cemented by possibly fault-transported fluids are then considered.

Faults within clean sandstones

Faults within clean sandstones (<5% clay at the time of deformation) occur either in isolation or in dense clusters. The latter are similar to the banded deformation bands identified in the field by Jamison and Sterns (1982). The microstructure of the faults varies depending upon the extent of grain-fracturing incurred during deformation and the extent of post-deformation quartz cementation. In addition, variations exist in the manner in which the grain-size distribution changes across individual faults. Two end-members were identified. The first, have a relatively constant cross-fault grain-size distribution (Fig. 1a). These occur either in isolation or in dense clusters and account for over 98% of the fault rocks examined from clean sandstones in the North Sea. The second, are composed of a gouge which itself has a banded sub-structure, containing discrete slip planes (such as described by Logan *et al.* 1992) along which strain was concentrated (Fig. 1b); these were only observed within the densely clustered faults.

Reservoir sandstones within the North Sea often have not experienced significant diagenetic alteration between the depths of 0.5 km and

(a)

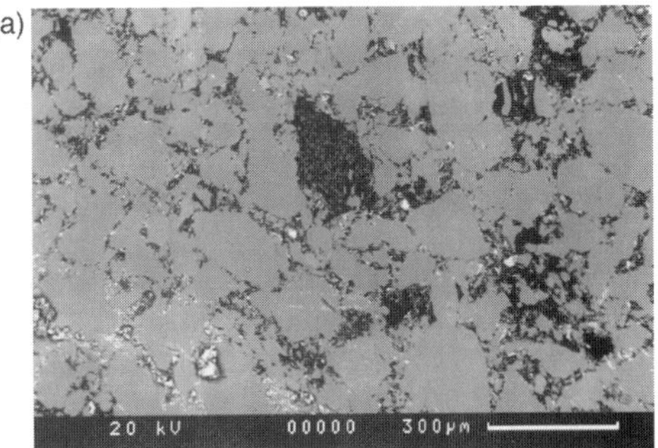

(b)

(c)

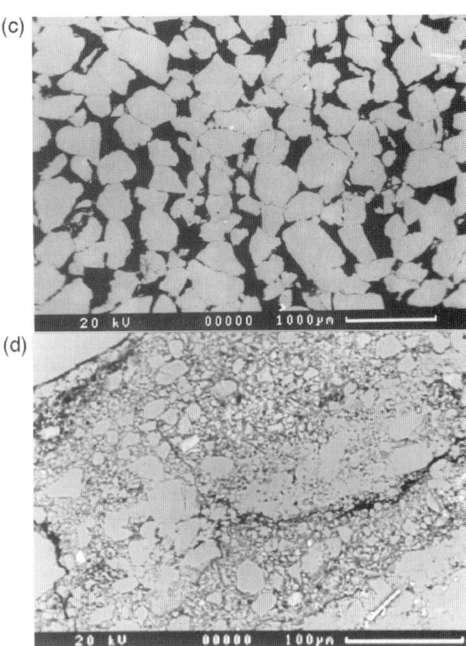

Fig. 1. BSE micrographs showing faults gouges in clean sandstones with (**a**) an homogeneous microstructure and a broad grain-size distribution, (**b**) discrete slip bands (arrows), (**c**) no grain-size reduction, (**d**) a large grain size reduction but a narrow grain size distribution.

~3 km. Using petrographic data alone, it is therefore difficult to accurately constrain the burial depth at which deformation occurred. The information available does, however, provide important clues as to the depth at which they formed. Namely, only homogeneous fault gouges were identified within sandstones that have been buried to <2.8 km. Fault gouges containing discrete slip bands were only identified in reservoirs that had experienced deeper burial.

The fault rocks identified ranged from those that experienced no grain fracturing (Fig. 1c) to those in which every grain was extensively fractured (Fig. 1d); a full range of intermediates also exist. Faults that experienced no grain-size reduction have an homogenous structure, similar to that of the host sediment. These features are termed disaggregation zones and deformation probably occurred by particulate flow under low effective stress conditions (Borradaile 1981). Fault rocks that experienced grain fracturing are termed clay-free cataclasites. Fault rocks that experienced a partial grain-size reduction also tend to have a homogenous macrostructure with a broad grain-size distribution due to the presence of large, mostly unfractured, grains and small fragments produced by cataclasis. The grain fragments within these fault rocks tend to form the matrix between the undeformed grains. Small elongate particles are also very common which probably formed as a result of spalling (see Knipe & Lloyd 1994).

Extensive grain-size reductions occur mainly within the banded faults. In the fault gouges that lack discrete slip surfaces, the fragments produced by grain fracturing are often not widely separated from their parent grain (Fig. 2). The fragments are, however, separated sufficiently from each other to produce a texture that appears dilational. It is important to realize that in porous sandstones the grain-fragments within these faults have collapsed into (and filled) macroporosity resulting in net compaction. The overall displacement along individual faults is often too small to be identified in hand specimen (i.e. ≪0.5 mm).

The most extensive grain-size reductions were confined to the discrete slip planes within the banded fault gouges, such as shown in Fig. 1b. The grains within some discrete slip planes are better sorted than those within the host sandstone. The fault gouge outside these discrete slip zones often has a broad grain-size distribution similar to the homogeneous faults.

In the majority of samples (>99%) examined, no evidence of distributed microcracking was observed within the host sediment adjacent to the fault. The only exception to this was adjacent to a small number of thrust faults which formed following the initiation of quartz overgrowth precipitation (see Knipe & Lloyd 1994).

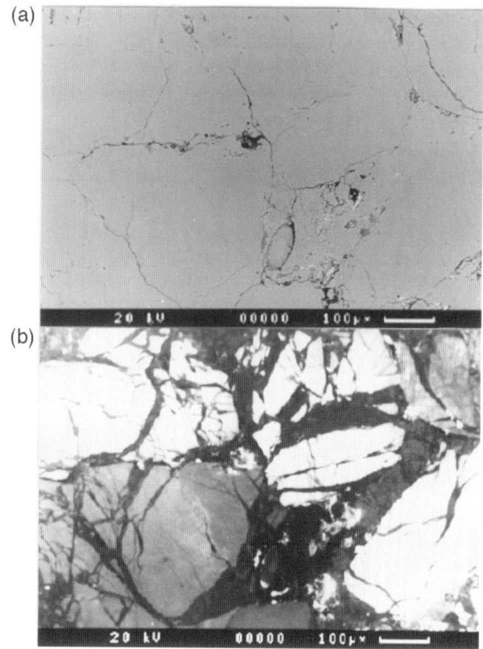

Fig. 2. BSE (**a**) and CL (**b**) images of the same area of a quartz-cemented cataclastic fault. Note the dark quartz cement in the CL image. Also, note that the fabric appears dilational because the grain fragments have separated from each other. In fact, the fragments have compacted into what was macroporosity.

Mesocrystalline quartz cement is concentrated within cataclastic faults in clean sandstones that had been buried to >3 km. These fault rocks often have very low porosities (≪3%) and, when viewed using BSE imagery, have the appearance of quartz-cemented veins (Fig. 2a). CL images reveal, however, that the faults are composed of fragments of detrital grains cemented by mesocrystalline quartz (Fig. 2b). Detrital fragments are not supported by the cement. Instead, the quartz appears to have preferentially precipitated between fragments in mutual contact and along intragranular fracture surfaces. All of the quartz-cemented cataclastic faults studied have lower intragranular volumes (see Houseknect 1987 for a definition) than their host sandstone, suggesting that they experienced a deformation-induced porosity collapse.

The amount of quartz cement within the cataclastic faults often appears to be related to the amount of grain-size reduction experienced. On the whole, faults that have experienced only very small amounts of grain fracturing appear

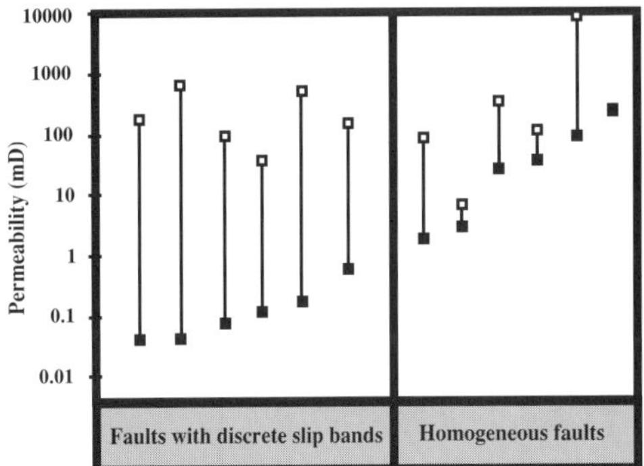

Fig. 3. Diagram showing the range of permeabilities obtained from cataclastic faults (solid symbols) from an Australasian oil field. These fault rocks have not experienced post-faulting quartz cementation or pressure solution. The results have been divided according to whether the fault gouges have an homogenous sub-structure or contain discrete slip planes. The permeability of their host sediment (open symbols) has been included for comparison.

less cemented than those that experienced pervasive cataclasis.

Cataclastic faults with a grain-size of >5 μm did not show evidence of having experienced enhanced post-faulting pressure solution compared to their host sediment. The microstructure of cataclasites with a grain-size of <5 μm could not, however, be resolved using CL. The relative amount of pressure solution and quartz cementation which contributes to the final microstructure of these very fine-grained cataclasites could therefore not be established.

Not surprisingly, the petrophysical properties of fault rocks within clean sandstones show enormous variation. Those that did not experience cataclasis have very similar petrophysical properties to their host sediment. Figure 3 shows a range of permeability values obtained from faults rocks within clean sandstones from an Australasian oil field in which cataclasis was the principle porosity reduction mechanism (a North Sea example was not used because most of the faults that have extensive grain-size reductions are also quartz cemented). The permeabilities obtained range from >100 mD to <0.05 mD. The principle factor which affects these permeability values is the final grain-size and grain-sorting of the gouge (see Knipe *et al.* 1997). The high permeability values are from faults that have experienced little grain-size reduction, whereas the low permeability values are from cataclasites that have experienced extensive grain-size reductions. The faults containing discrete slip bands have the lowest permeability values.

The permeabilities of cataclastic fault rocks within clean sandstones that have experienced post-faulting quartz cementation are on the whole far lower than the uncemented faults (Fig. 4). The extent to which the permeability of fault rocks is further reduced by quartz cementation depends both on the extent of grain-size reduction and the quantity of cement present. Cataclastic faults in clay-free sandstones from the North Sea have mercury–air threshold pressures of 15 psi to 2000 psi.

Fault rocks within sandstones containing low concentrations of clays

Fault rocks within sandstones containing low concentrations of phyllosilicates (5 to 15%) have a similar structure and range of grain-size distributions to those within clean sandstones. These fault rocks sometimes also contain discontinuous domains in which phyllosilicates partly fill the intragranular volume. These domains form by the deformation induced mixing of phyllosilicate material derived from the host sediment with the framework grains. In reservoirs that have been buried to >3 km, these domains usually show signs of enhanced pressure solution compared to their host sediment.

The volume of phyllosilicates within the host sandstone is insufficient for phyllosilicate-rich domains to be continuous along the entire fault plane. The faults therefore contain domains which have the same microstructure as observed within the phyllosilicate-free equivalents. The petrophysical properties of these fault rocks are therefore similar to those within the clean sandstones.

Fault rocks within impure sandstones

Faults within impure sandstones (those with 15–40% clay) also experienced a wide range of grain-size reductions, although the very large grain-size reductions experienced by the faults within the less clay-rich sandstones were not observed in the specimens studied. The microstructure and petrophysical properties of these fault rocks often differ markedly from their host sediment, regardless the extent of grain fracturing.

The permeability of impure sandstones is usually as not as high as clay-free sandstones from the same reservoir. Nevertheless, they can have good to moderate reservoir properties, because the phyllosilicates present often have a heterogeneous distribution leaving interconnected clay-free channels (e.g. Fig. 5a). Deformation of

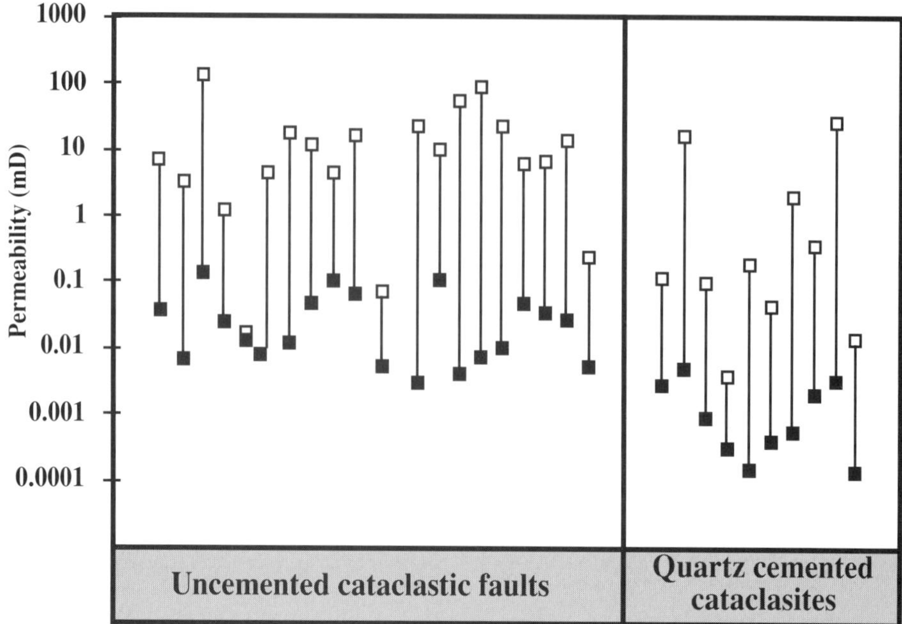

Fig. 4. Diagram showing the range of permeabilities obtained from cataclastic faults (solid symbols) from the North Sea. The fault rocks have been subdivided according to whether or not they have experienced post-deformation quartz cementation. The permeability of their host sediment (open symbols) has been included for comparison.

these impure sandstones results in the mixing of clays with framework grains and produces a fault rock that has a more homogenous distribution of clays than the host sediment (Fig. 5b). This process we have termed deformation-induced mixing and produces a distinct group of deformation features which have been termed phyllosilicate-framework fault rocks (Knipe *et al.* 1997)

In reservoirs that have been buried to >3 km, these phyllosilicate-rich fault rocks have often experienced enhanced pressure solution compared to the host sandstone (Fig. 5c & d). The extent of enhanced pressure solution often varies significantly both along and between individual faults within reservoirs. The faults which experienced most pressure solution are within sandstones with moderate concentrations (~15 to 25%) of heterogeneously distributed phyllosilicates. Faults within sediments which contain >25% intergranular clays, or quartz grains which have extensively developed clay coats, often do not experience significant enhanced pressure solution.

Sutured grain contacts, formed by enhanced pressure solution within fault rocks, frequently have preferred orientations. The grain contacts in most fault rocks examined are predominantly sub-horizontal (Fig. 5c). Fault rocks which have dominantly sub-vertical grain contacts have also been identified (Fig. 5d). Grain elongation as a result of pressure solution can be very significant. For example, quartz grains with aspect ratios of >5 have formed by enhanced pressure solution of what were originally sub-rounded grains.

Faults within impure sandstones that experience grain fracturing sometimes contain slightly more quartz cement than their host sediment. This tends to be concentrated along quartz fracture surfaces that are not coated by clay minerals. It should be emphasized that the extent of post-deformation quartz cementation of these fault rocks is far less than observed within the less clay-rich sediments.

The permeability of 53 fault rocks produced by the deformation of impure sandstones examined from the North Sea has been measured. The values obtained range from 0.9 mD to 0.2 μD. Their mercury–air threshold pressure varies between 75 and 1600 psi. The extent to which post-faulting pressure solution has contributed to the reduction in permeability of the fault rocks is difficult to assess. Nevertheless, some indication of this is gained by the observation that most of the framework-phyllosilicate fault rocks that have been buried to <2.5 km have Hg–air threshold pressures of <300 psi. However, most framework-phyllosilicate fault rocks buried to >3.5 km have threshold pressures of >350 psi.

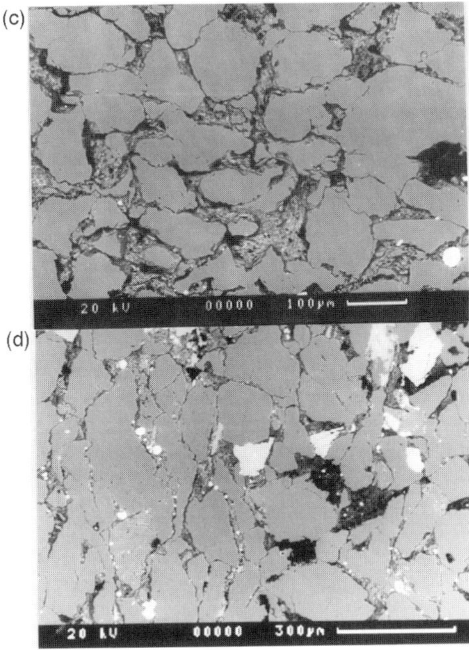

Fig. 5. BSE micrographs showing: (**a**) the undeformed phyllosilicate-rich sandstones close to (**b**) a fault rock that has experienced porosity collapse by disaggregation and deformation induced mixing; (**c**) fault rock that has experienced enhanced grain-to-grain dissolution along sub-horizontal grain-to-grain contacts; (**d**) fault rock that has experienced enhanced grain-to-grain dissolution along sub-vertical grain-to-grain contacts.

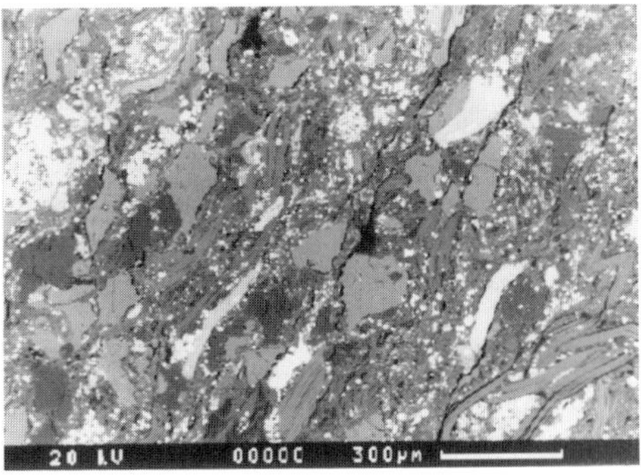

Fig. 6. BSE image showing the alignment of phyllosilicates within a clay smear.

Deformation features in clay-rich sediments

Sediments containing >40% phyllosilicates deform to produce fault rocks containing domains of aligned phyllosilicates (Fig. 6). These fault rocks frequently become smeared between the hangingwall and footwall cut-offs of phyllosilicate-rich horizons. In addition, faulting of some very phyllosilicate-rich horizons has resulted in the injection of a clay gouge into less clay-rich sediment. These fault rocks are clay smears and have been described elsewhere in the literature (e.g. Smith 1980; Bouvier *et al.* 1989; Sassi *et al.* 1992; Jev *et al.* 1993; Lindsay *et al.* 1993; Gibson 1994).

The microstructure and continuity of clay smears examined from hydrocarbon fields within the North Sea varies enormously. For example, discontinuous clay smears were identified along faults with displacements of less than twice the thickness of the clay-rich layer present. However, coherent clay smears were identified along faults with displacement of >15 times the thickness of the clay-rich layer present. The main difference between these cases was the grain-size of the phyllosilicates present. The continuous clay smears contained coarse-grained phyllosilicates (>20 μm), whereas the clay within the discontinuous smears was very fine-grained (<0.5 μm).

The permeability and threshold pressure of seventeen clay smears have been analysed during this study. Twelve yielded permeability values of <0.0005 mD. Clay smears with higher permeability values later proved to be discontinuous on the hand-specimen scale. The permeability value of 0.0005 mD must, however, be taken as a maximum, as it is close to the resolution of the water flow permeameter used. All phyllosilicate smears analysed had mercury-air threshold pressures in excess of 2000 psi, with fourteen of the measurements being over 10000 psi.

Cemented deformation features present within North Sea hydrocarbon reservoirs

A variety of cement types have precipitated along faults and fractures within the North Sea reservoirs, including: anhydrite, ankerite, barite, calcite, dolomite, kaolin, microcrystalline quartz, pyrite, siderite, sphalerite, and mesocrystalline quartz. These cements can be found in a range of structures including: cataclastic faults, veins, and fluid escape features. Mesocrystalline quartz is by far the most common cement concentrated within cataclastic faults in clean sandstones that have not experienced reactivation.

Anhydrite, ankerite, barite, calcite, dolomite, pyrite, siderite, and sphalerite occur as vein-filling cements within lithified sandstones and reactivated cataclastic faults. Calcite veins were found to be particularly common within calcite cemented sandstones (Fig. 7a). In the Rotliegendes of the southern North Sea, anhydrite, ankerite, siderite and barite are particularly common within fractures formed (Fig. 7b), and cataclastic faults reactivated (Fig. 7c), during inversion.

Microcrystalline quartz (Fig. 7d) and pyrite cement are concentrated within faults formed during very early burial associated with fluid escape. These deformation features often have very high minus cement porosities (>40%) suggesting that they experienced dilation during cement precipitation. Microcrystalline quartz-cemented deformation features generally occur within sandstones containing large concentrations of biogenic silica, such as sponge spicules, suggesting that the cement is a product of its dissolution.

Veins usually have extremely low permeabilities (<1 μD) and high mercury-air threshold pressures (>1500 psi). Higher permeability, and lower threshold pressure values were only obtained from veins which microstructural analysis revealed to be discontinuous. In the case of reactivated faults, cements are usually concentrated within dilational jogs. Portions of such faults that have not experienced extensive dilation remain uncemented. The ultimate sealing capacity of the fault is therefore controlled by the capillary pressure of the fault rock between cemented sections. For example, Fig. 8 shows a cataclastic fault which has been reactivated and cemented by barite. Barite is only present along the parts of the fault that experienced dilation, and the sealing capacity of the fault rock is controlled by the pore structure of the cataclasite between.

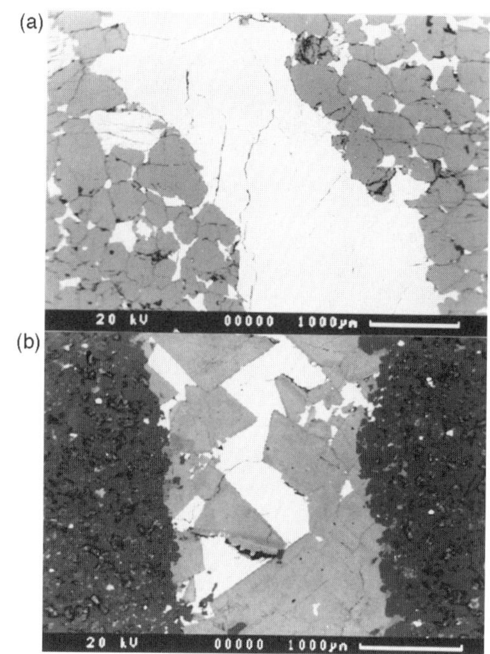

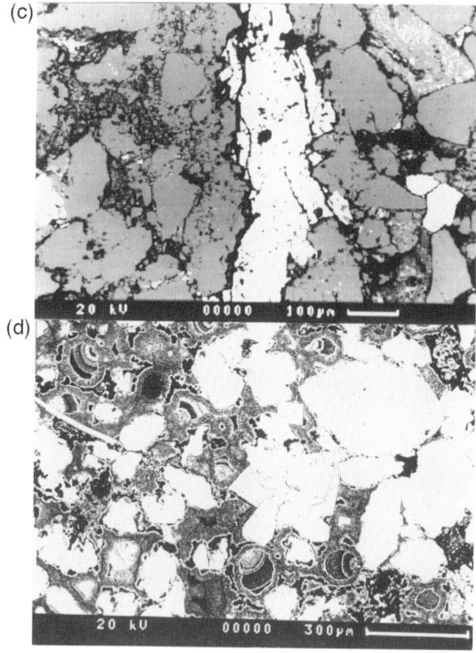

Fig. 7. BSE micrographs showing: (**a**) a calcite vein in a calcite cemented sandstone; (**b**) a dolomite and anhydrite vein in a lithified sandstones; (**c**) an anhydrite cemented reactivated cataclastic fault; (**d**) a microcrystalline quartz-cemented water-escape structure;

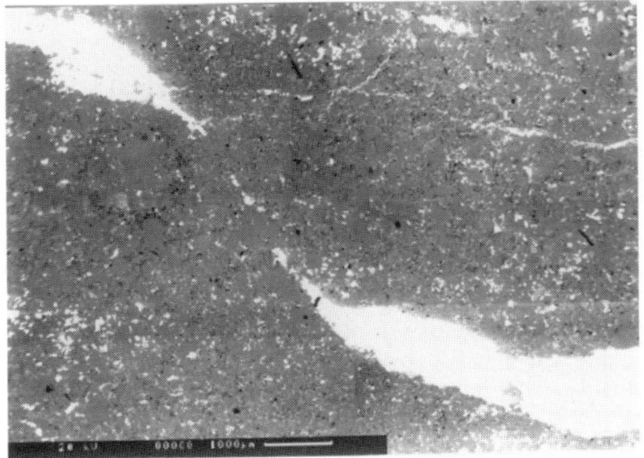

Fig. 8. BSE photomontage of a cataclastic fault that has been reactivated. The dilated areas produced during reactivation have been cemented by barite. Note that the barite cement is discontinuous therefore the sealing potential of the fault is still controlled by the pore structure of the cataclastic fault.

The permeability of microcrystalline quartz cemented water-escape structures varies dramatically (between 2 mD and <0.1 μD), depending upon cement morphology. Quartz cements within these deformation features range from highly porous networks of microcrystalline quartz to non-porous chert veins. These water escape structures are usually discontinuous; however, they often occur in high abundance and may greatly increase the overall tortuosity of fluid flow paths within sponge spicule-rich hydrocarbon reservoirs.

Discussion

Fault seal processes and their controls

The results presented earlier show that the porosity and permeability of fault rocks may be decreased and their threshold pressure increased by a number of processes, including:

(i) deformation induced porosity reduction by disaggregation and mixing;
(ii) pressure solution;
(iii) cataclasis;
(iv) cementation, and
(v) clay smearing.

This section discusses the results and identifies the main factors which control the extent to which these processes operate.

Porosity collapse by deformation induced mixing. The deformation of clean sandstones by particulate flow without grain-fracturing does not result in a long term change in their petrophysical properties, because the grain-size and grain-sorting of the fault is not significantly changed. Temporary dilation may occur due to changes in grain packing, however, if the fault rock is not quickly cemented, mechanical compaction usually results in a return of the grains to their former packing arrangement. Deformation of impure sandstones results in the mixing of fine-grained phyllosilicates with framework grains and the production of a more homogenous microfabric in which macroporosity is partly or wholly replaced by clay minerals and microporosity. This may occur during early burial by particulate flow without the need for grain-fracturing. Macroporosity destruction results in a reduction in permeability and average pore-aperture size. The most important factor controlling the ability of this process to create a continuous barrier to flow is the amount of fine-grained phyllosilicates present. Only sediments containing >15% phyllosilicates deform by particulate flow without grain-fracturing to produce a low permeability fault gouge in which all intragranular porosity is partly filled by fine-grained phyllosilicates. If grain fracturing occurs, less phyllosilicates are required to fill intergranular porosity.

Pressure solution and quartz cementation. The extent to which the fault rocks have experienced post-faulting quartz cementation or pressure solution depends on both their clay content and distribution at the time of burial, as well as on the depth/temperature history experienced. Faults within sandstones containing <5% phyllosilicates experienced enhanced quartz cementation but evidence of enhanced pressure solution was not observed. Faults within sandstones containing ~5 to 15% phyllosilicates experienced both enhanced quartz cementation and pressure solution. Faults within sandstones containing ~15 to 25% experienced enhanced pressure solution but not, in most cases, extensive quartz cementation. Faults in sandstones containing very large quantities of phyllosilicates (>25%) were usually not significantly affected by either of these processes.

The effect that clays have on the rate of quartz cementation and pressure solution in sandstones is well established (Heald 1955; Dewers & Ortoleva 1991; Bjørkum 1996; Oelkers *et al.* 1996). In particular, small concentrations of phyllosilicates at grain-contacts increase the rate of pressure solution (e.g. Heald 1955, 1959; Thompson 1959; de Boer *et al.* 1977; Chang & Yortsos 1994; Bjørkum 1996; Oelkers *et al.* 1996). In contrast, grain-coating clay minerals suppress quartz cementation (Cecil & Heald 1971; Tada & Siever 1989) by reducing the reactive quartz surface area and hence reducing the overall rate of quartz precipitation (Walderhaug 1996).

Theoretical considerations suggest that silica solubility and the rate of fluid flow in sandstones are not sufficient to account for the mobility of large volumes of silica by advection. For example, Bjørlykke (1994) showed that 3×10^9 pore volumes of fluid would be required to completely cement a single pore volume with quartz. Local processes are therefore required to explain the magnitude of quartz cementation and pressure solution observed in sandstones (Bjørlykke 1994). It seems likely that silica generated by pressure solution precipitates locally (i.e. at a distance of <1 m) as mesocrystalline quartz cement (Oelkers *et al.* 1996).

The link between quartz cementation and pressure solution is emphasized by the observation that faults rocks within the North Sea that have experienced these processes have been buried to >3 km. This finding reflects the fact that quartz cementation in sandstones, which is necessary for pressure solution, occurs at negligible rates below temperatures of 90 °C (Rimstidt & Barnes 1980; Giles 1997).

The susceptibility of a fault rock to pressure solution can also be related to variations in the textural distribution of phyllosilicates. Clays within many sandstones are either pore-filling or grain-coating, both of which inhibit quartz cementation; as they are not present between grain contacts, they do not increase the rate of pressure solution. During deformation, some of these clays become emplaced at quartz grain-to-grain contacts, rendering the quartz more susceptible to pressure solution. Faults in very clay-rich sandstones (>25% clay) do not experience enhanced pressure solution, because quartz cementation in the surrounding sediment is inhibited by the presence of the clays, which lowers the surface area available for quartz cementation.

It may be tempting to suggest that enhanced quartz cementation of cataclasites occurs due to the precipitation of silica from fluids focused along the fault. The quartz-cemented cataclastic faults identified during this study have, however, a far lower finite intergranular volume than their host sediment. There is therefore no reason why fluids would be focused along these faults as opposed to the more permeable host sandstone. This suggests that the faults sampled did not act as long term conduits for fluid flow and therefore cementation is unlikely to have occurred from non-local fluids focused along these features (Fisher *et al.* in press). The reason why cataclastic faults within clean sandstones often become preferentially quartz cemented requires an understanding of the kinetics of quartz cementation.

The rate of quartz cementation in most sandstones is inhibited either due to the presence of small quantities of clays (Cecil & Heald 1971) or other surface pollutants such as Al^{3+} (Iller 1979). For example, Mullis (1991) showed that the precipitation constant for silica was up to three orders of magnitude lower in the Jurassic Dogger Beta Formation from northern Germany than the value derived in laboratory experiments using clean sands. Cataclastic deformation of a clean sandstone produces a fault gouge that not only has a larger quartz surface area than the host sediment but also the newly formed fracture surfaces will contain less contaminants (clay etc.) than are present on the surface of the undeformed detrital quartz grains. Hence, quartz can precipitate faster in the cataclastic gouge than in the surrounding host sediment.

As advective processes cannot supply large volumes of silica required for quartz cementation in sandstones (Bjørlykke 1994; Giles 1997), it seems likely that the silica in the cataclasites was produced by processes such as pressure solution near the fault. Integrating the microstructural observations made during this study, it appears that faults in clean sandstones often act as sinks for silica generated by diagenetic processes in the host sediment. However, faults in impure sandstones act as sources of silica for the precipitation of quartz cement in the adjacent sediment.

The hypothesis that cataclastic fault gouges offer kinetically favourable sites for the precipitation of silica generated by pressure solution in the host sediment implies that the timing of faulting may be an important control on their sealing capacity. In particular, if deformation occurs below the temperature that quartz can rapidly precipitate (~90°C), the fracture surfaces generated during cataclasis may become polluted by the growth of clay minerals, etc. Indeed, the cataclasites examined during this study that experienced most quartz cementation are those which also experienced the highest grain-size reductions and probably deformed at the deepest depths.

Pressure solution and quartz precipitation may be suppressed by the emplacement of hydrocarbons (Griggs 1940) which lowers the rate of silica diffusion from grain contacts to precipitation sites (Worden *et al.* 1998). This implies that early hydrocarbon migration may reduce the capacity of faults above the hydrocarbon–water contact to seal by post-faulting pressure solution and quartz cementation.

Cataclasis. Cataclasis is the dominant process responsible for reducing the porosity and permeability and increasing the threshold pressure of porous, clay-free sandstones during faulting. These changes occur because cataclasis causes macroporosity to collapse, average grain-size to be reduced and grain-sorting to become poorer. It is difficult to determine the key factors which control the grain-size distribution and hence the petrophysical properties of cataclasites from microstructural observations of natural fault rocks alone. Integration of microstructural data with published accounts of triaxial deformation experiments (Engelder 1974; Zoback & Byerlee 1976; Logan 1992; Gu & Wong 1994; Wong *et al.* 1997; Zhu & Wong 1997) has, however, the potential to provide such information.

Sandstone deformation experiments show that low confining pressures favour brittle faulting in which failure occurs along single slip planes (Handin *et al.* 1963; Scott & Nielson 1991). High confining pressures favour ductile deformation by distributed cataclastic flow without the formation of discrete slip planes (Handin *et al.* 1963; Scott & Nielson 1991). A transitional regime exists by which deformation occurs along multiple slip planes (Scott & Nielsen 1991). The effective pressure at which this transition occurs is related to the porosity and grain-size of the sandstone (Wong *et al.* 1997). In particular, it has been shown that fine grain sizes (Zhang *et al.* 1990), low porosities (Rutter & Hadizaddeh 1991), and high cement contents (Ménendez *et al.* 1996) favour brittle faulting as the dominant deformation process. The relative manner in which these factors vary during burial can be complex. For example, confining pressures tend to increase during burial but porosity tends to decrease (e.g. Jamison & Stearns 1982).

The microstructure of fault gouges within clean sandstones is also dependant upon the amount of shear strain. For example, Logan (1992) showed that under high effective pressures (equivalent to >2000 m of overburden), deformation becomes concentrated along discrete slip planes within the gouge zones even at low shear strains. Strains of >100 are, however, required to form discrete shear planes within gouge zones under lower confining stress conditions.

The faults within the clean sandstones examined during this study can be subdivided according to whether they occur in isolation or in dense clusters, and whether the fault gouges themselves have a homogeneous grain size distribution or contain discrete slip planes. The micromechanical work presented by Ménendez *et al.* (1996) suggests that isolated faults are produced by single brittle faulting events at low confining pressures. In contrast, dense clusters of cataclastic faults probably deformed at higher confining pressures and are associated with more distributed brittle deformation. It seems likely that the structure of natural cataclastic faults is also controlled by the effective pressure conditions during deformation.

Discrete slip planes within fault gouges were observed mainly within the densely packed cataclastic faults. It is suggested that these also deformed under high effective pressures, although Logan (1992) suggests that these fault rocks may also have accommodated more strain than those with a more homogeneous sub-structure.

The permeability of sands and porous sandstones decreases as strain is increased during high pressure triaxial deformation experiments (Zoback & Byerlee 1976; Zhu & Wong 1997). The extent to which the permeability decreases varies depending upon whether the sandstone deforms in a localized, transitional or distributed manner. For example, triaxial experiments on the Berea sandstone (porosity = 21%), show that under low mean effective pressures (<10 MPa), porosity and permeability decrease slowly as strain is increased until failure occurs (Zhu & Wong 1997). At failure, the samples lose cohesive strength, porosity increases slightly but permeability continues to fall by up to an order of magnitude (Zhu & Wong 1997). Under high effective pressures (160 MPa), the porosity and permeability decrease slowly with increasing strain until a critical stress is reached, after which a dramatic fall in permeability of around two orders of magnitude occurs but the samples strain harden (Zhu & Wong 1997). Under intermediate effective pressures (~40 MPa) the permeability behaves in a similar manner to the high pressure experiments but at a critical pressure the samples strain softened (Zhu & Wong 1997). Integration of the microstructural observations of Ménendez *et al.* (1996) with the permeability results of Zhu & Wong (1997) suggest that the sudden fall in permeability in the samples deformed in a distributed and transitional manner occurs due to pervasive porosity collapse and grain fracturing.

Microstructural studies of fault gouges have shown that the particle size of cataclastic faults decreases with increasing shear strain and confining pressure (Engelder 1974; Sammis *et al.* 1987; Marone & Scholtz 1989). The non-quartz-cemented faults within the clay-free sandstones with the lowest permeabilities, were also the ones which experienced the largest grain size reductions. These fault rocks were usually from dense clusters or contained discrete slip planes that probably deformed under high effective pressures. It therefore seems likely that high effective stresses at the time of deformation favour the production of cataclastic faults with low permeabilities and high capillary pressures. It should, however, be emphasized that experimental work shows that low porosity sandstones are likely to deform by localized brittle deformation (Rutter & Hadizaddeh 1991). The porosity and permeability of sandstones decreases due to chemical compaction beyond around 90 °C. It is therefore possible that sandstones that have experienced severe porosity reduction due to burial diagenesis would undergo localized faulting even when deformed under very high effective stresses. The results presented by Zhu & Wong (1997) suggest that these would not produce the high permeability reductions as observed within the sandstones that have deformed in a transitional manner.

Experimental studies have not yet fully explored the effect that mineralogy has on sediment susceptibility to grain-size reduction and porosity collapse by cataclastic processes. Deformation of feldspar-rich rocks produces cataclasites with a lower average grain size than less feldspathic lithologies under similar conditions (Blenkinsop 1991; Antonellini *et al.* 1994). Microstructural work undertaken during this study also suggests that sediments containing concentrations of phyllosilicates tend to be less prone to cataclasis than clean sandstones. This is because clay-rich sediments can deform by grain-sliding, grain rotation, and the plastic deformation of phyllosilicates without the need for significant dilation and grain crushing.

Cementation. The microstructure of the cemented faults and fractures provides clues for identifying the origin of the cements, the role of exotic fluids in cementation, as well as conditions under which fault- and fracture-related fluid flow occurs. Except for the quartz-cemented cataclastic faults, cements are only concentrated within faults and fractures that had a higher intergranular volume than their host sediment following deformation. This suggests that prior to cementation they had a higher porosity and permeability than their host sediment. Indeed, if the faults were not more permeable that the host sediment there would be no reason why focused fluid flow and cement precipitation would occur.

The cemented faults and fractures that were identified during this study fall into two categories. Those that formed in lithified sediment and those which formed in poorly lithified sediment. In these two cases, fundamental differences may exist in the mechanism by which dilation was maintained during cementation.

In the case of the veins, the wall rock was sufficiently strong or the fluid pressure was sufficiently high, either due to early cementation or later lithification, to allow fractures to remain open following initial deformation (Bjørlykke & Høeg 1997). For example, all of the cemented fractures within the Rotliegendes sandstones of the southern North Sea studied appear to have formed during structural inversion after the sandstone had become well lithified. These veins had a higher permeability than their host sediment and were capable of acting as conduits for fluid flow. There are no obvious local sources of solutes for the precipitation of some vein-filling cements, suggesting that they precipitated from exotic fluids. For example, the anhydrite, ankerite, barite, and sphalerite cements in the Rotliegendes of the southern North Sea may have required external sources of solutes such as the overlying Zechstein evaporites or the underlying Carboniferous sediments (Sullivan *et al.* 1990; Gaupp *et al.* 1993). In other cases, such as the calcite veins in calcareous sandstones, local cement sources are available and the open veins may have merely provided porosity for cementation to occur.

The cemented water-escape structures have very high intergranular volumes, suggesting that they experienced significant fluid pressure-induced dilation, because the host sandstone was poorly lithified at the time of deformation. As with the veins, cementation in fluid escape structures may result from decreases in pore pressure and temperature, or due to the mixing of fluids.

Other than mesocrystalline quartz, cements are not concentrated within faults within porous sandstones that were normally pressured at the time of deformation. This observation suggests that under these conditions, faulting does not result in the creation of fluid flow conduits for sufficient lengths of time for cementation to occur. This is entirely consistent with the results of sandstone deformation experiments (e.g. Zhu & Wong 1997).

Continuous cemented faults and fractures have been shown to have very low permeabilities and high threshold pressures. Microstructural evidence shows however, that they are rarely continuous – even on the hand specimen scale. In the case of reactivated faults, cements are usually concentrated within dilational jogs. Regions along such faults that have not experienced dilation may be uncemented, in which case the ultimate sealing capacity of the fault may still be controlled by the petrophysical properties of the fault between cement zones.

The continuity of fractures may also be very limited. For example, Peacock *et al.* (this volume) show an example where an array of calcite veins are linked by pressure solution seams. In this example, the sealing capacity of the vein array is probably controlled by the pressure solution seams and not the veins.

Hydrocarbon field examples are also inconsistent with the idea that faults and fractures cemented by externally derived fluids are sufficiently continuous to control the sealing capacity of faults on a large scale. The capillary pressures measured during this study suggest that continuous, cemented faults would be capable of supporting hydrocarbon column heights of hundreds of metres. Such large column height differences are seldom seen across faults in hydrocarbon fields in areas of known cementation, suggesting that their sealing capacity is usually controlled by more permeable fault rock types.

Clay smear. Continuous clay smears have very low permeabilities and high threshold pressures, emphasizing that they provide extremely effective barriers to fluid flow. The continuity of clay smears has, however, been found to be highly variable. The more continuous clay smears identified during this study contain coarse-grained phyllosilicates (>20 μm), whereas the clay within the discontinuous smears are very fine-grained (<0.5 μm). It seems likely that the continuity of clay smears will be dependent upon a large number of factors, including: sediment lithification, effective stress, confining pressure, strain rate, mineralogy etc.

The prediction of the presence and continuity of clay smears in the sub-surface is usually based on parameters such as clay smear potential – CSP (Bouvier *et al.* 1989), shale smear factor – SSF (Lindsay *et al.* 1993), and clay gouge ratio – CGR (Gibson 1994). The large variations in the proportion of clay required to form an effective clay smear emphasizes that it is important to calibrate these techniques based on known hydrocarbon distributions, fluid pressures and production data from within the area of interest.

Fault rock types, fault sealing processes and their relationship to sediment composition

The results and discussion presented above illustrate that there is often a distinct link between sediment composition and the sealing processes that occur within faults both during and following faulting. In particular, faults within clean sandstones tend to seal by cataclasis and post-deformation quartz cementation. Faults within impure sandstones tend to seal by deformation induced mixing, cataclasis and post-faulting pressure solution. Faults within very clay-rich sediments tend to deform by disaggregation and grain-boundary sliding to produce clay smears. We have therefore divided fault rocks developed in siliciclastic sediments into three series (clean sand series, impure sand series and clay-rich series) according to the composition of the host sediment at the time of deformation (Fig. 9).

Each of these fault rock series can be subdivided, based partly on the classification of Sibson (1977) into a number of fault rock types according to the extent of cataclasis experienced. For example, Fig. 9, shows that the clean sand series can be divided into disaggregation zones, protocataclasites, cataclasites and ultracataclasites depending upon the proportion of fractured grains.

Each of the fault rocks can also be sub-divided according to the extent of post-deformation lithification using the scheme adopted by Knipe (1986). Again, it is important to note that the extent to which post-faulting processes may operate is a function of the fault rock composition. Cataclastic faults in clean sandstones are susceptible to quartz cementation. Faults in impure sandstones are susceptible to post-faulting pressure solution. Faults in ultra clay-rich faults become lithified by mechanical compaction and clay mineral diagenesis.

The classification scheme described is sometimes difficult to apply without the use of electron microscopes. For example, in ultra clay-rich sandstones it is impossible to establish the extent of grain fracturing without the use of transmission electron microscopes. We have therefore outlined a series of fault rock types (clay-free disaggregation zones, clay-free cataclasites, phyllosilicate-framework fault rocks, clay smears and cemented faults) which can, in the majority of circumstances, be used to classify fault rocks on both a hand-specimen and microstructural scale. These are described below:

(1) Clay-free disaggregation zones are formed during shallow burial; strain is accommodated by particulate flow without extensive grain fracturing. In clay-free sandstones, disaggregation zones do not contain sufficient phyllosilicates to render them susceptible to extensive porosity reduction, either by deformation induced mixing or by later pressure solution.
(2) Clay-free cataclastic faults have porosities and permeabilities reduced mainly by cataclasis and post-deformation quartz cementation.

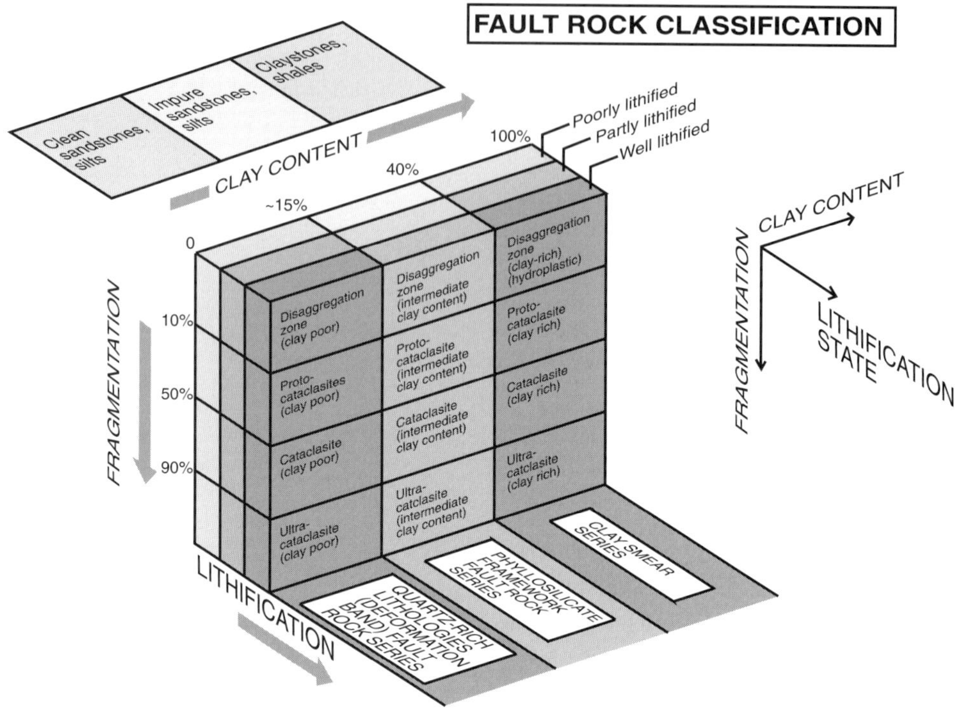

Fig. 9. Diagram showing the types of fault rocks developed in the North Sea and their relationship to the composition of the host sediment and the extent of grain-size reduction and post-deformation lithification experienced.

Cataclastic faults in clay-free sandstones identified within North Sea reservoir sandstones have permeability values of >100 mD to <1 μD and mercury–air threshold pressures of 15 psi and 2000 psi.

(3) Phyllosilicate-framework fault rocks are a new class of deformation feature (Knipe *et al.* 1997) introduced to describe fault rocks that are developed within impure sandstones (>15% phyllosilicates). The porosity and permeability of phyllosilicate-framework fault rocks are reduced mainly by deformation-induced mixing of framework grains with fine-grained phyllosilicates and often later pressure solution. Cataclasis may also reduce porosity and permeability of these fault rocks. Phyllosilicate-framework fault rocks are therefore used here to describe both clay-rich disaggregation zones and clay-rich cataclasites. Phyllosilicate-framework faults can be regarded as transitional to clay smears as the length and continuity of domains of aligned phyllosilicates increases. The phyllosilicate-framework fault rocks examined during this study have permeabilities of 0.9 mD to 0.2 μD and mercury-air threshold pressures of 75 to 1600 psi.

(4) Clay smears contain coherent domains of aligned phyllosilicates that form by the deformation of extremely phyllosilicate-rich sediments (i.e. >40% phyllosilicates).

(5) Cemented faults/fractures, are those in which the main porosity reduction mechanism is cementation. The features classified in this group are those which experienced significant dilation during deformation, resulting in the presence of open fractures along which cement could precipitate. Quartz cementation of cataclastic faults is an integral part of the evolution of a cataclasite, and therefore this type of fault rock is not grouped with the other cemented deformation features. The continuous cemented faults and fractures analysed during this study have permeabilities of <0.1 μD and threshold pressures of >1500 psi.

Prediction of fault sealing capacity

The relationship between fault rock type, fault sealing mechanisms and sediment composition suggests that, with a knowledge of the fault displacement distributions and sediment distribution within a field, it should be possible to predict the distribution of fault rock types (Knipe 1997; Knipe *et al.* 1997). The extent to which each of the fault sealing mechanisms operates for a given sediment composition depends upon the interaction between the stress, strain and temperature history of the reservoir. Only comprehensive knowledge obtained by microstructural and physical property analysis of faults within cored sections of reservoirs allows the sealing capacity of faults to be accurately predicted.

In the majority of individual fields examined, the relative importance and extent to which each of the fault sealing mechanisms operate (and hence the petrophysical properties of each fault rock type) is quite restricted. For example, faults present within clean sandstone in the Sleipner Field have permeabilities of ~0.4 to 43 mD (Ellevset *et al.* this volume). Only in five out of the thirty eight fields we have studied have the petrophysical properties of each fault rock type been highly varied, with permeabilities varying over four orders of magnitude. This is probably because faulting occurred throughout their burial history. Even in such cases, a knowledge of the range of petrophysical properties of each fault rock type may have an important impact on both exploration targets and production strategies.

Future work

The present paper has concentrated on the processes which reduce permeability and increase the threshold pressure of fault rocks. Microstructural and petrophysical analysis of core material can provide extremely important data on the likely properties of faults in the various lithologies. In addition to petrophysical properties of individual fault rocks types, the sealing capacity of field-scale faults will depend on other factors such as the distribution of open fractures in the subsurface and the continuity of fault rock types (particularly clay smears); these are important areas of future research.

Conclusions

(1) The principle fault sealing processes which operate within siliciclastic hydrocarbon reservoirs are: the deformation induced mixing of heterogeneously distributed fine-grained material with framework grains, pressure solution, cataclasis, clay smear, and cementation.

(2) The potential for fault sealing by each of these processes is dependent upon the sediment composition at the time of deformation as well as their pre- syn-, and post-deformation stress, strain and temperature histories. Clay-free sandstones tend to seal by cataclasis and post-deformation quartz cementation. Impure sandstones tend to seal by deformation induced mixing and post-faulting pressure solution.

(3) Fault rocks in siliciclastic sediments do not experience significant quartz cementation or pressure solution unless they are buried to >90 °C.

(4) The extent of cataclasis is particularly dependent upon the effective stress and the porosity at the time of deformation. Deformation of high porosity sandstones under high effective stress conditions leads to extensive porosity collapse and grain-size reductions.

(5) Excluding sediment composition, many of the principle factors which control the final petrophysical properties of a fault rock are common to an individual field as a whole (i.e. maximum temperature, stress state at the time of deformation, etc.). Microstructural and physical property analysis of deformation features within cores can therefore be used to reduce risks associated with fault seal analysis and can provide realistic fault rock properties for robust reservoir modelling.

The authors would first of all like to thank our industrial sponsors (Agip, Amoco, Arco, British Gas, British Petroleum, Conoco, Chevron, Elf, JNOC, Mobil, Phillips, Texaco and Statoil) who provided financial support and core material for this study. We would also like to thank M. Antonellini, E. Sverdrup and D. Peacock for providing thoughtful reviews which greatly improved this manuscript. B. Clennell, B. Kidd, A. Harrison, D. Condliffe and R. Jones are thanked for conducting the physical property analyses.

References

Antonellini, M. & Aydin, A. 1994. Effect of faulting on fluid flow in porous sandstones: petrophysical properties. *American Association of Petroleum Geologists Bulletin*, **78**, 335–377.

——, —— & Pollard, D. D. 1994. Microstructure of deformation bands in porous sandstones at Arches National Park, Utah. *Journal of Structural Geology*, **16**, 941–959.

Aydin, A. 1978. Small faults formed as deformation bands in sandstone. *Pure and Applied Geophysics*, **116**, 913–942.

Berg, R. R. 1975. Capillary pressure in stratigraphic traps. *American Association of Petroleum Geologists Bulletin*, **59**, 939–956.

—— & AVERY, A. H. 1995. Sealing properties of Tertiary growth faults, Texas Gulf coast. *American Association of Petroleum Geologists Bulletin*, **79**, 375–393.

BJØRKUM, P. A. 1996. How important is pressure in causing dissolution of quartz in sandstones? *Journal of Sedimentary Research*, **66**, 147–154.

BJØRLYKKE, K. 1994. Pore-water flow and mass transfer of solids in solution in sedimentary basins. *In*: PARNELL, A. (ed.) *Geofluids: Origin, Migration and Evolution of Fluids in Sedimentary Basins*. Geological Society, London, Special Publications, **78,** 127–140.

—— & HØEG, K. 1997. Effects of burial diagenesis on stresses, compaction and fluid flow in sedimentary basins. *Marine and Petroleum Geology*, **14**, 267–276.

BLENKINSOP, T. G. 1991. Cataclasis and processes of particle size reduction. *Pure and Applied Geophysics*, **136**, 60–86.

BORRADAILE, G. J. 1981. Particulate flow and the generation of cleavage. *Tectonophysics*, **72**, 306–321.

BOUVIER, J. D., SIJPESTEIJN, K., KLEUSNER, D. F., ONYEJEKWE, C. C. & VAN DER PAL, R. C. 1989. Three-dimensional seismic interpretation and fault sealing investigations, Nun River field, Nigeria. *American Association of Petroleum Geologists Bulletin*, **73**, 1397–1414.

CECIL, C. B. & HEALD, M. T. 1971. Experimental investigation of the effects of grain-coatings on quartz over-growth. *Journal of Sedimentary Petrology*, **41**, 582–584.

CHANG, J. & YORTSOS, Y. C. 1994. Lamination during silica diagenesis – effects of clay content and Oswald rippening. *American Journal of Science*, **294,** 137–172.

DE BOER, R. D., NAGTEGAAL, P. J. C. & DUYVUS, E. M. 1977. Pressure solution experiments on quartz sand. *Geochimica et Cosmochimica Acta*, **41**, 249–256.

DEWERS, T. & ORTOLEVA, P. J. 1991. Influences of clay minerals on sandstone cementation and pressure solution. *Geology*, **19**, 1045–1048.

ELLEVSET, S. O., KNIPE, R. J., OLSEN, T. S., FISHER, Q. J. & JONES, G. 1998. Fault controlled communication in the Sleipner Vest Field, Norwegian Continental Shelf; detailed, quantitative input for reservoir simulation and well planning. *This volume*.

ENGELDER, J. T. 1974. Cataclasis and the generation of fault gouge. *Geological Society of America Bulletin*, **85**, 1515–1522.

FISHER, Q. J., KNIPE, R. J. & WORDEN, R. In press. The microstructure of deformed and undeformed sandstones from the North Sea: its implications for the origin of quartz cement. *In*: MORAD, S. & WORDEN, R. (eds) *Quartz Cement: Origin and Effects on Hydrocarbon Reservoirs*. International Association of Sedimentology, Special Publication

GAUPP, R., MATTER, A., PLATT, J., RAMSEYER, K. & WALZEBUCK, J. 1993. Diagenesis and fluid evolution of deeply buried Permian (Rotliegende) gas reservoirs, northwest Germany. *American Association of Petroleum Geologists Bulletin*, **77**, 1111–1128.

GIBSON, R. G. 1994. Fault-zone seals in siliclastic strata of the Columbus Basin, offshore Trinidad. *American Association of Petroleum Geologists Bulletin*, **78**, 1372–1385.

GILES, M., 1997. *Diagenesis: a Quantitative Perspective, Implications for Basin Modelling and Rock Property Prediction*. Kluwer, Dordrecht.

GRIGGS, D. T. 1940. Experimental flow of rocks under conditions favouring recrystallisation. *Geological Society of America Bulletin*, **51**, 1001–1022.

GU, Y. & WONG, T-F. 1994. Development of shear localisation in simulated quartz gouge: effect of cumulative slip and gouge particle size. *Pure & Applied Geophysics*, **143**, 387–423.

HANDIN, J., HAGER, R.V., FRIEDMAN, M. & FEATHER, J. N. 1963. Experimental deformation of sedimentary rocks under confining pressure: pore pressure effects. *American Association of Petroleum Geologists Bulletin*, **47**, 717–755.

HEALD, M. T. 1955. Stylolites in sandstones. *Journal of Geology*, **63**, 101–114.

—— 1959. Significance of stylolites in permeable sandstones. *Journal of Sedimentary Petrology*, **29**, 251–253.

HOUSEKNECT, D. W. 1987. Assessing the relative importance of compaction processes and cementation to reduction of porosity in sandstones. *American Association of Petroleum Geologists Bulletin*, **71**, 633–642.

ILLER, R. K. 1979 *The Chemistry of Silica*. John Wiley and Sons, UK.

JAMISON, W. R. & STEARNS, D. W. 1982. Tectonic Deformation of Wingate sandstone, Colorado National Monument. *American Association of Petroleum Geologists Bulletin*, **66**, 2584–2608.

JEV, B. I., KAARS-SIJPESTEIJN, C. H., PETERS, M. P. A. M., WATTS, N. L. & WILKIE, J. T. 1993. Akaso field, Nigeria: use of integrated 3-D seismic, fault slicing, clay smearing, and RFT pressure data on fault trapping and dynamic leakage. *American Association of Petroleum Geologists Bulletin*, **77**, 1389–1404.

KATZ, A. J. & THOMPSON, A. H. 1986. Quantitative prediction of permeability in porous rock. *Physical Reviews*, **B34**, 8179–8181.

—— & —— 1987. Prediction of rock electrical conductivity from mercury injection measurements. *Journal of Geophysical Research*, **92**, 599–607.

KNIPE, R. J. 1986. Deformation mechanism path diagrams for sediments undergoing lithification. *In*: MOORE, J. C. (ed.) *Structural Fabrics in Deep Sea Drilling Project Cores From Forearcs*. Geological Society of America, Memoirs, **166**, 151–160.

—— 1992. Faulting processes and fault seal. *In*: LARSEN, R. M., BREKKE, H., LARSEN, B. T. & TALLERAAS, E. (eds) *Structural and Tectonic Modelling and its Application to Petroleum Geology*. Norwegian Petroleum Society, Special Publications, **1**, 325–342.

—— 1997. Juxtaposition and seal diagrams to help analyse fault seals in hydrocarbon reservoirs. *American Association of Petroleum Geologists Bulletin*, **81**, 187–195.

—— & LLOYD, G. E. 1994. Microstructural analysis of faulting in quartzite, Assynt, NW Scotland: implications for fault zone evolution. *Pure & Applied Geophysics*, **143**, 229–254.

—— FISHER, Q. J., JONES, G., CLENNELL, M. R., FARMER, A. B., KIDD, B., MCALLISTER, E., PORTER, J. R. & WHITE, E. A. 1997. Fault seal analysis, successful methodologies, application and future directions. *In*: MØLLER-PEDERSON, P. & KOESTLER, A. G. (eds) *Hydrocarbon Seals — Importance for Exploration and Production*. Norwegian Petroleum Society (NPF), Special Publications, **7**, 15–37.

LINDSAY, N. G., MURPHY, F. C., WALSH, J. J. & WATTERSON, J. 1993. Outcrop studies of shale smears on fault surfaces. International Association of Sedimentologists, Special Publication, **15**, 113–123.

LOGAN, J. M. 1992. The influence of fluid flow on the mechanical behaviour of faults. *In*: TILLERSON, J. R. & WAWERSIK, W. R. (eds) *Rock Mechanics*. Balkema, Rotterdam, 141–149.

MARONE, C. & SHULTZ, C. H. 1989. Particle-size distribution and microstructures within simulated fault gouge. *Journal of Structural Geology*, **11**, 799–814.

MENÉNDEZ, B., ZHU, W. & WONG T.-F. 1996. Micromechanics of brittle faulting and cataclastic flow in Berea sandstone. *Journal of Structural Geology*, **18**, 1–16.

MORROW, C. A., SHI, L. Q. & BYERLEE, J. D. 1984. Permeability of fault gouge under confining pressure and shear stress. *Journal of Geophysical Research*, **89**, 3193–3200.

MULLIS, A. M. 1991. The role of silica precipitation kinetics in determining the rate of quartz pressure solution. *Journal of Geophysical Research*, **96**, 10 007–10 013.

OELKERS, E. H., BJØRKUM, P. A. & MURPHY, W. M., 1996, A petrographic and computational investigation of quartz cementation and porosity reduction in North Sea sandstones. *American Journal of Science*, **296**, 420–452.

OLSON R. E. & DANIEL, D. E. 1981. Measurement of the hydraulic conductivity of fine-grained soils. *In*: ZIMMIE, T. F. & RIGGS, C. O. (eds) *Permeability and Groundwater Contaminant Transport*, American Society for Testing of Materials, STP 746, 18–64.

——, GILL, J. D., WILLDEN, A. T. & NELSON, K. R. 1991. Innovations in hydraulic conductivity measurements. In: Proceedings of the Transportation Research Board 70th Annual Meeting., Washington D.C. Paper 910367.

PEACOCK, D. C. P., FISHER, Q. J., WILLEMSE, E. J. M. & AYDIN, A. 1998. The relationship between faults and pressure solution seams in carbonate rocks, and the implications for fluid flow. *This volume*.

PITTMAN, E. D. 1981. Effect of fault-related granulation on porosity and permeability of quartz sandstones, Simpson Group (Ordovician), Oklahoma. *American Association of Petroleum Geologists Bulletin*, **65**, 2381–2387.

RIMSTIDT, J. D. & BARNES H. L. 1980. The kinetics of silica–water reactions. *Geochimica et Cosmochimica Acta*, **44**, 1683–1699.

RUTTER, E. H. & HADIZADEH, J. 1991. On the influence of porosity on the low temperature brittle–ductile transition in siliciclastic rocks. *Journal of Structural Geology*, **13**, 609–614.

SAMMIS, C. G., KING, G. & BIEGEL, R. 1987. The Kinematics of Gouge Deformation. *Pure and Applied Geophysics*, **125**, 777–812.

SASSI, W., LIVERA, S. E. & CALINE, B. R. P. 1992. Reservoir compartmentation by faults in Cormorant Block IV, U.K. northern North Sea. *In*: LARSEN, R. M., BREKKE, H., LARSEN, B. T. & TALLERAAS, E. (eds) *Structural and Tectonic Modelling and its Applications to Petroleum Geology*. Norwegian Petroleum Society, Special Publications, **1**, 355–364.

SCOTT, T. E. & NIELSON, K. C. 1991. The effects of porosity on the brittle–ductile transition in sandstones. *Journal of Geophysical Research*, **96**, 405–414.

SIBSON, R. H. 1977. Fault rocks and fault mechanisms. *Journal of the Geological Society, London*, **133**, 199–213.

SMITH, D. A. 1966. Theoretical consideration of sealing and non-sealing faults. *American Association of Petroleum Geologists Bulletin*, **50**, 363–374.

—— 1980. Sealing and non-sealing faults in Louisiana Gulf Coast salt basin. *American Association of Petroleum Geologists Bulletin*, **64**, 145–172.

SULLIVAN, M. D., HASZELDINE, R. S. & FALLICK, A. E. 1990. Linear coupling of carbon and strontium isotopes in Rotliegend Sandstone, North Sea: Evidence for cross-formational fluid flow. *Geology*, **18**, 1215–1218.

SVERDRUP, E. & BJøRLYKKE, K. 1992. Small faults in sandstones from Spitsbergen and Haltenbanken. A study of diagenetic and deformational structures and their relation to fluid flow. *In*: LARSEN, R. M., BREKKE, H., LARSEN, B. T. & TALLERAAS, E. (eds) *Structural and Tectonic Modelling and its Application to Petroleum Geology*. Norwegian Petroleum Society, Special Publications, **1**, 507–518.

TADA, R. & SIEVER, R. 1989. Pressure solution during diagenesis. *Annual Reviews of Earth and Planetary Sciences*, **17**, 89–118.

THOMPSON, A. 1959. Pressure solution and porosity. *In*: IRELAND, H. A. (ed.) *Silica in Sediments*. Society of Economic Paleontologists and Mineralogists Special Publications, **7**, 92–110.

UNDERHILL, J. R. & WOODCOCK, N. H. 1987. Faulting mechanisms in high porosity sandstones; New Red Sandstone, Arran, Scotland. *In*: JONES, M. E. & PRESTON, R. M. F. (eds) *Deformation of sediments and sedimentary rocks*. Geological Society, London, Special Publications, **29**, 91–105.

WALDERHAUG, O. 1996. Kinetic modeling of quartz cementation and porosity loss in deeply buried sandstone reservoirs. *American Association of Petroleum Geologists Bulletin*, **80**, 731–745.

WATTS, N. L. 1987. Theoretical aspects of cap-rock and fault seals for single and two phase hydrocarbon columns. *Marine and Petroleum Geology*, **4**, 274–307.

WONG, T.-F. DAVID, C. & ZHU, W. 1997. The transition from brittle faulting to cataclastic flow in porous sandstones: Mechanical deformation. *Journal of Geophysical Research*, **102**, 3009–3025.

WORDEN, R. H., OXTOBY, N. H. & SMALLEY, P. C. 1998. Can oil emplacement stop quartz cementation in sandstones? *Petroleum Geoscience*, **4**, 129–138.

Yielding, G., Freeman, B. & Needham, D. T. 1997. Quantitative fault seal prediction. *American Association of Petroleum Geologists Bulletin*, **81**, 897–917.

Zhang, J., Wong, T.-F. & Davis, D. M. 1990. Micromechanics of pressure-induced grain crushing in porous rocks. *Journal of Geophysical Research*, **95**, 341–352.

Zhu, W. & Wong T.-F. 1997. The transition from brittle faulting to cataclastic flow: Permeability evolution. *Journal of Geophysical Research*, **102,** 3027–3041.

Zoback, M. D. & Byerlee, J. D. 1976 Effect of high-pressure deformation on permeability of Ottawa Sand. *American Association of Petroleum Geologists Bulletin*, **60**, 1531–1542.

Experimental investigation of molecular transport and fluid flow in unfaulted and faulted pelitic rocks

B. M. KROOSS, S. SCHLOEMER & R. EHRLICH

Institute of Petroleum and Organic Geochemistry (ICG-4), Forschungszentrum Jülich GmbH, D-52425 Jülich, Germany

Abstract: Laboratory experiments have been performed to investigate the fluid transport properties of fine-grained (pelitic) sedimentary rocks with respect to their hydrocarbon sealing efficiency. The experiments comprised the molecular transport (diffusion) of gases and the pressure-driven volume flow (Darcy flow). Diffusion coefficients of methane and nitrogen in water-saturated rock samples were measured at 90°C from an equimolar mixture of the two gases (10 MPa gas pressure). Methane had lower effective diffusion coefficients but higher steady-state molecular transport rates than nitrogen. Permeability coefficients measured with water on selected unfaulted natural samples, under controlled effective stress up to 47 MPa, ranged between 10^{-22} and 10^{-19} m^2. The permeability reduction with increasing stress mostly followed an exponential relationship. Permeabilities of faulted mudstones from compressive fault zones ranged from 2 up to 20×10^{-21} m^2. No systematic relationship between permeability and microfault frequency within the samples or distance of samples from microfault planes could be identified. Permeability measurements were also performed on macroscopically homogeneous Carboniferous shales from a compressive tectonic stress field. These samples, which showed distinct sonic velocity anisotropies, had permeability coefficients in the nanodarcy (nDarcy; 10^{-21} m^2) range. Only in one case was a significant permeability anisotropy associated with the sonic velocity anisotropy.

Fluid transport in fine-grained clastic sedimentary rocks (pelitic rocks, mudstones, shales, claystones, siltstones) plays an important role in various aspects of the evolution of hydrocarbon systems. Porosity–permeability–effective stress relationships control the compaction behaviour, overpressure build-up and dissipation during burial of shale sequences. Primary migration and petroleum expulsion involves multiphase fluid flow which is induced and influenced by the conversion of solid organic matter with concomitant formation of fluid phases and breakdown of load-bearing structural elements.

One of the economically important aspects of shales and mudstones in sedimentary basins is their function as flow barriers in hydrocarbon systems. Shales and mudstones may act as top seals, lateral seals in clay smears of fault systems or as intraformational seals inhibiting the mixing of reservoir fluids and enclosing pressure compartments. Although the importance of pelitic rocks in hydrocarbon trapping and sealing is generally appreciated, the sealing efficiency and the potential rates of hydrocarbon leakage are up to now only barely quantified. Pressure driven volume flow is commonly attributed to play the dominant role among the conceivable transport mechanisms. Molecular transport (diffusion) has been advocated as a rate-controlling mechanism in primary migration (Stainforth & Reinders 1990) but its main role in petroleum migration appears to be restricted to the redistribution and dismigration (i.e. loss from reservoir accumulations) of hydrocarbon and non-hydrocarbon gases in sedimentary sequences.

While the fluid transport properties of reservoir rocks, due to the immediate economic relevance for production efficiency, are generally well established, information on fluid transport in mudrocks is limited (e.g. Brace 1980; Neuzil 1986, 1994; Katsube et al. 1991; Katsube & Williamson 1994; Faulkner & Rutter, *this volume*). Permeability data used in basin modelling are partly speculative or deduced from regional studies on overpressure evolution (Bredehoeft & Hanshaw 1968; Deming 1994). Although one must be aware that the permeability of sedimentary rocks is scale-dependent, laboratory measurements, although not immediately transferable to geologic systems, provide an important contribution to the understanding of fluid transport processes in the lithosphere.

This paper presents selected results of a systematic investigation of different types of pelitic rocks with the aim of providing an improved database for fluid flow modelling in the context of numerical basin analysis. The experimental work comprised both (single phase) pressure-driven volume flow (Darcy flow) and molecular transport (diffusion) of methane and nitrogen in water-saturated sedimentary rocks at elevated pressures and temperatures.

KROOSS, B. M., SCHLOEMER, S. & EHRLICH, R. 1998. Experimental investigation of molecular transport and fluid flow in unfaulted and faulted pelitic rocks. *In*: JONES, G., FISHER, Q. J. & KNIPE, R. J. (eds) *Faulting, Fault Sealing and Fluid Flow in Hydrocarbon Reservoirs*. Geological Society, London, Special Publications, **147**, 135–146.

Experimental methods

Experimental methods and equipment developed in our laboratory during recent years have been used to study fluid transport in fine-grained lithologies under simulated subsurface conditions. Measurements were performed on plugs with 28.5 mm diameter and a maximum length of 30 mm at elevated fluid pressures and temperatures. Using triaxial flow cells (Fig. 1) experiments can be carried out under controlled axial and confining stress. Confining pressures can range up to 50 MPa and axial loads up to 100 kN (corresponding to a total axial stress of 157 MPa) can be applied. Transport measurements (diffusion and volume flow) on sedimentary rocks are mostly performed at temperatures between room temperature and 150°C while maximum temperatures up to of 320°C can be reached to perform pyrolysis experiments under controlled stress conditions (Hanebeck 1995).

Using micropumps (Shimadzu LC6), in a constant pressure mode and a liquid phase (water or *n*-hexane), permeability coefficients can be measured down to the sub-nanodarcy level ($<10^{-21}$ m^2) by means of a constant flow technique. The fluid flow across the sample is monitored and recorded continuously via the pumping rate on the upstream side and by a calibrated burette on the downstream side. Once a steady-state flow has been established both flow rates are used for the assessment of the permeability coefficients. The pressure differences used for these permeability tests range between 3 and 10 MPa, corresponding to pressure gradients of 200–1000 MPa m^{-1}. Within this range of experimental conditions no measurable deviations from Darcy's law have been observed, i.e. pressure gradients and corresponding fluxes show a linear correlation. Taking into account the maximum errors for the assessment of temperature (affecting water viscosity), sample thickness and cross section, fluid volume measurements (upstream and downstream) and time, the maximum experimental error of the permeability values is estimated to range between 4 and 7%. Due to the natural heterogeneity of sedimentary rocks even on the cm scale, differences of a factor of two or more in the measured permeability coefficients of samples from the same core are common. During the individual experiments the steady-state fluid flow, measured over different time intervals, does not usually vary by more than 10%.

Diffusion measurements of hydrocarbon and non-hydrocarbon gases are performed with water-saturated rock samples in the absence of a pressure gradient. The source phase consists of a pure gas or a gas mixture brought into contact with one side of the rock sample at the start of the experiment. The diffusive flux is determined by exchanging, at regular intervals, the aqueous phase at the opposite side of the sample and

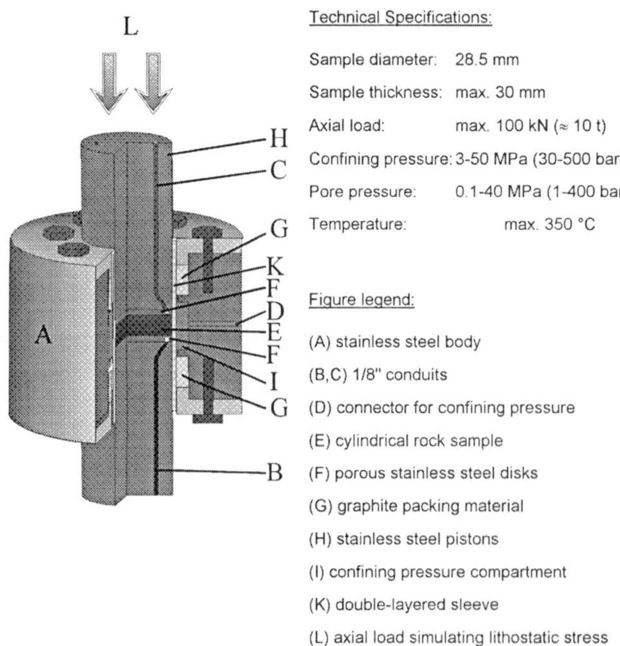

Fig. 1. Schematic view of the triaxial flow cell used for fluid flow studies.

analysing the gas content. These experiments can be conducted at temperatures up to 200°C. The evaluation of the experimental data to assess the effective diffusion coefficients followed the procedures described by Krooss & Schaefer (1987) and Krooss & Leythaeuser (1988).

Mercury porosimetry (up to 350 MPa) and specific surface area measurements (BET method) are used as standard methods for the petrophysical characterization of the rock samples. Furthermore, qualitative examination of the rocks has been performed by microscopic inspection of sample thin sections, SEM and on selected plugs by X-ray microtomography.

Samples

Fluid flow experiments have been performed on several sets of mudstones from different regions and geological ages. One of the main constraints on the experimental approach is the availability of coherent samples, preferably from drill cores with sufficient mechanical stability to withstand the sawing and grinding procedures necessary to produce cylindrical plugs of the required dimensions. This imposes severe limitations, particularly with respect to samples from fault zones. Investigations on rocks with open fractures which disintegrate during sample preparation are not possible on the laboratory scale. But even *in situ* measurements, which might be envisaged as a potential alternative, bear these limitations because of unavoidable formation damage during drilling or excavation works. An overview of the sample origins and the scope of the investigations of different studies is given in Table 1. Fluid flow measurements were usually performed perpendicular to the bedding plane, if not stated otherwise.

Detailed petrophysical and mineralogical information for the sample sets from Haltenbanken and the North German Basin are reported elsewhere (Schlömer & Krooss 1997). The mudstones from the Haltenbanken area had TOC contents between 0.3 and 5.5% and were partly laminated. The organic matter contents of the mostly red coloured homogeneous claystones from North Germany were very low (TOC <0.05%). The Haltenbanken shales had slightly higher quartz contents (average: 28% quartz) and, correspondingly, lower phyllosilicate contents (63%) than the Palaeozoic red claystones from North Germany (average: 21% quartz, 68 % phyllosilicates). Small percentages (2–9%) of mixed-layered clay minerals (smectite/illite with max 15% smectite) were found in some of the Mesozoic Haltenbanken samples. Porosities were very low for both sample sets, ranging mostly below 3 % and in many instances even below 1%.

The massively faulted shale samples investigated in this study originate from four wells in the Ruhr area that intersect compressive fault zones. The cores were taken at depths ranging between 700 and 1500 m and the selection was necessarily strongly influenced by criteria like sample integrity and coherence. Generally, it was attempted to select specimens with different degrees of microfaulting on the size scale of the plugs used in the experiments (diameter *c.* 3 cm, thickness *c.* 1.5 cm). The samples had matrix contents (grain size < 30 μm) between 49 and 59% and 4–6 % feldspar. They had average grain sizes of 60 μm and showed a fine lamination (mm scale) with grain sizes around 20 μm and 100–130 μm in the fine- and the coarse-grained laminae, respectively.

The samples from S. Limburg had been taken in 1972 from two coal mines at depths between

Table 1. *Overview of samples used in fluid flow experiments and scope of investigations*

Sample set	Depth	Scope of study	Types of measurements
Haltenbanken (Norway) Mesozoic mudstones (shale/siltstones)	3900–4700 m	Sealing efficiency of unfaulted pelitic rocks	Diffusion and permeability
North German Basin Palaeozoic red claystones	1500–5000 m	Sealing efficiency of unfaulted pelitic rocks	Diffusion and permeability
Ruhr area (Germany) Palaeozoic shales from compressive fault zones	700–1500 m	Transport in faulted pelitic rocks	Permeability
South Limburg mining area (The Netherlands) Dark Palaeozoic shales from a tectonic stress field	250–550 m	Transport in pelitic rocks from compressive tectonic stress fields	Permeability

250 and 550 m. Their matrix content ranged between 59 and 64% with 2–3% feldspar. Their lamination was less distinct than in the Ruhr area samples. Average grain sizes were around 20 μm, varying between 5 μm in the fine and 150 μm in the coarse laminae.

Selected results

Fluid transport in unfaulted pelitic rocks

Transport processes in unfaulted pelitic rocks were investigated to quantify the hydrocarbon sealing efficiency of fine-grained sedimentary rocks. The study comprised both diffusive transport and volume flow measurements in two sets of Mesozoic and Palaeozoic mudstones. The pore structure of the selected samples was characterized by mercury porosimetry. Detailed results of this research are published elsewhere (Schlömer & Krooss 1997).

Diffusive transport of methane and nitrogen. The potential role of diffusive transport of hydrocarbons, in particular with respect to the stability of natural gas reservoirs on the geologic time scale has been a matter of recurrent discussion and dispute for more than 40 years (Krooss & Leythaeuser 1996). While the problems and pitfalls of the estimation procedure for diffusive losses have been largely recognized (Krooss *et al.* 1992a,b; Nelson & Simmons 1992) the main source of controversy appears to reside in the assumptions on seal thickness and porosity, and reservoir size and geometry (productive area, reservoir content, thickness of pay zone; cf. Nelson & Simmons 1995, 1997; Krooss & Leythaeuser 1997).

Extensive experimental measurements on the diffusive transport of light hydrocarbons in various types of sedimentary rocks were carried out by our group during the past decade. Recently, molecular nitrogen, a major component in natural gas accumulations in some areas of the North German Basin, was included in the investigations. Following research on the sources of molecular nitrogen in the lithosphere (Krooss *et al.* 1995; Littke *et al.* 1995), these experiments were deemed necessary to provide a basis for the assessment of the role of transport-related fractionation processes in the formation of nitrogen-rich gas accumulations suggested by some workers (Boigk & Stahl 1970; Boigk *et al.* 1976; Müller *et al.* 1976).

Generally, the effective diffusion coefficients of molecular nitrogen (N_2) in water-saturated

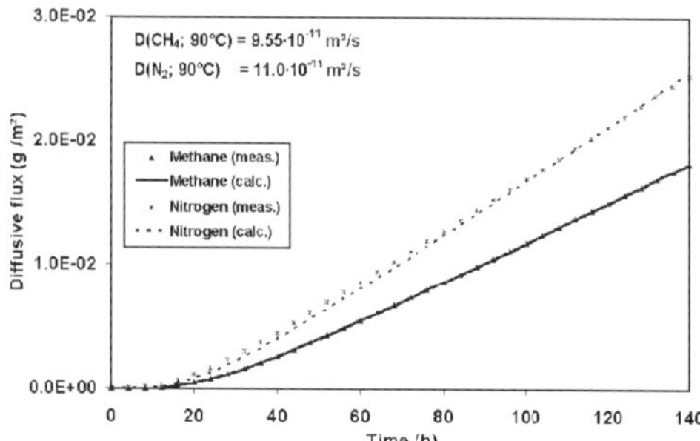

Fig. 2. Cumulative diffusion curves for methane and nitrogen in a shale at 90°C in mass units. The source phase consisted of a 50/50% (by volume) CH_4/N_2 mixture.

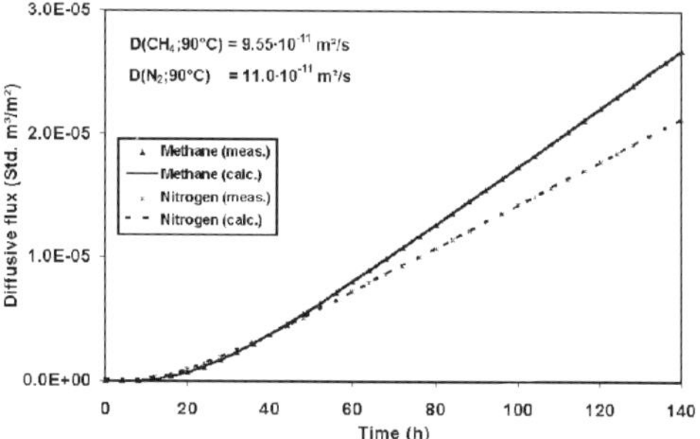

Fig. 3. Cumulative diffusion curves for methane and nitrogen in a shale at 90°C in volume units (Std. m^3). The source phase consisted of a 50/50% (by volume) CH_4/N_2 mixture.

sedimentary rocks measured at 90°C were higher than those of methane. This result conforms with diffusion coefficients of these two gases in pure water. According to experimental data (Tammann & Jessen 1929; Gertz & Löschcke 1954; Witherspoon & Saraf 1965; Bonoli & Witherspoon 1969; Sahores & Witherspoon 1970) the diffusion coefficient of N_2 in water is about 1.5 times higher than that of methane at temperatures between 21 and 37°C. The Wilke–Chang estimation method of binary liquid phase diffusion coefficients (Reid *et al.* 1987) predicts nitrogen diffusion coefficients in water to be higher by a factor 2 than for methane. The experimental diffusion curves for methane and nitrogen through a water-saturated rock sample at 90°C are shown in Fig. 2. This experiment was performed using a 50/50 vol.% methane/nitrogen gas mixture as source phase, at a fluid pressure of 10 MPa (100 bar) and an effective stress of 40 MPa. The curves represent the cumulative mass fluxes (in g m^{-2}) of the two gases per unit cross sectional area through a rock slice as a function of time. The evaluation of these cumulative curves yielded diffusion coefficients of 11.0×10^{-11} and 9.55×10^{-11} $m^2 s^{-1}$ for N_2 and CH_4, respectively, i.e. differing by a factor of 1.16. Evidently, the total *mass* of nitrogen diffused through the sample until the end of the experiment is higher than the corresponding value for methane. In contrast, conversion of the mass flux units to molar flux units ($mol\, m^{-2}$ or Std $m^3 m^{-2}$) shows that the molar flux of methane (molar mass: $16\, g\, mol^{-1}$) is higher than for nitrogen (molar mass: 28 g mol). This is demonstrated in Fig. 3 where the cumulative diffusion curves are plotted in units of Std. m^3. Consequently, during steady-state diffusion, the molecular transport efficiency of methane (i.e. the number of molecules passing through a unit cross-section under otherwise identical conditions), despite its lower diffusion coefficient, is larger than that of nitrogen. As natural gas reservoir contents are usually expressed in standard volumetric units, the molecular transport efficiency will be the appropriate parameter for the analysis of molecular diffusion effects on the composition of natural gases.

Diffusion experiments under lithostatic stress. No measurements of hydrocarbon diffusion under controlled effective stress conditions have been reported so far in the literature. Recent experimental developments (Krooss *et al.* 1993; Hanebeck 1995) have provided the possibility of investigating the effective stress-dependence of molecular transport in water-saturated sedimentary rocks. Selected results of these measurements for methane and nitrogen are shown in Fig. 4 and Fig. 5. For the Mesozoic silty shale used in this experiment, an increase in effective stress from 10 to 40 MPa results in a drop in the effective diffusion coefficient of methane by 38% from 7.5×10^{-11} to 4.6×10^{-11} $m^2 s^{-1}$. Under the same conditions the effective diffusion coefficient for nitrogen is reduced by 31% (from 14×10^{-11} to $9.6 \times 10^{-11} m^2 s^{-1}$). The steady-state diffusive fluxes of methane and nitrogen decrease by 70 and 60%, respectively.

The results indicate a two-fold control of the effective stress on the molecular transport process. The decrease in effective diffusion coefficients can be attributed in part to a corresponding increase in tortuosity, which reduces the mobility of the gas molecules in the water-saturated pore space. The decrease in steady-state diffusive flux exceeds this effect, indicating a constriction of the pore throats with concomitant reduction of the effective cross-sectional area of the flow paths.

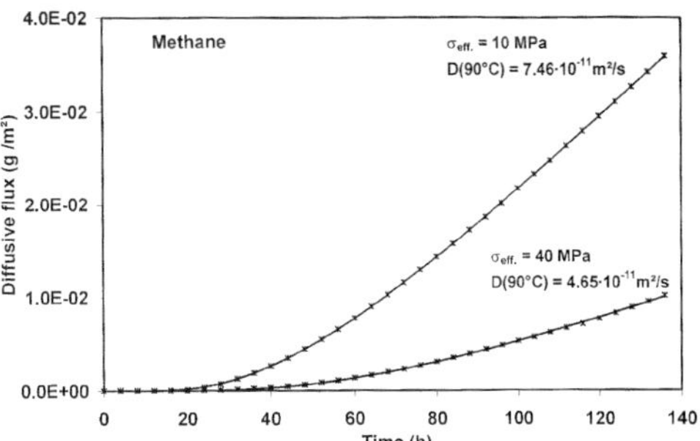

Fig. 4. Cumulative diffusion curves for methane at 90°C and effective stresses of 10 and 40 MPa.

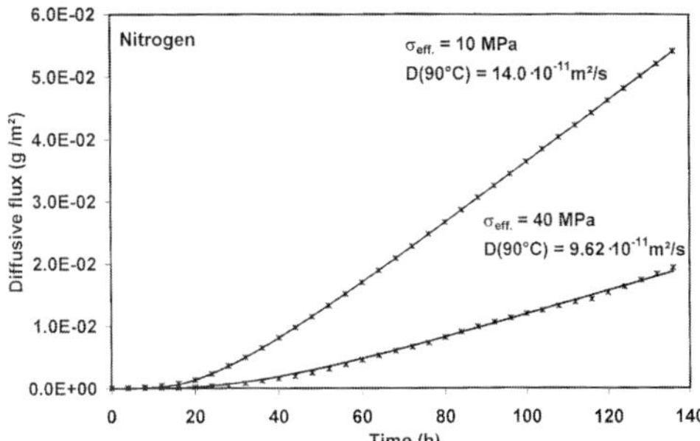

Fig. 5. Cumulative diffusion curves for nitrogen at 90°C and effective stresses of 10 and 40 MPa.

Permeability measurements under controlled effective stress. A comprehensive characterization of the sealing efficiency of pelitic rocks from Haltenbanken (Mesozoic shale/siltstones) and North Germany (Palaeozoic red claystones), comprised permeability measurements which were performed in a simple flow cell without application of controlled axial stress (Schlömer & Krooss 1997). These measurements were used as screening experiments to explore the range and variability of permeability coefficients of fine-grained sedimentary rocks. The permeability ranges found for the two sample sets are shown in Fig. 6 together with permeability data from faulted rocks and rocks from tectonic stress fields (see below).

Following the screening experiments, permeability measurements at controlled effective stresses up to 47 MPa have been performed on selected mudrock samples. These measurements were conducted at a temperature of 90°C with intermittent diffusion experiments at different stress levels. The results of permeability measurements on four different pelitic rocks are shown in Fig. 7. Although the present database is still too small for general conclusions, two different

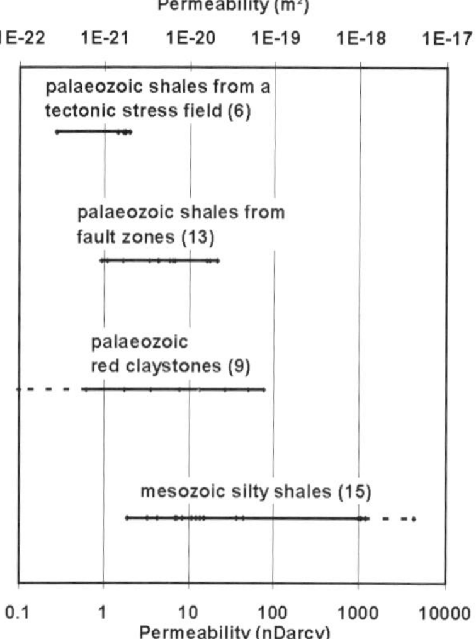

Fig. 6. Ranges of permeabilities of pelitic rocks investigated in this study (figures in brackets denote number of samples measured).

permeability-stress relationships can be distinguished: an exponential decrease (approx. linear trends in the semi-logarithmic plot) in permeability with effective stress and a power law (curved line in the semi-logarithmic plot for cemented silty claystone). To date, only very few permeability measurements on pelitic rocks have been performed under lithostatic stress conditions. It is interesting to note that our results correspond to the findings of Katsube *et al.* (1991) who report permeability measurements on two shale samples from the Scotian Shelf performed at effective pressures from 2.5 to 60 MPa. The measured permeabilities range from $1.6 \times 10^{-20} m^2$ (16 nDarcy) down to $1.3 \times 10^{-22} m^2$ (0.13 nDarcy). Katsube et al. (1991) found that one of their datasets can be described by an exponential permeability–effective stress relationship of the type:

$$k = k_0 \exp(-\alpha \Delta P)$$

with $k_0 = 1.18 \times 10^{-20} m^2$ and $\alpha = 0.074$ and where ΔP denotes the effective stress (in MPa). Their second dataset shows a better fit with a power law curve:

$$k = k_0 (\Delta P / P_0)^{-\alpha}$$

with $k_0 = 1.3 \times 10^{-19} m^2$ and $\alpha = 0.92$ and P_0 is atmospheric pressure. In a more recent study, Katsube & Williamson (1994) quoting Katsube & Coyner (1994), suggest a representation of the permeability–effective stress relationship by a system of three exponential relationships corresponding to:

(i) restoration of pores widened due to stress release of rock samples brought to the surface;
(ii) restoration of pores widened due to compaction-induced overpressure, and
(iii) reduction of intrinsic pore-sizes due to compaction.

Similarly, Morrow et al. (1984) find indications of varying relationships of permeability to effective stress at high, as compared to low, stress levels. Further investigations will have to show how the observed variations of permeability with effective stress relate to structural and petrophysical features of shales.

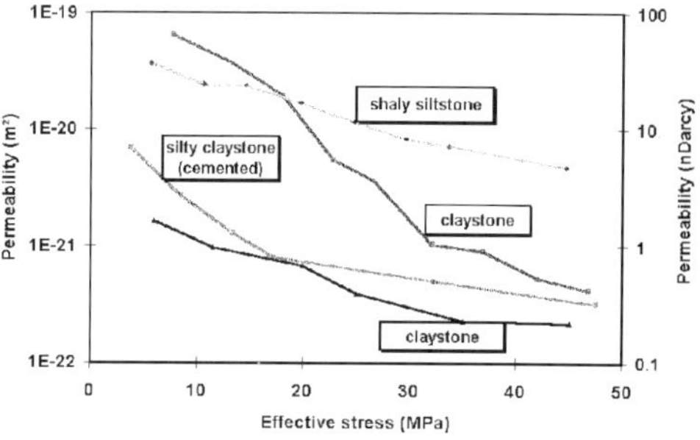

Fig. 7. Stress–permeability relationships for unfaulted pelitic rocks.

Fluid transport in faulted pelitic rocks

Fluid transport in fault zones is an important issue in the evaluation of hydrocarbon reservoir evolution. Fault zones and deformation bands in reservoir sandstones are known to impede the fluid exchange in hydrocarbon reservoirs. Antonellini & Aydin (1994, 1995) conducted field studies using a micropermeameter to investigate, on a statistical basis, the influence of these structures on fluid transport in sandstones.

The role of mudstones with respect to hydrocarbon migration is seen predominantly as hydrocarbon seals in stratigraphic traps or in smear gouges of fault systems (Smith 1980; Sibson 1995). The hydrocarbon sealing efficiency of pelitic rocks will depend to a considerable extent on the behaviour of the formation during faulting and when subjected to tectonic stress.

The experimental approach to fluid transport in faulted pelitic rocks meets some inherent technical limitations related to the low permeability values and the scale difference between laboratory and natural systems. Permeability measurements on pelitic rocks are very time-consuming, so that it is usually not possible to produce a database sufficiently large for statistical evaluation. Laboratory experiments can provide petrophysical information and flow parameters for specimens of cm size, whereas fault systems in geologic systems represent structural units in the range of tens of metres up to kilometres. *In situ* permeability measurements, both in unfaulted and faulted tight rock sequences, are technically very demanding, and have only been performed to a very limited extent (Neuzil 1986).

The experimental measurements on pelitic rocks from fault zones performed during this study were aimed at exploring the potential and the limitations of laboratory methods in this field. Due to the earlier-mentioned requirements with respect to the mechanical stability of rock samples to be used, the sample selection is necessarily biased. This fact must be kept in mind when interpreting the results. With respect to the experimental efforts associated with these measurements, the sample selection was undertaken with particular care.

The experimental investigation of fluid transport in pelitic rocks affected by tectonic events was carried out in a dual way referred to as the 'microscopic' and the 'macroscopic' approach, respectively. In the microscopic approach, permeability measurements were conducted on coherent plugs from drill cores transsecting massively faulted shale intervals with microfault structures on the cm–dm scale. The macroscopic approach involved permeability tests and petrophysical characterization of essentially homogeneous mudrocks from a tectonic stress field.

The microscopic approach. The aim of this investigation was to assess potential permeability differences in microfaulted cores in relation to the intensity and frequency of disturbances in the rock fabric (thrusting, cleats, fractures etc.) and as a function of distance to microfractures.

In order to address the former issue, an attempt was made to relate the permeabilities of laboratory-size (30 mm diameter) samples from a severely faulted (coherent) rock volume to the fault intensity and frequency and compare them to permeabilities of undisturbed reference samples of the same lithotype. This work comprised:

(i) preparation of samples from strongly disturbed core intervals and adjacent, presumably undisturbed intervals of the same lithological composition.
(ii) preparation of samples with varying intensities and frequencies of microfaults.

The sample selection was dictated, to a considerable extent, by the behaviour of the rock material during sample preparation. The assessment and judgement of microfault frequency and intensity relied essentially on visual inspection and must be considered to be qualitative rather than quantitative. X-ray microtomography measurements were perfomed on selected samples mainly to test the applicability of this new technique for characterization of microfaults in shales.

A deconvoluted sketch of the surface of a faulted core piece is shown in Fig. 8. The positions of the sample plugs used in permeability tests are marked together with the corresponding permeabilities ranging from 4 to 18 nD. Permeability values for the other plugs shown in this diagram could not be measured due to mechanical damage during sample preparation. For two of the samples permeability was measured perpendicular to, and for the other one (denoted as "12E2aIIIsrS") parallel to the bedding plane. This latter sample, quite unexpectedly, had the lowest permeability of all three samples from this core. This observation reflects the general finding that neither the samples shown in Fig. 8 nor those from other cores showed a systematic dependence of permeability on fault frequency, intensity and orientation.

Sample plugs from the largely undisturbed shale matrix were taken at different distances from microfaults to examine the effects of the microfaults on the permeability of the adjacent matrix. The deconvoluted sketch of the surface of a core piece with a small compressive fault is shown in Fig. 9. Permeability measurements were performed on three plugs from a small, laminated clay/silt interval upthrust by about 4 cm. Particular care was taken to prepare plugs from the same lithologic layer and to drill at identical angles with respect to the lamination. Secondary mineralization was found preferentially in the immediate vicinity of the fault plane and decreased slightly with increasing distance from the fault plane. The mineralization occurs mainly in the clay/siltstone layer and only to a lesser extent in the adjacent claystone. The permeabilities of all three plugs are very low, ranging between 1.4 and 3.5 nD, and show no systematic trend with distance from the microfault. Although the minimum and maximum values differ by a factor of two, they reflect, according to our experience with unfaulted shales, common fluctuations in the lithological composition rather than a fault-related effect. The unexpectedly low permeability coefficients of the clay/siltstone layer are attributed to secondary mineralization.

The generally low permeability coefficients observed in this study suggest that any permeability increases associated with faulting in

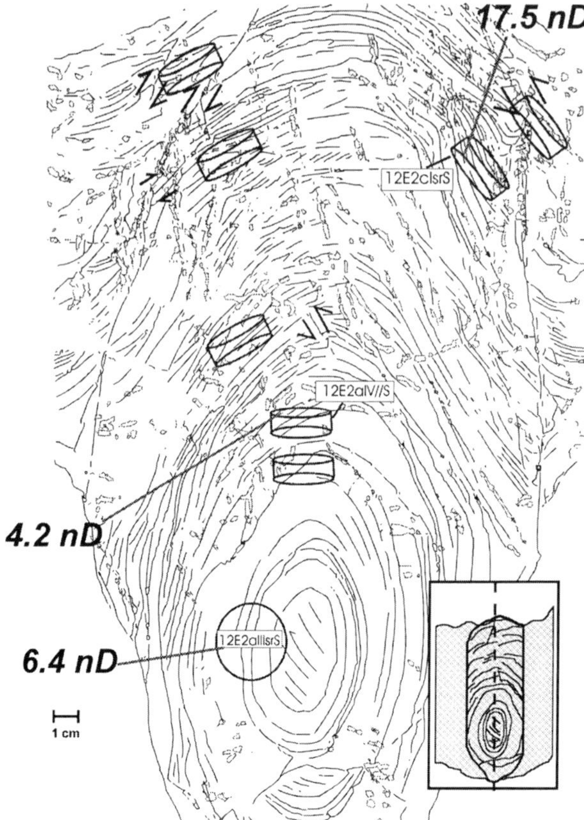

Fig. 8. Surface peel of a faulted core piece with approximate positions of sample plugs used to investigate the effects of microfault frequency on shale permeability. Measured permeability values (in nDarcy = 10^{-21} m^2) are indicated for the individual samples.

mudstones are rapidly reduced and eliminated by local secondary mineralization processes and/or ductile deformation of the shales.

Influence of tectonic stress fields on the permeability of pelitic rocks – the macroscopic approach. Acoustic anisotropies in sedimentary rocks can be produced by tectonic stress fields (Dahlen 1972). Anisotropies resulting from fractures and combinations of fractures and tectonic stress fields in mylonites and crystalline rocks were investigated by Anderson *et al.* (1974), Siegesmund *et al.* (1990) and Zang *et al.* (1989). In order to examine the influence of tectonic stress fields on the permeability of pelitic rocks, samples were selected from a set of approximately 500 drill cores from Upper Carboniferous layers of the South Limburg mining area in the Netherlands.

The geological evolution of this area was characterized by a compressive regime which has induced acoustic anisotropies in the horizontally layered Carboniferous shales. The direction of maximum velocity, oriented predominantly NW–SE, is interpreted as the direction of principal effective stress. The samples were selected according to homogeneity (by visual inspection) and the extent of the velocity anisotropy recorded after coring in 1972. Sound velocity measurements were repeated and showed that the anisotropies still persisted in some of the samples during storage whereas in others they were reduced. Four samples showing the highest degree of acoustic anisotropy, in combination with a largely homogenous structure and texture and without visible cracks or fractures, were used in the permeability tests.

For the sample with the strongest acoustic anisotropy (>4500 m s^{-1} in NNW–SSE direction, <3500 m s^{-1} in ENE–WSW direction) a substantial permeability anisotropy was found with 0.27 nD in the high velocity direction and

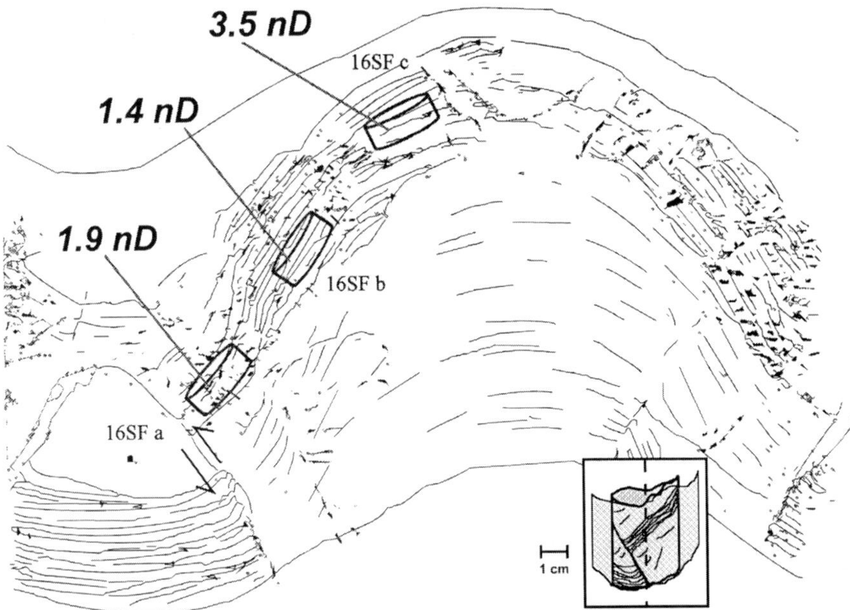

Fig. 9. Surface peel of a faulted core with approximate positions of sample plugs used to investigate the effect of distance from a micro-thrust fault on shale permeability. Measured permeability values (in nDarcy = 10^{-21} m^2) are indicated for the individual samples.

c. 2 nD in the low velocity direction (Fig. 10). The mercury capillary pressure curves for this sample measured on two plugs of 1 cm diameter oriented along the directions of maximum and minimum sonic velocities, also indicated a distinct anisotropy (Fig. 11). Additional measurements were performed on two other rock samples (403 and 405) from the same sequence exhibiting smaller acoustic anisotropies. The results listed in Table 2 show no permeability anisotropy in these cases. This also holds for the mercury porosimetry measurements.

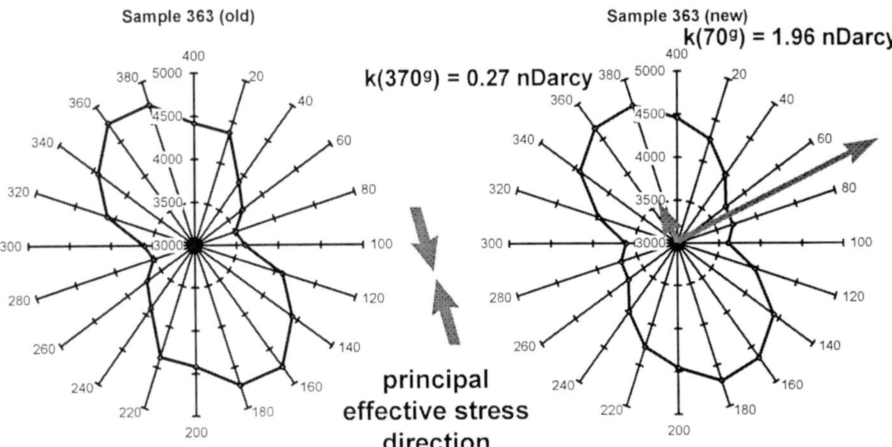

Fig. 10. Sonic velocity distribution for a shale sample from a tectonic stress field. Results of measurement after sampling in 1972 (left) and before permeability measurement in 1996 (right). The permeability coefficients (k) show a distinct anisotropy associated with the velocity anisotropy.

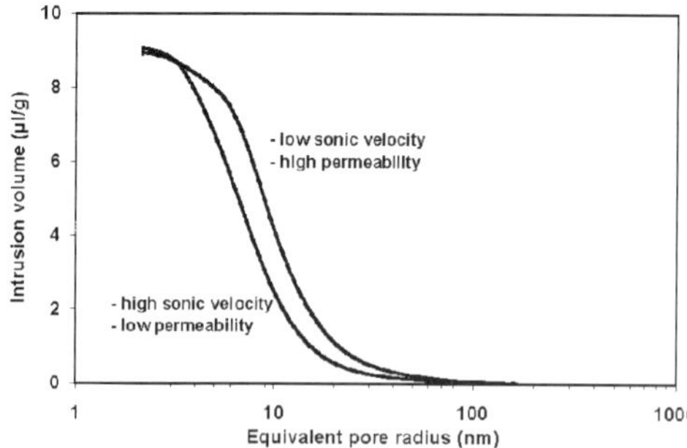

Fig. 11. Capillary pressure curves from mercury porosity on two orientated plugs from a core sample of the South Limburg mining area exhibiting strong sonic velocity anisotropy.

Table 2. *Seismic velocity data and permeability coefficients for mudrock samples from a tectonic stress field*

Sample	Sonic velocity (m/s)	Azimuth (Gon)	Permeability (nDarcy)
363 (slow)	3611.1	100	1.95
363 (fast)	4688.9	380	0.27
403 (slow)	4020.4	100	1.46
403 (fast)	4641.5	20	1.67
405 (slow)	3949.2	60	1.74
405 (fast)	4599.1	200	1.79

1 nDarcy = 10^{-21} m^2.

Conclusions

Reliable information on transport parameters in pelitic rocks is one prerequisite for the understanding and quantification of fluid flow in geologic systems. The scope of potential applications ranges from petroleum-related issues such as sealing efficiency, hydrocarbon leakage and pressure evolution to waste disposal and pollutant transport problems.

Despite inherent limitations with respect to sample size and properties, laboratory experiments represent an indispensable contribution to ongoing research efforts in this field and provide the basic data for subsequent upscaling procedures. Due to the technical problems associated with fluid transport measurements on tight lithologies, the number of experimental data will always be limited. In consequence, this approach requires a careful sample selection and measuring strategy.

Permeability measurements on consolidated mudrock samples yielded permeability coefficients ranging from the sub-nanodarcy up to the microdarcy range. It could be demonstrated that both pressure-driven volume flow (Darcy flow) and molecular transport in pelitic rocks is strongly dependent on effective stress. Fluid transport properties for these lithologies should therefore be measured under controlled effective stress conditions and at different effective stress levels.

The permeability measurements on micro-faulted pelitic rocks from compressive fault zones have delineated the limits of experimental work on this issue. Permeability values obtained for these samples were comparatively low, with values ranging between 2 and 20 nD, indicating a rapid elimination of any fault-induced permeability increases by mineralization or ductile deformation. Investigation of shale samples with sonic velocity anisotropies originating from a tectonic stress field showed a distinct associated permeability anisotropy only in one case.

The results presented in this contribution illustrate how, with a relatively small number of experiments, some key questions of fluid transport in pelitic rocks have been addressed. The results have immediate implications for the numerical modelling of molecular transport in geologic systems. Future studies will have to substantiate these results and attempt to establish relationships on the basis of the growing dataset. The success of this approach will also rely on the systematic integration of ancillary petrophysical, mineralogical, geochemical and rock-mechanical information.

This work was performed as part of the German–Norwegian Geoscientific Co-operation (Phase II). The authors acknowledge funding by Statoil, Deutsche Wissenschaftliche Gesellschaft für Erdöl, Erdgas und Kohle e.V. (DGMK), and the German Ministry of Education, Science, Research and Technology (BMBF, project No. ET6906B 'Hydrocarbon Movement and Trapping'). We thank the reviewers B. Clennell and D. Faulkner for valuable comments and suggestions which helped to improve the original manuscript.

References

ANDERSON, D. L., MINSTER, B. & COLE, D. 1974. The effect of oriented cracks on seismic velocities. *Journal of Geophysical Research*, **79**, 4011–4015.

ANTONELLINI, M. & AYDIN, A. 1994. Effect of faulting on fluid flow in porous sandstones: Petrophysical properties. *American Association of Petroleum Geologists Bulletin*, **78**, 355–377.

—— & —— 1995. Effect of faulting on fluid flow in porous sandstones: Geometry and spatial distribution. *American Association of Petroleum Geologists Bulletin*, **79**, 642–671.

Boigk, H. & Stahl, W. 1970. Zum Problem der Entstehung nordwestdeutscher Erdgaslagerstätten [On the problem of the formation on North-West German natural gas reservoirs]. *Erdöl und Kohle-Erdgas-Petrochemie*, **23**, 325–333 [in German].

——, Hagemann, H. W., Stahl, W. & Wollanke, G. 1976. Isotopenphysikalische Untersuchungen zur Herkunft und Migration des Stickstoffs nordwestdeutscher Erdgase aus Oberkarbon und Rotliegend [Isotope–physical investigations on the origin and migration of nitrogen in northwest German natural gases from upper Carboniferous and Rotliegend]. *Erdöl und Kohle-Erdgas-Petrochemie*, **29**, 103–112 [in German].

Bonoli, L. & Witherspoon, P. A. 1969. Diffusion of paraffin, cycloparaffin and aromatic hydrocarbons in water and some effects of salt concentration. *In*: Schenck, P. A. (ed.) *Advances in Organic Geochemistry Proc. 4th Int. Meet.*, 1968, Pergamon, Oxford, 373–384.

Brace, W. F. 1980. Permeability of crystalline and argillaceous rocks. *International Journal for Rock Mechanics and Mining Sciences & Geomechanics Abstracts*, **17**, 241–251.

Bredehoeft, J. D. & Hanshaw, B. B. 1968. On the maintenance of anomalous fluid pressures: I. Thick sedimentary sequences. *Geological Society of America Bulletin*, **79**, 1097–1106.

Dahlen, F. A. 1972. Elastic velocity anisotropy in the presence of an anisotropic initial stress. *Bulletin of the Seismological Society of America*, **62**, 1183–1193.

Deming, D. 1994. Factors necessary to define a pressure seal. *American Association of Petroleum Geologists Bulletin*, **78**, 1005–1009.

Faulkner D. R. & Rutter E. H. 1998. The permeability of clay-bearing fault gouges. *This volume*.

Gertz, J. H. & Löschcke, H. H. 1954. Bestimmung der Diffusions-Koeffizienten von H_2, O_2, N_2, und He in Wasser und Blutserum bei konstant gehaltener Konvektion. [Determination of the diffusion coefficients of H_2, O_2, N_2, and He in water and blood serum at constant convection]. *Zeitschrift für Naturforschung*, **9b**, 1–9 [in German].

Hanebeck, D. 1995. Experimentelle Simulation und Untersuchung der Genese und Expulsion von Erdölen aus Muttergesteinen. [Experimental simulation and investigation of the generation and expulsion of petroleum from source rocks]. PhD. Thesis, Rheinisch-Westfälische Technische Hochschule Aachen. *Berichte des Forschungszentrums Jülich* [in German].

Katsube, T. J. & Coyner, K. 1994. Determination of permeability(k)–compaction relationship from interpretation of k-stress data for shales from Eastern and Northern Canada. *Current research/(Geological Survey of Canada); 1994-D, Paper 94–1D*, 169–177.

—— & Williamson, M.A. 1994. Effects of diagenesis on shale nano-pore structure and implications for sealing capacity. *Clay Minerals*, **29**, 451–461.

——, Mudford, B. S. & Best, M. E. 1991. Petrophysical characteristics of shales from the Scotian shelf. *Geophysics*, **56**, 1681–1689.

Krooss, B. M. & Schaefer, R. G. 1987. Experimental measurements of diffusion parameters of light hydrocarbons in water-saturated sedimentary rocks: I. A new experimental procedure. *Organic Geochemistry*, **11**, 193–199.

—— & Leythaeuser, D. 1988. Experimental measurements of diffusion parameters of light hydrocarbons in water-saturated sedimentary rocks: II. Results and geochemical significance. *Organic Geochemistry*, **12**, 91–108.

—— & —— 1996. Molecular diffusion of light hydrocarbons in sedimentary rocks and its role in the migration and dissipation of natural gas. *In*: Schumacher, D. & Abrams, M. A. (eds) *Hydrocarbon Migration and Its Near-Surface Expression. American Association of Petroleum Geologists Memoir*, **66**, 173–183.

—— & —— 1997. Diffusion of methane and ethane through the reservoir cap rock: Implications for the timing and duration of catagenesis: Discussion. *American Association of Petroleum Geologists Bulletin*, **81**, 155–161.

——, Leythaeuser, D. & Schaefer, R. G. 1992a. The quantification of diffusive hydrocarbon losses through cap rocks of natural gas reservoirs – a reevaluation. *American Association of Petroleum Geologists Bulletin*, **76**, 403–406.

——, Leythaeuser, D. & Schaefer, R. G. 1992b. The quantification of diffusive hydrocarbon losses through cap rocks of natural gas reservoirs – a reevaluation: Reply. *American Association of Petroleum Geologists Bulletin*, **76**, 1842–1846.

——, Hanebeck, D. & Leythaeuser, D. 1993. Experimental measurement of the molecular migration of light hydrocarbons in source rocks at elevated temperatures. *In*: Doré, A. G., Augustson, J. H., Hermanrud, C., Stewart, D. J. & Sylta, Ø. (eds) *Basin Modelling: Advances and Applications*, NPF Special Publications **3**, 277–291.

——, Littke, R., Müller, B., Frielingsdorf, J., Schwochau, K. & Idiz, E. F. 1995. Generation of nitrogen and methane from sedimentary organic matter: implications on the dynamics of natural gas accumulations. *Chemical Geology*, **126**, 291–318.

Littke, R., Krooss, B. M., Idiz, E. F. & Frielingsdorf, J. 1995. Molecular nitrogen in natural gas accumulations: Generation from sedimentary organic matter at high temperatures. *American Association of Petroleum Geologists Bulletin*, **79**, 410–430.

Morrow, C. A., Shi, L. Q. & Byerlee, J. D. 1984. Permeability of fault gouge under confining pressure and shear stress. *Journal of Geophysical Research*, **89**, 3193–3200.

Müller, E. P., May, F. & Stiehl, G. 1976. Zur Isotopengeochemie des Stickstoffs und zur Genese stickstoffreicher Erdgase. [On the isotope geochemistry of nitrogen and the generation of nitrogen-rich natural gas]. *Zeitschrift für Angewandte Geologie*, **22**, 319–324. [in German].

Nelson, J. S. & Simmons, E. C. 1992. The quantification of diffusive hydrocarbon losses through cap rocks of natural gas reservoirs—A reevaluation: Discussion. *American Association of Petroleum Geologists Bulletin*, **76**, 1839–1841.

—— & —— 1995. Diffusion of methane and ethane through the reservoir cap rock: Implications for the timing and duration of catagenesis. *American Association of Petroleum Geologists Bulletin*, **79**, 1064–1074.

—— & —— 1997. Diffusion of methane and ethane through the reservoir cap rock: Implications for the timing and duration of catagenesis: Reply. *American Association of Petroleum Geologists Bulletin*, **81**, 162–167.

NEUZIL, C. E. 1986. Groundwater flow in low-permeability environments. *Water Resources Research*, **22**, 1163–1195.

—— 1994. How permeable are clays and shales? *Water Resources Research*, **30**, 145–150.

REID, R. C., PRAUSNITZ, J. M. & POLING, B. E. 1987. *The properties of gases and liquids*. (4th ed). McGraw-Hill, New York.

SAHORES, J. J. & WITHERSPOON, P. A. 1970. Diffusion of light paraffin hydrocarbons in water from 2°C to 80°C. *In*: HOBSON, G. D. & SPEERS, G. C. (eds) *Advances in Organic Geochemistry: 3rd International Congress Organic Geochemistry Proceedings*, 219–230.

SCHLÖMER, S. & KROOSS, B. M. 1997. Experimental characterisation of the hydrocarbon sealing efficiency of cap rocks. *Marine and Petroleum Geology*, **14**, 565—580.

SIBSON, R. H. 1995. Selective fault reactivation during basin inversion; potential for fluid redistribution through fault-valve action. *In*: BUCHANAN, J. G. & BUCHANAN, P. G. (eds) *Basin inversion*. Geological Society, London, Special Publications **88**, 3–19.

SIEGESMUND, S., KERN, H. & VOLLBRECHT, A. 1990. The effect of microcracks on seismic velocities in an ultramylonite. *Tectonophysics*, **186**, 241–251.

SMITH, D. A, 1980. Sealing and nonsealing faults in Louisiana Gulf coast salt basin. *American Association of Petroleum Geologists Bulletin*, **64**, 145–172.

STAINFORTH, J. G. & REINDERS, J. E. A. 1990. Primary migration of hydrocarbons by diffusion through organic matter networks, and its effect on oil and gas generation. *Organic Geochemistry*, **16**, 61–74.

TAMMANN, G. & JESSEN, V. 1929. Über die Diffusionskoeffizienten von Gasen in Wasser und ihre Temperaturabhängigkeit. [On the diffusion coefficients of gases in water and their temperature dependence]. *Zeitschrift für anorganische Chem.*, **179**, 125–144 [in German].

WITHERSPOON, P. A. & SARAF, D. N. 1965. Diffusion of methane, ethane, propane and *n*-butane in water from 25 to 43°C. *Journal of Physical Chemistry*, **69**, 3752–3755.

ZANG, A., WOLTER, K. & BERCKHEMER, H. 1989. Strain recovery, microcracks and anisotropy of drill cores from KTB deep well. *Scientific Drilling*, **1**, 115–126.

The gas permeability of clay-bearing fault gouge at 20°C

D. R. FAULKNER & E. H. RUTTER

Rock Deformation Laboratory, Department of Earth Sciences, University of Manchester, Oxford Road, Manchester M13 9PL, UK

Abstract: Permeability measurements of clay-bearing fault gouges have been made using the pulse-transient and pore pressure oscillation methods with argon as the pore fluid, and at effective pressures up to 160 MPa and at a constant pore pressure of 40 MPa. Samples were collected from the outcrop of a major transpressional fault in southeastern Spain, in three orthogonal directions relative to the planar fabrics developed within the fault zone. Measurements show the gouge to exhibit permeability anisotropies of up to 3 orders of magnitude. In addition, pressure cycling of the gouge reduced the permeability by 2 to 3 orders of magnitude after 5 pressure cycles, when minimum permeabilities slightly less than 10^{-21} m^2 were measured.

The collection of data on the permeability of fault rocks is of utmost importance for the modelling of gas, water and hydrocarbon flow through reservoirs. In the past few years, the importance of fault sealing has been recognized to have a two-fold effect on the development of hydrocarbon fields. First, larger faults may act as structural seals, able to sustain excess pressures over geological time, and secondly, smaller faults may inhibit the flow of hydrocarbons in a reservoir during its production life. Clay-bearing fault rocks may also play a special role in the behaviour of seismogenic fault activity, by permitting the impounding of high pore water pressures within the fault zone and thereby reducing the frictional resistance to slip. Engineered clay seals may also be used to control the flow of groundwater around a waste repository. The sealing capacity of a fault zone is a function of the thickness of the fault rock, the type of fluid permeating through it (e.g. oil, gas, water), the permeability of the fault rock, the pressure difference to be sustained and the time frame over which the seal must act (Watts 1987; Deming 1994).

Clay-bearing fault gouge is an obvious candidate to form a natural low-permeability seal owing to its fine-grained and highly-structured nature. The microstructure of clay-bearing fault gouge has been described in both natural and experimental samples (Logan *et al.* 1979; Rutter *et al.* 1986) and also in less clay-rich cataclastic faults (Chester *et al.* 1985). Figures 1 and 2 show the essential features of the microstructure, consisting of a primary P foliation, developed roughly perpendicular to the principal shortening direction, the R_1 Riedel shear orientation and Y shears that develop parallel to the shear zone boundary. These features are observable on a mesoscopic and microscopic scale. Clay-bearing fault gouges tend to form in large, mature fault zones when the protolith is rich in phyllosilicates and/or feldspars, and are not simply a product of phyllosilicate reorientation, but (in part) of recrystallization. Hence they may be viewed as a low-grade metamorphic rock, potentially able to act as a seal over geological time periods.

Although the measurement of the permeability of clay-bearing fault gouges is a key factor in modelling them as fault seals over geological time, very few laboratory measurements on such rocks have been made. In this paper we present new data on the permeability of clay-bearing fault gouges and compare these with previous data. We also consider the implications for the fault sealing process and the design of further investigations aimed at understanding permeation through such rocks.

Sample collection and preparation

All the samples measured in this work were collected from the Carboneras fault, a major transpressional fault zone contained within the Betic Cordilleras of southeastern Spain (see Rutter *et al.* 1986; Keller *et al.* 1995 for further information). All the material used was collected from a fault strand containing clay-bearing fault gouge derived from a protolith of graphitic mica schist (Mapa Topográfico Nacional de España, El Agua del Medio Sheet, 1031-IV, 989006). The mineralogy of this fault gouge, as determined by X-ray diffraction studies (Rutter *et al.* 1986), shows quartz, muscovite and illite (unresolved), chlorite, some biotite and, in contrast to other permeability studies conducted on clay-bearing fault gouges (Morrow *et al.* 1981, 1984; Chu *et al.* 1981), no discernible montmorillonite. Protolithic phyllosilicate grains are variably comminuted and

FAULKNER, D. R. & RUTTER, E. H. 1998. The gas permeability of clay-bearing fault gouge at 20°C. *In*: JONES, G., FISHER, Q. J. & KNIPE, R. J. (eds) *Faulting, Fault Sealing and Fluid Flow in Hydrocarbon Reservoirs.* Geological Society, London, Special Publications, **147**, 147–156.

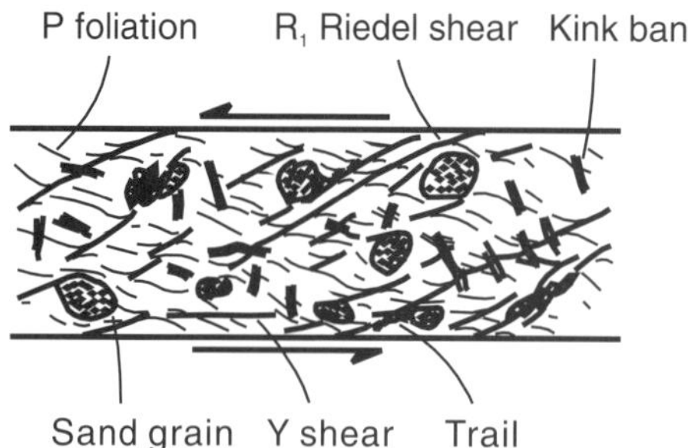

Fig. 1. Schematic diagram showing P foliation, R_1 Riedel shears, Y shears, kink bands and cataclastic fragment trails for an idealized, foliated, clay-bearing fault gouge (after Rutter *et al.* 1986).

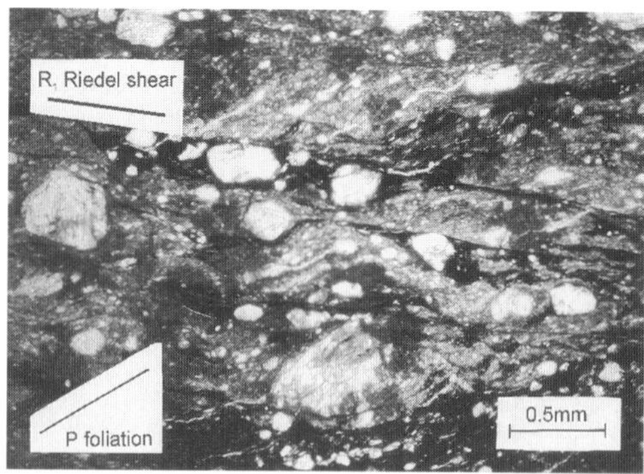

Fig. 2. Thin section photomicrograph (plane-polarized light) of the fault gouge studied. This microstructure is typical of the entire volume of the samples used. Rounded clasts of quartz and feldspar (transparent) are dispersed in a near opaque matrix of fine grained mica and clay, that display well-developed P and R_1 foliations (indicated).

altered and range in size from 1 to 50 microns and are decorated with oriented, submicron sized authigenic clay grains. Other silicates and protolithic clasts range from 10 microns up to 2 mm, and some pore spaces are assumed to be submicron in size (see Fig. 3).

One of the main reasons why more studies of the permeability of clay-bearing fault gouge have not been undertaken is perhaps due to the difficulty of recovering and dealing with the gouge in an undisturbed state. It is a highly friable rock, which does not lend itself well to coring or transportation. We have developed a new method for coring, transportation and subsequent testing of clay-bearing fault gouge. The outer surface of weathered 'crust' was cut away from the exposure until a substantial volume of rock reasonably free from macrocracks and major heterogeneities was exposed. Sharpened copper tubes, 20.3 mm in internal diameter, were hammered into this soft material, in directions relative to the main phyllosilicate foliations.

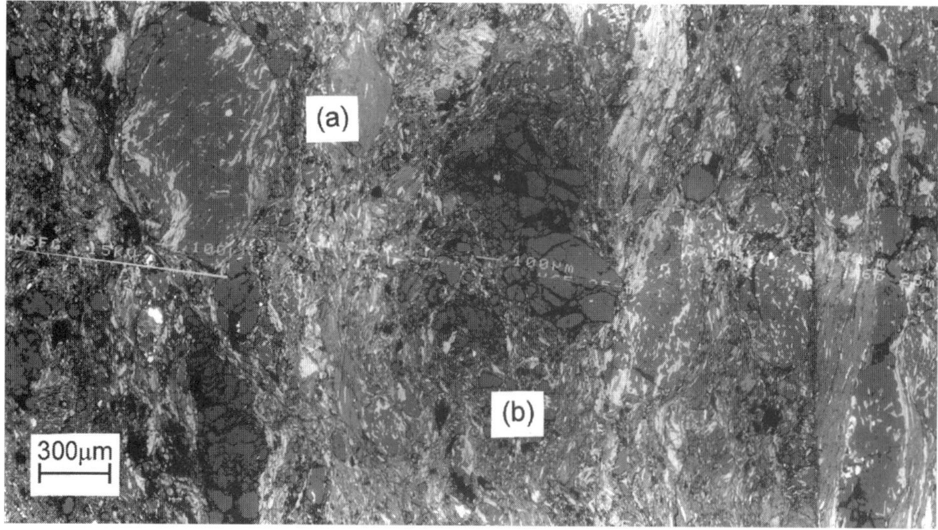

Fig. 3. SEM photomicrograph of a polished thin section of clay-bearing fault gouge taken in back scattered electron mode. A layered structure is apparent, with (**a**) fine-grained phyllosilicate-rich layers, with intense comminution of relict grains and formation of new authigenic clays (not visible at this magnification), (**b**) granular clastic layers, in which appears to reside most of the specimen porosity, other than that residing in collapsible macrocracks. Many of the cracks and voids visible here would collapse or close under pressure, and many voids may also be due to plucking of grains during specimen preparation for SEM.

From each coring site, three directions were cored, one along the intersection of the P foliation and the R_1 Riedel shear plane, one perpendicular to the acute bisector plane of the P foliation and R_1 Riedel shears (i.e. normal to the fault plane), and finally, one perpendicular to both the previous two cores, in the transport direction of the fault. At each site, two sets of cores were collected in this way, so that assessment of the heterogeneity of the gouge could be made from the measurement of each pair of cores collected in the same direction from the same site.

After the cores had been collected, both ends of the tube were filled and sealed with fast-hardening (20 min) epoxy resin. Transportation of the samples back to the laboratory could then be achieved without fear of damage. In the laboratory, the walls of the copper tubes (originally 1 mm in thickness) were thinned to 0.5 mm, and the epoxy resin coated ends were removed such that the sample length was nominally the same as the diameter. The copper sleeve could not be removed completely, or the sample would crumble. Through the sealing of the samples inside the copper tubes, the water content of the sample remained as it was in the field. Owing to the arid climate and the fact that the samples were all collected in high summer when there is typically no rain, the samples contain no free liquid water. Differential thermal analyses of the gouge showed a weight loss of 0.15% over a temperature range of 0 to 100°C. This is inferred to be attributable to the driving off of any free water, but also a contribution may be present in this weight loss due to incipient clay breakdown. It is not known how great this effect is. Vacuum impregnation with low viscosity resin and longitudinal sectioning of a gouge core showed that there was no discernible disruption of the periphery of the core by friction against the core tube.

The porosity of this fault gouge was determined to be typically about 20% when collected. This was estimated from density measurements of the intact and disaggregated gouge. From microstructural studies, it is clear that this porosity is largely in the form of open cracks, produced as a result of removal of overburden. The application of confining pressure causes instantaneous and largely irrecoverable collapse of nearly all the porosity (to *c.* 1 to 3%), based on measurement of sample dimensions after removal from the pressure vessel. Measurements of the storage capacity of the gouge samples as part of the permeability measurement indicated porosities of the same order under pressure, residing mainly in clastic-rich lenses (Fig. 3), and these did not change significantly with pressure cycling.

The specimen, in its 0.5 mm wall thickness copper jacket, was inserted into a PVC jacket with bronze sintered disks of known porosity at each end, and placed in the sample assembly as shown in Fig. 4. During preparation of the samples, it is possible that some preferred orientation of phyllosilicates occurred at the ends of the samples (i.e. smearing). However, this effect was assumed only to affect the top few grains of the sample, and hence any error introduced, although unavoidable, was small.

The upstream and downstream ends of the sample were in turn connected to the pore fluid system. In these experiments, argon gas was used as the pore fluid. It is anticipated that permeability to water may be lower than with inert argon, owing to the probability of clay swelling, leading to constriction (possibly irrecoverable) of pore throats, and electroviscous interactions between water and the clay surfaces. Owing to its inert nature, argon will adsorb onto the clay surfaces to a much lesser degree than water. Adsorption is favoured by high surface areas, higher pressures and very low temperatures (much less than room temperature). This effect was considered to introduce very small errors into the results (D. Dewhurst, personal commun., 1997) and hence was discounted. Effective pressure on the samples was varied by changing the confining pressure at a constant pore pressure, to avoid difficulties in the interpretation of the results that would arise from a complicated pressure history. The sensitivity of pore pressure to ambient temperature variations is far less with gas pore fluid than with water.

Measurement techniques

Permeability measurements were all made at room temperature (kept at 20°C) and 40 MPa pore pressure, and confining pressures were varied between 100 and 200 MPa. When pressurizing the sample up to initial values of 40 MPa pore pressure and 100 MPa confining pressure, care was taken to increase both in small increments, to allow temperature changes induced by pressure changes to dissipate and to avoid exposing the sample to effective pressures greater than the initially desired effective pressure of 60 MPa.

Two methods of permeability measurement were used for each sample, the pulse transient method (Brace *et al.* 1968) and the pore pressure oscillation method (Krantz *et al.* 1990; Fischer 1992). Transducers used to measure the pressure variations in the upstream and downstream reservoirs of the sample were amplified, and

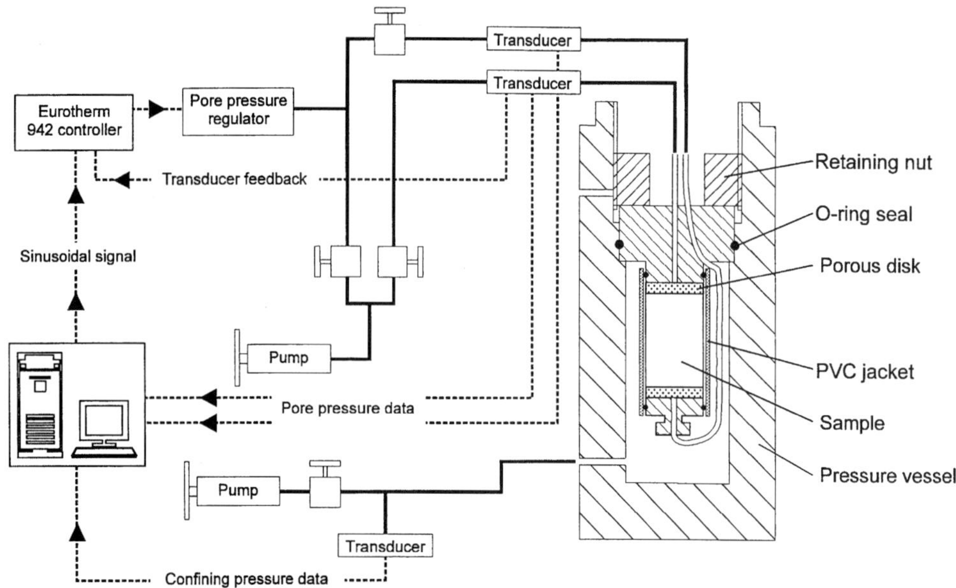

Fig. 4. Schematic diagram of the experimental apparatus used. The solid lines indicate the plumbing of the system, and the dotted lines indicate electrical connections. During permeability measurements using the pore pressure oscillation technique, the Eurotherm 942 controller receives the sinusoidal signal from the computer and the transducer feedback. The controller attempts to equalize these signals by driving the pore pressure regulator.

produced sensitivities consistently better than 0.01 MPa.

The pulse transient method utilizes the decay of a pressure transient introduced into the upstream end of the sample. The exponential decay of this transient may be used to calculate the permeability. The method assumes the storage capacity of the sample to be zero in order to obtain a solution to the diffusion equation, and therefore is only useful to measure the permeability of low porosity rocks.

The pore pressure oscillation technique employs a forced sinusoidal pressure oscillation of known amplitude and wavelength in the upstream reservoir of the sample. The downstream reservoir is then monitored for the transmission of this oscillation through the sample. From the attenuation and phase shift of the wave in the downstream reservoir with respect to the upstream oscillation, both the permeability and the storage capacity may be calculated, thereby providing a complete solution to the diffusion equation for the flow of fluids through porous media. Driving waveforms were varied in period from 100 s to 5000 s, according to the range of permeability being measured and the optimal amplitude ratio sought. These considerations are discussed in detail by Fischer (1992). Amplitudes and phases of the upstream and downstream waveforms were obtained from Fast Fourier Transform of the raw data. It was determined experimentally that the upstream waveform amplitude used (*c*. 2 MPa p-p) yielded permeability values that were largely independent of the range of wave periods used, and there was excellent agreement between the results obtained using the oscillation and pulse transient techniques (Fig. 5).

A number of tests were conducted using impervious nylon plugs surrounded by an annulus of remoulded and compacted gouge and contained within a 20 mm internal diameter copper jacket. These were performed to determine whether flow took place preferentially between the sample and the inside of the jacket, perhaps due to crimping of the jacket as a result of differential compaction of the sample with respect to the copper jacket under pressure. Different diameters of nylon cores were used, thereby changing the ratio of the cross-sectional surface area of the gouge available for pore fluid permeation relative to the cylindrical surface area. If there was a 'short-circuiting' effect along the inside of the jacket, then this would become evident in a predictable way as the cross-sectional area of the gouge provided for permeation was reduced. Therefore, if the permeability of the gouge was calculated for each

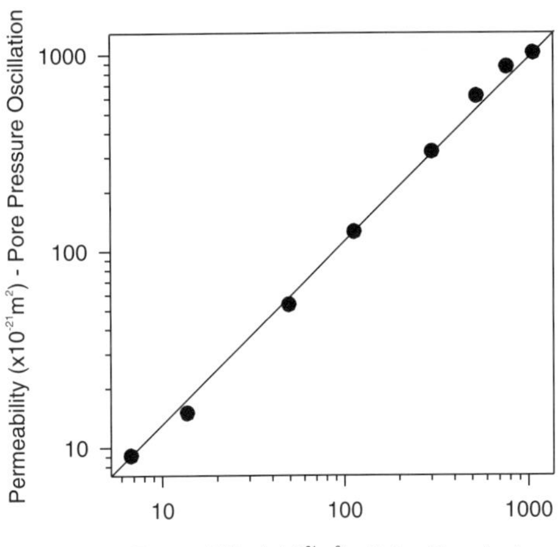

Fig. 5. Correlation between permeabilities measured using the oscillation (PPO) and pulse transient methods.

surface area, apparently higher permeabilities would be measured for the samples containing the largest plug of nylon. If there was to be no significant effect, then the apparent permeabilities in all the samples should be the same. These results are illustrated in Fig. 6, and show that the effect on the permeability of the gouge of any enhanced flow between the specimen and the copper jacket is negligible.

Permeability results

Effect of pressure cycling

The effect of pressure cycling on a sample cored perpendicular to the fault zone plane is shown in Fig. 7. The sample was pressurized and permeabilities measured after 20 MPa effective pressure increments up to 200 MPa total confining pressure (160 MPa effective pressure). The sample was then depressurized down to 100 MPa confining pressure, and the cycle repeated. The second run shows a marked decrease in the permeability of the sample. Pressure cycles 3, 4 and 5 show the reduction in permeability after a complete depressurization of the sample (both pore and confining pressure). From these data, it can be seen that the permeability is further decreased with each pressure cycle. An asymptote would probably eventually be attained, but this was not seen after typically 4 or 5 cycles applied in the present study. The results indicate that the amount of depressurization, whether complete or partial, has an effect on the subsequent permeability history with effective pressure change.

It was considered that the reduction in permeability with each cycle might be due to time-dependent compaction of the sample. To test this hypothesis, permeability was measured at the beginning and end of a hold period of at least 50 h between cycles 3 and 4, and 4 and 5

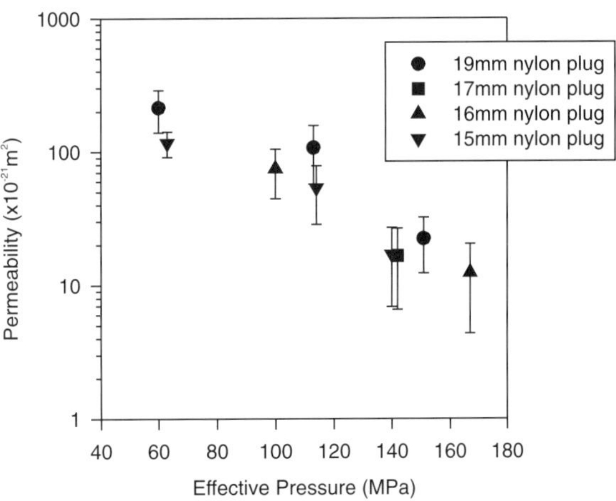

Fig. 6. Results of experiments conducted with a succession of nylon plugs of different diameters surrounded by an annulus of remoulded and compacted gouge contained within a copper jacket. Details of the experiment may be found in the text. Owing to the smaller cross sectional areas of the specimens, errors of measurement (1 standard error) are larger than for full-sized gouge specimens. Only the thinnest gouge annulus seems to show consistently lower apparent permeabilities, as might be expected if there was a short-circuit fluid pathway between the gouge and copper sleeve. However, these data are not significantly different from measurements made with thicker gouge annuli.

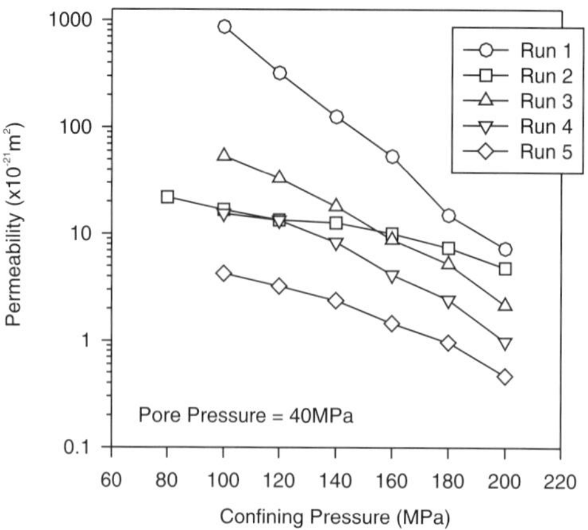

Fig. 7. The effect of pressure cycling on the permeability of clay-bearing fault gouges at different total confining pressures. In all cases pore pressure was 40 MPa. Run 2 represents permeability variation after the total pressure had been dropped to 100 MPa. For subsequent cycles both the confining and pore pressures were reduced to zero before starting the next cycle. Successive cycles continue to reduce permeability, and it is not yet known how many cycles would be needed to produce behaviour independent of further cycles.

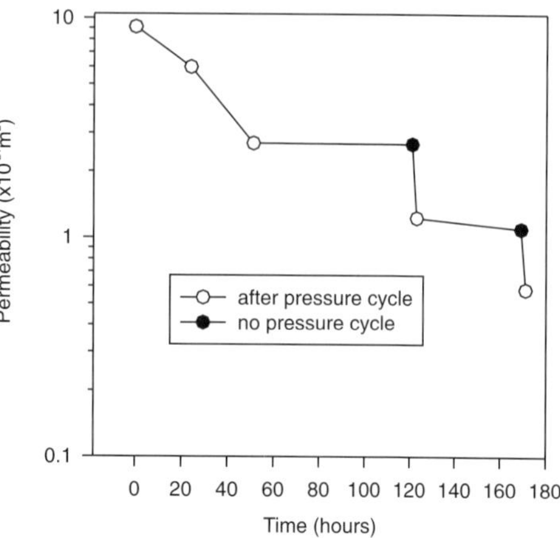

Fig. 8. The relative contribution of pressure cycling and time dependent compaction to the reduction in permeability of clay-bearing fault gouge. Measurements were made normal to foliation at 200 MPa total confining pressure and 40 MPa pore pressure after a succession of pressure cycles (diamonds). The squares indicate measurements made after the specimen had been held at pressure for more than 50 hours following the previous pressure cycle and before the next one. Whereas the permeability falls after successive pressure cycles, it does not fall significantly after merely being held at pressure for times comparable with the duration of each cycle.

(Fig. 8). It was observed that there was little or no reduction in permeability during the extended hold times between pressure cycles, indicating that the reduction is primarily due to the mechanical working during pressure cycling, rather than to time-at-pressure.

It might be argued that the permeability variations observed with pressure cycling are attributable, at least in part, to the progressive drying of the sample through the passage of dry argon gas. At the time of writing, we know that the permeability of this rock with water as pore fluid is lower than with argon, hence it would be expected that drying of the gouge will increase permeability, and may to some extent offset the observed decrease of permeability with pressure cycling.

Permeability anisotropy

The results of permeability measurements made on two sets of three cores collected from the same site in orthogonal directions relative to the planar fabrics are shown in Fig. 9 as a function of effective pressure. These show that the permeability of the sample perpendicular to the acute bisector of the P foliation and R_1 Riedel shears (i.e. normal to the fault plane) is up to three orders of magnitude lower than the sample which was collected parallel to both of these fabrics. The sample in the transport direction of the fault shows an intermediate value for permeability. Estimated errors in the measurement of these permeability values are about the size of the symbol used to represent them. Figure 9 also shows the variability between the two sets of cores. This is an indication of the heterogeneity of the gouge on a scale of a few metres. The variability in a given direction is less than the variability between directions relative to the fabric elements.

Discussion

Comparison with previous work

Three main studies of the permeability of clay-bearing fault gouges have previously been conducted. Morrow *et al.* (1981) measured the permeability of clay-bearing fault gouge collected from a borehole on the San Andreas fault in California. The mineralogy included montmorillonite, mixed-layer clays, illite, kaolinite and chlorite. The samples, however, did not retain their original fabrics, and were ground up and mixed into a slurry with distilled

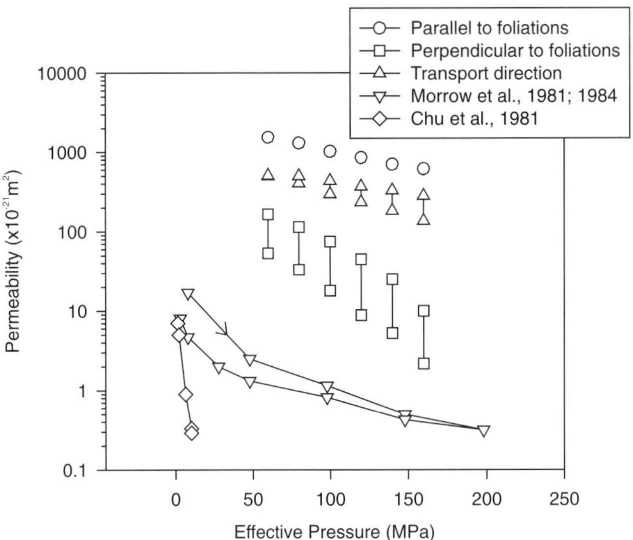

Fig. 9. Results of permeability measurements showing permeability anisotropy as a function of effective pressure of the clay-bearing fault gouge after four pressure cycles. Data are shown for two sets of samples collected from within a few metres of each other, apart from the second core collected parallel to the foliations, which suffered damage during preparation, rendering it useless for measurement. The variability between different orientations is greater than the variability between similarly oriented specimens. In all cases pore pressure was 40 MPa. Results are also shown for previous permeability measurements made on remoulded clay-bearing fault gouge by Morrow *et al.* (1981 and 1984), and an unoriented core of clay-bearing fault gouge by Chu *et al.* (1981). The line joining the data points of Morrow *et al.* indicates the path of the pressure cycle followed.

water. This mixture was then placed in a 30 degree sawcut in Berea sandstone, and its permeability measured as a function of effective pressure (confining minus pore pressure) up to 199 MPa (Fig. 9). Permeability measurements on a wider range of fault rocks, including those containing more non-phyllosilicate clastic material, were made by Morrow *et al.* (1984) but the same basic procedure was followed and the results for the phyllosilicate-rich samples were comparable to results from the earlier work (Fig. 9).

Chu *et al.* (1981) conducted tests on a core of San Andreas fault gouge that did retain its original fabric, but no indication of the orientation of the core with respect to the internal rock fabric was given. The sample measured contained montmorillonite, mixed-layer clays, kaolinite and small quantities of illite and chlorite. The sample was preconsolidated at 200 MPa effective pressure for 200 hours, and permeability measurements were made at up to 11 MPa effective pressure. Results reported are reproduced in Fig. 9.

The main problems with these previous studies were that the gouge was not preserved in its original state, and when it was, no indication of the orientation was given. From Fig. 1 it is clear that clay-bearing fault gouge is a highly structured rock, and this will affect the permeability in different directions through the gouge. Also, only a small number of tests were made, during which the effective pressure on the sample was increased up to its maximum, and then reduced (i.e. measurements were made during one pressure cycle only).

If the data collected during the present study are compared with those of Morrow *et al.* (1981, 1984) and Chu *et al.* (1981; see Fig. 9), important differences may be seen. The lowest permeability data we collected are comparable to those of Morrow *et al.* (1984) in terms of a permeability versus effective pressure trend, but do not compare with those of Chu *et al.* (1981). However, Chu *et al.* preconsolidated their samples to a much higher effective pressure than used in the present study.

One of the most important considerations to make when comparing our data with those of others is the nature of the pore fluid used. In all the other studies mentioned, permeability was measured using water as pore fluid. By using high-purity argon at this stage in our study, we ensure that there is minimal interaction of our pore fluid with the gouge (i.e. no electrostatic forces, only van der Waals forces, which are small due to the mass of the argon molecule). We consider that it is essential to provide an initial standard of reference in this way, so that effects of other physicochemical interactions, between the rock and the fluid, on permeability can be evaluated. Lower absolute permeabilities measured by Morrow *et al.* (1981, 1984) and Chu *et al.* (1981) may have been produced by the interaction of water with swelling clays contained within their gouge (Moore *et al.* 1982). Finally, the source of our samples may have produced an effect on the permeability. It has been shown that samples collected from boreholes provide lower permeabilities than samples from the same lithologies collected from the surface (Morrow & Lockner 1994). However, it is possible that pressure cycling progressively causes recovery from the effects of the damage caused by total depressurization at the surface, provided there has been no significant chemical weathering. Studies of the variation of permeability with depth will be required to assess these effects.

Pressure cycling and anisotropy

This work is the first to clearly demonstrate a significant dependence of the permeability of clay-bearing fault gouges on pressure cycling. Similar effects were, however, found for crystalline rocks by Morrow *et al.* (1986) and Bernabé (1987), although to a lesser degree. They also found that permeability was reduced with cycling. The data of Morrow *et al.* (1981) on fault gouge (reproduced in Fig. 9) also indicate that a reduction in permeability due to pressure cycling exists that is similar to that described here. The effect in crystalline rocks was explained by better closure of microcracks and reduction of the mean hydraulic radius in the rocks, whereas in clay-bearing fault gouges, the effect may be due to reshuffling of the phyllosilicates with respect to one another, producing enhanced compaction with each pressure cycle. The differences in permeability with complete or partial depressurization of the sample may be explained by opening or re-opening of microcracks (during complete depressurization), but it is assumed that this effect is negligible at higher pressures.

Permeability variations in anisotropic materials have only been demonstrated previously in deformed wet sediments (Arch & Maltman 1990; Brown & Moore 1993). Results from those studies have shown permeability anisotropies of up to two orders of magnitude along the foliation as opposed to across the foliation. These values for the anisotropy are comparable to, although lower than those found in the present study. They related the anisotropy to the square of the tortuosity ratio of the fluid flow path, following England *et al.* (1987). Wilkinson & Shipley (1972) found that for permeability development in the consolidation of muds, a maximum achievable tortuosity ratio was in the order of 2.6, which indicates a maximum permeability anisotropy of just under $\times 7$. These values are clearly rather lower than the results from our study. The reasons for this are obvious when the microstructure of clay-bearing fault gouge is observed in the scanning electron microscope (SEM) (Fig. 3). Microlayering, on the scale of 400 to 500 mm can be seen, with fine-grained phyllosilicate-rich layers interspersed with higher porosity granular clastic layers, composed of variably broken down protolith. These layers will greatly enhance the fluid flow parallel to the foliation. The lower values of permeability anisotropy obtained by Arch & Maltman (1990) in comparison to our study, may be due to these observed differences in fabric and also the higher effective pressures used in our experiments. The anisotropy of our material was smaller at lower pressures (see Fig. 9). Arch & Maltman used confining pressures of only up to 0.3 MPa. In addition, the fabric anisotropy of the materials used in the previous studies was produced predominantly by mechanical grain rotation during shear. In clay-bearing fault gouges these processes also occur, but additionally, recrystallization of many new phases has occurred, so that these rocks may be viewed as a low-grade metamorphic rock that suffered both mineralogical and mechanical modification. Hence, in the fine-grained phyllosilicate-rich layers, permeability will be reduced even further by these effects, thereby increasing the anisotropy.

Other factors: temperature and deviatoric stress

The effectiveness of clay bearing fault gouge as a fluid pressure seal is likely to be high, although discussion of the detailed implications of this qualitative statement lie beyond the scope of the present paper. We have measured permeabilities in the order of 10^{-21} m^2 at the highest effective pressures used (160 MPa), normal to the foliation and after several pressure cycles, without a constant value being attained. As indicated above, permeabilities are likely to be lower still when water is used as the pore fluid. Additionally, elevated temperatures (between 100 and 200°C), are likely to promote further compaction and consolidation of the gouge, either through neomineralization or through thermally activated plastic creep of the phyllosilicate grains. In nature, fault rocks are likely to be subjected to non-hydrostatic stresses. For example, the San Andreas fault, California, is currently subjected to a regional stress regime in which the greatest principal stress is almost normal to the fault zone (Zoback & Healy 1992). Both elastic and inelastic porosity reduction are facilitated by non-hydrostatic stress relative to purely hydrostatic stress, and this is likely to be reflected in further lowering of permeability. It is our intention to progressively evaluate these effects on the permeability of the clay-bearing fault gouges of SE Spain.

Conclusions

(1) Comparisons of previous studies of the permeability of clay-bearing fault gouge have shown a conflicting and confusing picture of their transport properties. This study shows that the room temperature gas permeability of undisturbed Carboneras fault zone gouge is similar to, although higher than, the water permeability of remoulded San Andreas fault gouge reported by Morrow *et al.* (1981).

(2) The permeability of clay-bearing fault gouge was found to be highly anisotropic in principal directions through the gouge, relative to the main planar and linear fabrics developed. Comparisons with other work on deformed wet muddy sediments shows that greater permeability anisotropies are found in clay-bearing fault rocks. These are due to a layered microstructure, observable in the SEM, involving fine-grained phyllosilicate-rich layers and porous granular clastic layers. Differential compaction of these layers may lead to higher anisotropies at increased pressures.

(3) Pressure cycling of clay-bearing fault gouge reduces initial permeabilities by two to three orders of magnitude after three to four pressure cycles, and more cycles appear to be required to reach a steady-state condition. Similar studies on crystalline rocks have also shown permeability reductions but of much less than one order of magnitude with pressure cycling.

(4) The results of this study and the anticipated effects of further factors (temperature, differential stress, the nature of the pore fluid) mean that clay-bearing fault gouges probably can make effective pressure seals.

D.R.F. acknowledges with thanks a N.E.R.C. research studentship (ref: GT4/94/216/G) held during the duration of this study. This work was supported partially though grant aid from the Royal Society of London. The Spanish Ministry of the Environment kindly gave permission for our studies in the area of the Nijar-Cabo de Gata Natural Park area. Construction and maintenance of the experimental apparatus was carried out with the help of R. F. Holloway and R. Mason. Helpful reviews by B. Kroos, B. Clennell and D. Dewhurst are gratefully acknowledged. In particular, discussions with B. Clennell and B. Kroos have helped to improve the content of this paper.

References

ARCH, J. & MALTMAN, A. 1990. Anisotropic permeability and tortuosity in deformed wet sediments. *Journal of Geophysical Research*, **95**, 9035–9045.

BERNABÉ, Y. 1987. The effective pressure law for permeability during pore pressure and confining pressure cycling of several crystalline rocks. *Journal of Geophysical Research*, **92**, 649–657.

BRACE, W. F., WALSH, J. B. & FRANGOS, W. T. 1968. Permeability of granite under high pressure. *Journal of Geophysical Research*, **73**, 2225–2236.

BROWN, K. M. & MOORE, J. C. 1993. Comment on "Anisotropic permeability and tortuosity in deformed wet sediments" by J. Arch & A. Maltman. *Journal of Geophysical Research*, **98**, 17859–17864.

CHU, C. L., WANG, C. Y. & LIN, W. 1981. Permeability and frictional properties of San Andreas fault gouges. *Geophysical Research Letters*, **8**, 565–568.

CHESTER, F. M., FRIEDMAN, M. & LOGAN, J. M. 1985. Foliated cataclasites. *Tectonophysics*, **111**, 139–146.

DEMING, D. 1994. Factors necessary to define a pressure seal. *Bulletin of the American Association of Petroleum Geologists*, **78**, 1005–1009.

ENGLAND, W. A., MACKENZIE, A. S., MANN, D. M. & QUIGLEY, T. M. 1987. The movement and entrapment of petroleum fluids in the subsurface. *Journal of the Geological Society, London*, **144**, 327–347.

FISCHER, G. J. 1992. The determination of permeability and storage capacity: pore pressure oscillation method. *In*: EVANS, B. & WONG, T.-F. (eds) *Fault Mechanics and Transport Properties of Rocks*. Academic Press, 187–212.

KELLER, J. V. A., HALL, S. H., DART, C. J. & MCCLAY, K. R. 1995. The geometry and evolution of a transpressional strike-slip system: the Carboneras fault, S.E. Spain. *Journal of the Geological Society, London*, **152**, 339–351.

KRANTZ, R. L., SALTZMAN, J. S. & BLACIC, J. D. 1990. Hydraulic diffusivity measurements on laboratory rock samples using an oscillating pore pressure method. *International Journal of Rock Mechanics, Mineral Science and Geomechanical Abstracts*, **27**, 345–352.

LOGAN, J. M., FRIEDMAN, M., HIGGS, N., DENGO, C. & SHIMAMOTO, T. 1979. Experimental studies of simulated gouge and their application to studies of natural fault zones. *In*: Proceedings of Conference VIII, Analysis of Actual Fault Zones in Bedrock, U.S.G.S. Open File Report, **79**, 305–343.

MOORE, D. E., MORROW, C. A. & BYERLEE, J. D. 1982. Use of swelling clays to reduce permeability and its potential application to nuclear waste repository sealing. *Geophysical Research Letters*, **9**, 1009–1012.

MORROW, C. A. & LOCKNER, D. A. 1994. Permeability differences between surface-derived and deep drill-hole core samples. *Geophysical Research Letters*, **21**, 2151–2154.

——, SHI, L. Q. & BYERLEE, J. D. 1981. Permeability and strength of San Andreas fault gouge under high pressure. *Geophysical Research Letters*, **8**, 325–328.

——, —— & —— 1984. Permeability of fault gouge under confining pressure and shear stress. *Journal of Geophysical Research*, **89**, 3193–3200.

——, ZHANG BO-CHONG & BYERLEE, J. D. 1986. Effective pressure law for permeability of Westerly granite under cyclic loading. *Journal of Geophysical Research*, **91**, 3870–3876.

RUTTER, E. H., MADDOCK, R. H., HALL, S. H. & WHITE, S. H. 1986. Comparative microstructures of natural and experimentally produced clay-bearing fault gouges. *Pure & Applied Geophysics*, **124**, 3–30.

WATTS, N. L. 1987. Theoretical aspects of cap rock and fault seals for single- and two-phase hydrocarbon columns. *Marine and Petroleum Geology*, **4**, 274–307.

WILKINSON, W. B. & SHIPLEY, E. L. 1972. Vertical and horizontal laboratory permeability measurements in clay soils. Developments in Soil Science **2.** International Symposium on Fundamentals of Transport in Porous Media. Elsevier, Amsterdam.

ZOBACK, M. D. & HEALY, J. H. 1992. In situ stress measurements to 3.5 km depth in the Cajon Pass scientific research borehole: implications for the mechanics of crustal faulting. *Journal of Geophysical Research*, **97**, 5039–5057.

Numerical simulation of departures from radial drawdown in a faulted sandstone reservoir with joints and deformation bands

S. K. MATTHÄI[1*], A. AYDIN[2], D. D. POLLARD[2] & S. G. ROBERTS[2]

[1] *Rock Fracture Project, Department of Geological and Environmental Sciences, Stanford University, Stanford, CA 94305, USA*

[2] *Mathematics Department, Australian National University, GPO Box 4, Canberra, ACT 2601, Australia*

[*] *Present Address: Departement Erdwissenschaften, Swiss Federal Institute of Technology, ETH Zentrum, CH-8092 Zürich, Switzerland*

Abstract: Field measurements constrain the fluid flow characteristics of an analogue hydrocarbon reservoir in the faulted Entrada sandstone, Arches National Park, Utah. These data comprise maps of the geometry, inhomogeneous permeability, and porosity of fault zones, joints, and deformation bands in a region where two discontinuous normal faults overlap. Two-dimensional computer simulations of drainage of this analogue reservoir identify normal faults with highly permeable slip planes as the most important reservoir inhomogeneities. These faults compartmentalize fluid pressure over timespans greater than years while fluid can be drained on the kilometre scale along their highly permeable slip planes. Joints induce the second most important distortions of radial drawdown, influencing the timespans over which fault signatures are observed in pressure decline curves. The joints often extend to the boundaries of the reservoir. This also reduces the time before the rate of pressure decline accelerates due to boundary interaction. Zones of deformation bands less than 25 cm wide with a spacing $\geq$30 m have little effect on radial drawdown in our single phase fluid flow simulations. When drawdown spreads with time over the deformation structures in the analogue reservoir, the different structures simultaneously influence the change of pressure at the wellbore (pressure derivative). This temporal overlap prohibits an analysis of the effects of individual structures. Drawdown does not 'recover' to radial flow after an inhomogeneity is encountered.

The prediction of the flow rate and optimal placement of oil production and environmental remediation wells in fractured sandstones represents a difficult problem for engineers, production geoscientists and hydrologists. There are implications for the drainage efficiency and the time it takes until a steady-state pumping pattern is obtained such that an effective engineering strategy can be formulated. As the size of hydrocarbon accumulations is usually limited, decisions on well placement also influence the risk of water coning and mixing of oil and water during production. It is therefore important to predict the flow paths in the reservoir, since they may determine which parts of the reservoir are drained by the well over time and which regions are by-passed by the flow. Thus, for fluid flow during hydrocarbon production, Tyler & Finley (1988, 1991) have reported that the ability to recover the oil stored in a specific reservoir decreases rapidly with increasing geological complexity.

Predictions of the flow paths are also needed to design strategies for toxic waste remediation by pumping (Gailey & Gorelick 1993). In this context, Hsieh & Shapiro (1994) and Shapiro *et al.* (1995) have recently shown that the predictability of the pathways of fluids degrades with increasing magnitude of spatial variations in the permeability of the drainage region. This reflects the need for determining those aspects of formation heterogeneity that are critical for fluid flow, such as the continuity and connectivity of highly permeable zones.

Today, many predictions in reservoir and aquifer engineering are made using flow simulations in which geological inhomogeneity is represented by a host of approaches (Wang 1991; Wheatcraft & Cushman 1991; Hewett 1992; Savioli *et al.* 1995; Deutsch *et al.* 1996). Commonly, inhomogeneities in hydrocarbon reservoirs are identified on the large scale in seismic data and are measured only on the small scale in selected intervals of boreholes with both direct (televiewer fracture mapping etc., e.g. Barton *et al.* 1991) and indirect methods (e.g. gamma ray logs to determine porosity, transient well testing determining permeability, Horne 1990, 1996). Statistical properties of these data are then extrapolated into a model of the reservoir (e.g. Hewett 1992;

MATTHÄI, S. K., AYDIN, A., POLLARD, D. D. & ROBERTS, S. G. 1998. Numerical simulation of departures from radial drawdown in a faulted sandstone reservoir with joints and deformation bands. *In*: JONES, G., FISHER, Q. J. & KNIPE, R. J. (eds) *Faulting, Fault Sealing and Fluid Flow in Hydrocarbon Reservoirs*. Geological Society, London, Special Publications, **147**, 157–191.

Wei *et al.* 1995). Thus, comparatively small samples are used to constrain the properties of large rock volumes. Additionally, because of the large degree of property variability in borehole samples, reservoir heterogeneity is often represented by effective parameters (Journel *et al.* 1986; Deutsch 1989). However, simulations based on effective properties cannot be expected to reproduce the geometry and magnitude of the flow in the actual reservoir, the tortuosity of the flow, and the spatial variations in flow velocities. The lack of geometric detail in such models does not permit the development of critical flow paths and percolations networks (cf. Gueguen *et al.* 1991) which are characteristic of hydrocarbon flow in nature (England *et al.* 1987).

A major cause of inhomogeneity in sandstone reservoirs is brittle deformation. The geometries of deformation structures are controlled by underlying fracture-mechanical processes (Aydin 1978; Dyer 1983; Pollard & Segall 1987; Cruikshank *et al.* 1991). At the top of a hierarchy of deformation structures in hydrocarbon reservoirs are km-scale faults (e.g. Fig. 1). There is a growing body of evidence that faults strongly influence fluid flow during production (Smith 1980; Knott 1993; Gibson 1994). We have previously identified (Antonellini & Aydin 1994, 1995) the inhomogeneous permeability of normal faults in the Entrada sandstone at Arches National Park, Utah. These faults are generally composed of ensembles of highly permeable slip planes at the boundary of the hanging wall blocks and low permeability zones of cataclastic deformation bands in the footwall (Fig. 2). The faults are discontinuous on most scales and often overlap, forming relay structures. The fault relays also contain small scale deformation structures such as joints and deformation bands, that provide a scale-dependent permeability structure that is intrinsically complex.

The flow characteristics of faults have been studied in heuristic models of isolated and simple fault geometries for which analytical solutions describe the flow (Horner 1951; Cinco *et al.* 1976; Yaxley 1987; Kuchuk & Habashy 1992). For these models, geometric patterns of flow that could guide our intuition in the analysis of real geometries have not been published.

We propose that quantitative structural datasets, like those from the Arches National Park, can be used to build geometrically realistic models of reservoir heterogeneity for flow simulations and provide an alternative approach to models based on extrapolated *in situ* data. Outcrop data can comprise complete information on the continuity of flow zones and their connectivity, which is otherwise poorly constrained in statistical models based on data obtained from well logs. We will show such information to be critical for the prediction of fluid flow paths in reservoirs.

Joints (Pollard & Aydin 1988), deformation bands which are thin zones of grain crushing accompanied by a collapse of the pore space (Aydin 1977, 1978), and faults are ubiquitous inhomogeneities in sandstones. In the recent past, a large amount of new quantitative structural geological data has been collected on these inhomogeneities (Underhill & Woodcock 1987; Antonellini *et al.* 1994; Antonellini & Aydin 1994, 1995; Knipe & Lloyd 1994; Forster *et al.* 1993, 1995). Studies in fracture mechanics provide us with a framework for understanding the controls on the geometry of these structures (Aydin & Johnson 1983; Pollard & Aydin 1988; Cruikshank *et al.* 1991).

The normal faults and related structures in the Arches National Park, Utah, cross-cut sandstones with typical reservoir properties. These are continuously exposed over large areas (Fig. 1) and were mapped in detail by Dyer (1983), Zhao (1991), Cruikshank *et al.* (1991) and Antonellini & Aydin (1994, 1995) who measured the permeability and porosity of deformation bands and altered fault rocks, and estimated the apertures of joints and slip surfaces. These parameters can be used directly in numerical reservoir models (Fig. 3). Due to the similarity between the deformation structures in the Arches National Park and many oil fields in fractured sandstones, insights gained into the flow behaviour of this analogue should be relevant to questions raised with regard to reservoir prediction.

Following this rationale, we attempt to incorporate the full structural complexity of the fault relays in the Arches National Park into numerical simulations of drawdown around a well. This is not possible with simulators such as ECLIPSE, because they do not permit the discrete representation of small scale heterogeneities like joints: the grid spacing would need to correspond to their narrow width in models with km-scale dimensions. Instead we use an algebraic multigrid finite element method (Roberts & Matthäi 1996) and highly refined two-dimensional finite element meshes to simulate fluid flow in this continuously characterized analogue reservoir.

By means of numerical simulation, we determine whether one should be able to detect fault responses in spite of the presence of small-scale inhomogeneities, and whether unperturbed

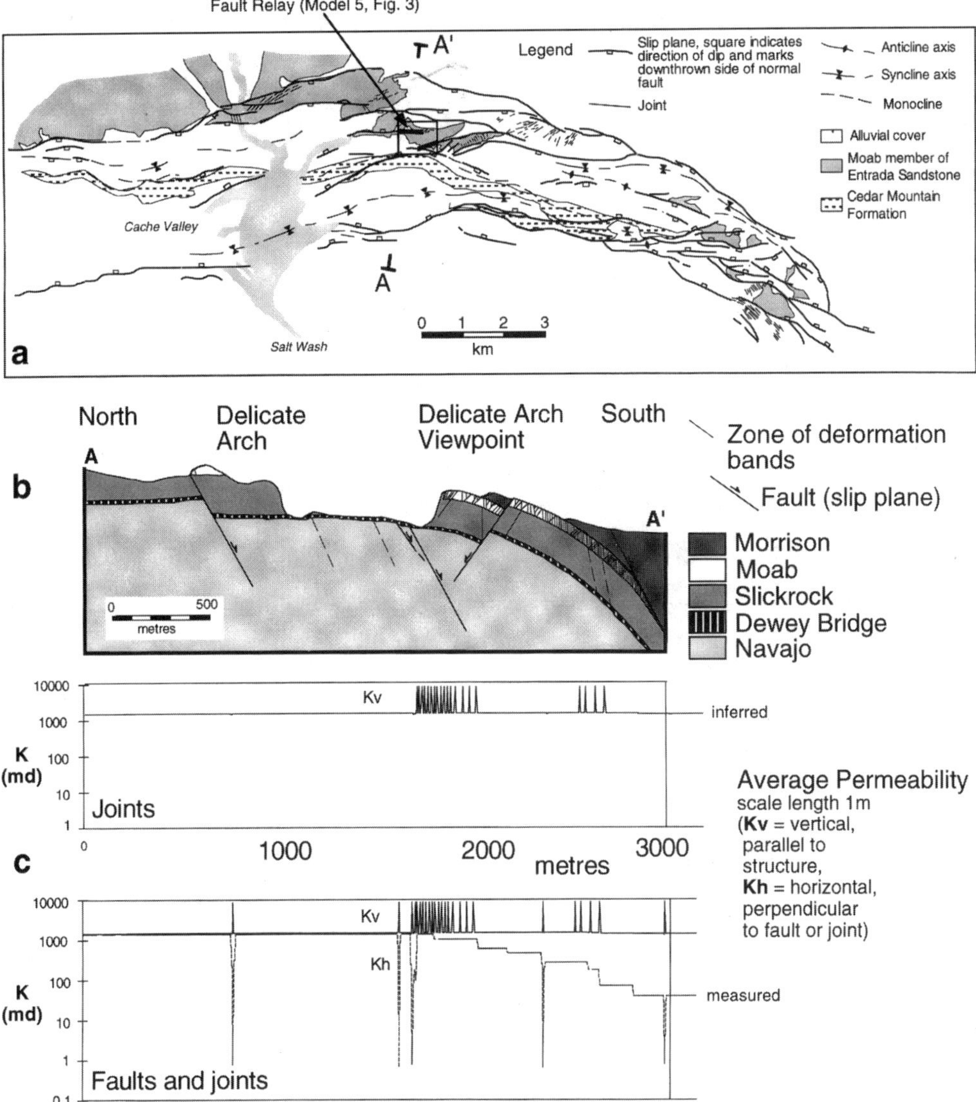

Fig. 1. Geological map and cross-section of the Cache Valley, Arches National Park, Utah (adapted from Antonellini & Aydin 1994, 1995). (**a**) Map of normal faults with segmented slip planes and the salt anticline in the Entrada Formation in the Cache Valley. (**b**) Cross-section along the profile A—A′ through the Cache Valley. (**c**) The graphs below the profile show measured and inferred permeabilities of joints and faults, respectively.

radial drawdown periods occur in their presence. Such periods would allow one to invert well-test results for the sandstone matrix permeability. We also evaluate the relative magnitude of signatures of faults and small-scale inhomogeneities in pressure derivative curves, by examining how they spatially distort the radial drawdown pattern. With this analysis, we try to identify a hierarchy of importance of the deformation structures for reservoir simulations, and we investigate how their signatures originate in transient well tests in inhomogeneous reservoirs. A by-product of these simulations is an assessment of the effectiveness of reservoir compartmentalization by fault relays, i.e. the likelihood that reservoir domains will not be drained or that water coning will occur.

Matthäi *et al.* (1998) show that pressure perturbations can travel with a speed of kilometres per month along highly permeable slip planes of

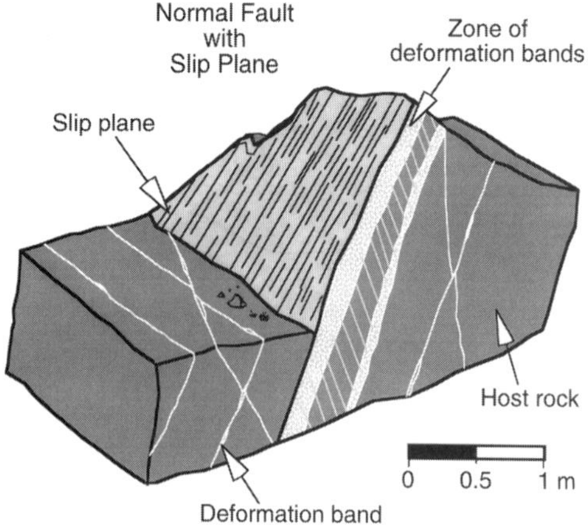

Fig. 2. Topology of a normal fault in sandstone at Arches National Park (after Antonellini & Aydin, 1994).

normal faults. In contrast, such pressure perturbations travel at hundreds of metres per month within the reservoir sandstone (Matthäi & Fischer 1996; Matthäi *et al.* 1998). For reservoir production scenarios, it is commonly assumed that a steady-state fluid pressure distribution is obtained. For the hydraulic diffusivities constrained by the field data from Arches National Park, we investigate whether this assumption is valid and how reservoir fluid flow paths can vary with time.

First, we describe our numerical model of transient single phase fluid flow and review the field and laboratory data it is based on. Secondly, we present some simple small scale results that elucidate the basic characteristics of flow in sandstone with joints and deformation bands. Then we present results from km-scale models of the selected structural settings in the Arches Park (Fig. 3).

Simulation technique for transient single phase flow

We use the algebraic multigrid finite element method (Ruge & Stueben 1987; Matthäi & Roberts 1996; Roberts & Matthäi 1996) to simulate single phase fluid flow into vertical wells that are underpressured relative to the two-dimensional flat lying sandstone reservoir. To quantify the fluid diffusivity in the reservoir, we need to specify fluid viscosity, μ, permeability, k, and storage capacity, S (storativity), of the

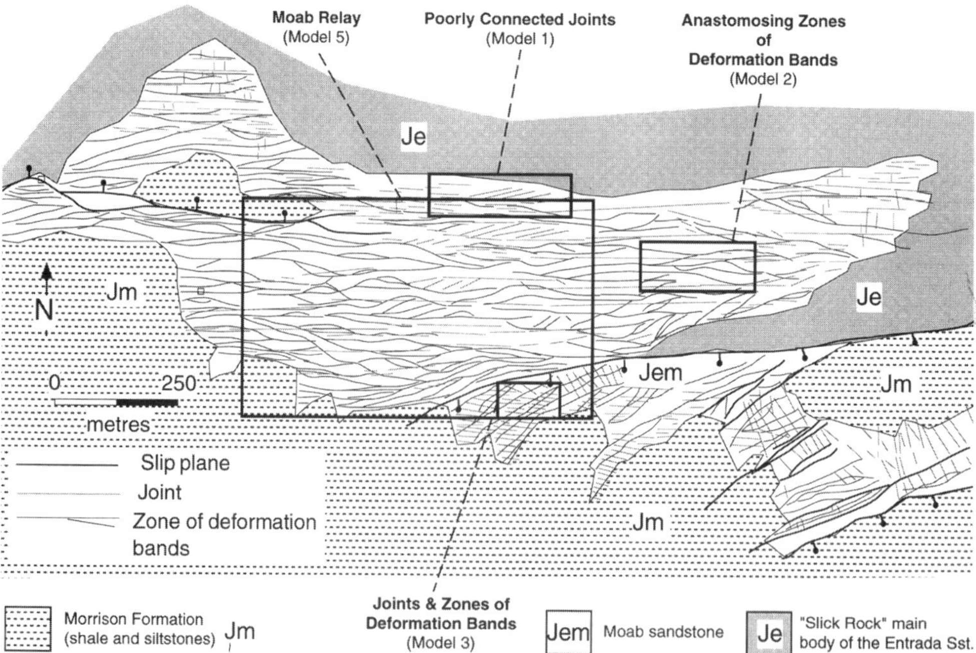

Fig. 3. Map of a fault relay in the Delicate Arch area, Arches National Park, Utah (after Antonellini & Aydin 1995). The map shows the hierarchy of structural inhomogeneities: normal faults with slip planes (bold black lines with halve dumpbells), joints (stippled lines), and deformation bands (solid anastomosing lines). Rectangles mark the regions which were represented in two dimensional fluid flow models of this analogue reservoir.

heterogeneous rock. S describes the change with fluid pressure, P, of the fluid volume, V_f, stored in the porosity, Φ, of a unit volume of reservoir rock. We calculate S from the porosity measured in the field and sand matrix and fluid compressibilities, α and β respectively, for a constant matrix mass.

$$\frac{\partial(V_f)}{\partial P} = ((1-\phi)\alpha + \phi\beta) = S. \quad (1)$$

This constitutive equation represents a simplification of the formulation of Brace *et al.* (1966) which also incorporates the compressibilities of mineral grains. Since the present day compressibility of the Entrada sandstone may have been modified during exhumation and compressibility varies with confining stress (Fischer & Paterson 1992), we use oilfield sandstone compressibility measurements under *in situ* conditions. Dvorkin & Nur (1996) measured α of cemented Oseberg sandstone sampled from oil fields in the North sea. At a porosity of 25% and an effective pressure of 30 MPa, the matrix compressibility, α, of the Oseberg sandstone is approximately 25 GPa, corresponding to a storativity of $1.68 \times 10^{-7}\,\mathrm{m^3\,Pa^{-1}}$, for a light oil (API = 45) as a pore fluid.

Given the measurement of Poisson's ratio of the Oseberg sandstone of 0.25 by Dvorkin & Nur (1996), the storativity of joints in the reservoir, S_j can also be calculated. It is equivalent to the joint-volume change with fluid pressure which can be derived as an extension of the analysis of mode-I fractures presented in Pollard & Segall (1987). Under the assumption that the joints are straight and the joint walls are not in contact along the joint plane

$$S_j = \frac{\partial V_j}{\partial P} + V_j\beta\,\Delta P \quad (2)$$

$$V_j = \frac{-\pi a^3(v-1)(P+\sigma_{yy})}{G} \quad (3)$$

$$\frac{\partial V_j}{\partial P} = \frac{-\pi a^3(v-1)}{G} \quad (4)$$

where V_j, v, G, and σ_{yy} are the joint volume, Poisson's ratio, shear modulus, and the joint normal stress, respectively; a is the half length of the joint. This joint storativity increases with V_j and joint aspect ratio (length/width). With equation (4) we calculate length-dependent joint storativities for the oil compressibility constrained below. Joint storativities calculated from equation (4) for joints that extend more than 100 m are slightly greater than that of the reservoir sandstone.

The algebraic-multigrid finite-element code (Roberts & Matthäi 1996) calculates the spatial and temporal change in fluid pressure from the pressure diffusion equation using Darcy's law

$$S\frac{\partial P}{\partial t} = \frac{k}{\mu}\nabla^2 P. \quad (5)$$

The viscosity, μ, of the hydrocarbon fluid has to be interpolated from experimental data for selected reservoir conditions. Since we limit this analysis to single phase flow of oil, we only consider a petroleum liquid above the 'bubble' point. For a reservoir fluid pressure of 30 MPa, a temperature of 50°C, a gas–oil ratio (GOR) of 0.5, and an oil density of $800\,\mathrm{kg\,m^{-3}}$, we calculate a viscosity of about $4 \times 10^{-3}\,\mathrm{Pa\,s}$ with the commercial program PetroTools (PetroSoft Inc., San Jose). The compressibility of this oil at reservoir conditions is approximately $7.1 \times 10^{-10}\,\mathrm{Pa^{-1}}$ (Batzle & Wang 1992).

While we can take the permeability, k, of the reservoir sandstone, fault zones, and contained zones of deformation bands directly from the minipermeameter measurements of Antonellini & Aydin (1994, 1995), we have to calculate the oil viscosity-dependent transmissivity of joints and slip planes, $K(=k/\mu)$ from their aperture and length data. The lengths of the joints in the analogue reservoir are obtained from the maps (Fig. 3). Joint aperture measurements and calculations by Cruikshank *et al.* (1991) are discussed in the field data section. The transmissivity of joints, K, is commonly treated as analogous to the transmissivity of parallel plates (Witherspoon *et al.* 1980; Neuzil & Tracy 1981) leading to the cubic relationship

$$K = \frac{d^3}{12\mu} \quad (6)$$

where d is the plate spacing i.e., mean fracture aperture (Krantz *et al.* 1979). Brown (1987) compared the transmissivity predicted by this approximation with calculations using Reynolds equation to simulate fluid flow through rough-walled fractures with contacting asperities. His results indicate that equation (6) approximates the flow within a factor of 2. For wider fractures ($d \geq 100$) this factor further decreases towards 1.0 (Brown 1987; Renshaw 1995).

The described treatment of fluid flow in the reservoir relies on the following assumptions:

(1) Porosities and permeabilities measured in outcrop are similar to *in situ* values;
(2) Three dimensional reservoir flow can be approximated with a two dimensional model;
(3) Porosity and permeability of the sandstone matrix are set as homogeneous $\phi = 25\%$,

$k = 10^{-12}\ m^2$ whereas variations ≥ 1.5 orders of magnitude occur in the field;
(4) The fluid can be modelled as a slightly compressible single phase petroleum liquid above the 'bubble point', GOR 0.5, density $800\ kg\ m^3$, dynamic viscosity 4.0×10^{-3} Pa s (50°C, 30 MPa, analogue reservoir conditions);
(5) Fluid flow in the joints and along the slip planes of the faults is laminar.

A two-dimensional model of the analogue reservoir (assumption 2) seems appropriate in view of its planar field geometry. The cross-section through the reservoir unit (Fig. 1b) also shows that the faults, joints and deformation bands are at a high angle to the bedding and more or less continuous from the top to the base of the permeable sandstone unit. Thus, a planar representation of the analogue reservoir seems appropriate. Since we focus primarily on reservoir-scale effects of the structural heterogeneities in the reservoir, we limit this analysis to single phase fluid flow of oil.

Model implementation

To implement the reservoir models, the selected regions of the faulted Moab Sandstone in plan view (Fig. 3) are represented by regularly gridded, two-dimensional finite element meshes. These meshes vary in size from 700,000 to 4 million triangular elements. The investigated regions (rectangles in Fig. 3) are selected to capture the structural variability within the analogue reservoir.

Calculated storativities, measured permeabilities, and an initially uniform fluid pressure are assigned to the meshes. The well is represented by 2 triangular elements such that it has a square shape and a fixed pressure is assigned to the nodes of the well elements. Since the hydraulic conductivity of joints, zones of deformation bands, and slip planes in our model does not depend on fluid pressure, the pressure differential applied at the well has no effect on the propagation velocity of fluid-pressure perturbations in the reservoir. We normalize this differential to 100 such that the change in reservoir pressure is equivalent to a percentage change of the difference between the initial pressure in the reservoir and the fixed drawdown pressure in the well. The effect of a pressure-dependent joint and slip plane permeability is discussed after presentation of the modelling results.

Temporal and spatial variations in fluid pressure are computed with exponentially increasing time-step size ranging from 60 s to 3 days. This approach is taken because initial (local) pressure changes in the reservoir occur rapidly but subsequent (regional) changes occur more slowly. This is a consequence of the proportionality between diffusion velocity and driving gradient (cf. equation 5).

In spite of the high resolution of the finite element meshes, we cannot represent the slip-planes and the joints with their correct thickness. Therefore, effective permeabilities are applied to compensate for the over-representation of slip plane and joint thickness in the models (see later). These effective permeabilities are calculated with the power averaging method of Deutsch (1989) as anisotropic and directional averages over the cubic-law permeabilities calculated from the correct joint apertures, and the permeability of the Moab sandstone that would occupy respective model regions in an exact representation. This treatment reduces joint-parallel transmissivities calculated from equation (6) by one order of magnitude and joint-normal transmissivities by up to three orders of magnitude.

The analogue reservoir: fluid flow properties

We employ Antonellini & Aydin's (1994, 1995) field maps and porosity and permeability measurements of the faulted and fractured Moab and Slickrock sandstones in the Delicate Arch area (Figs 1 & 3) to build a model of the analogue reservoir for our simulations. These sandstones are extensively jointed and intersected by arrays of deformation bands, representing large positive and negative deviations from the permeability and porosity of the undeformed sandstone, respectively. Additionally, the Delicate Arch area contains a relay structure of normal faults dipping away from the relay centre (Fig. 3). This spatial coincidence of critical structures permits their relative influence on fluid flow to be investigated in a single model.

Sandstone porosity and permeability. The Moab sandstone is a cross-bedded, well-rounded, and well-sorted dune sandstone (Doelling 1985) with porosity variations between 4 and 28% (Antonellini & Aydin 1994). The majority of porosity measurements cluster around 22%. As determined with an air flow minipermeameter, the permeability of the undeformed Moab sandstone varies over 3 orders of magnitude between 10^{-11} and $10^{-14}\ m^2$ (Antonellini & Aydin 1994). Commonly, this permeability range is narrower, between 10^{-11} and $10^{-12}\ m^2$. The Moab sandstone conformably overlies the 'Slickrock' sandstone which is another permeable unit with roughly the same properties (permeability 10^{-12} to $10^{-13.5}$ m^2). The combined thickness of the Moab and Slickrock sandstones varies between 80 and 200 m and the thickness of the Moab varies between 20 to 40 m in the Delicate Arch area (typically 30 m, Antonellini & Aydin 1995). The Moab and Slickrock sandstones occur between the much less permeable Dewey Bridge and the Morrison Shale formations, which can be considered as impermeable boundaries to the analogue sandstone reservoir. We treat sandstone permeability and porosity as uniform over the area of the models, using values of 25% and 10^{-12} m^2, respectively (Table 1).

Normal faults with slip planes. Antonellini & Aydin (1994, 1995) identify these faults as zones of deformation bands that are associated with slip surfaces with offsets of metres to a few tens of metres. These faults are segmented with the individual segments being semi-continuous for kilometres along strike (Fig. 1a). In cross-section (Fig. 1b), they extend into the stratigraphy above and below the Moab and Slickrock sandstones. The slip planes form 300–500 m segments along fault strike and occur, most-commonly, at the interface between the fault zone and the relatively undeformed hanging wall of the normal faults (Figs 2 & 4c). The two surfaces of a typical slip plane are smooth and polished with thin lineations (slickensides). These surfaces are wavy with a wavelength of 2–15 cm and an amplitude of 1 cm or less. Present day apertures of the slip planes are equal to or less than a few mm. Locally, *in situ* (sub-surface) dilatation of slip planes by up to 10–15 cm is indicated by open-space fillings of calcite. These are considered as evidence for preferential fluid circulation paths in the faults. In view of those characteristics, and the growing body of evidence for open fractures at several km depth with apertures in the cm range (e.g. Engelder & Scholz 1981; Barton & Zoback 1992), it appears reasonable to infer *in situ* apertures, at depth, of a few mm to centimetres (cm) for the slip planes of the normal faults (Fig. 2). Such apertures agree well with Cruishank *et al.*'s (1991) calculated apertures of sheared joints in the Entrada sandstone in the Garden area of Arches National Park during the time of the deformation (at a burial depth of $\geq$1 km). These apertures imply a high slip-plane transmissivity (Pittmann 1981).

The porosity in the footwall rock adjacent to well-developed slip planes is very low ($\geq$1%) due to grain crushing, recrystallization, and perhaps diagenesis. In the Slickrock and the Moab Sandstone, the permeability of slip-plane wallrock is around one millidarcy (10^{-15} m^2) and less than or equal to 10^{-18} m^2 (the detection limit of the permeameter), respectively (Fig. 5).

The width of the fault zones as defined by the extent of highly clustered deformation bands generally correlates to fault displacement (Antonellini & Aydin 1995). In the Delicate Arch area, this width ranges from a few metres to about 20 metres. Over this width, the fault-normal permeability is reduced to the spatial average of the minipermeameter measurements.

Zones of deformation bands. Deformation bands in porous sandstones are mm-wide tabular bodies in which the porosity is reduced and the grains may be crushed (Aydin 1977, 1978; Figs 3 & 4). Typically, the porosity is reduced to a few percent in the centre of the deformation band. In the Moab sandstone in the Delicate Arch area, deformation bands rarely occur individually, but they tend to cluster in zones that may reach a few decimetres in thickness.

In contrast with the rectilinear pattern of joints in the Delicate Arch area, the zones of deformation bands form a curvilinear pattern and anastomose (Fig. 3). In most cases, the deformation bands are cross-cut by the joints. This indicates an earlier faulting episode and later jointing (Antonellini & Aydin 1995), which is important for the permeability structure. The zones of deformation bands extend for tens of metres along strike and cut the Moab sandstone from top to base at a predominant dip angle of approximately 50°. In cross-section, the zones of deformation bands are restricted to the Moab and Slickrock sandstones and form an apparent conjugate pattern, compartmentalizing the sandstone into lozenges.

In accordance with the Kozeny–Carman relation between grain size and permeability (Bear 1972), most deformation bands are less permeable than the undeformed parent rock. A summary of the minipermeameter measurements of Antonellini & Aydin (1994) is given in Fig. 6. For two types of deformation bands (without and with cataclasis) the permeability of the host sandstone is reduced by 1 and 3 orders of magnitude, respectively (Antonellini & Aydin 1994). In some cases, the permeability is reduced to a tenth of a millidarcy (10^{-16} m^2). Where deformation bands in the Moab sandstone contain calcite cement, permeabilities down to the detection limit of the minipermemeter (10^{-18} m^2) were observed.

Joints. Joints in the Moab sandstone and Slickrock are straight and continuous over tens to

Fig. 4. Field characteristics (photographs) versus model implementations of joints (**a**), zones of deformation bands (**b**), and faults with slip planes (**c**) in the Delicate Arch area. (a) Joints (subvertical discontinuities with vegetation within them) are represented by regular meshes with a minimum width of two triangular finite elements. (b) Anastomosing zones of deformation bands with oblique boundaries (relative to the regular mesh) are approximated by vertical and 45° element boundaries. (c) Faults with slip planes (cross-sectional photograph with the slip plane on the left), are represented by an assembly of slip-plane elements and footwall elements. The footwall elements are assigned average transmissivities calculated in Antonellini & Aydin (1995).

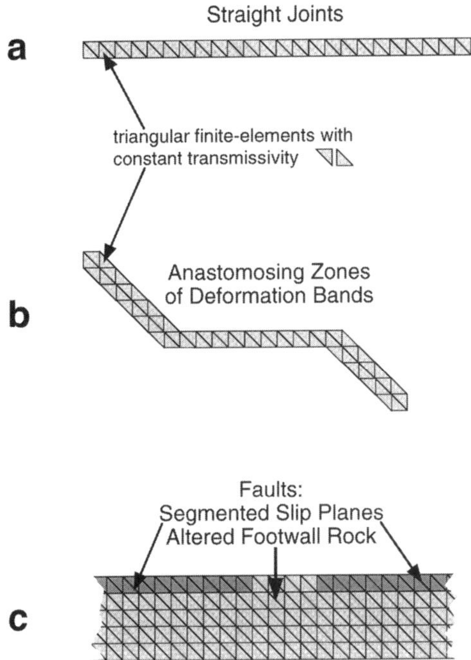

Fig. 4. Continued.

hundreds of metres (Fig. 3). In some cases, larger joints consist of arrays of m-long en echelon joint segments. Some joints have been sheared and tail cracks are present at their terminations (Dyer 1983; Cruikshank *et al.* 1991). The joints dip steeply ($\geq 50°$), penetrating the Moab and Slickrock sandstone layers from top to base, but not beyond. Their vertical extent is thereby limited to the thickness of the Moab and Slickrock units (approx. 50 m and 100 m, respectively, in the Delicate Arch area except for sandstone layers that are separated by fine-grained material and interdune deposits). The joints can be continuous along strike over distances greater than several tens of metres.

While most of the outcropping joints are open fractures, there are joints that are filled with mm to cm thick calcite precipitates. These mineral precipitates indicate *in situ* apertures from less than a mm to a few cm. Cruikshank *et al.* (1991) calculated the initial aperture of plain and sheared joints in the Entrada sandstone in the Garden area, Arches Park, Utah. From the magnitude of joint-parallel displacement, stress ratios deduced from the angle of splay cracks, and the mechanical properties of the sandstone, they constrained minimum apertures of 4 to 12 mm during mode-I joint formation and later mixed mode-I/mode-II growth of tail cracks on tens-of-metres-long joints.

Independently of these fracture-mechanical constraints on joint aperture, the minimum average aperture of joints in the Moab sandstone is constrained to about $\frac{1}{2}$ of the size of the sand

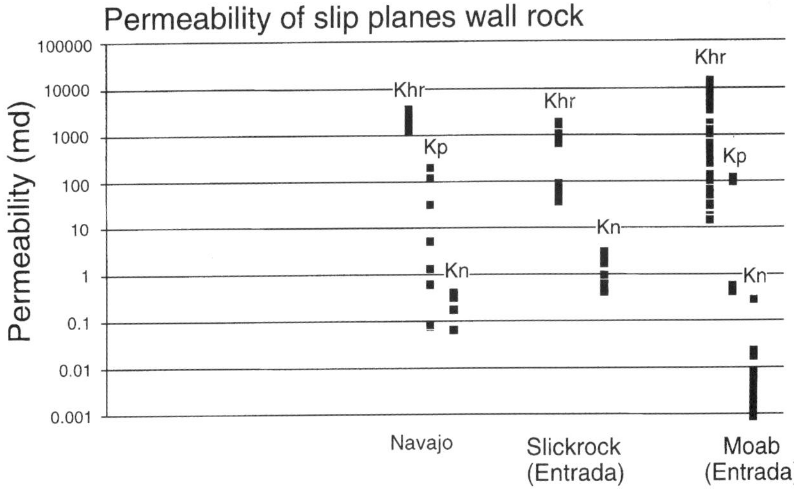

Fig. 5. Permeability of the wall rock of slip planes at Arches National Park (adapted from Antonellini & Aydin 1994). The lower limit of the slip plane perpendicular permeability of the rock in the footwall of the slip planes is unconstrained, because it lies below the detection limit of the minipermeameter. K_{hr}, K_p, K_n are the permeability of the host rock, and the deformation band parallel and normal permeability, respectively.

Table 1. *Summary of input data and characteristics of the fluid flow models*

General model properties			
Analog reservoir sandstone			
Porosity			0.25
Storativity			1.68E – 07 m^3 Pa^{-1}
Bulk modulus			2.40E + 10 Pa
Poisson's ratio			0.2
Fault properties			
Slip-plane storativity			5.71E – 07 m^3 Pa^{-1}
Footwall storativity			1.00E – 09 m^3 Pa^{-1}
Oil properties			
Viscosity			4 cp
Density			800 kg m^3
API gravity			45
GOR			0.1
Compressibility			7.14E – 10 Pa^{-1}
Poorly interconnected joints model			
Dimensions	83 m	×	289 m
Resolution			0.25 m
Joint width			0.25 m
Effective joint transmissivity			2.50E – 06 m s^{-1}
Anastomosing zones of deformation bands model			
Dimensions	95 m	×	229 m
Resolution			0.25 m
Deformation-band thickness			0.25 m
Deformation-band permeability			1.00E – 15 m^2
Morrison formation permeability			1.00E – 15 m^2
Joints & deformation bands model			
Dimensions	64 m	×	125 m
Resolution			0.165 m
Joint width			0.165 m
Effective joint transmissivity			8.00E – 06 m s^{-1}
Deformation-band thickness			0.165 m
Deformation-band permeability			1.00E – 15 m^2
Fault relay model			
Dimensions	410 m	×	1100 m
Resolution			1 m
Slip-plane thickness			1 m
Slip-plane transmissivity			2.50E – 05 m s^{-1}
Fault thickness			5 m
Avg. footwall permeability			1.00E – 16 m^2
Complex fault relay in delicate arch area model			
Dimensions	433 m	×	757 m
Resolution			0.75 m
Slip-plane thickness			0.75 m
Slip-plane transmissivity			8.00E – 05 m s^{-1}
Fault thickness			3.75 m
Avg. footwall permeability			1.00E – 16 m^2
Joint width			0.75 m
Effective joint transmissivity			2.50E – 06 m s^{-1}
Deformation-band thickness			0.75 m
Deformation-band permeability			1.00E – 15 m^2

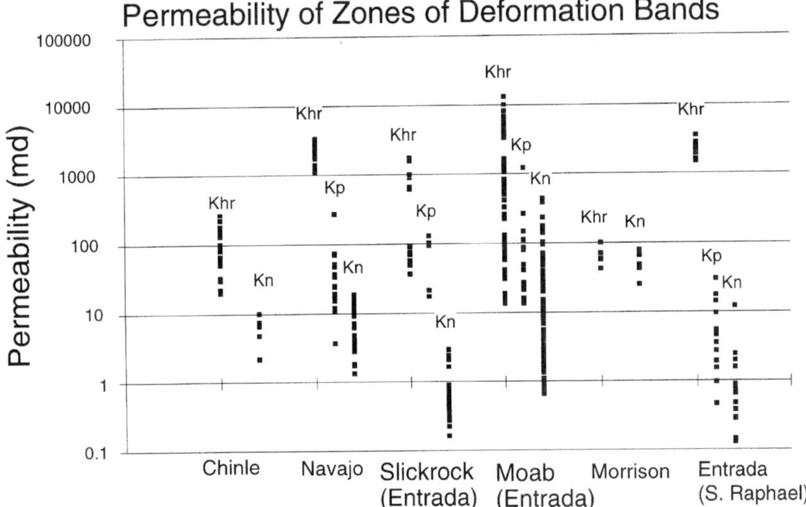

Fig. 6. Summary of permeability measurements for zones of deformation bands, Arches National Park and San Raphael desert (adapted from Antonellini & Aydin 1994). K_{hr}, K_p, K_n are the permeability of the host rock, and the deformation band parallel and normal permeability, respectively.

grains (approx. 1 mm in average for the well-rounded grains, cf., Fig. 2, Antonellini & Aydin 1994). This applies because jointing occurred along the cemented grain contacts and many of the joints show very small amounts of later slip, such that the joint walls are slightly offset relative to one another. Consequently, surface irregularities do not match across the joint walls.

In view of these constraints, we implement transmissivities for the tens-of-metre-long joints that are calculated for an intermediate joint aperture of 5 mm (Table 1). Joint transmissivities for the aperture minimum of 0.5 mm would be 2.5 orders of magnitude lower than the implemented values. Since model resolution dictates that we represent joints with a width of a few tens of cm (see Table 1), effective transmissivities calculated with the power-averaging method of Deutsch (1989) are used to compensate for the joint thickness over-representation. A dependence of joint aperture on the ambient stress state, fluid pressure and burial depth has been demonstrated in a range of studies (Makurat *et al.* 1990; Barton *et al.* 1995; Zhang & Sanderson 1996*a*,*b*). We do not incorporate such dependence in our models. Implications of this limitation are discussed after the presentation of the modelling results.

Results

Among many, Cinco *et al.* (1976), Yaxley (1987), Grader & Horne (1988), Kuchuk & Habashy (1992), and Matthäi *et al.* (1998) have investigated how linear high and low permeability zones affect the production rate and drainage pressure in a well. Highly permeable joints more rapidly conduct pressure changes in a well into the surrounding reservoir which is expressed spatially as a positive deviation from radial drawdown. This is illustrated in Figure 7 which shows pressure contours (Pa) at 2.7, 17.5 h and 4.8 days (Fig. 7a), and a corresponding pressure/derivative plot for this example simulation (Fig. 7b), respectively. Since fluid is supplied more rapidly to the well, the pressure decline at the well decelerates while the pressure front travels along the joint. The normalized pressure derivative is therefore lower than for radial drawdown ($\log t\, \partial P/\partial t \leq 0.5$). The magnitude of this deviation of the derivative scales with the joint matrix permeability ratio, joint length, and orientation relative to the radial drawdown front (Cinco *et al.* 1976).

Linear low permeability zones have the opposite effect on radial drawdown (Fig. 7a, b). They retard the radial spreading of the drawdown and reduce the rate of production which is reflected in an increase in the pressure derivative ($t \log \partial P/\partial t \geq 0.5$, cf., Kuchuk & Habashy 1992). As a result of this direct translation of spatial deviations from radial drawdown to derivative curves, and also because transient flow fields around a well are rarely presented in the literature, we focus on the display and discussion of the spatial evolution of deviations from radial drawdown and fluid-flow paths in the reservoir.

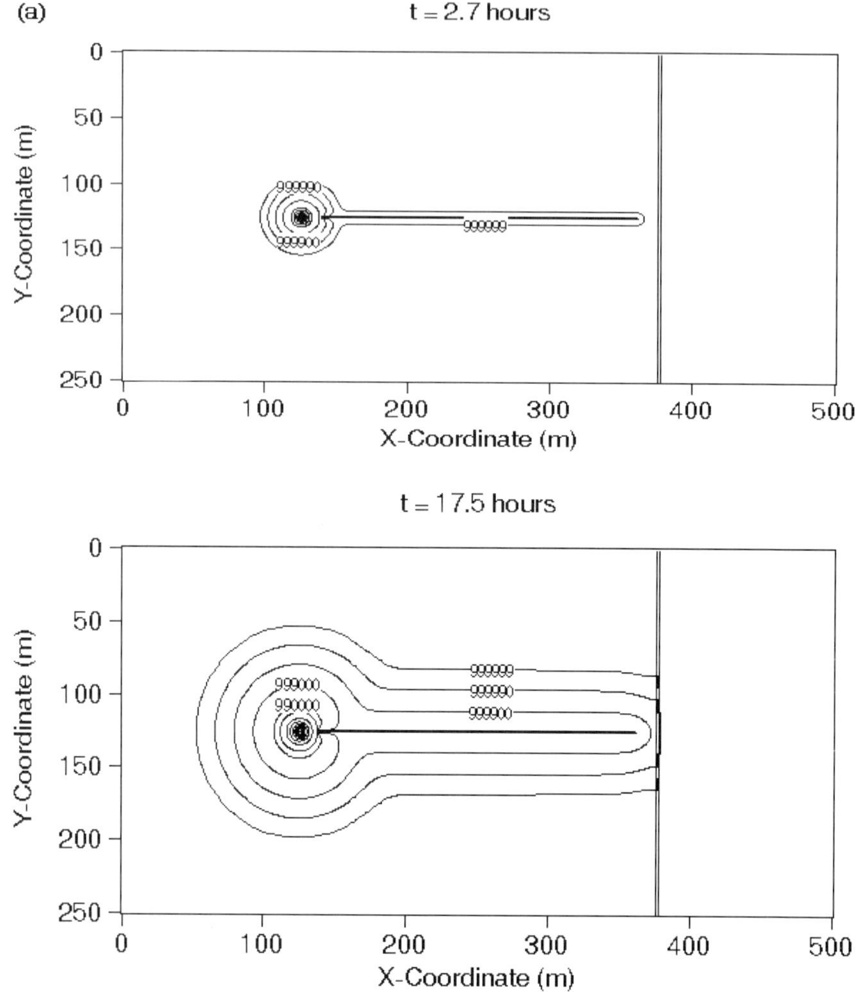

Fig. 7. Example simulation of the spatial variation in drawdown pressure with time (**a**) and its signature in a pressure derivative plot (**b**). Drawdown interacts sequentially with a highly permeable joint (10^{-8} m s^{-1}) and a low permeability fault (10^{-16} m^2). These are embedded in a homogeneous medium (10^{-12} m^2). (a) Fluid pressure contours (Pa) at 2.7, 17.5 h, and 4.8 days, showing the perturbation of the initially radial drawdown by the joint and the fault. Initial reservoir pressure is uniform (1 MPa). Drawdown spreads faster along the joint but is retarded by the fault. (b) In the derivative plot, interaction of the drawdown with the highly permeable joint decelerates pressure change below 0.5 (radial drawdown). However, as this interaction develops, the pressure front also reaches the low permeability fault and the pressure derivative finally increases above 1.

The figures we employ (e.g. Figs 8–15), show deviations from radial drawdown using a grey-scale scheme in which departures from grey toward white represent regions of slower pressure decay relative to radial drawdown in a homogeneous model. Regions that are darker than grey indicate that fluid pressure decayed faster than in the homogeneous model. These deviations are calculated by subtracting fluid pressure computed for a homogeneous and isotropic model from the fluid pressure in the inhomogeneous model.

Overlain on the grey-scale images are fluid pressure contours (isobars of fluid pressure), which illustrate how the pressure front propagates and interacts with reservoir inhomogeneities over time. Additional contour lines depict spatial variations in the hydraulic conductivity of the model. However, the model resolution is too fine to permit the thickness of joints and deformation

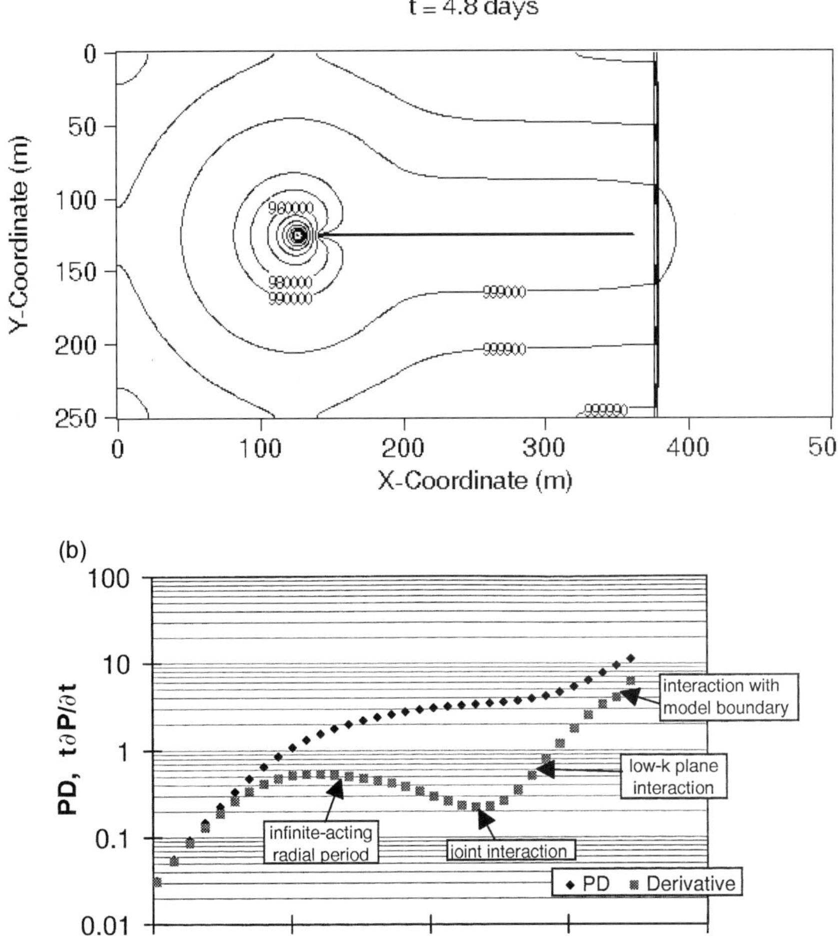

Fig. 7. Continued.

bands to be correctly represented in the images. The contours have been made wider such that they can be discerned in the images.

Note that the spatial extent of deviations from radial drawdown (as shown in the grey-scale images) has a more direct bearing on the shape of the pressure derivative, than the shape of the outer pressure contours. These contours depict small changes in pressure which pre-date the development of a deviation in the derivative curve.

In some of the figures (Figs 9, 12 & 16), fluid flow directions are illustrated with vectors. They are used instead of streamlines, because streamlines apply only if there are no fluid sources along the flow path. They are not useful here, because the fluid flow is unsteady and fluid flux variations along streamlines would therefore not be equal to zero. The length of the vectors scales with transport velocity (Darcy velocity), but vectors are omitted along joints as their disproportionally greater length (greater magnitude) would obscure matrix flow. A logarithmic length scaling of the vectors would remedy this, but is not used because it is too insensitive to variations in the transport velocity in the sandstone.

Poorly connected joints

Joints in the Delicate Arch area are often oriented parallel to the normal faults and, by contrast to the Garden area (Cruikshank *et al.*

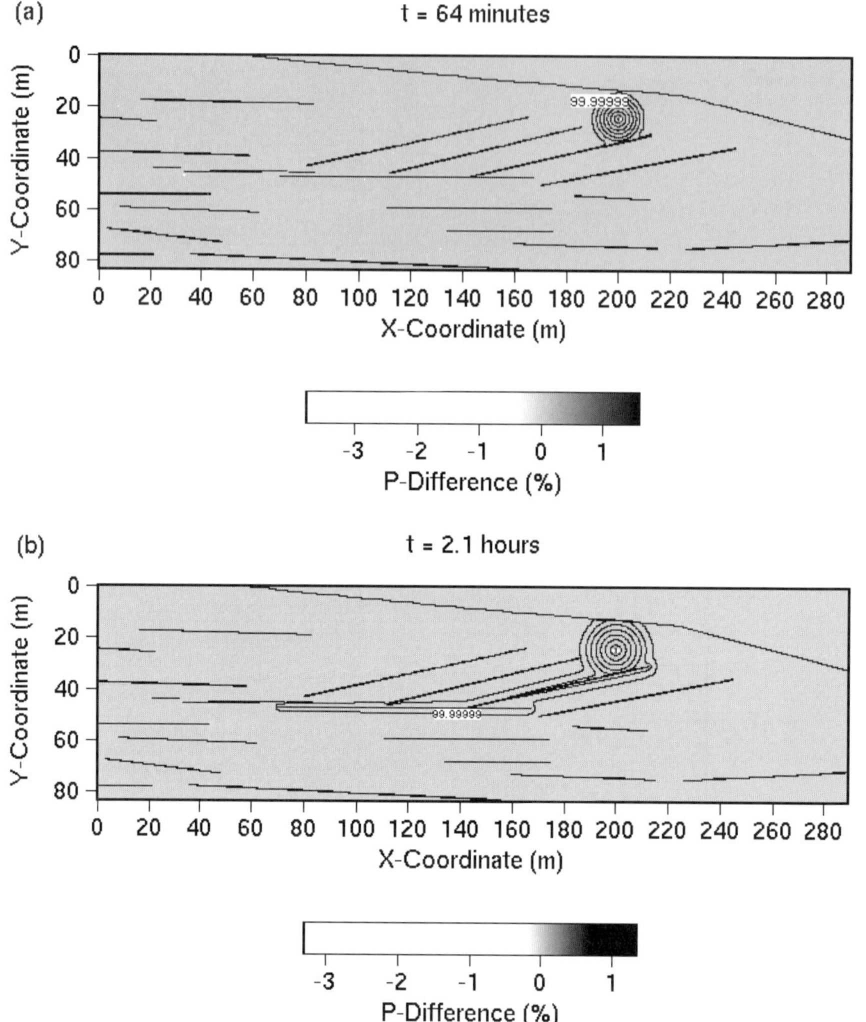

Fig. 8. (**a–e**). Model of poorly-connected joints in Moab sandstone. (joint parallel transmissivity 2.5×10^{-6} m s^{-1}, sandstone permeability 10^{-12} m^2, low permeability region in upper right 10^{-15} m^2). Deviations from radial drawdown are shown in grey-scale (white = less drainage, dark = increased drainage). Fluid pressure contours in percent of the difference between initial reservoir pressure and fixed drawdown pressure at well head. Fluid pressure contour intervals: 99.99999, 99.9999, 99.999, 99.99, 99.9, 99 and 90 %. Directions of the transient flow are indicated by vectors. Snapshots of the evolution of drawdown at 64 min, 2.1, 8.5, 68.3 h, and 11.4 days (**a–e**, respectively).

1991), they are poorly interconnected. Figure 8 shows an example of this joint pattern from the northeast of the Delicate Arch area that has been selected as the first flow model (Model 1, Fig. 3). The well is placed close to the tip of one of the WSW–ENE striking joints in the upper part of the model. This well placement ensures that radial drawdown interacts both with joints oriented radially to the well and joints that are oriented tangentially.

North of the well, where the Moab sandstone has been eroded, a low permeability region has been added to Model 1 (Figs 3 & 8). In this region the permeability is reduced by three orders of magnitude relative to the Moab sandstone, such that it is exactly the same as that of zones of deformation bands in the Moab sandstone. This low permeability region is included to illustrate how much slower the pressure front propagates for a lower matrix permeability.

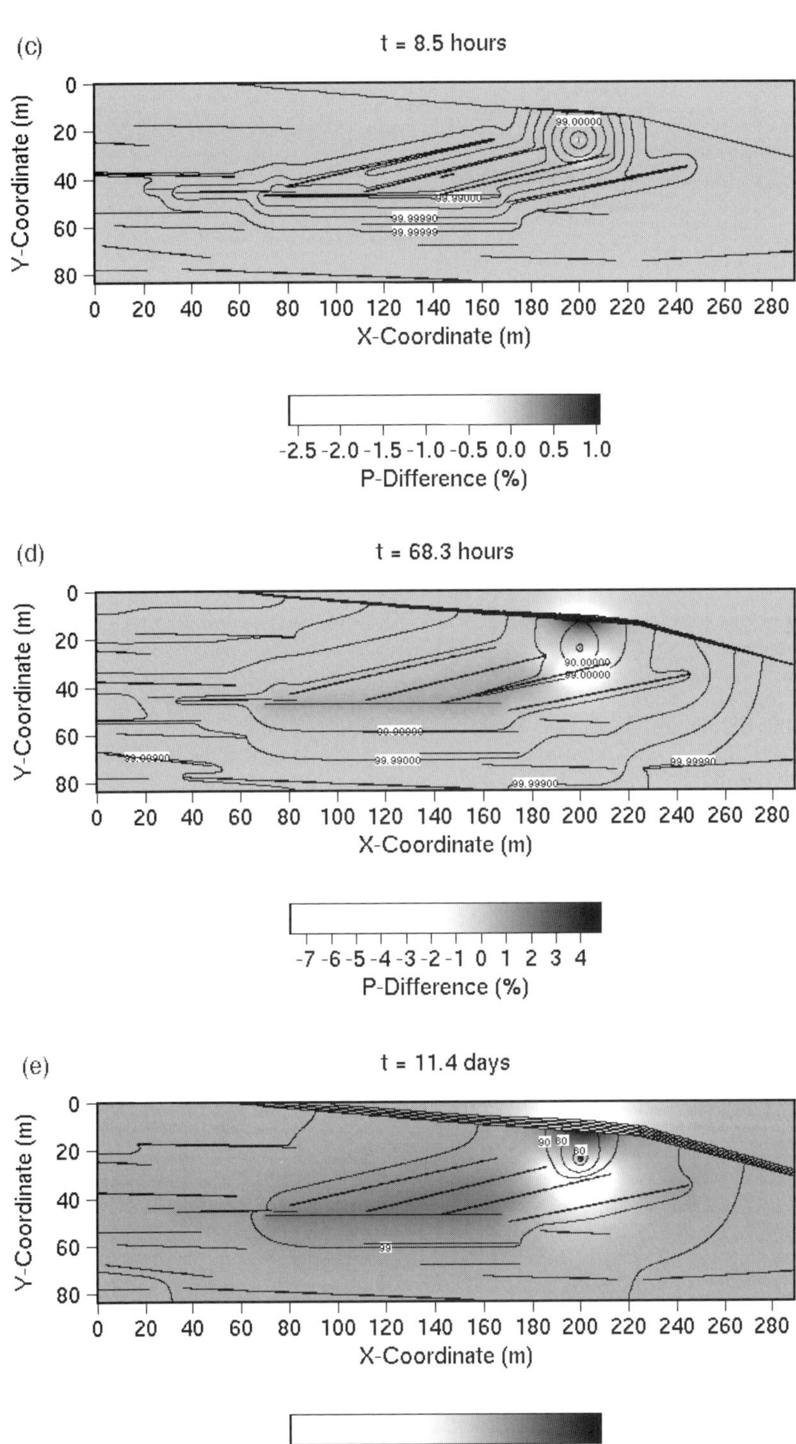

Fig. 8. Continued.

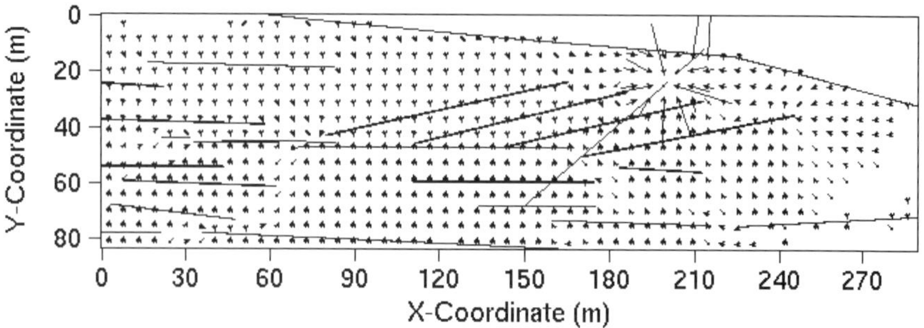

Fig. 9. Transient flow directions in the model of poorly connected joints in Moab sandstone at $t = 68.3$ h. The flow to the well is focused by the poorly-connected joints and where these are not connected, the flow follows a path that minimizes the travel distance through rock bridges among joints.

The evolution of drawdown from 64 min to 6 months after the application of the fixed pressure differential at the well is shown in Fig. 8. The 99.99999% contour first departs from radial at about 60 min ($= \Delta t$), rapidly spreading along the joint that is closest to the well (joint 1). It also spreads into the country rock along joint 1, at the same speed that it spread from the well to this joint. This is illustrated in the snapshot at 2.1 h (Fig. 8b) that shows the analogue reservoir just before interaction commences of the 99.99999% contour with the second closest joint to the well (joint 2). Interaction of the drawdown with another joint (joint 4) located in the west of the horizontal joint (joint 3) that is connected to joint 1, is also just about to occur. At the same time, the spreading of radial drawdown has stopped along the formation boundary north of the well.

At 8.5 h, the distorted drawdown front has reached the western model boundary having spread over all the joints in the vicinity of the well. The 10^{-6}% pressure perturbation indicated by the 99.99999% contour, has spread asymmetrically towards the west for over 7.5 times the distance that it would have spread in a homogeneous medium. The pressure gradients along the joints are minimal and the fluid that is supplied to the well along the poorly connected joint network counteracts the pressure decay near the well. Therefore, a positive pressure anomaly, relative to radial drawdown, develops where joint 1 is closest to the well. A similar anomaly develops in the low permeability region north of the well, where drawdown occurs at a slower rate than in the homogeneous model.

After 68.3 h, the positive and negative pressure deviations from radial drawdown shown in the grey-scale image (Fig. 8d) have developed in magnitude up to 7 and 4 %, respectively. In spite of the additional fluid supply along the joints from a much larger area, the analogue reservoir next to the joints 1–3 is depleted more rapidly than a homogeneous reservoir. The distant fluid supply is reflected in the flow field (Fig. 9) and in the stabilized fluid pressure at the tips of joints 1 and 2 that changed very little as compared with the 8.5 h snapshot. A steep pressure gradient has built up within the low permeability region north of the well where it borders the Moab sandstone. Beyond this boundary zone, the pressure in the low permeability region has not been altered. Elsewhere, significant interaction with the no-flow boundaries of the model has occurred.

After 11.4 days (Fig. 8e), the relative fluid pressure in large areas of the model has decreased by more than 1%, but it has been altered appreciably only at a 4–8 m wide margin of the low permeability region. The fluid pressure near the well, which determines the pressure derivative, is under strong influence of joints 1 and 2. This indicates a strong negative deviation of the pressure derivative from 0.5, in spite of the simultaneous interaction of the well with the low permeability region.

In summary, in the poorly-connected joints model, over a 1–2 h initial period, flow is radial and the analogue reservoir acts as though it were infinite in extent for the chosen well placement. After 2 hours, the drawdown starts to interact simultaneously with the joints that are south of the well, but also with the low permeability region in the north of the well. Subsequently, the influence of the poorly-connected joint network on the drawdown increases, leading to a decline in the pressure derivative below 0.5. This decline occurs in spite of the continuous interaction with the low permeability region in the northeast of the model. In spite of the poorly-connected nature of the joint network, the fluid that enters the well in the investigated period of time is derived largely from the western central part of the model. This origin would be severely misinterpreted under the assumption of radial drawdown. Also, the mean deviation in the hydraulic conductivity of the Moab sandstone fraction of the model relative to intact Moab sandstone is small ($\leq 0.1\%$). The flow, however, is strongly influenced by the joints which would not be captured by an effective permeability based simulation.

Anastomosing zones of deformation bands

Anastomosing zones of deformation bands are the most common inhomogeneities in the Moab/Slickrock sandstones. Antonellini & Aydin (1994, 1995) have shown that these zones are typically less than 2 cm wide. Model 2 overlaps with region 5 (Antonellini & Aydin 1995, fig. 24) in the east side of the Delicate Arch (Fig. 3). It over-represents the thickness of the zones of deformation bands by a factor of 20. In spite of this over-representation, the measured permeability (10^{-15} m^2) is applied to emphasize the effect of the deformation bands on radial drawdown. Two joints that occur in the represented analogue reservoir region are omitted from the model such that they do not interfere with the deformation bands. The well is located in a narrow region, where two zones of deformation bands coalesce.

The spatial variation in fluid pressure at time $t = 34.1$ h after the pressure differential was applied to the well is shown in Fig. 10. Drawdown is radial with the exception of small offsets in the pressure contours across the deformation bands in the west of the well. The grey-scale image does however show a fluid pressure difference of a few percent relative to the homogeneous model, indicating a higher fluid pressure in the west of the well and a lower fluid pressure in the east as compared with the homogeneous model. These differences indicate the inhomogeneity introduced with the zones of deformation bands, and their polarity reflects reduced fluid flow normal to the deformation bands, next to the well. The overall drawdown behaviour is close to homogeneous, which we ascribe to the narrowness of the zones of deformation bands. This narrowness prohibits the deformation bands from having a significant effect on drawdown.

This result indicates that the behaviour of Model 2 could be reproduced by an effective

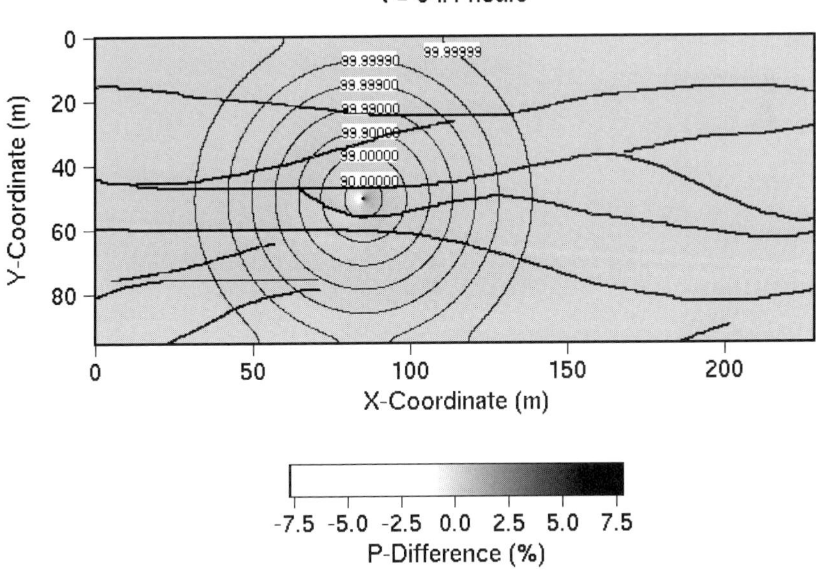

Fig. 10. Model of anastomosing zones of deformation bands in Moab sandstone, time-step 34.1 h (deformation band permeability 10^{-15} m^2, sandstone permeability 10^{-12} m^2). Deviations from radial drawdown shown in grey-scale (white = less drainage, dark = increased drainage). Fluid pressure contours in percent of the difference between the initial reservoir pressure and the fixed drawdown pressure at the well head. Contour intervals: 99.99999, 99.9999, 99.999, 99.99, 99.9, 99 and 90 %. Directions of the transient flow are indicated by vectors.

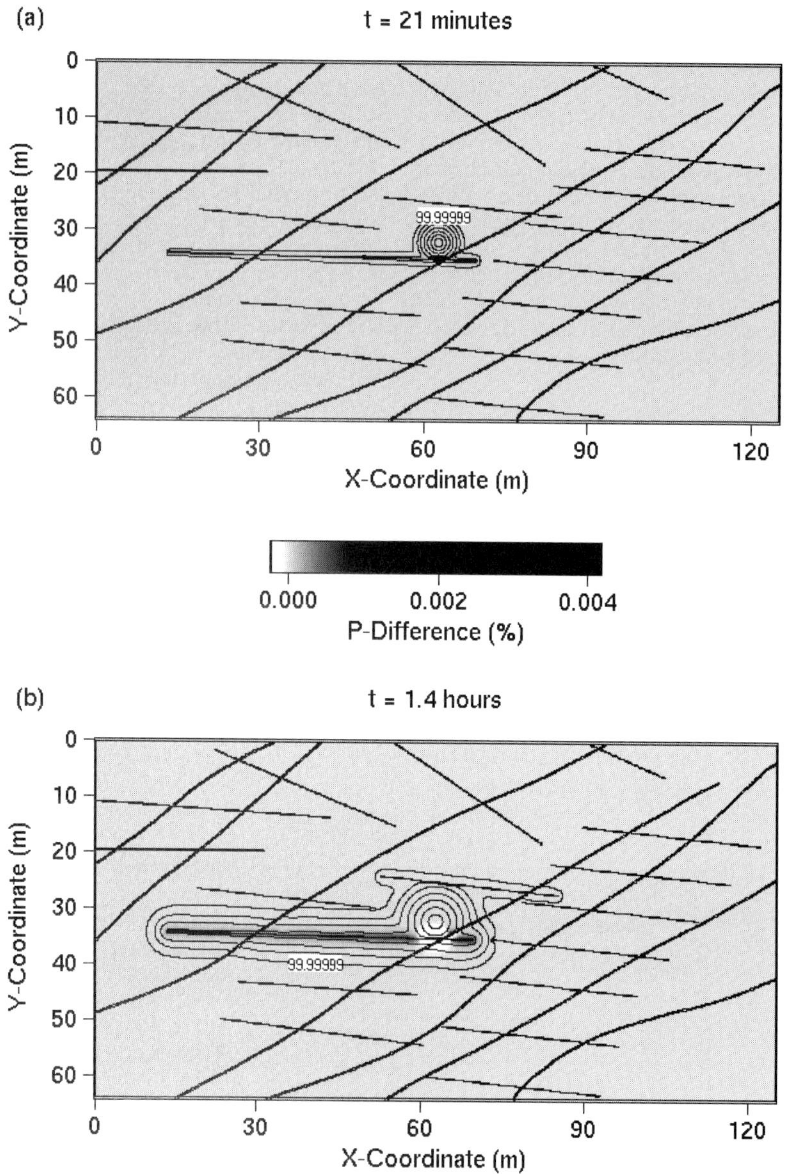

Fig. 11. Model of joints intersecting deformation bands in the Moab sandstone (joint parallel transmissivity 2.5×10^{-6} m s^{-1}, sandstone permeability 10^{-12} m^2, zones of deformation bands 10–15 m^2). Deviations from radial drawdown shown in grey-scale (white = less drainage, dark = increased drainage). Fluid pressure contours in percent of the difference between initial reservoir pressure and fixed drawdown pressure at well head. Fluid pressure contour intervals: 99.99999, 99.9999, 99.999, 99.99, 99.9, 99 and 90 %. (a) to (e): Snapshots of the evolution of drawdown at 21 min, 1.4, 5.6, 45 h, and 60 days.

permeability model (e.g. Deutsch 1989). The small volume of zones of deformation bands does not appreciably alter the average hydraulic conductivity of the represented region (mean log $K = -9.64933$ v. -9.6). In the derivative plot, however, the small-scale heterogeneities may be detectable as is indicated by the grey-scale image (Fig. 10) showing a difference between radial and model drawdown, which persists throughout the simulation. Only one time

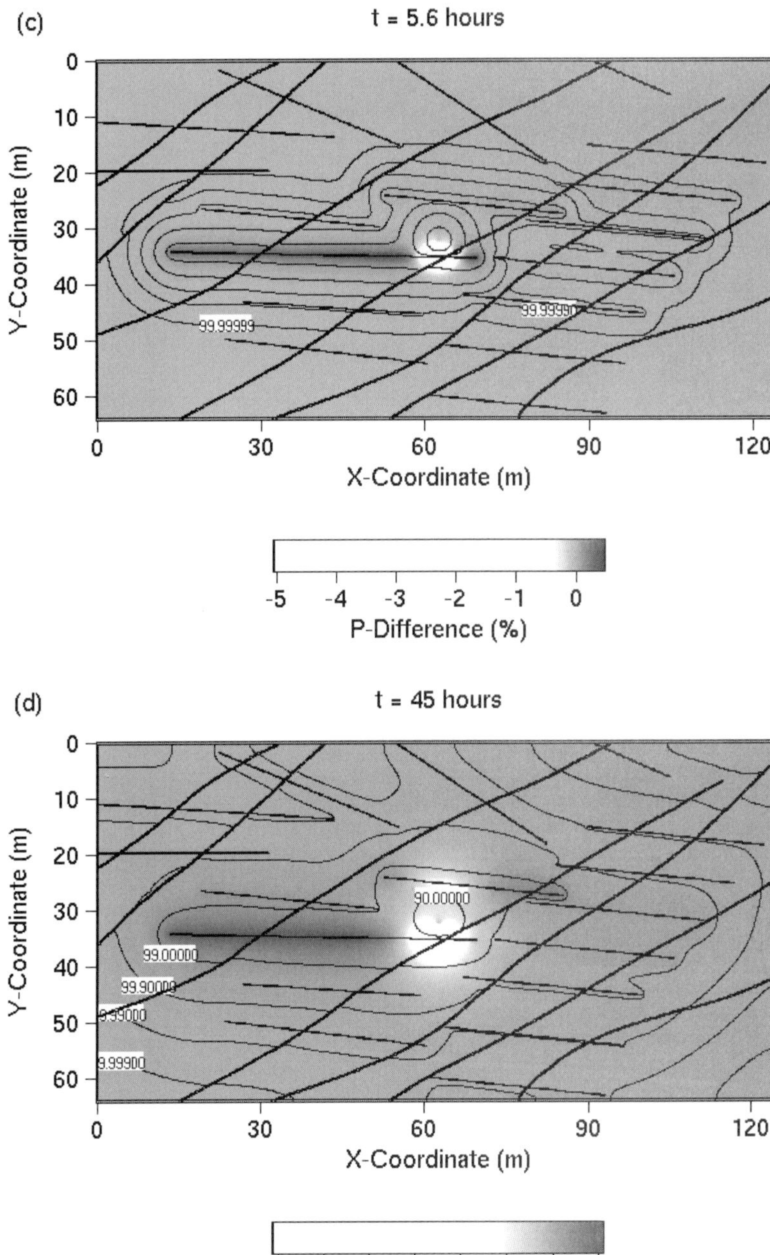

Fig. 11. Continued.

slice is shown here because drawdown remains very close to radial from the onset of the simulation to the interaction with the model boundaries shown in Figure 10.

In a multiphase flow scenario, deformation bands should impede flow of a non-wetting fluid or even act as capillary seals due to their high capillary pressure relative to the undeformed

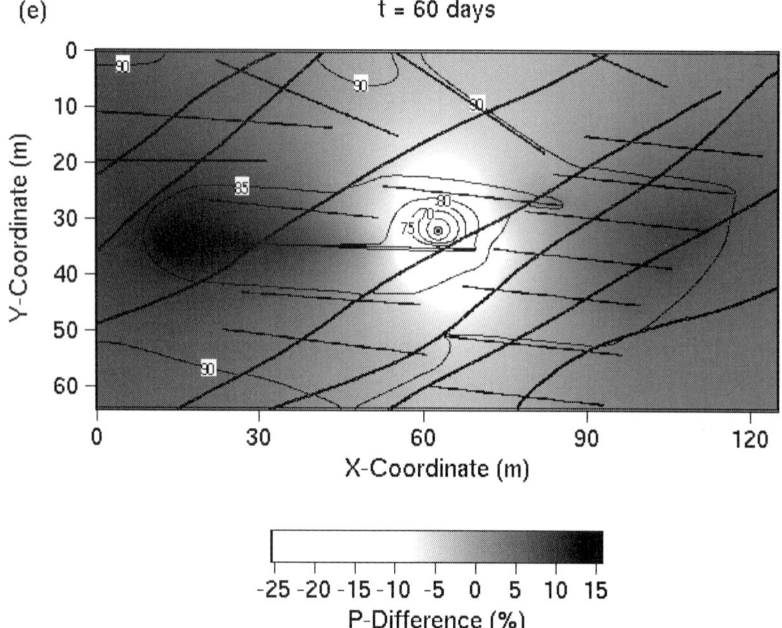

Fig. 11. Continued.

sandstone (Antonellini & Aydin 1994). Thus, our conclusions are only meaningful for single phase fluid flow.

Joints intersecting deformation bands

In the southeast of the Delicate Arch area in the hanging wall of the relay fault (Fig. 3, Region 8; fig. 24, Antonellini & Aydin, 1995), the curvature of the Moab sandstone layer increases while joint spacing decreases. Curvature-related joints intersect regularly spaced deformation bands at a greater angle than within the fault relay that bounds the analogue reservoir. These overprinting relationships between joints and deformation bands are represented in the flow model 'joints and deformation bands' (Model 3; Fig. 3; Table 1).

Model 3 develops a drawdown pattern that is relatively similar to the poorly-connected joints model. Drawdown is radial only for the first 15 min of the simulation, after which interaction with the joint south of the well and the deformation band in the east of the well occurs (Fig. 11a). This interaction is associated with a decrease in the pressure derivative as indicated by the white spot at the joint near the well. Drawdown spreads quasi-instantaneously along the joint and perpendicular to it, into the sandstone. The convergence of the 99.99999% contour to the joint with increasing distance from the well records the time it took for the pressure perturbation to propagate along the length of the joint. Flow in the joint is bilinear as described in Cinco *et al.* (1978) for finite-conductivity fractures. The effect of the zone of deformation bands on radial drawdown is very small.

After 1.4 h (Fig. 11b), the magnitude of the joint-induced pressure perturbation in the reservoir has increased to 0.5% and there is distinct darkening along the joint, indicating a faster pressure decrease where it drained fluid from the country rock. Drawdown has affected a rectangular region parallel to the strike of the joints, in spite of the presence of the deformation bands. The fluid pressure next to the well is stabilized by the rapid fluid supply along the joint and the pressure derivative would therefore be smaller than 0.5. Other joints close to the well start to distort the drawdown. There is little variation along strike in the distance of the pressure isobars from the joints. This indicates linear flow within the joints (Cinco *et al.* 1978). The joints near the well, that are not yet intersected by the pressure contours, deflect these toward the well, indicating that they already supply fluid to the well, stabilizing fluid pressure. There are small pressure discontinuities associated with the zones of deformation bands, as is indicated by

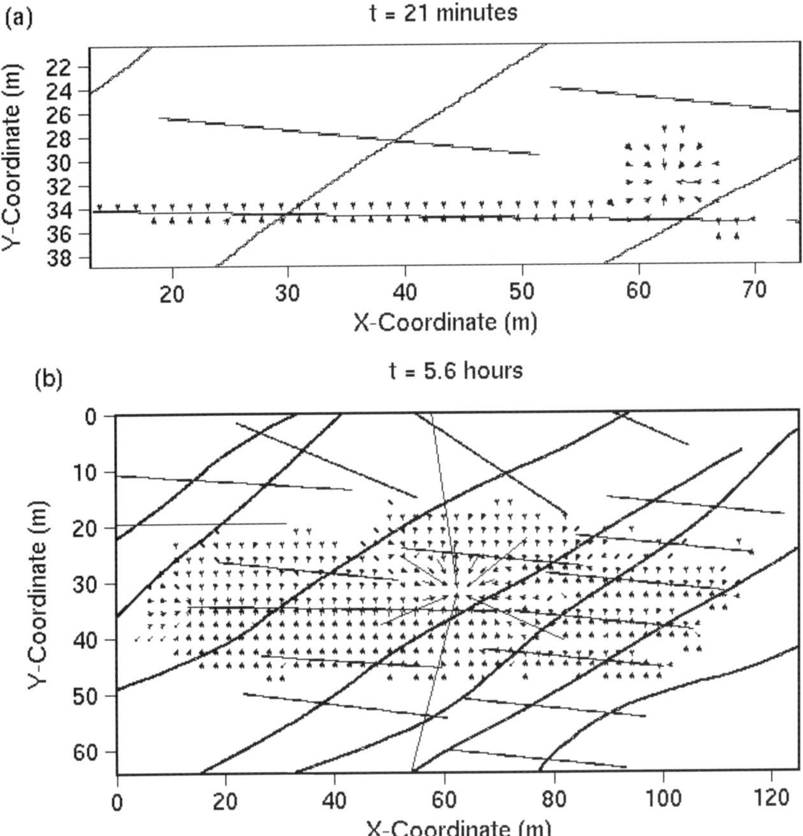

Fig. 12. Transient flow directions in the model of joints intersecting deformation bands in the Moab sandstone at $t = 21$ min and 5.6 h. (**a**) 21 min: Magnified central region of the model; the flow to the well is focused in the horizontal joint below the well in which flow is bilinear (Cinco *et al.* 1978). (**b**) 5.6 h: A complex flow pattern has developed; flow is focused into the joints that extend to near the well. This focusing is enhanced where the joints cross the zones of deformation bands.

the offset in the pressure isobars. These offsets are most pronounced where the long horizontal joint intersects the deformation band in the west of the well. The contours curve toward the deformation band, because it impedes cross-flow. Fluid flows with greater ease along the deformation band perpendicular to the pressure contours. The joints do, however, provide for good flow connections across the deformation bands.

After 5.6 h (Fig. 11c), the magnitude of the pressure deviations from radial drawdown has further increased to 5%. This means that the pressure near the well has declined appreciably less in the presence of the joints which supply fluid from a great distance. This supply is drawn from the jointed sandstone in a heterogeneous pattern that is best illustrated by the pressure island which formed near the second deformation band to the east of the well. Transient flow is away from this pressure high into the joints, as is shown in the vector plot (Fig. 12). The pressure front that spreads to the east of the well (99.99999% contour) is wavy due to the influence of the joints. In the west (on the left between 20 and 30 m Y-coordinate), this contour is offset significantly across the deformation band indicating its diminishing effect on pressure diffusion.

After 45 h (the maximum length of a typical drawdown test, cf. Horne 1996) the pressure in about a third of the model region has decreased by more than or equal to 0.1%. The small joint in the north of the well has developed a distinct inflow and outflow region as mirrored in the grey-to-white-to-grey transition along its strike. The flow pattern around the joint in the south of the well has changed relative to the 5.6 h snapshot: it now mainly supplies fluid from the

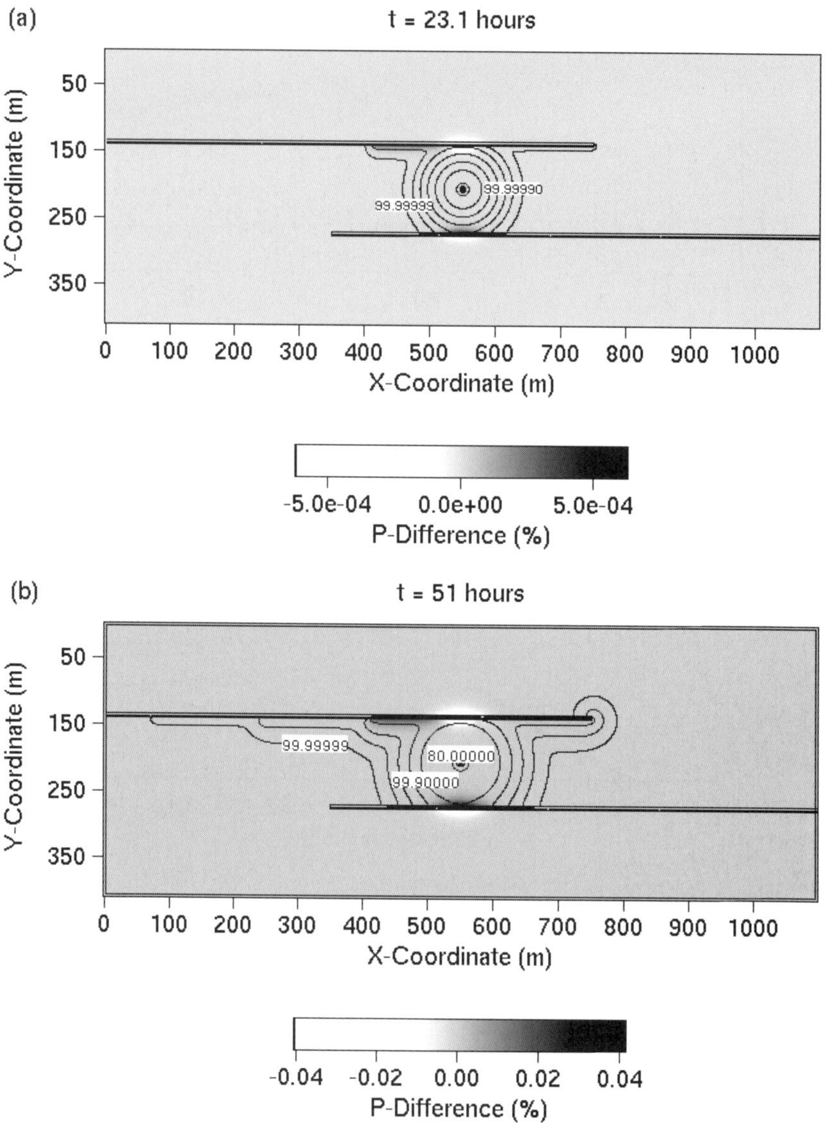

Fig. 13. Model of a relay of normal faults with highly-permeable segmented slip planes on the south side of the faults. Deviations from radial drawdown are shown in grey-scale (white = less drainage, dark = increased drainage) and dimensionless fluid-pressure is visualized with contour lines. Thickness-adjusted slip-plane parallel transmissivity 2.0×10^{-5} m s^{-1}, fault zone permeability 10^{-16} m^2, sandstone permeability 10^{-12} m^2. The permeability structure of the faults is illustrated in Fig. 6(c). (**a**)–(**f**): 23.1, 51 h, 4.7, 7.9, 13.3, 29.3 days, respectively.

west, while the joint north of the well is the main conduit for fluids from the east. The pressure deviation from radial drawdown has increased to 18% near the well while the pressure away from the well is only up to 3% lower than for radial drawdown. Computer-aided data analysis shows that this asymmetry is a measure for the relative sizes of the fluid supply regions of the joints relative to their discharge region next to the well; fluid is collected from six times the area over which it is discharged to the well. Flow to the joints has also progressed toward radial (Cinco *et al.* 1978) with the growth of the charge regions of the joints.

Sixty days after the start of the simulation (Fig. 11e), strong interaction with the no-flow

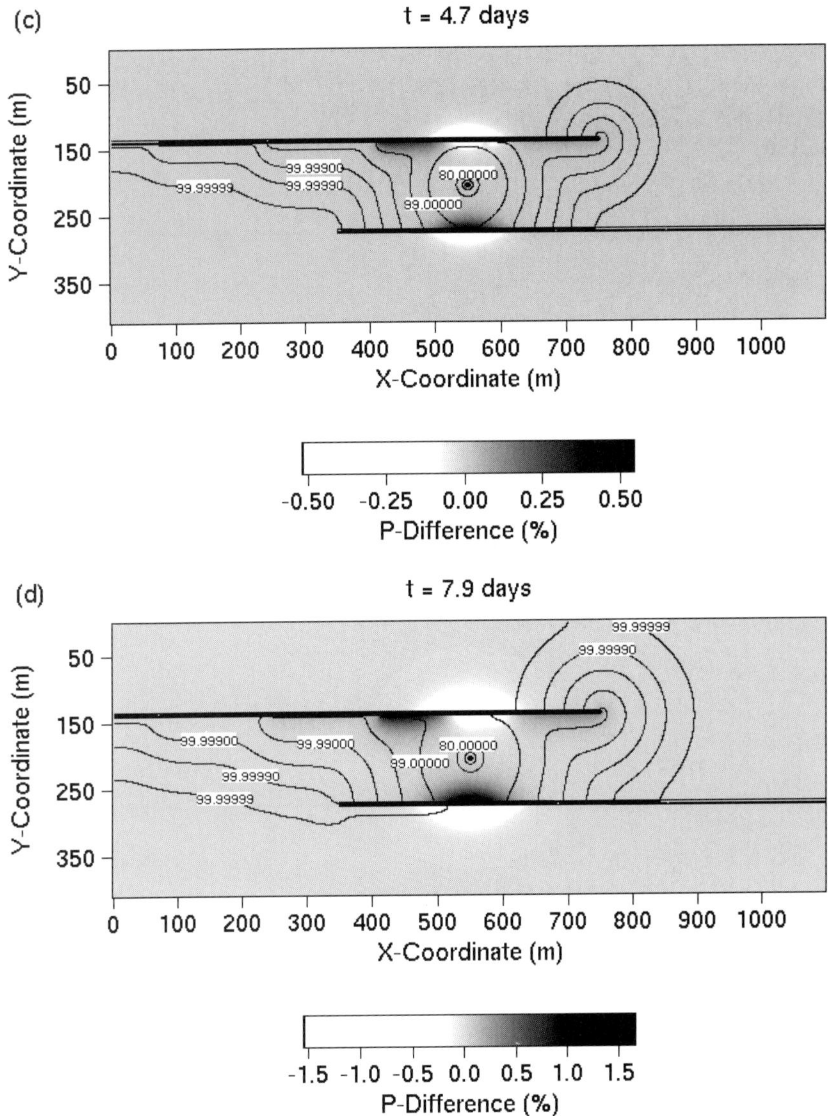

Fig. 13. Continued.

boundaries of the model has lead to a distinct polarization of the deviations relative to radial flow: Pressure is 25% higher near the well and up to 15% lower in its periphery. Fluid pressure near the joint tip to the west of the well is distinctly lower than to the east of the well indicating that the western region supplied most of the fluid. As compared with a homogeneous medium, the behaviour of the joints and deformation bands model is again very different and drawdown is controlled by the poorly interconnected joints. In spite of their over-represented thickness, the deformation bands play a subordinate role for single phase flow.

In summary, after a very short radial drawdown period, the pressure derivative at the well in Model 3 would decrease with time, as more and more joints become involved in drawdown-induced fluid flow to the well. These joints simultaneously decrease the pressure derivative and they induce a contraction of the time sequence in which the deformation bands influence the pressure derivative. Most of the fluid volume that is supplied to the well first flows out of the

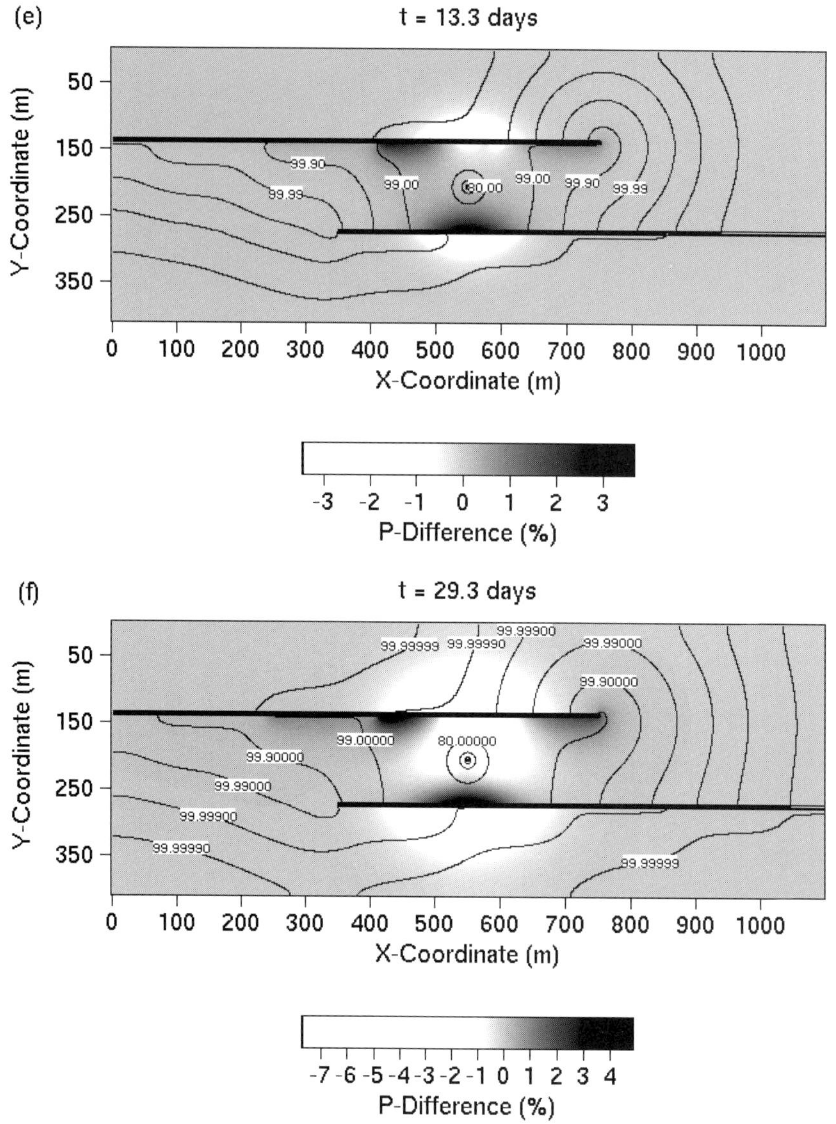

Fig. 13. Continued.

sandstone into these joints, then to those parts of the joints which are nearest to the well, and then through the sandstone to the well. Therefore, the zones of deformation bands are by-passed by the tortuous flow through the joints which cross-cut the zones of deformation bands.

Discontinuous overlapping faults: fault relay

On the top level of the hierarchy of deformation structures in the Arches National Park are the normal faults with slip planes which form fault relays (Fig. 1). The placement of a well within such a relay represents an interesting case of a semi-bounded reservoir. In this scenario, the overlap of the faults along strike and the location of the highly permeable slip planes in the faults are important for the evolution of drawdown. Since the highly permeable slip planes are commonly located in the hanging wall of the faults, the dip direction of the overlapping faults determines whether the slip plane or the low permeability footwall zone faces the inside of the

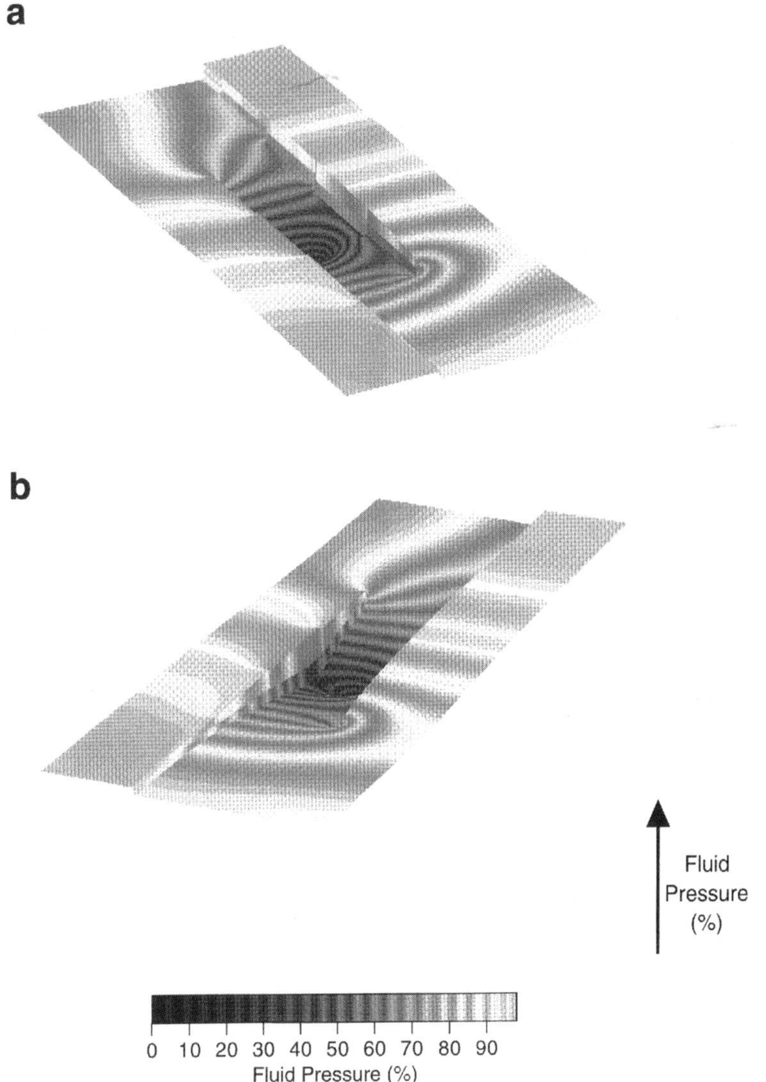

Fig. 14. Fluid pressure gradients across the relay bounding faults after 1a of production from the well in the centre of the relay. (**a**) Pressure gradient across the fault that faces the well with its slip plane. (**b**) Pressure gradient across the fault that faces the well with its low permeability footwall.

relay. Both cases are captured by the idealized flow model of a fault relay of normal faults that dip in the same direction (Fig. 13).

In this model the faults are 100 m apart from one another and the well is placed in the centre of the relay. The fault thickness is 5 m and the slip planes are represented with a thickness of 1 m. This over-representation of slip-plane thickness is compensated for by using an effective transmissivity of $2.0 \times 10^{-5}\,\mathrm{m\,s^{-1}}$ (Table 1). The slip planes of both faults dip to the south (bottom of the model). The fault overlap is three times the fault spacing, as is common in strike-slip faults (Aydin & Nur 1982). The overlap of normal faults in relay structures tends to be smaller, ranging from negative values to values larger than 1.0 depending on the degree of development of the fault pattern. The slip plane consists of 300-m-long segments separated by 5-m-wide low permeability zones. This degree of continuity is compatible with the common segment geometries observed along the normal faults in the Arches National Park (Fig. 1).

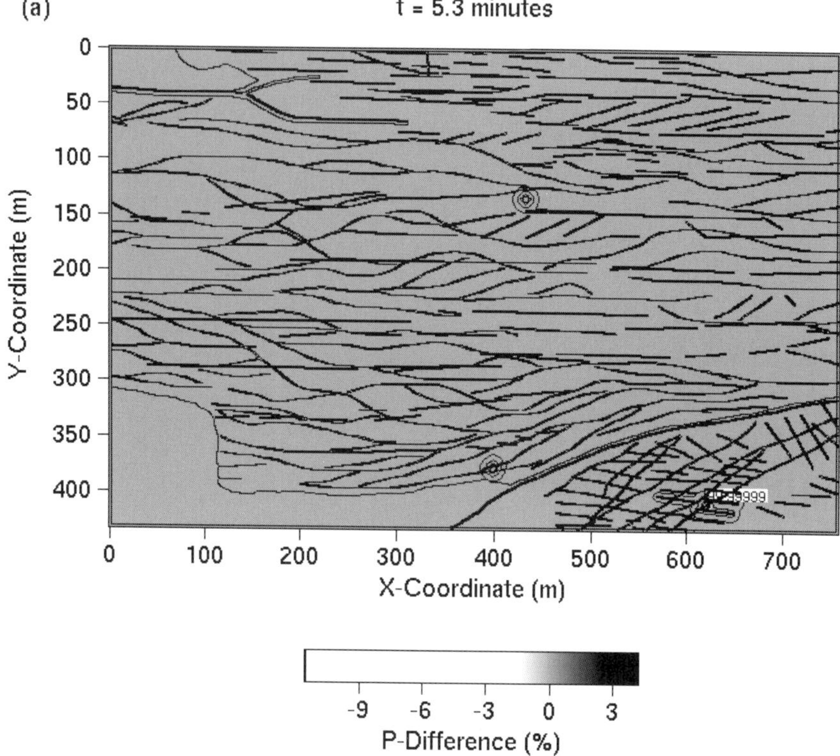

Fig. 15. Model of the fault relay in the Delicate Arch area, Arches National Park as mapped by Antonellini & Aydin (1995, fig. 24), fault permeability 10^{-16} m^2, simulated slip plane transmissivity 1.0×10^{-5} m s^{-1}, joint parallel transmissivity 2.5×10^{-6} m s^{-1}, sandstone permeability 10^{-12} m^2, zones of deformation bands 10^{-15} m^2, Morrison formation 10^{-15} m^2. Deviations from radial drawdown shown in grey-scale (white = less drainage, dark = increased drainage). Fluid-pressure contours in percent of the difference between initial reservoir pressure and fixed drawdown pressure at well head. Fluid pressure contour intervals: 99.99999, 99.9999, 99.999, 99.99, 99.9, 99 and 90%. (**a**)–(**e**) Snapshots of the evolution of drawdown at 5.3 min, 1.6, 12.8 h, 8.5 days and 3 months. Drawdown is simulated for 3 wells: Well 1 [440 m(x), 140 m(y)], well 2 (400 m, 370 m), and well 3 (630 m, 410 m).

The time sequence in Fig. 13 shows that when the radial pressure drawdown first interacts with the highly permeable slip plane of the northern normal fault, it does not increase in magnitude, but is propagated rapidly along the fault strike (see perturbed pressure contours, 23.1 h). This signifies rapid movement of fluid along the slip plane of the fault. This movement is reflected in the darkening in the drawdown periphery and the lightening where the slip plane is closest to the well. On both sides of the relay, the low permeability footwall of the fault impedes flow such that there is less pressure decay than for radial drawdown in a homogeneous medium. This pressure shielding by the faults becomes more and more pronounced with time. With regard to the incipient signature of the faults in a derivative plot, the simultaneous effects of the slip plane and the low permeability footwall counteract each other, thereby reducing the fault signal.

With time increasing (51 h to 29.3 days), a strong asymmetry develops in the shape of the drawdown region in the fault relay. Even after 30 h, the smallest displayed pressure perturbation reaches the tip of the upper fault and subsequently starts to affect its far side. Interestingly, a radial drawdown pattern develops subsequently around this fault tip. The fluid pressure deviation from radial drawdown next to the slip plane near the fault tip stays small, in contrast with the region along the slip plane in the inside of the relay.

After 4.7 days (Fig. 13c), the smallest displayed pressure perturbation also reaches the tip of the lower relay fault and thereafter rapidly propagates along the segmented slip plane in the hanging wall of this fault. In the meantime, the

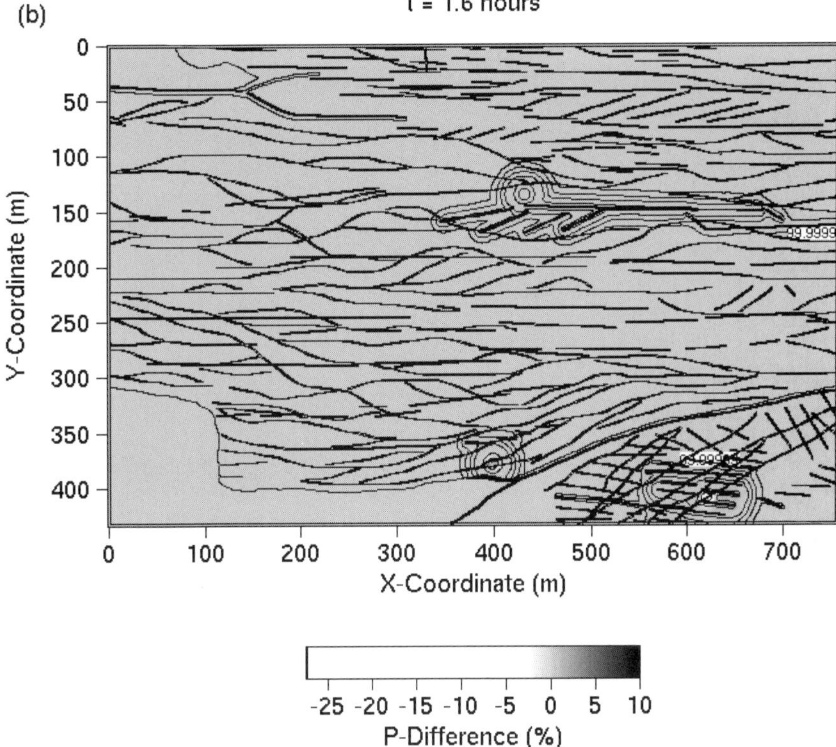

Fig. 15. Continued.

pressure at the faults inside the relay has changed by a few percent of the drawdown pressure differential applied at the well. After 7.9 days, the regions outside the relay have undergone no pressure change (whitening) and the region near the well along the southern fault facing the well with its footwall has been drained more rapidly (black half circle). The segmentation of the slip plane in the northern fault leads to a pronounced stabilization of pressure in the slip plane segment closest to the well. This stabilization is supported by the laterally adjacent slip plane segments that draw fluid from the reservoir sandstone, such that it becomes more rapidly depleted than if the fault was not present. Thus, pressure in the reservoir at a given distance from the well changes more rapidly near the conductive slip plane.

Since the fault relay (Fig. 13) is asymmetric with regard to the location of the high permeability zones relative to the well, the pressure differentials that build up across the two faults differ in magnitude. The northern fault (that faces the well with its highly permeable hanging wall) does develop a much smaller cross-fault pressure differential, because the hanging wall pressure is sustained via the effective pressure communication along the fault strike. The highly permeable slip plane acts similar to a constant head boundary, because fluid can be supplied along it at a rate that is faster than the flow through the reservoir to the well. At the southern fault (that faces the well with its low permeability footwall) a distinct pressure low develops where the distance between the fault and the well is smallest. This different fault behaviour is displayed in the fluid pressure surface plots (Fig. 14a, b).

Fault relay in the Delicate Arch area, Arches National Park

In the Delicate Arch area, a relay of antithetically dipping faults is exposed (Model 5, Figs 1 & 3). The slip planes of the bounding faults dip away from the centre of this structure. Compared to the previous model, all small-scale heterogeneities that are concentrated in the relay have been added. This complex structural assemblage of joints and zones of deformation bands differs significantly from a homogeneous isotropic unit and it represents the full permeability spectrum measured in the Moab sandstone analogue

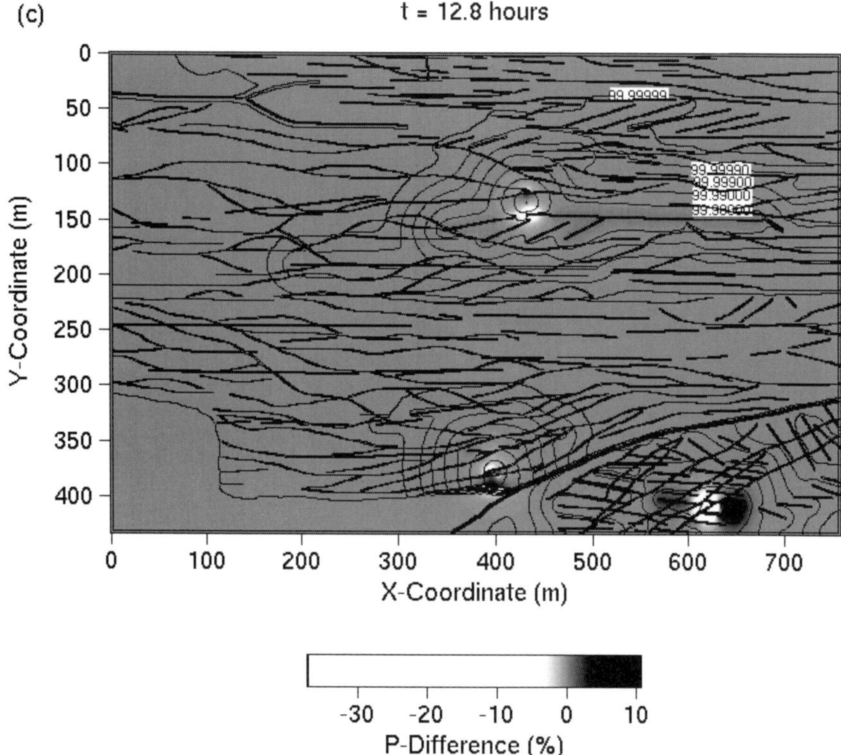

Fig. 15. Continued.

reservoir at the flank of the Cache Valley salt structure (Fig. 1). The joints are represented as continuous high permeability zones (Table 1; cf. Antonellini & Aydin, 1995, fig. 24, table 2). Zones of deformation bands are represented with a thickness equal to the resolution of the model (75 cm) and with a low permeability (10^{-15} m^2). The fault zones are represented with a thickness of 5 m and an average process-zone permeability as calculated in Antonellini & Aydin (1995). The slip planes of the faults have a high effective permeability (Table 1) and are located at their margins in the hanging wall facing directly against unaltered country rock (Fig 6). The slip plane of the normal fault in the southeast of the model consists of 100-m segments separated by 15-m gaps.

Three wells are placed in the Delicate Arch fault-relay model: Well 1 is placed east-southeast of the tip of the northern normal fault; Well 2 is placed at a distance of 25 m from the footwall of the southeastern normal fault and 15 m from the boundary between the Moab sandstone and the Morrison Formation; Well 3 is placed in a region of joints and deformation bands south of the southeastern normal fault. These well locations are chosen to span the range of structural settings one may encounter by drilling the analogue reservoir. Importantly, each of the 3 wells experiences a different evolution of drawdown pressure, and for all three wells the drawdown pattern differs appreciably from radial. Five minutes after the start of the simulation, inhomogeneities in the analogue reservoir are encountered by the drawdown and it begins to diverge from radial (Fig. 15a).

After 1.6 h (Fig. 15b), the 99.99999% pressure contour has spread over a distance of 350 m from Well 1 to the eastern boundary of the model. However, it has travelled only about 50 m to the north and 80 m to the west. Most of the joints near the well have begun to supply fluid to Well 1. Drawdown of Well 2 has diverged from radial as it has begun to interact with the formation boundary of the Morrison Formation in the south, and with the footwall of the normal fault in the east-southeast. Well 3 has developed a rectangular drawdown region, the geometry of which is controlled by the joints near the well.

12.8 h after the start of drawdown (Fig. 15c), all wells have developed distinct pressure anomalies relative to radial drawdown, as is indicated in

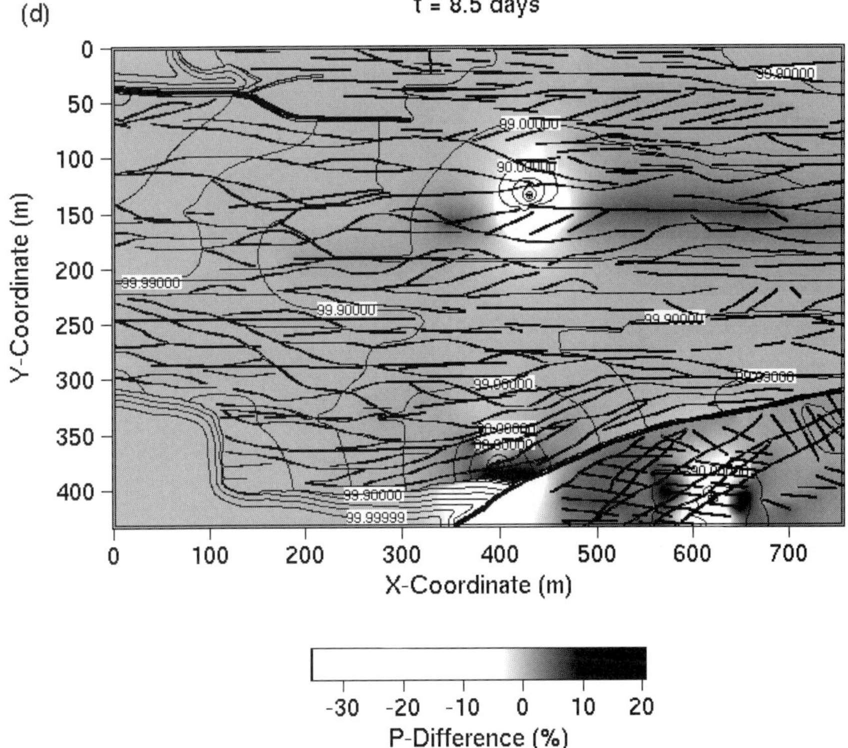

Fig. 15. Continued.

the grey-scale image. The flow patterns in the reservoir (Fig. 16) are complex and controlled by the flow focusing and defocusing around joints similar to Model 1. In the surroundings of Wells 1 and 3, the pressure has decreased less, as compared with radial drawdown such that the derivative would be less than 0.5 (=radial flow). The pressure near Well 2 is strongly influenced by the interaction with the Morrison Formation and the normal fault. From this direction, little fluid is supplied to the well as is indicated by the brightening of the grey-scale image in these areas. To the west and the east of Well 3, regions of accelerated pressure decline have grown along the joints that terminate near the well. In this hanging wall block of the normal fault, drawdown has spread widely and it has begun to interact with the highly-permeable slip plane of the fault. Due to drawdown interaction with the model boundaries and the fault, the pressure in the region around Well 3 has decayed the most in the model. There is no communication of pressure with the footwall of the fault.

After 8.5 days (Fig. 15d), the pressure in the entire Delicate Arch model has decayed by $\leq 0.1\%$ with the exception of the Morrison formation in the southwest. Along its boundary a 20–40 m-wide zone has developed that accommodates the pressure difference to the more permeable sandstone. Around all wells, deviations in pressure from radial drawdown have exceeded 20% and grey-scale expressions of inflow and outflow regions of joints have developed. The pressure perturbation diffusing around Well 1 has interacted with the slip plane of the normal fault in the northeast, which conducted it along strike more rapidly than the sandstone in the footwall. Therefore, a pressure differential has developed across this normal fault. As is indicated by the grey-scale image, the pressure derivatives in Wells 1 and 3 are below 0.5 and that in Well 3 is above 0.5. These deviations reflect cumulative influences of the structures near the wells. The signature of the normal fault in Well 2 should be discernible, but it appears impossible to invert the distance to the fault from a drawdown test at Well 2 in the presence of the small-scale heterogeneities in the Moab sandstone.

After 4 months ($\frac{1}{3}$ a, Fig. 15e), the pressure distribution in the Delicate Arch model is strongly

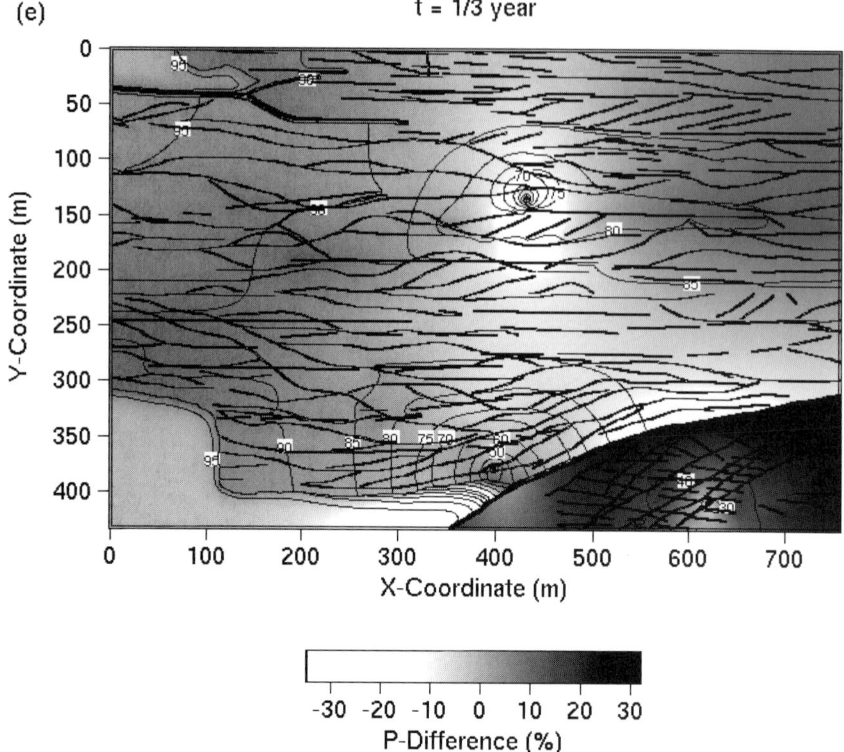

Fig. 15. Continued.

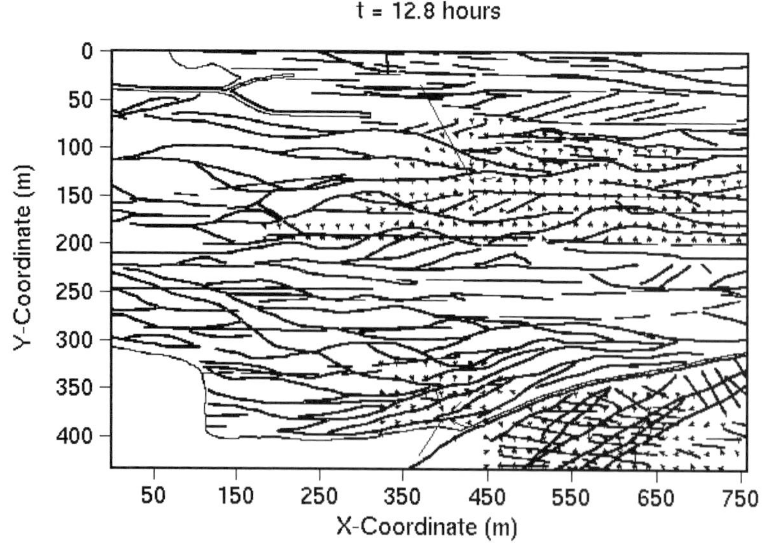

Fig. 16. Transient flow directions in the model of the Delicate Arch fault relay at $t = 12.8$ h.

influenced by the no-flow model boundaries. This time-step is shown mainly to illustrate the impervious nature of the normal fault in the southeast which compartmentalizes fluid pressure in this part of the model and thereby stands out among all the permeability inhomogeneities. South of this fault, the pressure has declined to 50% of the initial reservoir pressure such that the pressure differential across the fault is up to 35% of the drawdown pressure. Under the influence of the model boundaries, the pressure differences throughout the Delicate Arch model have leveled out.

The Delicate Arch analogue-reservoir simulation indicates that radial drawdown is strongly distorted by the joints that are restricted to the reservoir sandstone and which largely determine the spatial extent and direction of spreading of the drainage region. These distortions occur in spite of the poorly-connected nature of the joints and should make it difficult to impossible to isolate the properties of the reservoir sandstone from a drawdown test. The deformation bands play a subordinate role. The most important heterogeneities are the normal faults which distort the radial drawdown by rapid pressure communication along the highly permeable slip planes and by impeding flow perpendicular to them. In the presence of the poorly-connected joints, the transient flow paths in the reservoir are highly tortuous (Fig. 16) and throughout the simulation, there is no convergence of the pressure or flow patterns toward a steady state. While this conclusion is tentative, because the interaction with the model boundaries significantly affects the pressure evolution in the analogue reservoir after $t \geq 10$ days, formation boundaries with similar effects may also be encountered in nature, leading to a moderation of heterogeneity-induced pressure gradients in the reservoir.

These results indicate a hierarchy of influence on the pattern of fluid flow: faults > joints ≫ deformation bands. The zones of deformation bands have a negligible effect on single phase fluid flow but they may be more important for multiphase flow because of their distinctly high capillary pressure.

Discussion

The model representation of joints and slip planes with an exaggerated thickness and a correspondingly reduced hydraulic conductivity using power-law averaging (Deutsch 1989) implies that pressure pertubations spread slower but more uniformly along them than in a natural system. However, joints in the Moab sandstone occasionally consist of en echelon arrays of smaller joints. This segmentation should have a diminishing effect on fracture flux, joint storativity and propagation velocity of pressure perturbations, but it could not be resolved in the fluid flow models. Precipitates on the joint and slip plane surfaces or calcite impregnations in their country rock could shield joint and slip-plane conduits, further reducing fracture flux in a natural system (cf. Matthäi *et al.* 1998).

As an additional simplification, the hydraulic conductivity of joints and slip planes in the flow models does not depend on fluid pressure or stress state in the reservoir. This misrepresents fractured sandstone reservoirs in which the hydraulic conductivity varies with drawdown pressure (e.g. Anderson *et al.* 1994) and/or which are situated in an anisotropic regional stress field. Zhang & Sanderson (1996a, b) and Zhang *et al.* (1996) show, by means of numerical simulations, that the flux through interconnected fracture networks depends on the stress state of the fractured rock. Also, Barton *et al.* (1995) identified that fractures associated with thermal anomalies due to fracture flow tend to be oriented favourably relative to the regional stresses. Thus, for unfavourably oriented stresses under *in situ* conditions, the poorly-connected parallel joints and slip planes in the Moab analogue reservoir may have been poorly conductive with apertures controlled by contacting asperities (Renshaw 1995). However, the occasional cm-wide calcite fillings in the joints and slip planes in the Moab analogue reservoir are consistent with in situ apertures in the cm-range. Also, in many cases, normal faults in hydrocarbon reservoirs are oriented favourably relative to the regional stresses.

The analysis of mode I fractures by Pollard & Segall (1987) permits us to calculate the dependence of fracture aperture / hydraulic conductivity on fluid pressure. Assuming a reservoir burial depth of 3 km and mechanical sandstone properties similar to the Oseberg sandstone, we calculate only small aperture changes ($\leq 5\%$) for the joints represented in the Moab sandstone when fluid pressure is varied between hydrostatic and 0.7 lithostatic. Since the hydraulic joint matrix conductivity ratio is greater than the aspect ratio of the joints (aperture over length), these changes should have little effect on the fluid flux through the isolated joints (cf. Phillips 1991, chapter 3.8, fig. 3.13).

A more detailed analysis of the pressure dependence would need to account for the effects on joint aperture and stress state of faults, of production-induced sandstone compaction and subsidence (Pennington *et al.* 1986; Segall 1992). This was beyond the scope of the present analysis. The preceding discussion highlights that reservoir specific factors must be taken into account when the presented results are applied to develop an understanding of a natural reservoir.

The field data used in our simulations indicate that there is no gradual variation of structures with scale. Instead there are faults on one characteristic length scale, joints on another. These deformation structures have a characteristic size and play a distinct role in creating deviations from radial drawdown. Our results show that these structures cannot be substituted by grid blocks with effective properties for the purpose of predicting drainage patterns and the rate of fluid flow from wells through reservoir simulations except for models of single phase flow with only deformation bands. Statistical descriptions of reservoir inhomogeneities may be appropriate as long as the approximated structures are small relative to the scale of the drainage region. In this case, these descriptions may correctly model their composite effects on production. However, the joints, deformation bands, and faults in the Arches Park analogue reservoir are structures that have the same scale as the drainage area. Thus, even if a correct statistical representation for them could be found, it would not reduce the need for an exact representation and location of these structures in a reservoir.

The orientation, spacing, aperture and the important cross-cutting relationships among the investigated deformation structures are controlled by the mechanical properties of the sandstone, its mode and conditions of deformation, and its tectonic history at large. These factors can be constrained by detailed field documentation, mechanical modelling, geological interpretation of seismic data, and the geological history of a particular region. In the case of the analogue reservoir at Arches National Park, the joint spacing is controlled by the curvature and thickness of the flexed Moab and Slickrock sandstone layers, and the joint spacing and degree of shear along the joints provide information on the connectivity of the joints (Wu & Pollard 1995; Cruikshank *et al.* 1991, respectively). The tectonic history may elucidate why deformations bands formed first and were later cut by the joints. Potentially, this reflects the gradual induration of the Moab and Slickrock sandstones. The listed factors exert a strong control on the characteristics of inhomogeneities in sandstone reservoirs suggesting that if they can be better constrained through process analysis, we may improve our ability to predict reservoir production characteristics.

Conclusions

Time-dependent numerical simulations of drainage around wells in two dimensional models of a fractured sandstone reservoir analogue include all the structural inhomogeneities that were mapped in the Delicate Arch area of Arches National Park. Each structure has a discrete representation in these models. The simulations identify faults with highly permeable slip planes as the most important among all reservoir inhomogeneities, because they influence flow on the km scale and they form boundaries of flow domains over production time-spans.

Poorly-connected joints with a small storage but a high permeability are the second most important structures since they strongly distort radial drawdown. These joints also distort the timespans over which fault signatures are observed in drawdown tests and they become automatically lumped into estimates of matrix permeability as obtained from transient well-testing. As the joints often extend to the boundaries of the reservoir which, when encountered by the pressure front, have an accelerating effect on the rate of pressure decline, boundary effects should further impact permeability estimates obtained from well tests of faulted reservoirs. The timespan over which the properties of the reservoir sandstone itself are sampled by a well test is therefore limited to the time before the drawdown encounters the first joint that connects to a reservoir boundary.

As radial drawdown spreads with time over the deformation structures in the reservoir, various structures simultaneously leave their trace in the type curve. This overlap in time prohibits the identification of signatures of individual structures if multiple inhomogeneities are present.

Less than 25-cm-wide zones of deformation bands with a spacing of ≥ 30 m only have a very small flux-reducing effect for single phase fluid flow.

Pressure gradients across faults in relay structures can reach transient maxima and then decay again. In our simulations, the analogue reservoir does not reach a steady state behaviour on the km scale.

The identified drainage patterns indicate that the well placement is extremely important in a fractured sandstone reservoir: If a high permeability domain exists which does not extend significantly beyond the margin of the oil pool, the well should be placed within it.

This research was supported by the Stanford Rock Fracture Project, Stanford University. The authors wish to thank B. Biondi and J. Clearbout (SEP,

Stanford University) for gratuitous supply of CPU time on their SGI Power Challenge computer, and R. Horne, X. Zhang and an anonymous reviewer for their constructive and helpful comments.

References

ANTONELLINI, M. & AYDIN, A. 1994. Effect of Faulting on Fluid Flow in Porous Sandstones: Petrophysical Properties. *American Association of Petroleum Geologists Bulletin*, **78**, 355–377.

—— & —— 1995. Effect of Faulting on Fluid Flow in Porous Sandstones: Geometry and Spatial Distribution. *American Association of Petroleum Geologists Bulletin*, **79**, 642–671.

——, ——, POLLARD, D. D. & D'ONFRO, P. 1994. Petrophysical study of faults in sandstone using petrographic image analysis and X-ray computerized tomography. *Pageoph Topical Volumes*, **143**, 181–201.

AYDIN, A. 1977. *Faulting in sandstone*. PhD Thesis, Stanford University, Stanford, California.

—— 1978. Small faults formed as deformation bands in sandstone. *Pure and Applied Geophysics*, **116**, 913–930.

—— & NUR, A. 1982. Evolution of pull-apart basins and their scale independence. *Tectonics*, **1**, 91–105.

—— & JOHNSON, A. M. 1983. Analysis of faulting in porous sandstones. *Journal of Structural Geology*, **5**, 19–31.

BARTON, C. A. & ZOBACK, M. D. 1992. Self-similar distribution and properties of macroscopic fractures at depth in crystalline rocks in the Cajon Pass scientific drill hole. *Journal of Geophysical Research*, **97**, 5181–5200.

——, TESLER, L.G. & ZOBACK, M. D. 1991. Interactive image analysis of borehole televiewer data, *In*: PALAZ, I. & SENGUPTA, S. K. (eds) *Automated Pattern Recognition in Exploration Geophysics*. Springer, New York, 223–248.

——, ZOBACK, M. D. & MOOS, D. 1995. Fluid flow along potentially active faults in crystalline rock. *Geology*, **23**, 683–686.

BATZLE, M. & WANG, Z. 1992. Seismic properties of pore fluids. *Geophysics*, **57**, 1396–1408.

BEAR, J. 1972. *Dynamics of Fluids in Porous media*. Elsevier Scientific Publications, New York.

BRACE, W. F., ORANGE, A. S., & MADDEN, I. M. 1966. The effect of pressure on the electrical resistivity of water-saturated crystalline rock. *Journal of Geophysical Research*, **70**, 5669–5678.

BROWN, S. 1987. Fluid Flow through rock joints: The effect of surface roughness. *Journal of Geophysical Research*, **92**, 1337–1347.

CINCO, L. H., SAMANIEGO, V. F. & DOMINGUEZ, A. 1976. Unsteady state behavior for a well near a natural fracture. Society of Petroleum Engineering, **SPE6019**, presented at the SPE Annual Technical Conference and Exhibition, New Orleans.

——, —— & —— 1978. Transient pressure behaviour for a well with a finite-conductivity vertical fracture. Society of Petroleum Engineering Journal, August 1978, 253–264.

CRUIKSHANK, K. M., ZHAO, G. & JOHNSON, A. M. 1991. Analysis of minor fractures associated with joints and faulted joints. *Journal of Structural Geology*, **13**, 865–886.

DEUTSCH, C. 1989. Calculating effective absolute permeability in sandstone/shale sequences. *Society of Petroleum Engineering Formation Evaluation*, **4**, 343–348.

DEUTSCH, C. V., SRINIVASAN, S. & MO, Y. 1996. Geostatistical reservoir modeling accounting for the scale and precision of seismic data. *Society of Petroleum Engineering*. **SPE36497**, prepared for the 1996 SPE Annual Technical Conference and Exhibition held in Denver, Colorado, USA, 6–9 October 1996.

DOELLING, H. 1985. *Geology of Arches National Park*. Utah Geological and Mineral Survey Publications, to accompany map **74**.

DVORKIN, J. & NUR, A. 1996. Elasticity of high-porosity sandstones: Theory for two North sea datasets. *Geophysics*, **61**, 1–8.

DYER, J. R. 1983. *Jointing in sandstones, Arches National Park, Utah*. PhD Thesis, Stanford University, Stanford, California.

ENGELDER, T. & SCHOLZ, C. H. 1981. Fluid flow along very smooth joints at effective pressures up to 200 megapascals. *In*: CARTER, N. L., FRIEDMAN, M., LOGAN, J. M. & STEARNS, D. W. (eds) *Mechanical Behaviour of Crustal Rocks*, Washington D. C. 147–152.

ENGLAND, W. A., MACKENZIE, A. S., MANN, D. M. QUIGLEY, T. M. 1987. The movement and entrapment of petroleum fluids in the subsurface. *Journal of the Geological Society of London*, **144**, 327–347.

FISCHER, G. & PATERSON, M. 1992. Measurement of permeability and storage capacity in rocks during deformation at high temperature and pressure. *In*: EVANS, B. & WONG, T.-F. (eds) *Fault Mechanics and Transport Properties of Rocks*. New York, Academic Press, 213–252.

FORSTER, C. B., GODDARD, J. V. & EVANS, J. P. 1993. *Permeability structure of a thrust fault*. United States Geological Survey Open File Report **94-228**. Menlo Park, California, 216–223.

——, BRUHN, R. L. & CAINE, J. 1995. *Field and laboratory study of the spatial and temporal variability in hydromechanical properties of an active normal fault zone, Dixie valley, Nevada*. United States Geological Survey Open File Report **95-210/1**, 71–84.

GAILEY, R. M. & GORELICK, S. M. 1993. Design of optimal, reliable plume capture schemes: Application to the Gloucester landfill groundwater contamination problem. *Ground Water*, **31**, 107–114.

GIBSON, R. G. 1994. Fault-zone seals in siliciclastic strata of the columbus basin, offshore Trinidad. *American Association of Petroleum Geologists Bulletin*, **78**, 1372–1385.

GRADER, A. & HORNE, R. N. 1988. Interference testing: Detecting an impermeable or compressible subregion, *Society of Petroleum Engineering Formation Evaluation*, 428–437.

Gueguen, Y., David, C. & Gavrilenko, P. 1991. Percolation networks and fluid transport in the crust. *Geophysical Research Letters*, **18**, 931–934.

Hewett, T. A. 1992. *Modeling reservoir heterogeneity with fractals*. 4th International Geostatistics Conference, Troia, Portugal, September 1992, 274–289.

Horner, D. R. 1951. Pressure Buildup in Wells. Proceedings of the Third World Petroleum Congress, Chapter II, Brill, E. J. (ed.) Leiden.

Horne, R. N. 1996. *Modern Well Test Analysis: A Computer-Aided Approach*, 2nd edn (1st edn 1990), Petroway Inc., Palo Alto, California.

Hsieh, P. A. & Shapiro, A. M. 1994. Hydraulic characteristics of fractured bedrock underlying the FSE well field at the Mirror Lake site, Grafton County, New Hampshire. *In*: Morganwalp, D. W. & Aronson, D. A. (eds) United States Geological Survey Toxic Substances Hydrology Program—Proceedings of the Technical Meeting, Colorado Springs, Colorado, September 20–24, 1993. United States Geological Survey Water-Resources Investigations Report **94-4015**.

Journel, A. G., Deutsch, C. V., & Desbarats, A. J. 1986. Power averaging for block effective permeability. *Society of Petroleum Engineering* **SPE 15128**, presented at 1986 SPE California Reg. Meeting, Oakland, April 2–4.

Knipe, R. J. & Lloyd, G. E. 1994. Microstructural analysis of faulting in quartzite, Assynt, NW Scotland: Implications for fault zone evolution. *Pageoph Topical Volumes*, **143**, 229–254.

Knott, S. D. 1993. Fault Seal Analysis in the North Sea: *American Association of Petroleum Geologists Bulletin*, **77**, 778–792.

Krantz, R. L., Frankel, A. D., Engelder, T. & Scholz, C. H. 1979. The permeability of whole and jointed Barre granite. *International Journal of Rock Mechanics and Mining Science*, Geomechanics Abstracts, **16**, 225–234.

Kuchuk, F. J. & Habashy, T. M. 1992. Pressure behavior of laterally composite reservoirs. *Society of Petroleum Engineering*, **SPE24678**, 67th Ann. Conference and Exhibition, SPE, Washington, DC, October 4–7th, 171–186.

Makurat, A., Barton, N. & Rad, N. S. 1990. Fracture conductivity variation due to normal and shear deformation. Proceedings International Symposium on Rock Fractures, Loen, Norway, 535–540.

Matthäi, S. K. & Roberts, S. G. 1996. The influence of fault permeability on single phase fluid flow near fault–sand intersections: Results from steady-state high-resolution models of pressure-driven fluid flow. *American Association of Petroleum Geologists Bulletin*, **80**, 1763–1779.

—— & Fischer, G. 1996. Quantitative modeling of fault-fluid-discharge and fault-dilation-induced fluid-pressure variations in the seismogenic zone. *Geology*, **24**, 183–186.

——, Aydin, A. & Pollard, D. D. 1998. Simulation of transient well-test signatures for geologically realistic faults in sandstone reservoirs. *Society of Petroleum Engineering Journal*, **SPE 38442**.

Neuzil, C. E., & Tracy, J. V. 1981. Flow through fractures. *Water Resources Research*, **17**, 191–199.

Pennington, W. D., Davis, S. D., Carlson, S. M., Dupree, J. D., & Ewing, T. E. 1986. The evolution of seismic barriers and asperities caused by the depressuring of fault planes in oil and gas fields of south Texas, *Bulletin of the Seismological Society of the Americas*, **76**, 939–938.

Phillips, O. W. 1991. Flow and Reactions in Permeable Rocks. Cambridge University Press, Cambridge.

Pittmann, A. 1981. Effect of fault-related granulation on porosity and permeability of quartz sandstones, Simpson Group (Ordovician), Oklahoma. *American Association of Petroleum Geologists Bulletin*, **65**, 2381–2387.

Pollard, D. D. & Segall, P. 1987. Theoretical displacements and stresses near fractures in rock: with applications to faults, joints, veins, dikes, and solution surfaces. *In*: Atkinson, B. K. (ed.) *Fracture Mechanics of Rock*. Academic Press, London, 277–349.

—— & Aydin, A. 1988. Progress in understanding jointing over the past century. *Geological Society of America Bulletin*, **100**, 1181–1204.

PetroTools Vs. 2.3, 1996. *Seismic Rock Properties Software*. Petrosoft, 4320 Stevens Creek Blvd., Suite 282, San Jose, California, 95129.

Renshaw, C. E. 1995. On the relationship between mechanical and hydraulic apertures in rough-walled fractures. *Journal of Geophysical Research*, **100**, 24,629–24,636.

Roberts, S. G. & Mattha[umlaut]i, S. K. 1996. *High-resolution potential flow methods in oil exploration*. Mathematics Research Report **MRR003-96**, Centre for Mathematics and its Applications, Australian National University, Canberra.

Ruge, J. W. & Stueben, K. 1987. Algebraic Multigrid, *In*: McCormick, S. F. (ed.), Multigrid Methods. *SIAM Frontiers in Applied Mathematics*, **3**, 73–130.

Savioli, G. B., Bidner, M. S. & Jacovkis, P. M. 1995. The influence of heterogeneities on well test pressure response: A sensitivity analysis. *Society of Petroleum Engineering*, **SEP26985**, 67–72.

Segall, P. 1992. Induced stresses due to fluid extraction from axisymmetric reservoirs. *Pageoph Topical Volumes*, **139**, 535–560.

Shapiro, A. M., Hsieh, P. A. & Winter, T. C. 1995. *The mirror lake fractured-rock research site – a multidisciplinary research effort in characterizing ground-water flow and chemical transport in fractured rock*. United States Geological Survey Fact Sheet **FS-138-95**.

Smith, D. A. 1980, Sealing and non-sealing faults in Lousiana Gulf Coast Salt Basin: *American Association of Petroleum Geologists Bulletin*, **64**, 145–172.

Tyler, N. & Finley, R. J. 1988. Reservoir architecture; a critical element in extended conventional recovery of mobile oil in heterogeneous reservoirs. *American Association of Petroleum Geologists Bulletin*, **72**, 255.

—— & —— 1991. Architectural controls on the recovery of hydrocarbons from sandstone reservoirs. *Concepts in Sedimentology and Paleontology*, **3**, 1–5.

UNDERHILL, J. R. & WOODCOCK, N. H. 1987. Faulting mechanisms in high-porosity sandstones; New Red Sandstone, Arran, Scotland. *In*: JONES, M. E. & PRESTON, M. F. (eds) *Deformation of Sediments and Sedimentary Rocks*. Geological Society, London, Special Publications, **29**, 91–105.

WANG, J. S. Y. 1991. Flow and transport in fractured rocks. *Reviews of Geophysics*, Supplement, U.S. Natl. Rep. Intl. Un. Geodesy and Geophysics, April 1991, 254–262.

WEI, Z. Q., EGGER, P. & DESCOEUDRES, F. 1995. Permeability predictions for jointed rock masses. *International Journal of Rock Mechanics and Mining Science & Geomechanical Abstracts*, **32**, 251–261.

WHEATCRAFT, S. W. & CUSHMAN, J. H. 1991. Hierarchical approaches to transport in heterogeneous porous media. *Reviews of Geophysics*, Supplement, U.S. Natl. Rep. Intl. Un. Geodesy and Geophysics, 263–269.

WITHERSPOON, P. A., WANG, J. S. Y., IWAI, K. & GALE, J. E. 1980, Validity of the cubic law in a deformable rock fracture. *Water Resources Research*, **19**, 1016–1024.

WU, H. & POLLARD, D. D. 1995. An experimental study of the relationship between joint spacing and layer thickness. *Journal of Structural Geology*, **17**, 887–905.

YAXLEY, L. M. 1987. Effect of a partially communicating fault on transient pressure behavior. *Society of Petroleum Engineering Formation Evaluation*, December 1987, 590–598.

ZHANG, X. & SANDERSON, D. J. 1996a. Numerical modeling of fault slip on fluid flow around extensional faults. *Journal of Structural Geology*, **18**, 109–119.

—— & —— 1996b. Effects of stress on the two-dimensional permeability tensor of natural fracture networks. *Geophysics Journal International*, **125**, 912–924.

——, ——, HARKNESS, R. M. & LAST, N. C. 1996. Evaluation of 2-D permeability tensor of fractured rock masses. *International Journal of Rock Mechanics and Mining Science, Geomechanics Abstracts*, **33**, 17–37.

ZHAO, G. 1991. *Joints and faults in garden area of Arches National Park, Utah*. PhD Thesis, Purdue University, Indiana, USA.

An integrated approach for characterizing fractured reservoirs

P. S. D'ONFRO[1], W. D. RIZER[2], J. H. QUEEN[3], E. L. MAJER[4], J. E. PETERSON[4], T. M. DALEY[4], D. W. VASCO[4] A. DATTA-GUPTA[5] & J. C. S. LONG[6]

[1] *Conoco Inc., P.O. Box 2197, Houston, Texas, 77252–2197, USA*
[2] *Houston Advanced Research Center, 4800 Research Forest Drive, The Woodlands, Texas 77381, USA*
[3] *Conoco Inc., P.O. Box 1267, Ponca City, Oklahoma 74602–1267, USA*
[4] *Ernest Orlando Lawrence Berkeley National Laboratory, Berkeley, California 94720, USA*
[5] *Texas A & M University, Department of Petroleum Engineering, College Station, Texas 77843–3116, USA*
[6] *Mackay School of Mines, University of Nevada, Reno, Nevada 89557-0047, USA*

Abstract: Experience has shown that fractures and faults within a given array are not all equally conductive or well-connected. To investigate new techniques for locating conductive fracture flow paths, a series of high resolution (1 to 10 kHz) crosswell and single well seismic surveys and interference tests were conducted in a shallow five spot well array penetrating a fractured limestone formation. Two inverse approaches for constructing fracture flow models were applied to the interference test data. Both approaches successfully reproduced the transient pressure behaviour at the pumping and observation wells and indicated a preferential fracture flow path between two wells aligned in an east-northeast direction, the dominant direction of fracturing mapped in the area. Crosswell and single well seismic experiments were performed before and after air injection designed to displace water from the fracture flow path and increase seismic visibility. The crosswell experiments showed that replacement of water with gas produces significant changes in the seismic signal. The single well reflection surveys were able to precisely locate the position of the fracture flow path. This location was confirmed by core from a slant well which intersected a single open fracture at the targeted depth.

Over the last several years, the United States Department of Energy (DOE), with the kind participation from industry partners, has funded significant amounts of research aimed at improving fracture characterization in naturally fractured gas reservoirs (Watts 1996). One piece of work in particular, a collaborative effort between the Ernest Orlando Lawrence Berkeley National Lab (LBL), Conoco and Amoco, has shown the potential for providing high resolution information on the location of gas filled, natural fractures using developing borehole seismic techniques such as single well reflection profiling and crosswell methods.

We present the results of one part of this collaborative project, a series of single and crosswell seismic experiments and interference tests performed at Conoco's Borehole Test Facility (CBTF) in Kay County, Oklahoma. This work was focused on developing geological, geophysical and hydrological methods for detecting and characterizing natural fracture systems and their effect on fluid flow. The ultimate goal is to develop an integrated methodology that can be used to plan field development, for example, determine optimum infill well locations, offset poor wells to intersect fractures, or design effective water floods.

Geology of the Conoco Borehole Test Facility

The Conoco Borehole Test Facility is located in Kay County, north-central Oklahoma, on the Nemeha Ridge, a tectonic feature which, in Kay County, consists of differentially uplifted blocks bordered by northeast and northwest trending faults active in the Palaeozoic (Fig. 1). Surface rocks consist primarily of limestones and shales of the Lower Permian Chase Group (Fig. 2) and are described by Toomey (1992). Surface structure in the area is subtle, characterized by regional dips less than one degree west-southwest (Fig. 3).

The seismic experiments and interference tests were performed in the Fort Riley Limestone which is approximately 15.25 m thick and lies 15 m below the ground surface (Fig. 2). At the CBTF, the Fort Riley Limestone contains two orthogonal sets of near-vertical fractures, the dominant set striking east-northeast and the

D'ONFRO, P. S, RIZER, W. D., QUEEN, J. H. *et al.* 1998. An integrated approach for characterizing fractured reservoirs. *In*: JONES, G., FISHER, Q. J. & KNIPE, R. J. (eds) *Faulting, Fault Sealing and Fluid Flow in Hydrocarbon Reservoirs*. Geological Society, London, Special Publications, **147**, 193–208.

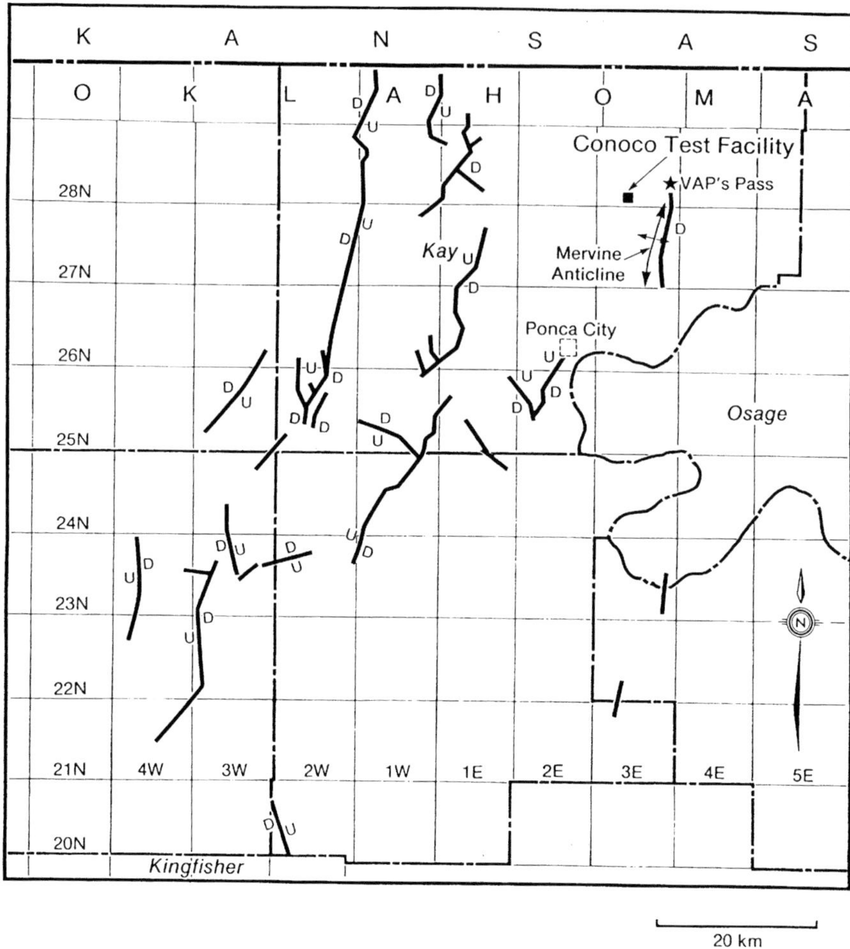

Fig. 1. Structure map for north-central Oklahoma showing the location of the Conoco Borehole Test Facility. Faults shown cut the basal Pennsylvanian unconformity (modified after Luza & Lawson 1981). D and U represent the downthrown and upthrown sides of faults, respectively.

secondary set striking north-northwest. These sets are parallel to fractures mapped in detail in a quarry exposure 5 k east-northeast of the CBTF (Fig. 4) by Queen & Rizer (1990) and are likely part of a regional system of fractures mapped in eastern Oklahoma by Melton (1929).

The CBTF contains six deep (914 m) and five shallow (45.7 m) wells used for geophysical and hydrological testing (Fig. 5). The shallow wells, designated GW-1 to GW-5 in Fig. 5, penetrate the Fort Riley Limestone and were drilled in a skewed '5-spot' pattern to provide maximum azimuthal coverage for both seismic and hydrological experiments. The GW wells were completed with polyvinylchloride (PVC) casing that is slotted in the Fort Riley interval (Fig. 6). A permeable sand pack was emplaced between the slotted casing and the wellbore and bentonite grout was emplaced between the casing and the wellbore above and below the Fort Riley Limestone to ensure hydraulic isolation (Fig. 6). The Fort Riley Limestone is overlain and underlain by the Doyle and Matfield Shales which act as confining beds.

Results of the interference tests

Two interference tests were performed in the GW well array, one of which will be discussed here. The results of both interference tests are presented in a paper by Datta-Gupta *et al.* (1995). Precision quartz pressure tranducers were installed in the GW wells. During the test, water

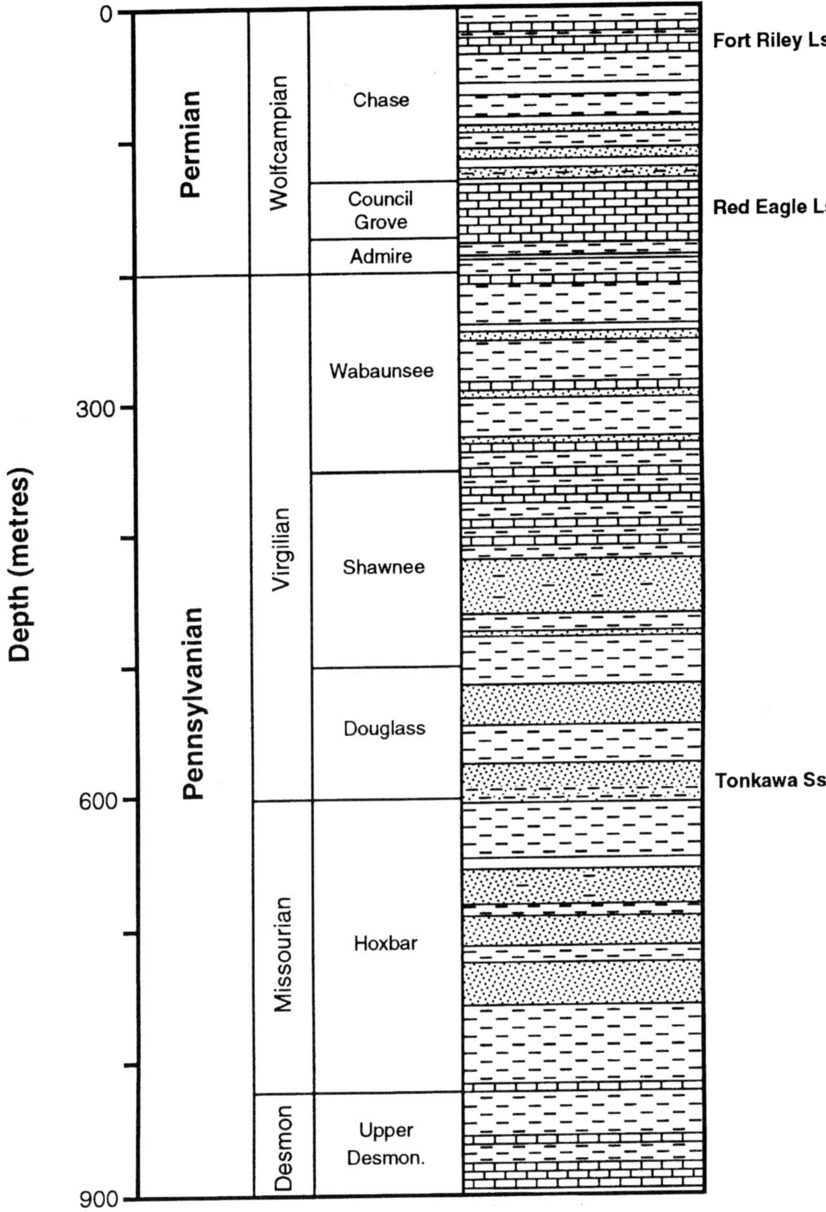

Fig. 2. Generalized stratigraphy at the Conoco Borehole Test Facility. The block pattern represents limestone, the dashed lines represent shale, and the stippled pattern represents sandstone.

was pumped from well GW-5 at approximately 0.5 gallon min^{-1} and the pressure responses in all five GW wells were recorded continuously by an automated data acquisition system controlled by a desktop computer. Background data collected before the beginning of the test indicated that the wells were recovering from rainfall when the test was started; hence the drawdown data were corrected for rainfall before further analysis. This correction was accomplished by simply subtracting the additional drawdown resulting from the recovery from the rainfall.

Figure 7 shows the final drawdown and recovery curves. The plot was generated by converting pressure (bars) to pressure head (m). The drawdown curves show that GW-2 responds much earlier to water production from GW-5 than wells GW-1, GW-3, and GW-4, indicating that

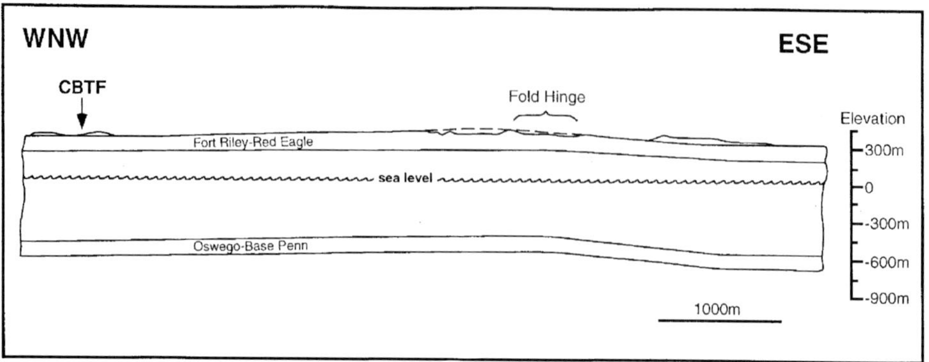

Fig. 3. WNW–ESE section across the Mervine anticline. The arrow indicates where the Conoco Borehole Test Facility projects on the section (modified after Gasteiger 1980).

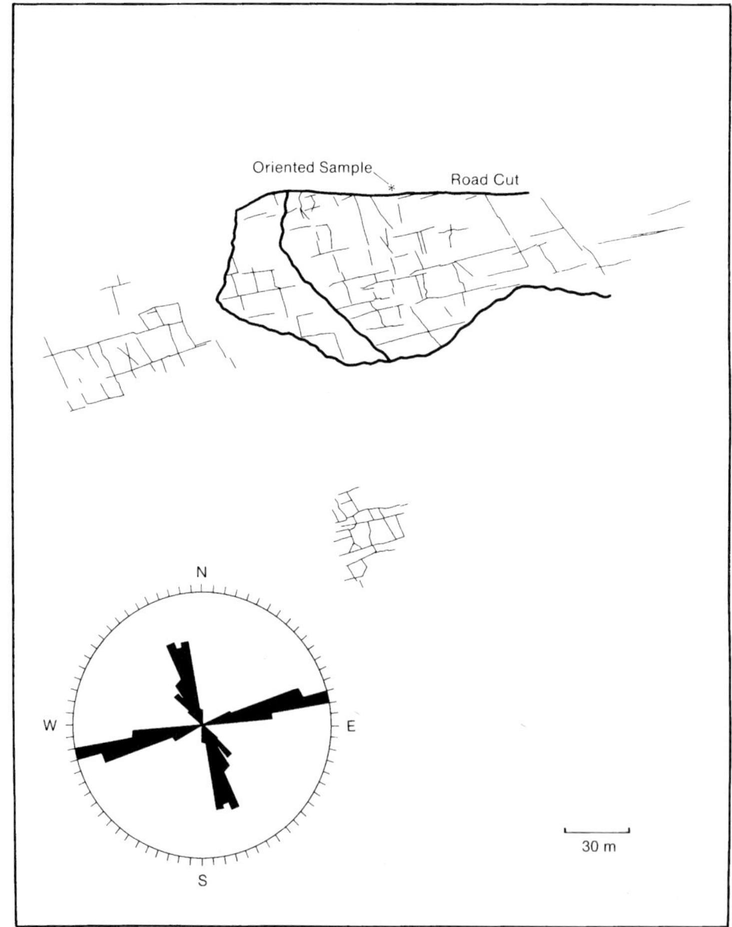

Fig. 4. Map and rose plot of surface fracture pattern in the Fort Riley Limestone in an abandoned quarry at Vap's Pass located 5 km east-northeast of the Conoco Borehole Test Facility. See Fig. 1 for the location of Vap's Pass.

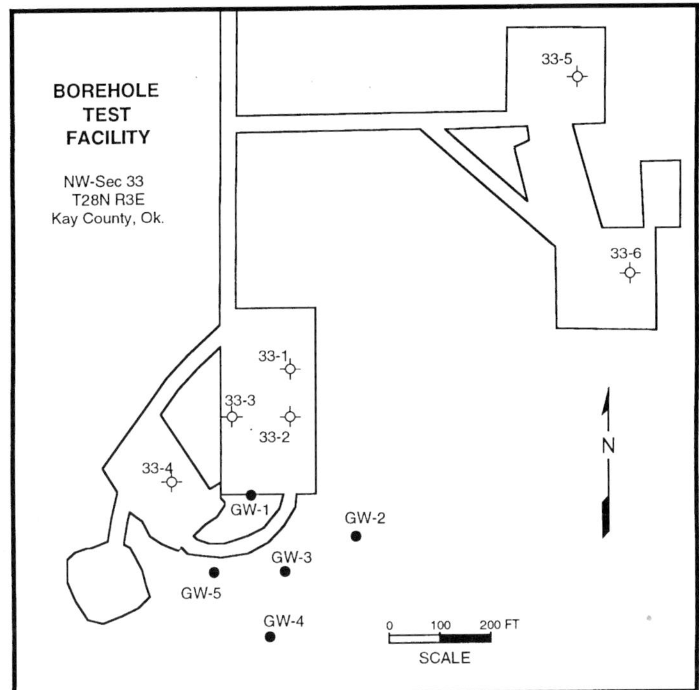

Fig. 5. Map of the Conoco Borehole Test Facility showing the six deep (914 m) 33 series wells and the five shallow (45.7 m) GW series wells. The seismic experiments and the interference tests were performed in the GW well array.

there is a fast flow path between GW-5 and GW-2. This result and the fact that GW-2 and GW-5 are aligned east-northeast, the dominant direction of fracturing in the Fort Riley Limestone, suggest the two wells may be connected by a conductive fracture or fracture network.

Fracture system characterization using inverse modelling approaches

The interference test data were analysed using inverse modelling methods. Inverse methods provide a means to derive fracture flow networks using fluid flow and transport information, such as pressure transients from interference tests and tracer breakthrough curves (Datta-Gupta *et al.* 1995). Unlike traditional discrete fracture network modelling approaches that rely on knowledge of fracture geometry to reproduce flow and transport behaviour, the inverse methods directly incorporate hydrologic test data to derive the fracture networks and thus emphasize the underlying features that control fluid flow and transport.

Two inverse modelling approaches were applied to the interference test data to characterize the fracture flow network in the Fort Riley Limestone. The first approach creates equivalent-discontinuum models that conceptualize the fracture system as a partially filled lattice of constant aperture conductors which are locally connected or disconnected to reproduce the observed pressure transient data. The second approach creates variable-aperture lattice models that represent the fracture system as a fully connected network composed of conductors of varying apertures or hydraulic conductivities. Detailed discussion of inverse modelling methods is beyond the scope of this paper. However, the two methods applied in this study are discussed more thoroughly by Datta-Gupta *et al.* (1995).

Equivalent discontinuum models

To simulate the well interference tests, we represented the fracture system in the Fort Riley Limestone as a network of 1D conductors having fixed apertures. The matrix permeability is assumed to be negligible (zero) relative to the fracture permeability. The fracture conductivity is related to the third power of fracture aperture

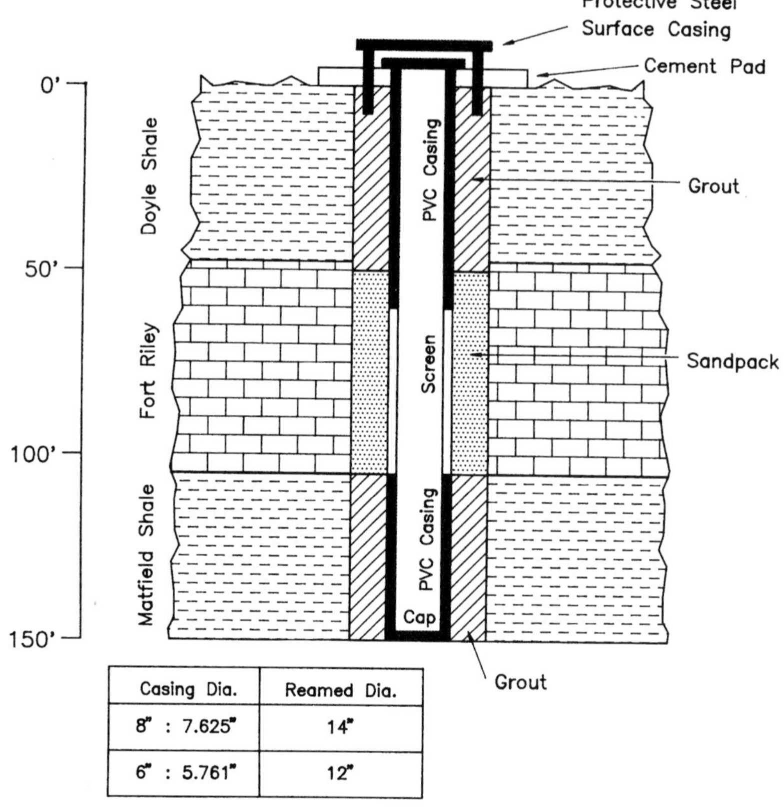

Fig. 6. Schematic diagram of a GW well completion. GW-3 was completed with 203 mm (8 inch) diameter PVC casing and the other four GW wells were completed with 152 mm (6 inch) diameter PVC casing.

through the widely used 'cubic law' (Gelhar 1993). The flow field was obtained using a Galerkin finite element method. Figure 8 shows the basic template of conductors used for inversion of the GW well tests. It consists of an inner

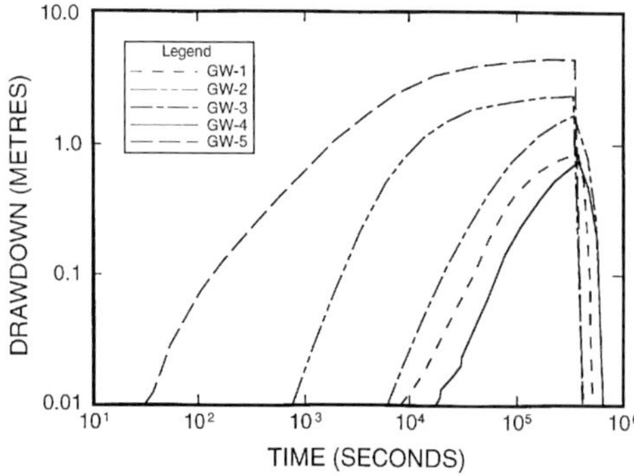

Fig. 7. Drawdown and recovery data for the interference test. The test duration was approximately eleven and a half days.

dense region with an element spacing of 7 m to obtain greater resolution of the fluid flow field and a coarse outer region with a 14 m element spacing. A sensitivity study was performed to determine the template dimensions required to minimize the impact of the boundary conditions on the flow field. A constant head boundary condition was then imposed on all four sides. During the inversion, a criterion was imposed whereby the probability of altering an element decreased exponentially with distance beyond the inner region, where the wells are located.

Equivalent discontinuum models were derived from the GW interference test data by starting with the template or lattice of conductors shown in Fig. 8. The discontinuum approach involves changing the configuration of conductors until it satisfies the well test data. A function was defined to reflect the misfit between the observed pressure response and the predicted response for all wells (Datta-Gupta *et al.* 1995). A technique called simulated annealing was used for efficiently finding the minumum value

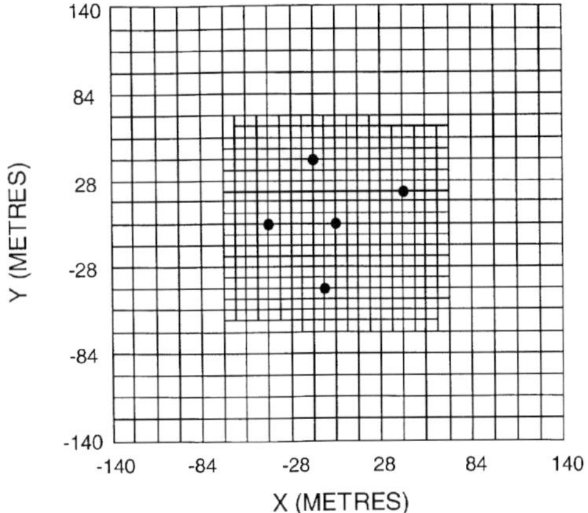

Fig. 8. Basic template of conductors used for inversion of the interference test data. The template consists of an inner dense region with an element spacing of 7 m and a coarse outer region with a 14 m element spacing. The solid black dots represent the locations of the GW wells.

of the function. The simulated annealing algorithm operates by choosing a lattice element at random and if the element is present or 'on', it is turned off and vice versa. The change in misfit is calculated for each change in the lattice. If the misfit decreases, the lattice change is accepted; otherwise the change is accepted with a probability $P(\Delta E) = \exp(-\Delta E/T)$ where E is an objective function that reflects the misfit between the data and the predicted response, and T is analogous to temperature in the Gibbs distribution. By accepting changes that result in an increase in energy, the simulated annealing technique provides a mechanism to escape from local minima. Changes in the lattice are made until the misfit function is reduced to a suitably small value which usually requires less than ten thousand iterations for the GW well interference test data. The final configuration of conductors represents a fracture model that satisfies the well test data. It is important to note that this model in non-unique and there are other possible configurations that will satisfy the well test data.

Figure 9 shows the match between the observed drawdowns (black dots) in the GW wells and the calculated drawdowns. For reference, the dashed line represents the drawdown calculated using the initial template shown in Fig. 8. The solid line represents the calculated drawdown for the final configuration of conductors that provided a low value of misfit between the calculated and observed pressure responses. For the producing well GW-5, early time data were excluded from matching because they were effected by wellbore storage. Overall, the fracture network derived from the inversion is able to reproduce the drawdown data reasonably well.

Figure 10(a) shows the final configuration of conductors obtained after inversion that produced the match in Fig. 9. Figure 10(a) focuses on the inner dense region of the template where the GW wells are located. The pattern suggests the presence of a direct fracture path extending from GW-5 to GW-2 to the north of GW-3,

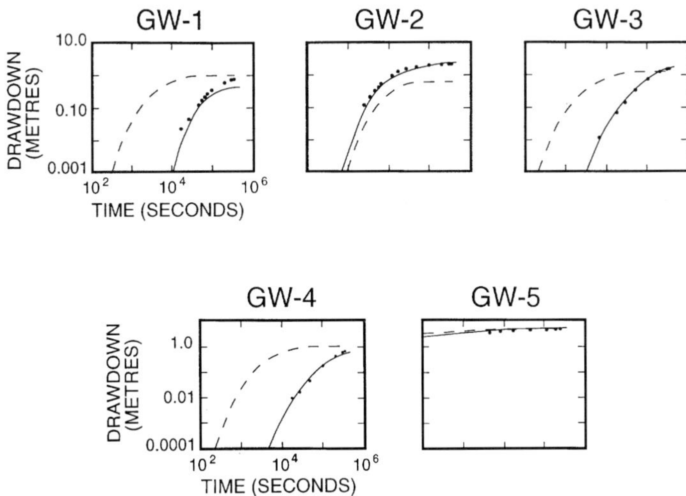

Fig. 9. Match between the drawdown calculated from the equivalent discontinuum model and the observed drawdown during the interference test. The dashed lines represent the calculated drawdown for the initial template shown in Fig. 8.

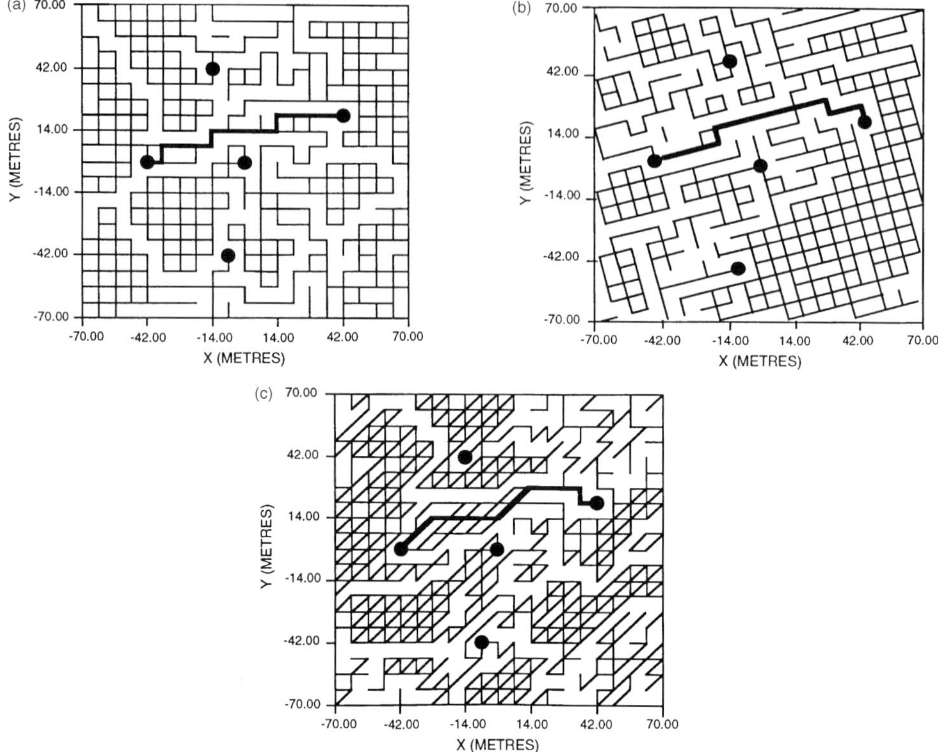

Fig. 10. The final configuration of conductors after inversion using the equivalent-discontinuum approach for three grid selections. (a) Grid oriented north–south and east–west. (b) Grid aligned east-northeast and north-northwest, the directions of fracturing in the Fort Riley Limestone. (c) Triangular grid.

accounting for the early response of GW-2 to production from GW-5. Figures 10(b) and (c) show the final configurations of conductors for two additional grid selections, one of which (Fig. 10b) is aligned with the mapped fracture pattern (Fig. 4). These models were run to determine the impact of grid selection on the results. Although the fracture patterns differ in details, in all three cases a direct fracture flow path exists between GW-5 and GW-2 to the north of GW-3.

Variable aperture lattice models

The variable aperture lattice approach seeks to find a spatial pattern of fracture apertures that satisfies the interference test data. The steps are similar to the ones used in the equivalent discontinuum approach. Hovever, instead of turning conductors on and off, conductors are assigned apertures sampled uniformly from a specified aperture distribution. In this study, we used a log normal distribution of apertures with a mean aperture of 0.00065 m and a log aperture variance of 0.5. Figure 11 shows the spatial pattern of apertures obtained from the inversion, where the mean aperture has been removed, hence the negative scale in the figure. In Fig. 11, the hot colours (e.g. yellow, orange and red) represent regions with fracture apertures less than the mean aperture and the cool colours (green and blue) represent apertures greater than the mean. The model obtained from inversion appears noisy and a preferential flow path connecting wells GW-5 and GW-2 is not apparent in Fig. 11.

Better results were obtained with an ensemble analysis of the test data. Ensemble analysis involves generating a collection of inverse models that satisfies the well test data within acceptable limits. We then determined properties that are shared by the ensemble of models and their associated uncertainties to create a model of the fracture system (Datta-Gupta *et al.* 1995). Specifically, several simulated annealing runs were conducted to generate multiple variable aperture lattice models and concluded when the misfit was reduced to a specified level.

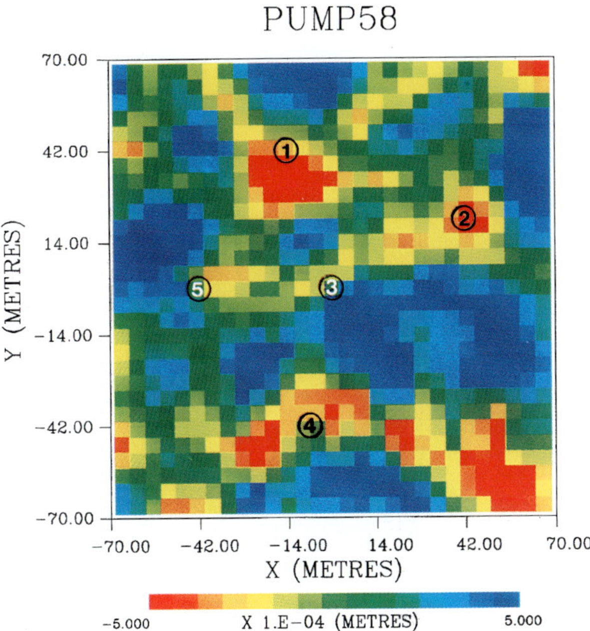

Fig. 11. Spatial pattern of fracture apertures obtained from inversion of interference test data using the variable aperture lattice approach. The 'hot' colours (red, orange and yellow) represent fracture apertures less than the mean aperture, whereas the 'cold' colours (dark green and blue) represent apertures greater than the mean.

When a sufficient number of models was generated, various statistical quantities were extracted from the ensemble. Figure 12 shows the ensemble median model from variable aperture annealing of the interference test data. The median model was chosen because it is less sensitive to outliers. Like the discontinuum models, the ensemble median model shows a preferential flow path between GW-5 and GW-2 to the north of GW-3.

Results of the seismic experiments

After the interference tests were conducted, a series of single well and crosswell seismic experiments were performed in the GW well array. The primary goal of these experiments was to test whether single well reflection profiling and crosswell methods could be used for imaging fractures in tight gas reservoirs. Specifically, we hoped to precisely locate the fracture suggested by the inverse models. Air was injected into the Fort Riley Limestone between packers placed in well GW-5. The concept was to perform seismic imaging experiments before, during, and after air injection to determine the effect of air in a fracture. Care was taken to keep the air injection pressure below the parting pressure of the Fort Riley. During the air injection, a pump was installed in the bottom of GW-2 to keep the water level below the bottom of the limestone. Our intent was to create a pressure sink in GW-2 and further encourage the flow of air along the fracture or fracture network connecting GW-5 and GW-2. It was thought that injecting air might increase the reflectivity and attenuation properties of the fracture.

Crosswell experiments

Crosswell experiments were performed between GW-3 and each of the outer GW wells before and after air injection. In this paper, we present the results related to the fracture detection. More detailed discussions of the crosswell and the single well seismic experiments are provided in two papers by Majer *et al.* (1996, 1997). The source–receiver configuration for the crosswell experiments is shown in Fig. 13. A piezoelectric source (cylindrical bender) was placed in GW-3 and an 8-channel receiver string (hydrophones) was placed in one of the outer wells. The source generates a swept sine wave using frequencies between 1000 to 10 000 Hz over a 50 ms time window and a recording time of 80 ms at 50 000 samples per second. Data were acquired using a 16-bit, 12-channel system, capable of recording

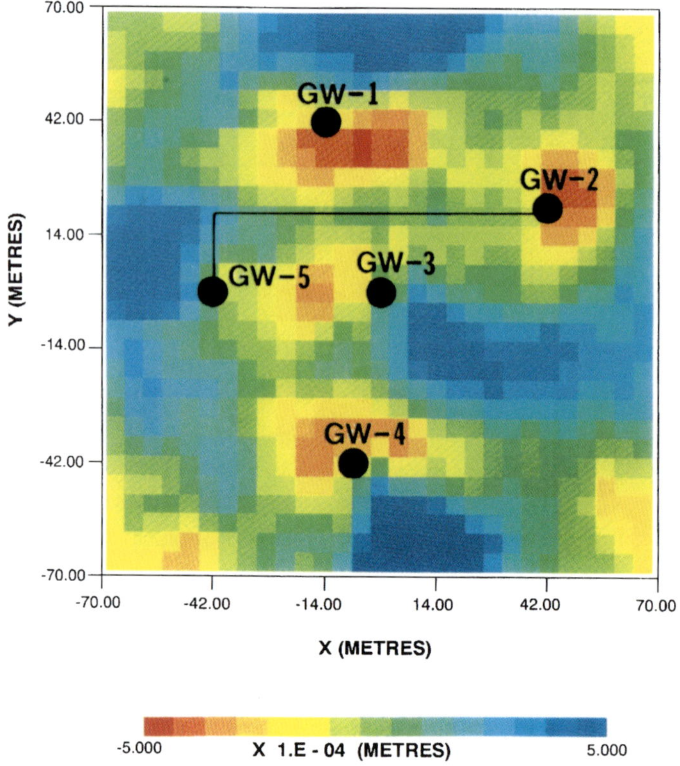

Fig. 12. Ensemble median model obtained from inversion of the interference test data using the variable aperture lattice approach. A fast fracture flow path between GW-5 and GW-2 is delineated by the east–west trending row of green and blue pixels north of GW-3 and is also marked by the thin black line.

100,000 samples per second per channel, including power electronics developed at LBL to deliver up to 8000 V peak to peak at several amps into a cable of up to 1 microfarad capacitance from 500 to 15,000 Hz. The source was

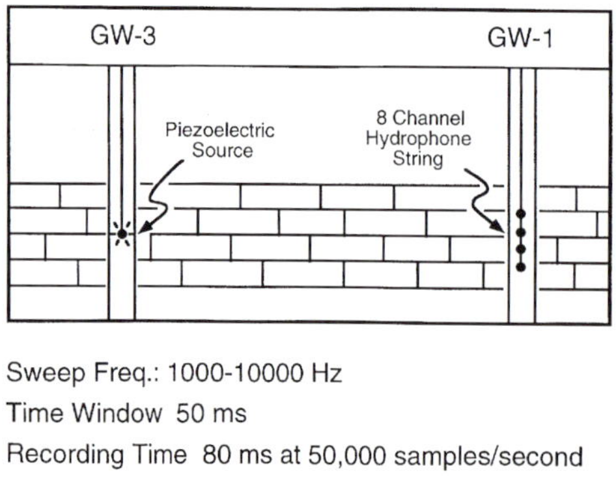

GW-3
GW-1
Piezoelectric Source
8 Channel Hydrophone String

Sweep Freq.: 1000-10000 Hz

Time Window 50 ms

Recording Time 80 ms at 50,000 samples/second

Fig. 13. Schematic diagram of the source–receiver configuration used for the crosswell seismic experiments.

positioned in GW-3 such that it was directly across from the centre of the receiver string. The procedure for crosswell measurements was to move the source and the receiver string in 0.25 m increments up the holes concurrently. The starting and ending positions of the source and the centre of the receiver string were the bottom and the top of the Fort Riley Limestone, respectively.

Crosswell experiments were performed between four well pairs, however we focus on the results from two well pairs: GW-3/GW-1 and GW-3/GW-4. The crosswell results from both well pairs are quantified in time–amplitude plots shown in Fig. 14. These plots were created by calculating a summed spectral amplitude over 4000 to 6000 Hz in 0.08 ms time steps along each trace at each depth. The 4 to 6 kHz band was chosen because it was the one with the most power. Figure 14 shows time–amplitude plots between the two well pairs before and after air injection. Red and blue represent high and low amplitudes, respectively. Note that almost all traces in well pair GW-3/GW-1 show a sharp

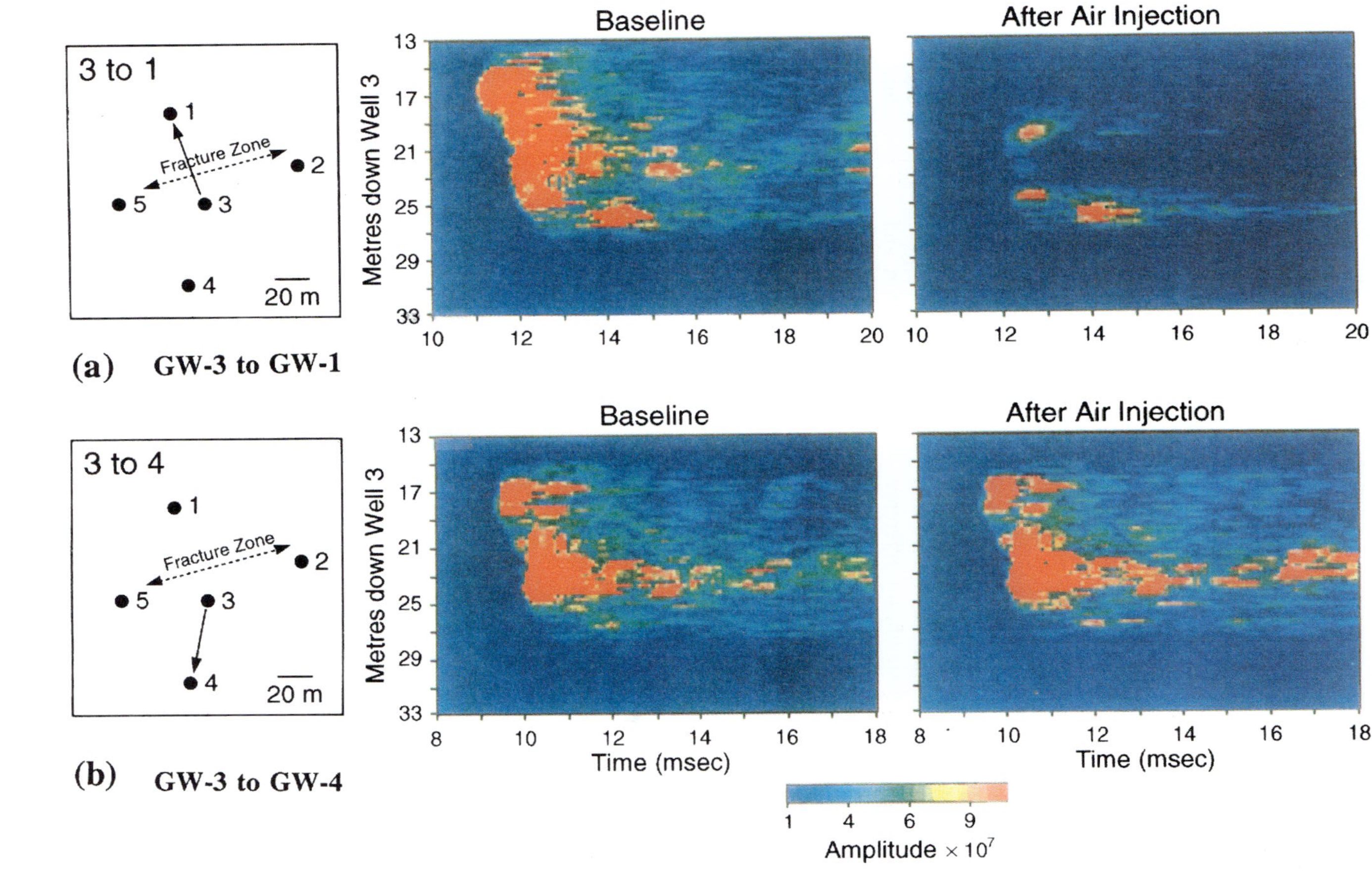

Fig. 14. Time–amplitude plots for crosswell pairs GW-3/GW-1 and GW-3/GW-4 before and after air injection.

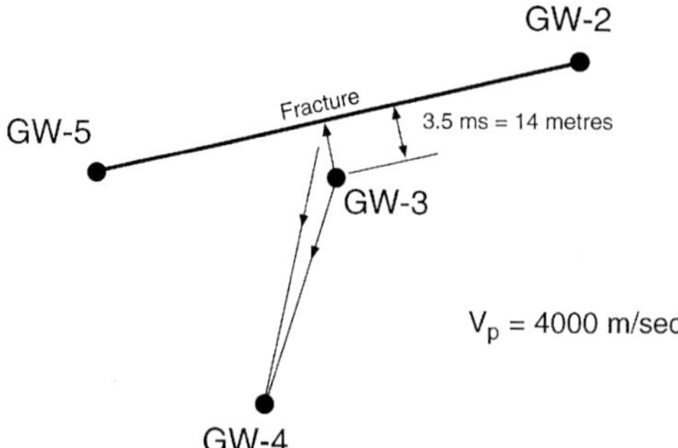

Fig. 15. Schematic diagram showing the position of a vertical fracture relative to GW-3 and the ray paths for a direct arrival between GW-3 and GW-4 and a reflection from the vertical fracture.

decrease in amplitude of the first arrival (between 11 and 12 ms depending on depth) after air injection whereas the first arrival signals in well pair GW-3/GW-4 look nearly identical before and after air injection. The only significant difference in the before and after plots for well pair GW-3/GW-4 is the increase in amplitude of a secondary arrival at 17 ms (7 ms after the first arrival). We interpreted the increase in energy to be a reflection from the suspected vertical fracture between GW-5 and GW-2. If a large drop in amplitude was caused by a fracture being filled with air, then it is likely that there would be reflected energy from this fracture. This situation is illustrated schematically in Fig. 15. In Fig. 15 the fracture is positioned 3.5 ms north of GW-3 to satisfy the observation that a secondary arrival reached GW-4 7 ms^{-1} after the direct arrival between GW-3 and GW-4. The P-wave velocity of the Fort Riley limestone is 4000 m/second, therefore a one-way travel time of 3.5 ms would place the fracture approximately 14 m from GW-3.

Single well reflection profiling

Single-well seismic reflection experiments were performed in wells GW-1 and GW-3 (Majer *et al.* 1996, 1997); however we focus here on the results from GW-3. The source–receiver configuration is illustrated schematically in Fig. 16. The 8-element receiver string with 0.25 m intervals between hydrophones was hung vertically in GW-3 with the piezoelectric source. As the receiver string was held in place, the source was moved from 1 m below the bottom hydrophone to the bottom of the Fort Riley Limestone at 0.25 m intervals. The receiver string was then moved up 0.25 m and the procedure repeated until the entire Fort Riley interval was covered. This procedure was then repeated with the source above the receiver string. The result was a multi-fold imaging dataset using a split spread configuration. The data were then processed as a common depth point (CDP) reflection survey.

Figure 17 shows the resulting CDP stacks in GW-3 before and after air injection in GW-5. The two sections are almost identical, except for a strong reflector at 7 ms in the after air injection section. This reflector comes in at the same

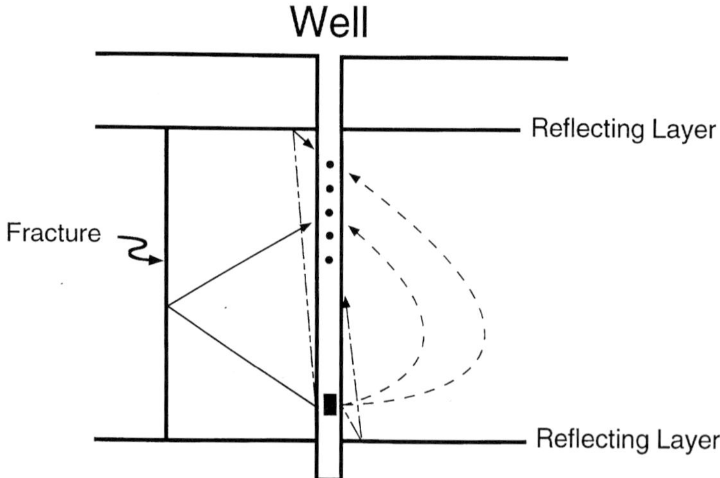

Fig. 16. Schematic diagram of the source–receiver configuration used for the single well seismic reflection surveys. The solid black dots represent the receiver string and the solid black rectangle represents the seismic source.

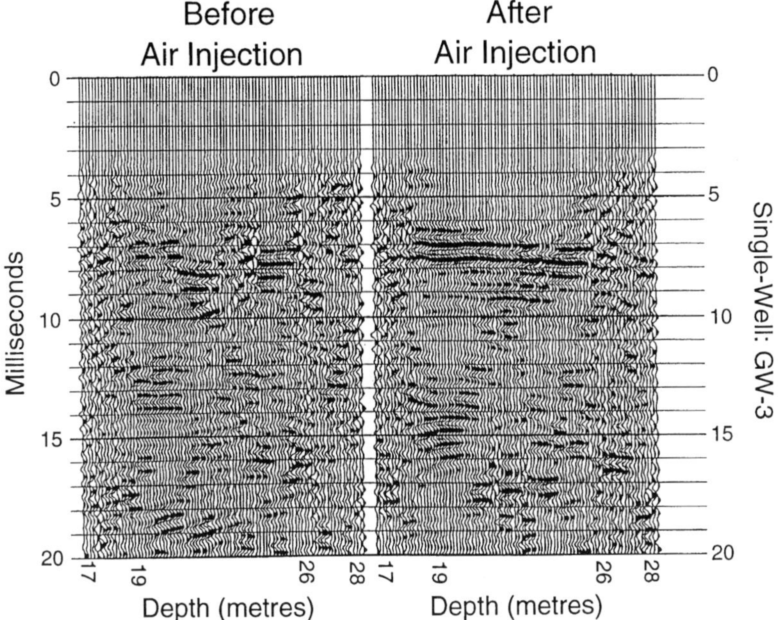

Fig. 17. A CDP stack of the single well reflection data from GW-3 before and after air injection. Note the increase in the reflected energy at 7 ms when air fills the fracture zone.

time as the proposed reflector seen in the crosswell data between GW-3 and GW-4. The reflection is strong from a depth of about 19 to 26 m which is a region of relatively high velocity as indicated on a P-wave velocity log in GW-3. The reflector appears to extend to a depth of 28 m with some evidence of reflected energy shallower than 19 m.

In summary, the single and crosswell results indicated the presence of a vertical fracture 14 m north of GW-3 along a line between GW-3 and GW-1 and at a depth below the surface between 19 to 28 m.

Results of slant well drilling

In order to verify the results of the seismic experiments, we drilled a slant well (GW-6) designed to penetrate the vertical reflector thought to be a fracture. Using the reflection arrival and average velocity of the Fort Riley Limestone, we picked a target 24 m below the surface and 14 m north-northwest of GW-3. A commercial air drilling rig was used to drill the slant well at 30° from the vertical. A schematic diagram of the well is shown in Fig. 18. The surface location of the well was 4.6 m west-southwest of GW-3 and its bearing was parallel to a line connecting GW-3 and GW-1. NX-size core (diameter = 84 mm) was taken in the Fort Riley Limestone from 24 to 33 m drilling depth (true vertical depth (TVD) between 20.8 and 28.6 m) with a recovery of nearly 100%. Only one natural fracture was intersected during drilling between 24.9 and 25.1 m TVD. This fracture was located less than 1 m from the targeted fracture depth of 24 m.

A photograph of the fracture is shown in Fig. 19. Three pieces of evidence indicate that the fracture is natural and not drilling induced. First of all, the fracture is planar and oriented 30° to the core axis. This orientation is consistent with an interpretation that the fracture is vertical. Vertical or near-vertical natural fractures are commonly observed in nearby outcrops of Fort Riley Limestone (Queen & Rizer, 1990) and one was observed in the Fort Riley core from well GW-5. Second, we examined the fracture surface under an optical microscope and observed euhedral calcite crystals (dog-tooth spar) and framboidal pyrite. Their occurrence indicates that the fracture was open in the subsurface enabling euhedral mineral crystals to form on the fracture surface. Third, there was significant water influx into the borehole during drilling immediately after 24.9 m TVD. This increase in the influx of water was manifested by a significant and abrupt increase in the volume of water blown out of the well whilst air drilling.

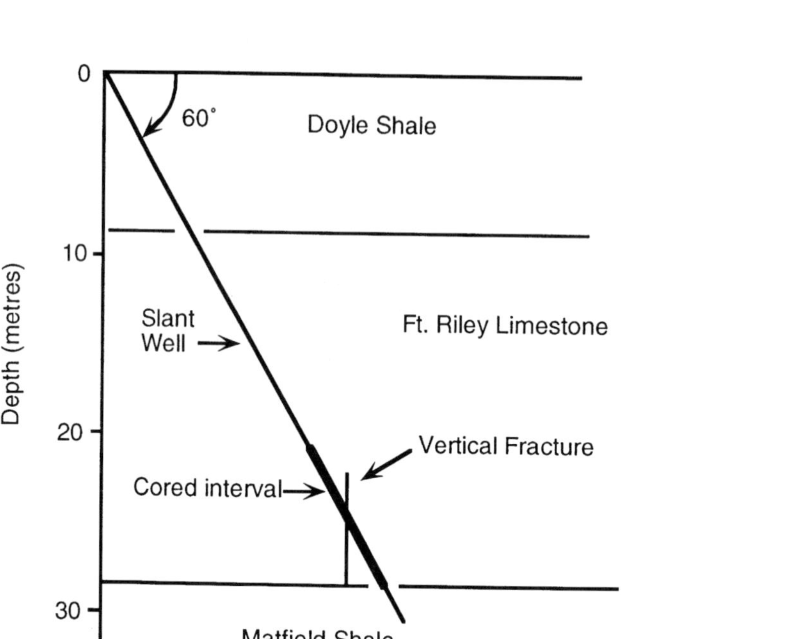

Fig. 18. Schematic diagram of the slant well, GW-6, showing targeted fracture zone and cored interval.

Recently, a series of single well and crosswell seismic experiments were performed in the GW-6 slant well. In these experiments air was injected into the Fort Riley Limestone between packers that straddled the fracture in GW-6. The data are being analysed and the interpretation of the results is not yet completed. However, water was observed flowing out of wells GW-2 and GW-5 during the air injection in GW-6 providing additional evidence that the fracture in GW-6 is part of the preferential fracture flow path between the GW-2 and GW-5 wells. Interestingly, there was only a slight water level change in nearby well GW-3 during the air injection in GW-6.

Discussion

Hydrological experiments

The results and application of the inverse modelling performed in this study are affected in a few significant ways by the assumption that the matrix permeability of the Fort Riley Limestone is zero. Actually, the matrix permeability ranges between 0.11 and 6.06 mD with a mean of 1.23 mD based on laboratory liquid permeability measurements on 29 horizontal core plugs from well GW-2. The fracture models derived from the inverse modelling of the interference tests (Fig. 10) have different spatial characteristics from the fracture system in the Fort Riley Limestone (Fig. 4) mapped by Queen & Rizer (1990). The fractures in the inverse models are, in general, more closely spaced and better connected than in the outcrop (Fig. 4). We suspect that a more dense and better connected fracture network was required to match the interference test data in the absence of matrix permeability. Consequently, the connections between the pumping well, GW-5, and wells GW-1, GW-3 and GW-4 with sluggish pressure responses (Fig. 9) were modelled as long, tortuous fracture flow paths embedded in impermeable rock when, in fact, the flow paths likely comprise partially connected fractures and low permeability matrix.

Another way to evaluate the effect of the zero matrix permeability assumption is to consider what consequences moving the locations of either or both wells GW-2 and GW-5 would have on the inverse modelling results. Oriented core from GW-5 contains an open, sub-vertical natural fracture in the lower 6 m of the Fort Riley Limestone. If GW-5 had been drilled a few to several metres away from its present location, missing the fracture, would inverse

Fig. 19. Natural fracture in GW-6 core between 24.9 and 25.1 m true vertical depth. The top of the core section is at the top of the photograph. The core piece on the top has a planar, natural fracture oriented 30° to the core axis. The other side of the fracture has been broken into rubble. The scale is 10 cm (4 in.).

modelling of the interference test have detected the east-northeast preferential fracture flow path north of GW-3? For future work it would be worthwhile testing what effect matrix permeability has on the ability of inverse modelling to detect fracture flow paths and generate fracture networks that more closely resemble observed data (e.g. outcrop).

Seismic experiments

There are aspects of the seismic data acquisition and processing and the well completion design at the CBTF that account for the success of the seismic experiments. The single well data presented here were characterized by a lack of tube waves, but contained large shear-wave energy. The tube waves may have been attenuated by the sand packing around the boreholes, and it must be anticipated that strong tube waves could exist in other single well surveys performed in wells with oil-field type completions. We believe that our success was a combination of careful attention to electronic noise reduction, the use of high frequency data, and the well bore conditions. Tube waves could have a strong complicating effect on the processing, but the shear-wave energy was easy to remove with *f–k* filtering so it can be assumed that the tube wave could just as easily be removed. In the worse case one would design the survey such that the arrival times of interest would not be in the same time window as the tube waves.

Conclusions

There are several significant results from this work:

(1) The inverse modelling methods were successful in reproducing the transient-pressure behaviour at the pumping and observation wells and may be viable methods for characterizing fractured reservoirs.

(2) Both inverse approaches (equivalent discontinuum and variable aperture lattice) were able to resolve a preferential fracture flow path in the Fort Riley Limestone.

(3) Single well reflection surveys can provide useful information on vertical features at tens to possibly hundreds of metres from the well. Single well surveys show high potential for characterizing fine scale reservoir heterogeneity, but due to operational issues (e.g. tube waves, horizontal velocity gradients and lack of commercial systems) the method has not been extensively used.

(4) Fractured reservoir characterization requires the use of high frequency energy in a combination of crosswell and single well surveys.

This work was supported by the Office of Oil Gas and Shale Technologies, U.S. Department of Energy (DOE) under Contract No. DE-AC03-76SF00098. We are grateful to Conoco Inc. for their support of this project and to H. Tan for Amoco's support and interest in this work. All computations were carried out at the Center for Computational Seismology and the field work was supported by the Geophysical Measurement Facility at the Ernest Orlando Lawrence Berkeley National Laboratory, both supported by DOE's Office of Energy Research Geosciences Program.

References

DATTA-GUPTA, A., VASCO, D. W., LONG, J. C. S., D'ONFRO, P. S. & RIZER, W. D. 1995. Detailed characterization of a fractured limestone formation by use of stochastic inverse approaches. *Society of Petroleum Engineers, Formation Evaluation*, **10**, 133–140.

GASTEIGER, C. M. 1980. *Strain analysis of a low amplitude fold in north-central Oklahoma using calcite twin lamellae*. M. S. thesis, University of Oklahoma, Norman.

GELHAR, L. W. 1993. *Stochastic Subsurface Hydrology*. Prentice Hall Inc., New York City.

LUZA, K. V. & Lawson, Jr, J. E. 1981. *Seismicity and tectonic relationships of the Nemaha Uplift, part III*. Special Publication **81–3**, Oklahoma Geological Survey.

MAJER, E. L., DATTA-GUPTA, A. PETERSON, J. E., VASCO, D. W., MYER, L. R., DALEY, T. M., KAELIN, B., QUEEN, J. H., D'ONFRO, P. S., RIZER, W. D., COX, D. & SINTON, J. 1996. Utilizing crosswell, single well and pressure transient tests for characterizing fractured gas reservoirs. *The Society of Exploration Geophysicists, The Leading Edge*, **15**, 951–956.

——, PETERSON, J. E., DALEY, T., KALEN, B., QUEEN, J. H., D'ONFRO, P. S. & RIZER, W. D. 1997. Fracture detection using crosswell and single well surveys. *Geophysics*, **62**, 495–504.

MELTON, R. A. 1929. A reconnaissance of the joint systems in the Ouachita Mountains and central plains of Oklahoma. *Journal of Geology*, **37**, 729–746.

QUEEN, J. H. & RIZER, W. D. 1990. An integrated study of seismic anisotropy and the natural fracture system at the Conoco Borehole Test Facility, Kay County, Oklahoma. *Journal of Geophysical Research*, **95**, 11255–11273.

TOOMEY, D. F. 1992. Microfacies correlation of the Early Permian Barneston Limestone, Conoco Test Facility to Vap's Pass, Kay County, Northern Oklahoma. *Oklahoma Geological Survey Bulletin* **145**, 193–219.

WATTS, R. W. 1996. Objectives of the U.S. DOE's research. *Society of Exploration Geophysicists, The Leading Edge*, **15**, 906.

Simulating polyphase faulting with a tensorial 3D model of fault growth

B. MAILLOT[1*], P. COWIE[1] & D. LAGUE[2]

[1] *Grant Institute, University of Edinburgh, West Mains Road, Edinburgh, EH9 3JW, UK*
[2] *Géosciences Rennes, UPR 4661 du CNRS, Campus de Beaulieu, 35042 Rennes Cedex, France*
[*] *Present address: Département de Géologie, Université de Cergy-Pontoise, Le campus, 8, Av. du Parc, 95 033 Cergy-Pontoise Cedex, France*

Abstract: Fault growth in brittle media has previously been extensively studied via a numerical approach using scalar representation of the stress and strain fields. More realistic simulations and further investigations of fault array evolution demand a fully tensorial three dimensional (3D) representation of these fields. In this paper we present a tensorial model of the spontaneous birth and growth of faults in a 3D medium. The medium is elastic and attenuating up to a stress threshold, determined by the Mohr–Coulomb criterion, where brittle failure is modelled by a partial shear stress drop. Elastic radiations generated by the rupture are explicitly extrapolated in time by a finite-difference scheme of the equation of dynamics until the static state is reached. We do not consider the dynamic process explicitly; further ruptures can only be triggered by the resulting static stress field, or by the imposed straining of the medium. Preliminary simulations of 3D straining of a 2D plate show how a pre-existing fault set (appearing as a perturbed stress field)can influence the development of a second fault set. We believe that our model provides a valuable tool for the study of fault development and in particular for the assessment of the effects of anisotropic stresses around faults on strain accumulation and the spatial organization of crustal deformation.

Since 1990 there has been a gradual yet fundamental shift in ideas about the growth and evolution of faults. Previously, growth models primarily focused on the propagation of a single isolated fault and the influence of rock properties on controlling displacement–length ratios for individual faults (Walsh & Watterson 1988, 1987; Cowie & Scholz 1992; Gillespie *et al.* 1992). The shift in perspective has been to consider instead the evolution of a population of faults forming during a particular tectonic episode (Davy *et al.* 1990; Sornette, A. *et al.* 1990; Cowie *et al.* 1995; Cartwright *et al.* 1995). It is clear from field observations and analogue modelling that crustal deformation, however small the net strain, is accommodated by large numbers of faults rather than a single structure. Moreover, the deformation occurs at all scales from microcracks to major faults that may penetrate the entire crust. Focusing on isolated faults alone has resulted in useful and robust results (e.g. Dawers *et al.* 1993), but it is now widely recognised that faults interact at both short-range (i.e. with nearest neighbours, see for a summary Willemse *et al.* 1996) and at long-range. These interactions are thought to play a key role determining the spatial organization of the deformation, i.e. the concentration of strain along fault zones with large displacements separated by regions which are much less deformed (Sornette, D. *et al.* 1990; Sornette, A. *et al.* 1990; Cowie *et al.* 1993, 1995).

The spatial organization is indicated by characteristic fractal scaling properties, now widely documented for natural fault patterns in continental areas (see Special Issue of *Journal of Structural Geology*). The proposed explanation for these scaling properties is based on the idea of coeval nucleation of many small faults each of which perturbs the stress field in the surrounding crust. Interaction between the developing faults leads naturally to enhancement of stress in some areas (which localizes the deformation) and stress shielding in other areas. The deformation develops a spatially correlated structure as a consequence of the stress field interference phenomenon (Sornette A. *et al.* 1990). There are two key assumptions in this explanation: the first is that the shallow crust (i.e. <10–15 km) can support elastic strains, the second is that the crustal properties are heterogeneous. (That the crust can support stress of low amplitude over geologic time is evident from the occurrence of induced seismicity.) Adopting these ideas, Cowie *et al.* (1993, 1995) presented a simple scalar model for crustal deformation which showed the development of a fractal fault pattern even though the material properties of the crust were set to be randomly heterogeneous and there was no preferential weakening of active

MAILLOT, B., COWIE, P. & LAGUE, D. 1998. Simulating polyphase faulting with a tensorial 3D model of fault growth. *In*: JONES, G., FISHER, Q. J. & KNIPE, R. J. (eds) *Faulting, Fault Sealing and Fluid Flow in Hydrocarbon Reservoirs*. Geological Society, London, Special Publications, **147**, 209–216.

faults. In their model, the simulated fault populations exhibited fractal scaling properties indistinguishable from natural examples. Following Sornette, A. *et al.* (1990), Cowie *et al.* (1995) argued that spontaneous self-organization of the deformation is a consequence of the long range nature of the elastic strain field described by the Laplace equation ($\nabla^2\varepsilon = 0$). In detail, the evolution of the fault pattern appears to be due to two competing effects:

(i) the nucleation of new faults primarily controlled by material heterogeneity; and
(ii) the degree to which the stress fields around developing faults may dominate and dictate fault propagation and linkage between existing faults. The first effect generally leads to more distributed deformation while the second effect leads to the development of longer faults and strain localization. Note that Davy *et al.* (1995) and Heimpel and Olson (1996) have shown that including ductile lower layers can modify this basic picture.

The results obtained by (Cowie *et al.* 1995) were derived using a scalar model of the deformation which is solely valid in a 2D anti-plane geometry. Hence, the conclusions to which they lead can only be taken as qualitative, and they should be seen as guides for further investigations of more realistic cases. Other 2D geometries (i.e. in-plane), and indeed the 3D case, demand a tensorial formulation of strains and stresses. Poliakov *et al.* (1994) present a 2D in-plane tensorial elasto-plastic model for shear band development. They found fractal scaling properties for the deformation although they analysed the strain rate, as opposed to the accumulated fault strain, so their results are not directly comparable with other studies. Other computer simulations of fault growth include the cellular automaton model of An & Sammis (1996). However the starting assumptions required for the automaton do not permit spontaneous development of the fault pattern.

Any future investigation on the behaviour of fault sets requires a tensorial formulation, because it is the only way to capture the intrinsically anisotropic nature of the stress and strain fields around faults. Sornette, D. *et al.* (1990), and Sornette & Virieux (1992) discuss the importance of this anisotropy in their theoretical model for crustal deformation as a self-organized critical phenomenon. Self-organized criticality has been proposed as an explanation for the Gutenberg–Richter relationship universally observed for earthquakes (Bak & Tang 1989; Sornette & Sornette 1989). However, the role of the tensorial strain anisotropy in the development of fault patterns and fault scaling relationships has not yet been explicitly investigated.

This paper presents a numerical method to calculate the distribution of tensorial stresses as faults nucleate and evolve in a heterogeneous 3D medium. The first section of the paper describes the model. As in the model of Cowie *et al.* (1995), we do not take account of the dynamics of the rupture process, and no *a priori* fault planes are assumed (faults grow spontaneously throughout the medium). The major difference with previous models is a switch from a scalar to a fully tensorial description of the strain field in three dimensions. The method used is based on that of Nielsen & Tarantola (1992), which was originally developed to study the dynamics of seismic rupture.

In the present model, the medium under tectonic loading, is elastic and attenuating up to a stress threshold determined by the Coulomb criterion. Rupture is modelled by a partial loss of shear strength, and the corresponding static stress field is obtained by integrating, with time, the equation of dynamics. Then, broken regions heal instantaneously, so that elements can re-rupture and accumulate shear displacement. However, the use of a finite-difference scheme on an undeformable grid imposes the constraint that accumulated displacements stay within a few per cent of a lattice element size. In other words, the ratio of displacement to length that can be modelled is of the order of a few per cent. Thus we do not attempt to model lithospheric scale faulting for which thermo-mechanical models on deformable grids are more appropriate (Beaumont & Quinlan 1994; Govers & Wortel 1995). However, one advantage of the present modelling approach is that both the initiation of faulting and the subsequent development of the fault array with increasing strain can be studied. The small-strain limitation is also a reasonable approximation for studying reservoir scale faulting in sedimentary basins. Since the mathematical formulation of the model is fully tensorial in three dimensions, we can impose any type and orientation of tectonic strains. In particular, we can superimpose two-dimensional in-plane and anti-plane deformation phases.

The second section of the paper presents several examples of simulation, illustrating how the model can be used to study the growth of a fault set superimposed on a pre-existing fault set resulting from a different deformation phase. We do not attempt a quantitative study of these effects, nor an assessment of the effects of anisotropy on strain localization, but rather we aim to demonstrate a method for incorporating realistic tensorial physics into models of faulting.

The model

In this section, we present in detail our model which calculates the stress and strain fields in an elastic–brittle medium under imposed straining. The medium is discretized by a regular rectangular 3D lattice.

Dynamics

The medium obeys the fully tensorial equation of dynamics

$$\rho \frac{\partial v_i}{\partial t} = \nabla_j \sigma_{ij}, \qquad (1)$$

where v_i is the ith component of the velocity of displacement, ρ the medium density, and σ_{ij} the stress. Equation (1) is discretized in space and time by a finite-difference scheme with a staggered grid (Virieux 1986).

Rheology

Where not broken, the medium is elastic, i.e. the stress is related to the strain via the generalized Hooke's law. Rupture occurs according to the Coulomb criterion, but is not restricted to a predefined fault plane, i.e. we solve for any point in the medium where the maximum shear stress τ and the mean normal stress σ are such that $|\tau| - \nu\sigma \geq S$, where ν is the internal friction, and S the cohesion of the medium. The parameter S is allowed to vary from point to point, and is the only heterogeneous property of the medium. The rupture process itself is then modelled by an amount of shear stress drop proportional to τ at the broken points, while the normal stress is kept constant. (This partial shear stress drop, although not as elegant, is more general than the viscous rheology adopted by Nielsen and Tarantola (1992), which leads to a total stress drop.) The shear stress drop generates elastic radiations. Since we are primarily interested in the final static strain and stress fields, we absorb these radiations by adding to the right hand side of (1) a force

$$f_i = -\alpha |\nabla_j \sigma_{ij}| \operatorname{sign}(v_i) \qquad (2)$$

opposite to the velocity of displacement. The optimal absorption coefficient is $\alpha = 0.47$. It was determined by comparing static stress and displacement fields around a single broken element obtained with various values of α, $\alpha = 0.01$ being the reference. Above 0.47, the static state is not accurate (i.e. relative errors on displacements become greater than 0.1%), below 0.47, the convergence is slower without substantially improving the accuracy.

Boundary conditions

The medium is uniformly loaded with applied strain rates that represent the imposed deformation. The resulting applied stress

$$\sigma^P_{ij}(t) = \sigma^P_{ij}(0) + f(i,j) c_{ijk\ell} \int_0^t \dot{\varepsilon}_{k\ell}(\tau)\, \mathrm{d}\tau \qquad (3)$$

is a sum of the initial pre-stress $\sigma^P_{ij}(0)$ (e.g. lithostatic pressure) and of stresses due to the strain rates $\dot{\varepsilon}_{ij}(\tau)$ applied during a time t. Since the prestress is uniform in space, we do not take account of the lithostatic pressure gradient due to gravity. The strain rates $\dot{\varepsilon}_{ij}(t)$ are uniform in space, but can vary with time, hence allowing the superposition of different deformation events. $c_{ijk\ell}$ is the elastic stiffnesses, and $f(i,j)$ is a function with values 1 or 0 specifying which component $\dot{\varepsilon}_{ij}$ of the loading strain rates are imposed.

The axes on which the components of the applied stress (3) are defined can have any orientation with respect to the axes of the discretization lattice. This is an important point since it allows us to rotate the imposed straining during a simulation.

A more general equation than (3) could be used, for instance to take account of visco-elastic relaxation of the stresses, such as that proposed by Tarantola (1988). By adopting equation (3), we are here in the extreme case of a permanent build up of stresses due to the imposed strain, that can be relieved only by fracturing the medium. We discuss this assumption below.

Finally, the boundary conditions are periodic. This means that the medium is infinite and spatially periodic, with periods equal to the size of the lattice along each spatial direction. In other words, a fracture reaching an edge of the lattice, may continue its growth from the opposite edge. This particular geometry allows us to perform two-dimensional simulations merely by setting one of the axes of the lattice to be of length one. This corresponds conceptually to a 3D medium where all spatial derivatives are null along the axis of length one. The displacement field remains, however, three dimensional, reflecting both 'in-plane' and 'anti-plane' types of fractures. All the examples we show here are such 3D calculations performed on a 2D plate.

The full 3D case is numerically much more expensive. Finally, it should be noted that such boundary conditions forbid, by definition, the simulation of free surfaces, or no-strain surfaces.

Loading and rupture cycles

We start by loading the medium (i.e. applying imposed strain rates) until a single node reaches failure. The imposed straining is then stopped. The elastic radiations generated by the shear stress drop at the node are integrated in time with the equation of dynamics and the absorbing force (2) to obtain the static state

$$\nabla_j \sigma_{ij} = 0. \quad (4)$$

In practice, the static state is achieved when the maximum force $|\nabla_j \sigma_{ij}|$ over the whole lattice has decayed by a factor 10^3 (a factor 10^4 makes no substantial difference).

The broken points then heal instantaneously (i.e. they fully recover their static strength; see next section for a discussion on this issue). We then test for further ruptures that may occur from stress concentration around the previously broken point, and compute the new static state (there may be several nodes breaking at once). This rupture process is repeated until no further ruptures occur. In the following text, these successive ruptures occuring under a fixed straining of the medium will be referred to as the 'rupture cycle'. Once a rupture cycle terminates, the straining of the medium resumes (this will be referred to as the 'loading cycle'). We hence ignore the effects of the imposed strain rates *during* a rupture cycle, because we assume that the rupture time scale is negligible with respect to that of the strain rates (which is the case for tectonic processes).

Healing and the static case

When slip occurs on a fault, the friction changes from a static to a dynamic value. Healing is the process by which a fault recovers its static strength after the end of a slip event. In a fully dynamic rupture process, healing may start on one part of a fault as other parts are still slipping (e.g. Nielsen *et al.* 1995; Cochard & Madariaga 1994). In the present model we consider the static case whereby additional slip can only be triggered once a ruptured element has finished slipping, and started to heal. Hence, healing occurs instantaneously, within the rupture cycle defined above. By making this assumption, we isolate one process of fault (and slip) propagation: the propagation by concentration of static stresses as opposed to dynamic stresses. The reason for this assumption is that we are primarily interested in the long time scale behaviour of the deformation, rather than in its dynamic aspects. We believe that, in this static limit, we are still capturing the essential physical phenomenon by which fault sets develop. In fact, it is not yet clear how to perform healing in the 3D dynamic case, although it was modelled in the 2D anti-plane case by Nielsen *et al.* (1995). Finally, the static limit has the advantage of providing a much faster numerical algorithm.

There is another possible way to model healing in the static case by healing only at the end of the rupture cycle, i.e. when the medium is in complete equilibrium with the imposed strain. It is, however, not our goal here to investigate the effect of variable healing rates. All the simulations presented here were performed with the first model of healing above. These simulations are indeed aimed at showing the possibilities of investigation opened by a 3D tensorial representation of the strain field in a brittle medium.

Examples of simulations

The following set of examples is intended to show some of the capabilities of the model. We simulate the birth and growth of fault sets under different applied strains. The first example is an anti-plane shear (Fig. 1a). The second example is a pure shear of the medium at -15° from the grid axes (Fig. 1b), which results in a conjugate set of faults. The third example is similar to the second, but superimposed on the last stage of the first (i.e. the medium is first sheared in the anti-plane geometry of Fig. 1a, and then deformed in pure shear as in Fig. 1b).

All simulations were performed on a 2D medium (grid size: $128 \times 128 \times 1$), using a 3D formulation. The model was programmed in Fortran 77 using MPI (Message Passing Interface: a set of standard subroutines allowing the programmer to spread the data and control data fluxes within a parallel architecture of processors). All computations were performed on the Cray-T3D at EPCC (Edinburgh Parallel Computing Center). Parameters common to all examples are as follows: coefficient of internal friction: 0.6; density: $2500\,\mathrm{kg\,m^{-3}}$; shear stress drop of half the shear stress at onset of rupture;-isotropic P- and S- wave velocities, respectively: $3500\,\mathrm{m\,s^{-1}}$ and $2000\,\mathrm{m\,s^{-1}}$. All these parameters are constant in space and time. The only heterogeneous parameter is the cohesion (or the 'strength of the medium'), constant in time but randomly

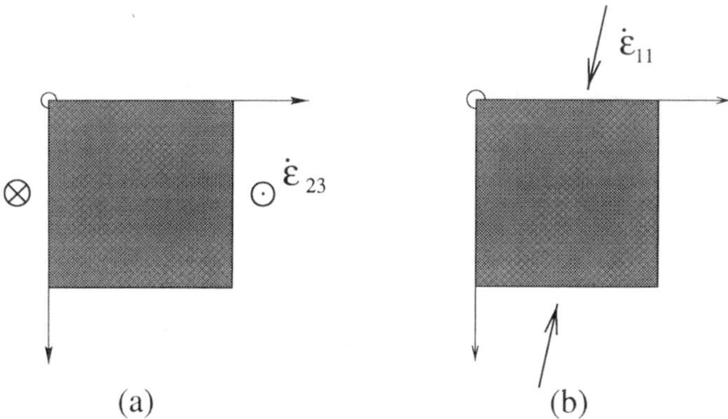

Fig. 1. (**a**) Imposed anti-plane strain rates for the first example. (**b**) Imposed in-plane strain rates for the second and third examples. All components of the strain rates which are not mentioned are left free. The shaded region represents the medium.

varying between 0.5×10^6 Pa and 1.5×10^6 Pa, without correlation from point to point. Its distribution is identical in all examples. The imposed strain rates (Fig. 1) were $10^{-9}\,a^{-1}$. However, the absolute values of the strain rates are irrelevant in our model because there is no time dependency in the rheology of the medium, which is only sensitive to the net amount of strain imposed. Should the medium be made visco-elastic [by changing the form of equation (3)], the net imposed strain and the time would no longer be linearly related and the actual values of strain rates would become relevant.

Figures 2 to 4 are plots of the rupture patterns for increasing imposed strain. Each plot contains four copies of the calculation grid in order to show the periodic nature of the boundary conditions.

In Figs 2 & 3, ruptures follow the expected orientations, forming linear features in the anti-plane deformation (Fig. 2), and a conjugate set in the in-plane deformation (Fig. 3). Both of these simulations started from an initial homogeneous stress field.

The third simulation (Fig. 4), was exactly identical to the second one, but started from the perturbed stress field of the last stage of the first simulation: the resulting ruptures start occuring at much lower net strains ($\varepsilon_{11} = 2.1 \times 10^{-5}$, as opposed to $\varepsilon_{11} = 1.27 \times 10^{-4}$ for the second

Fig. 2. Ruptures in the lattice in the first example (see Fig. 1a): anti-plane deformation $\varepsilon_{23} = 5.9110^{-5}$. Line thickness is proportional to the number of ruptures.

Fig. 3. Ruptures in the lattice in the second example (see Fig. 1b): in-plane deformation $\varepsilon_{11} = 1.5310^{-4}$. Line thickness is proportional to the number of ruptures.

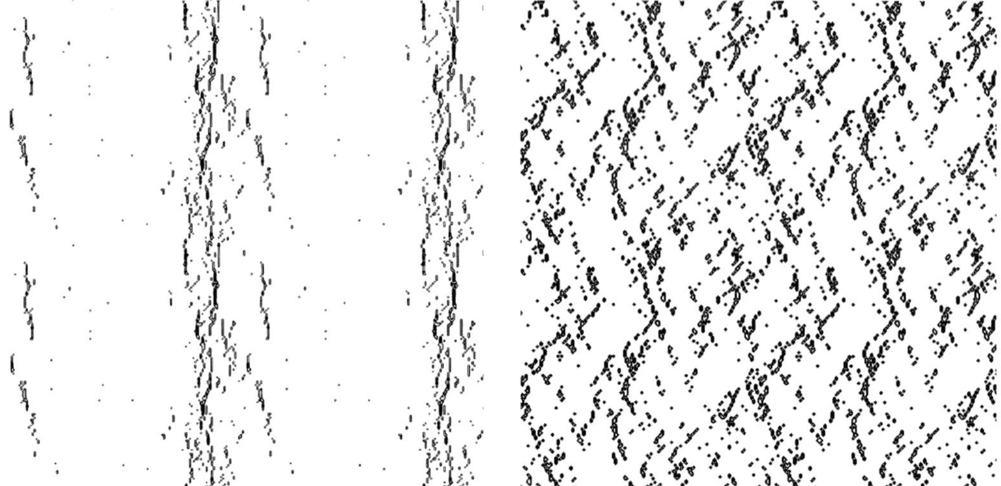

Fig. 4. Ruptures in the lattice in the third example (see Fig. 1): in-plane deformation superimposed on the perturbed stress field of the anti-plane fault set of Fig. 2. Line thickness is proportional to the number of ruptures. Left: ruptures at $\varepsilon_{11} = 3.910^{-5}$. Clearly, the presence of the first anti-plane fault set (Fig. 2) has strongly influenced the development of this second in-plane set, especially in its earlier stages. Right: ruptures occurring from $\varepsilon_{11} = 8.58 \times 10^{-5}$ to $\varepsilon_{11} = 1.04 \times 10^{-4}$. At this level of strain, the previous anti-plane deformation seems not to be felt any more: ruptures develop to form the expected conjugate set.

simulation), and clearly follow the inherited direction of the anti-plane ruptures. In fact these 'inherited' trends distort the formation of the conjugate set which is closest to that orientation. At further strains, the inherited stress field is overprinted and ruptures follow the conjugate directions (Fig. 4, right). However the conjugate pattern is not as strongly developed and the deformation is more diffuse.

This illustration of the effect of an inherited stress field is an extreme case because the medium is purely elastic. Visco-elasticity would result in a smoothing of the inherited stress field and hence reduce the magnitude of the effect. However, the wide spread occurrence of induced seismicity at relatively low fluid injection pressures demonstrates that elastic stresses are only partially relaxed.

Figure 5 shows the 3D displacement field in the third simulation for a net strain $\varepsilon_{11} = 4.5 \times 10^{-5}$, following the anti-plane deformation $\varepsilon_{23} = 5.9110^{-5}$. Here we have plotted just the calculation grid 128×128 in order to show the deformation pattern in detail. These plots should thus be compared to only one quadrant of the rupture plots shown in Figs 2 to 4. The top plot shows displacements resulting from the in-plane deformation. Note that they are complicated by the inherited stress field. The bottom plot is the component of the displacement perpendicular to the page, i.e. resulting from the initial anti-plane deformation.

Assumptions and further developments

Some of the main assumptions of the model, and therefore some of the possible improvements, are:

(1) Static limit, although the fully dynamic case should require only minor changes to the model.
(2) Periodic boundary conditions, which do not permit free surfaces or rigid walls to be modelled. Various ways to implement such boundaries within a finite-difference algorithm are widely available in the literature.
(3) Small strains, because we use Eulerian coordinates which do not follow the medium. The inclusion of large strains would require a Lagrangian formulation and a deformable computational grid, i.e. a completely different numerical method from that used here.
(4) Direct transition from elastic to brittle behaviour. Other tests not shown here have demonstrated that the inclusion of a Maxwell visco-elasticity is straightforward.
(5) No gravity effects (i.e. no lithostatic pressure gradient).
(6) No pore fluid effects. However, assuming that the fluid has the same bulk modulus as the rock matrix, previous work (Main *et al.* 1995) has demonstrated that a diffusion pore fluid pressure model (Maillot and

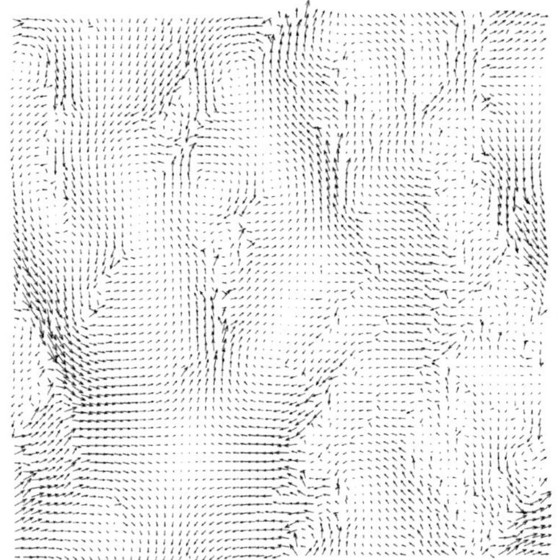

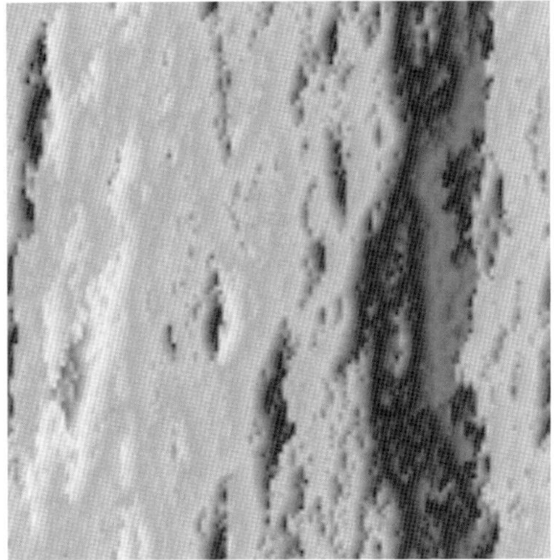

Fig. 5. Displacement field in the example 3, after a net applied strain $\varepsilon_{23} = 5.91 \times 10^{-5}$ (Fig. 1a) followed by $\varepsilon_{11} = 4.5 \times 10^{-5}$ (Fig. 1b). Top: in-plane displacements,bottom: displacements perpendicular to sheet (dark: inward (i.e. away from reader); light: outward (i.e. towards reader)).

Main 1996) can be coupled to the present model to capture poro-elastic effects: the rupture criterion is then applied to the effective stress, and the fluid pressure is coupled to the volumetric strains.

Conclusion

The model we have presented is an attempt to include realistic tensorial physics into the simulation of fault sets. Ruptures grow spontaneously as a response to applied strains on an elastic (2D or 3D) medium according to a rupture criterion (here, the Mohr–Coulomb criterion). After each rupture, the new static stress and strain fields are computed by extrapolating in time the velocity field according to the equation of dynamics with a damping term. Hence, the full displacement, stress and strain fields are known everywhere in the medium at any time of a simulation. The method has the potential to yield the seismic waveforms associated with rupture, but in this paper we calculate static stress changes and focus on long-time scale fault growth. This is a general and simple approach which makes no assumptions regarding the geometry of faults; they evolve spontaneously within the deforming medium.

The numerical implementation of the model uses a classical finite-difference algorithm to solve the equation of dynamics on a regular grid. In that respect, it is easier to implement than other techniques such as finite elements, distinct elements, and boundary elements which all meet a grid generation problem. The main limitation of this approach is that it is applicable only to small strains when tectonic rotations are negligable.

Using this model we can investigate the impact of an anisotropic stress field in the spatial organization of faulting during progressive deformation. The anisotropy is a fundamental property of crustal stress distributions yet its role in determining the scaling relationships of fault populations has not yet been explored. Previous scalar models established that if the material properties are randomly heterogeneous without spatial correlations and the boundary conditions remain unchanged throughout the deformation, then there is a straightforward transition from an early regime where nucleation of new faults dominates to a regime where growth and linkage of existing faults takes over. The present model permits further investigations: we can now address the question of how a set of faults formed during an early tectonic phase can influence the development of a later set of faults formed in a different tectonic episode. In particular, we have shown an example of how a fault set development may be perturbed by an inherited stress field due to a pre-existing fault set.

We thank S. Nielsen and A. Poliakov for their constructive comments. We thank Shell Research and Technology Services – SIEP for sponsoring this work, and the Edinburgh Parallel Computing Center for their technical support. For one of us (B. Maillot) this work was carried out as part of a Community training project (Marie-Curie Fellowship) funded by the European Commission.

References

AN, L.-J. & SAMMIS, C. G. 1996. Development of strike-slip faults: shear experiments in granular materials and clay using a new technique. *Journal of Structural Geology*, **18**, 1061–1077.

BAK, P. & TANG, C. 1989. Earthquakes as a self-organized critical phenomenon. *Journal of Geophysical Research*, **94**, 15 635–15 637.

BEAUMONT, C. & QUINLAN, G. 1994. A geodynamic framework for interpreting crustal-scale seismic-reflectivity patterns in compressional orogens. *Geophysical Journal International*, **116**, 754–783.

CARTWRIGHT, J., MANSFIELD, C. & TRUDGILL, B. 1995. The growth of faults by segment linkage: Evidence from the Canyonlands grabens of S. E. Utah. *Journal of Structural Geology*, **17**, 1319–1326.

COCHARD, A. & MADARIAGA, R. 1994. Dynamic faulting under rate-dependent friction. *Pure & Applied Geophysics*, **142**, 419–445.

COWIE, P. A. & SCHOLZ, C. H. 1992. Physical explanation for displacement-length relationship for faults using a post-yield fracture mechanics model. *Journal of Structural Geology*, **14**, 1133–1148.

——, MAILLOT, B. 1996. Numerical modelling of brittle deformation in rock: growth of fault and fracture arrays, abstract at E.G.S. Conference, Strasburg.

——, VANNESTE, C. & SORNETTE, D. 1993. Statistical Physics Model for the Spatio-temporal Evolution of Faults. *Journal of Geophysical Research*, **98**, 21809–21822.

——, SORNETTE, D. & VANNESTE, C. 1995. Multifractal scaling properties of a growing fault population. *Geophysical Journal International*, **122**, 457–469.

DAVY, P., SORNETTE, A. & SORNETTE, D. 1990. Some consequences of a proposed fractal nature of continental faulting. *Nature*, **348**, 56–58.

——, HANSEN, A., BONNET, E. & ZHANG, S.-Z. 1995. Localization and fault growth in layered brittle–ductile systems: Implications for deformation of the continental lithosphere. *Journal of Geophysical Research*, **100**, 6281–6294, 1995

DAWERS, N. H., ANDERS, M. H. & SCHOLZ, C. H. 1993. Growth of normal faults: displacement–length scaling. *Geology*, **21**, 1107–1110.

GILLESPIE, P. A., WALSH, J. J. & WATTERSON, J. 1992. Limitations of displacement and length data for single faults and the consequences for data analysis and interpretation. *Journal of Structural Geology*, **14**, 1157–1172.

GOVERS, R. & WORTEL, M. J. R. 1995. Extension of stable continental lithosphere and the initiation of lithospheric scale faults. *Tectonics*, **14**, 1041–1055.

HEIMPEL, M. & OLSON, P. 1996. A seismodynamical model of lithosphere deformation:development of continental and oceanic rift networks. *Journal of Geophysical Research*, **101**, B7, 16 155–16 176.

Journal of Structural Geology, Special Issue, **18**, No 2/3.

MAILLOT, B. & MAIN, I. G. 1996. A lattice BGK model for the diffusion of pore fluid pressure, including anisotropy, heterogeneity, and gravity effects. *Geophysical Research Letters*, **23**(1), 13–16.

MAIN, I. G., MAILLOT, B. & NIELSEN, S. 1995. A Numerical model of seismic rupture in fluid infiltrated rocks. *EOS Transactions*, **76**, 46.

NIELSEN, S. & TARANTOLA, A. 1992. Numerical model of seismic rupture. *Journal of Geophysical Research*, **97**, 15 291–15 295.

——, KNOPOFF, L. & TARANTOLA, A. 1995. Model of earthquake recurrence: Role of elastic wave radiation, relaxation of friction, and inhomogeneity. *Journal of Geophysical Research*, **100**, No B7, 12 423–12 430.

POLIAKOV, A. N. B., HERRMANN, H. J., PODLADCHIKOV, Y. Y. & ROUX, S. 1994. Fractal plastic shear bands. *Fractals*, **2**, 567–581.

SORNETTE, A. & SORNETTE, D. 1989. Self-organized criticality and earthquakes. *Europhysics Letters*, **9**, 197–202.

——, DAVY, P. & SORNETTE, D. 1990. Growth of fractal fault patterns. *Physics Review Letters*, **65**, 2266–2269.

SORNETTE, D., DAVY, P. & SORNETTE, A. 1990. Structuration of the lithosphere as a self-organised critical phenomenon. *Journal of Geophysical Research*, **95**, 17 353–17 361.

—— & VIRIEUX, J. 1992. Linking short-time scale deformation to long-time scale tectonics. *Nature*, **357**, 401–403.

TARANTOLA, A. 1988. Theoretical background for the inversion of seismic waveforms, including elasticity and attenuation. *Pageoph*, **128**, 365–399.

VIRIEUX, J. 1986. *P–SV* wave propagation in heterogeneous media – velocity-stress finite-difference method. *Geophysics*, **51**, 889–901.

WALSH, J. J. & WATTERSON, J. 1987. Distribution of cumulative displacement and of seismic slip on a single normal fault surface. *Journal of Structural Geology*, **9**, 1039–1046.

—— & —— 1988. Analysis of the relationship between displacements and dimensions of faults. *Journal of Structural Geology*, **10**, 239–247.

WILLEMSE, E. J. M., POLLARD, D. D. & AYDIN, A. 1996. Three-dimensional analyses of slip distributions on normal fault arrays with consequences for fault scaling. *Journal of Structural Geology*, **18**, 295–309.

Thermal effects of fluid flow in steep fault zones

C. G. FLEMING*, G. D. COUPLES† & R. S. HASZELDINE‡

Department of Geology and Applied Geology, University of Glasgow, Glasgow G12 8QQ, UK

** Present address: Babtie Geotechnical, 95 Bothwell St, Glasgow G12 7HX, UK*

† Present address: Department of Petroleum Engineering, Heriot-Watt University, Edinburgh EH14 4AS, UK

‡ Present address: Department of Geology and Geophysics, University of Edinburgh, Edinburgh EH9 3JW, UK

Abstract: Two-dimensional, porous-medium, steady-state, coupled fluid- and heat-transport models are used to investigate some of the hydrogeological and thermal consequences of steeply dipping fault damage zones in a normally pressured basin setting. Simple geometries and conservative petrophysical properties can produce large-scale buoyancy-driven circulation, both outside and within the fault zone. An average (homogeneous) basin permeability of only 7 mD k_h (horizontal permeability) and 0.07 mD k_v (vertical permeability) results in a free convection cell of this type, with the upflow being localized by a 300 m wide fault zone (50 mD k_h, 0.5 mD k_v). Steady-state temperature anomalies as large as 15°C at the top of the fault zone can be produced by this arrangement. Smaller values of basin permeability still result in a similar circulation pattern, but at flow rates which produce temperature anomalies that are below detection levels. When the basin fill is more heterogeneous (layered), higher permeabilities can exist in some layers without large-scale convection occurring, because of the dampening effect of other, lower permeability units. In realistic geometrical configurations that are similar to the North Sea Central Graben, the fluid flow system is dominated by within-fault-zone convective circulation that produces local (<10 km half-wavelength), high amplitude (50°C) temperature anomalies which are comparable to the largest of those actually observed in the subsurface.

Subsurface fluid flow is critically dependent on the petrophysical properties of the rock framework, the fluid properties, and the energy of the fluid system. Faults are an important factor in determining the petrophysical characteristics of the rock framework simply because they can disrupt the continuity of rock bodies. A further complexity in a faulted system is the presence of fault damage zones (Knipe 1993; Knipe *et al.* 1997) which are regions of deformed rock (constituting new material types) close to faults. Damage zones represent a further petrophysical heterogeneity that will affect the behaviour of the hydrogeological system.

Predicting the nature of fault damage zones is not simple, although progress is being made (Knipe *et al.* 1997). Depending on the rock types and the conditions of deformation, and on a knowledge of the history of fault movement, it is possible to predict (or at least explain after the fact) changes in fault zone porosity and permeability as a consequence of displacement. Some fault damage zones may exhibit both increases and decreases in their permeability characteristics: an increase in permeability parallel to the fault and a concurrent decrease across it (Foxford *et al.* 1998). This situation can arise both because of clay smears which cause flow barriers, and fault-parallel fractures which create flow paths.

Rather than focusing on faulting processes, this paper considers the hydrogeological ramifications of 'open', major, steep fault zones in a typical mature rift basin setting. We illustrate the changes in behaviour of such a large-scale flow system by emphasizing the thermal consequences of flow which is affected by the fault damage zone. We compare the magnitude and shape of the temperature disturbances resulting from such a situation to modern temperature anomalies observed in the North Sea. Our focus is on length scales that are rather larger than those usually necessary for modelling reservoir performance. However, this study is relevant to issues of fault-zone (and wider) diagenesis (Wood & Hewitt 1984; Haszeldine *et al.* 1992; Knipe 1993; Bjørlykke 1994; Fowles & Burley 1994; Mullis & Haszeldine 1995; Macaulay *et al.* 1997), hydrocarbon migration and the related concerns of sealing/leaking (Whelan *et al.* 1993; Grauls & Baliex 1994; Anderson *et al.* 1995), and the thermal history of basins (Smith & Chapman 1983; Person & Garven 1994; Le Carlier *et al.* 1994; Jessop & Majorowicz 1994; Wieck *et al.* 1995).

FLEMING, C. G., COUPLES, G. D. & HASZELDINE, R. S. 1998. Thermal effects of fluid flow in steep fault zones. *In*: JONES, G., FISHER, Q. J. & KNIPE, R. J. (eds) *Faulting, Fault Sealing and Fluid Flow in Hydrocarbon Reservoirs*. Geological Society, London, Special Publications, **147**, 217–229.

Model idealization

Porous medium models have proven beneficial in terms of their revelations concerning subsurface hydrogeological processes in basins (Bethke 1985; Garven 1995 and additional references therein; Person *et al.* 1996). However, a question concerning the applicability of these models arises when faults are a significant component of the geological system: Can faults be treated within the porous medium paradigm, or must open fracture modelling techniques be used? Open fracture modelling methods often assume that the matrix material does not participate in the flow, and that the entire system can be idealized with set(s) of partly or completely interconnected 'fractures' that have specified apertures and impervious walls (Cacas *et al.* 1990 *a*, *b*; Thomas & La Pointe 1995; Mourzenko *et al.* 1996). More complex 'dual-porosity' models have also been constructed (Pedersen & Bjørlykke 1994). Although such fractured medium models may not capture all of the elements of interest (de Marsily 1986), they give interesting insights and inspire further study. Still another class of model is familiar to those who use reservoir simulators that include discontinuities: in this type of approach, faults or fractures are assigned transmissibilities, but there is generally no consideration of specific fault zone width or properties (Foley *et al.* 1998; Manzocchi *et al.* 1998; Walsh *et al.* 1998).

In contrast, there has been some success in treating fault zones as porous materials in large-scale hydrogeological systems (Ge & Garven 1994; Haszeldine & Mckeown 1995; Mullis & Haszeldine 1995; Russell *et al.* 1995; Lewis 1996; Person *et al.* 1996; Garven pers. comm. 1997). Indeed, Garven (1995) argues that the hydrogeological effects of fractures and faults can be simulated successfully if the scale of the discretization is larger than the length of the majority of the discontinuities. Similarly, the determination of 'effective' flow properties for fractured media is the goal of many studies (D'Onfro *et al.* this volume; Ellevset *et al.* this volume; Knai & Knipe this volume; Matthäi *et al.* this volume; Steen *et al.* this volume; Townsend *et al.* this volume), with the effects of individual discontinuities 'homogenized' or up-scaled (Walsh *et al.* 1998) into a single bulk value. We assume that the small-scale deformation ('small' is relative to the scale of the entire zone) that characterizes fault damage zones can also be homogenized in this fashion, and that we can draw inferences about fault zone flow from the use of porous medium models that do not explicitly include any open fractures.

For our simulations, we use a numerical formulation which includes coupled fluid flow and heat transport (both conductive and advective), as described in detail by Garven (1989, 1995). There is only a single fluid phase (water: its properties are dependent on salinity, pressure, and temperature), and we ignore time-dependent changes. The adoption of this steady-state approach is quite deliberate: the flow systems, and especially the associated thermal disturbances which we simulate, could be effective for an indefinite period of time, and could therefore be important relative to (possibly) slow basinal processes.

The 2D simulator (OILGEN: Garven 1989) is implemented via finite element techniques (linear, triangular elements), with fluid flow and heat transport calculations alternating until the temperature change between iterations falls below a given criterion (here, a 1% temperature difference at any node since the last iteration). Properties for the rock framework (porosity, permeability, thermal conductivity) are fixed for each finite element, and so the construction of a model requires both a grid to represent the geometry, and the material properties to be distributed across the elements defined by that grid. Although the formulation allows variable salinities to be specified, we assume that the water is everywhere 'fresh', since this means that we do need to consider solute concentration differences that will arise as flow occurs (comparisons not included here show that there is very little difference between otherwise similar models with fresh water and those with a fixed salinity distribution). The sides and base of the model are closed to fluid flow, while its upper surface is the water table (open to water flux, at a constant head). The temperature along the top of the model is specified at a constant value (10°C), as is the heat flux on the base (60 mW m^{-2}); the sides of the model are thermally insulated, so there is no heat flux across them.

The work from which this paper is drawn (Fleming 1996) is focused on the modern temperature distribution in the North Sea Central Graben – a mature stage rift (Fig. 1). Although hydrogeological considerations are essential for understanding the present-day thermal state of this region (Fleming *et al.* 1997), the project has necessarily also been concerned with down-hole temperature measurements and the calibration of in situ thermal conductivities and heat flow. The details of the temperature data and the conductivity analysis (Fleming 1996) are to be the subject of a separate paper; here we include only an extract of that work to illustrate the presence of localized temperature anomalies –

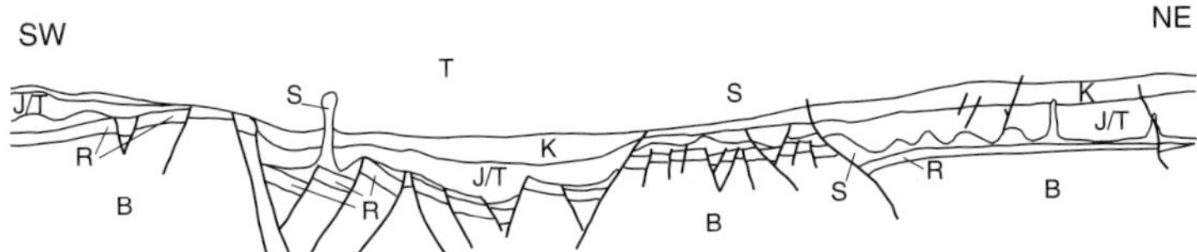

Fig. 1. Generalized cross section across the Central Graben showing structural style of mature rift basin. Based on interpreted seismic section provided by British Petroleum, public domain well data, and interpretations by Ziegler (1982). B = basement, R = Permian (Rotliegend), S = salt, J/T = Jurassic/Triassic, C = Cretaceous, T = Tertiary. Length of section *c.* 850 km. Height of section *c.* 10 km.

high amplitude (>40°C), short-wavelength (<10 km), and depth limited, which are spatially associated with major steep faults (Fig. 2; see also Andrews-Speed *et al.* 1984). All reasonable explanations for these temperature anomalies have been considered, including thermal conductivity variations, heat flow patterns, and advection of heat by flowing fluids, but only localized fluid flow in 'open', steep fault zones has proven adequate to explain the observed data. In this paper we emphasize some important generalities of fault zone hydrogeology through reference to simplified models which allow the basic behaviours to be ascertained.

In our simulations, fluid flow is driven by differences in hydraulic head (potential energy) and by buoyancy. We do not include here a mechanism to account for flows driven by overpressure (Darby *et al.* 1996; Wilkinson *et al.* 1997). We also ignore other situations with

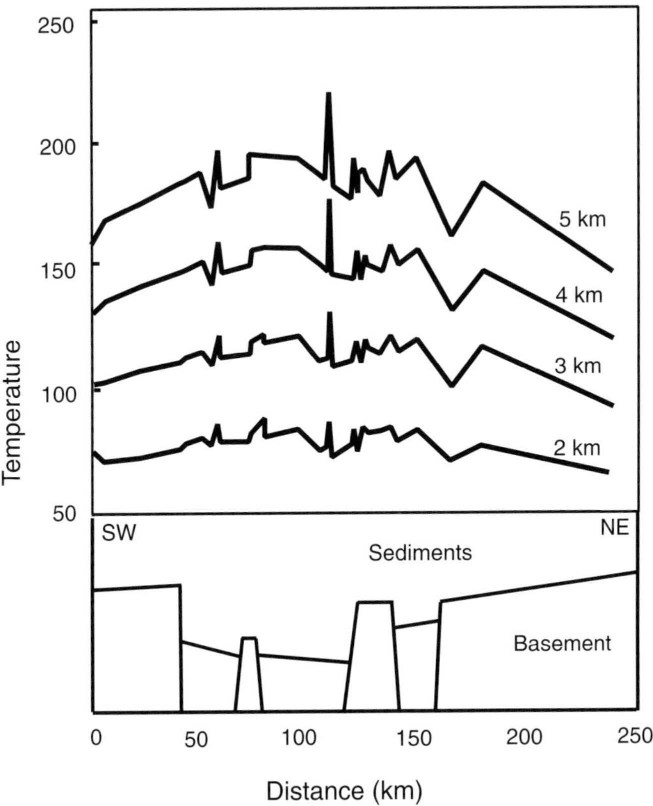

Fig. 2. Temperature profiles across the Central Graben. Drawn from dataset of corrected bottom hole temperatures interpolated to depths of 2, 3, 4 and 5 km. Scale at base gives distance (km) from left end of profile (located on rift terrace). Note particularly the broad anomalies (200+ km half-wavelength), with superimposed short wavelength anomalies. Schematic cross-section below (after Fleming 1996).

porosity changes. For example, it is widely accepted that compaction-driven fluid flows (related to porosity collapse: Bethke 1985; Audet & McConnell 1992; Waples & Couples 1998; and most commercial basin models) in the absence of significant overpressure disequilibrium, are too slow to have thermal effects. However, active deformation, and its associated poroelastic behaviour (Kümpel 1991), can induce rapid and voluminous fluid movement (Ge & Garven 1994). Related hydrogeological concepts involving porosity change (e.g. volumetric strain) include transient flows driven by earthquake-induced deformation (Rojstaczer & Wolf 1992; Muir-Wood & King 1993; Sibson 1994).

Volumetric strain (porosity change) may well, therefore, be causally associated with fluid flows in active fault zones. However, we are seeking to understand the general characteristics of flow in fault zones, including non-active ones, and especially the thermal effects of that flow. The use of only 'subtle' driving energies means that conclusions about the thermal effects of fluid flow in our hydrogeological models are more broadly applicable than might be the case for models related to a specific flow mechanism (e.g. overpressure or earthquakes) that may not occur everywhere, or that may occur only at limited times in a basin's history (active faulting).

Results

In order to illustrate the hydrogeological impact of inactive fault damage zones, we use two model geometries. The first is a simplified configuration which allows us to demonstrate how the primary parameters affect the behaviour of the system; this is referred to as the 'simple grid'. The second type of model is one which encapsulates the geometry of a mature rift basin (such as the North Sea), although here we only show a portion of such a model which, in this case, contains a single, steeply-dipping fault zone. In all of our models, the vertical permeability (k_v) of the country rocks is one hundred times smaller than the horizontal permeability (k_h), except for the fault zones, where the opposite holds (fault $k_v = 100\,k_h$).

Prior to conducting the simulations, we expected that the presence of a fault zone would permit simple upward flow of pore fluids responding to a minor hydraulic gradient caused by the topographic expression of the basin margin and the arrangement of basin fill materials. If the average linear flow rate up the fault were in excess of about 1 ma^{-1}, then a thermal disturbance should be observed (at smaller flow rates, conductive transfer dissipates the advected heat and no 'anomaly' is apparent). Our prediction was that there would be a simple relationship between the petrophysical properties of the fault zone and the resulting temperature anomaly. However, we were surprised by the complexity of the flow systems which developed, with both flushing and circulation aspects.

We use several illustrations to depict the behaviours of the simulations. 'Stream Function' indicates the flux of fluid, and is especially useful since its contours represent flow lines permitting a visualization of the flow paths. 'Temperature' contours illustrate the thermal consequences of rapid fluid flow that advects heat. We also use plots of temperature change (relative to a case without a hydrogeologically different fault zone) v. position, where the 'position' refers to the lateral locations of model nodes located at the same depth (usually at the top of the fault zone).

Simple grid

Our basic geometric prototype is the steer's head shape of a mature rift basin, although we only consider half of the basin (Fig. 3; note in this and several subsequent illustrations the extreme scale exaggeration). The landward edge of the model depicts a slight topographic rise which, because it reflects an elevated water table, imparts a small lateral hydraulic gradient to the system (although much of this energy is dissipated before it affects the basin proper). The sedimentary rocks of the idealized basin are in a simple, flat-lying, layer-cake arrangement. In the rift portion of the model, but away from both the rift margin and the other edge of the model, there is a depth limited (3.5–5 km) fault zone that is 300 m wide.

As our purpose here is to discover the fundamental hydrogeological effects which can be attributed to a fault damage zone, and not to explain the phenomena affecting a particular setting, it is appropriate to first consider a very simple configuration with three materials only: basement rocks ($k_h = 10^{-4}$ mD), sedimentary rocks, and the fault damage zone. For a fault zone with $k_v = 50$ mD ($k_h = 0.5$ mD), an approximately 5°C temperature disturbance occurs (Fig. 3B) at sediment permeabilities of $k_h = 5$ mD ($k_v = 0.05$ mD), but a 40°C disturbance occurs (Fig. 3C) with sediment permeabilities of $k_h = 10$ mD ($k_v = 0.1$ mD) for the same fault permeability. This bifurcation behaviour is similar for a range of fault zone permeabilities (Fig. 4). The smaller temperature anomalies in

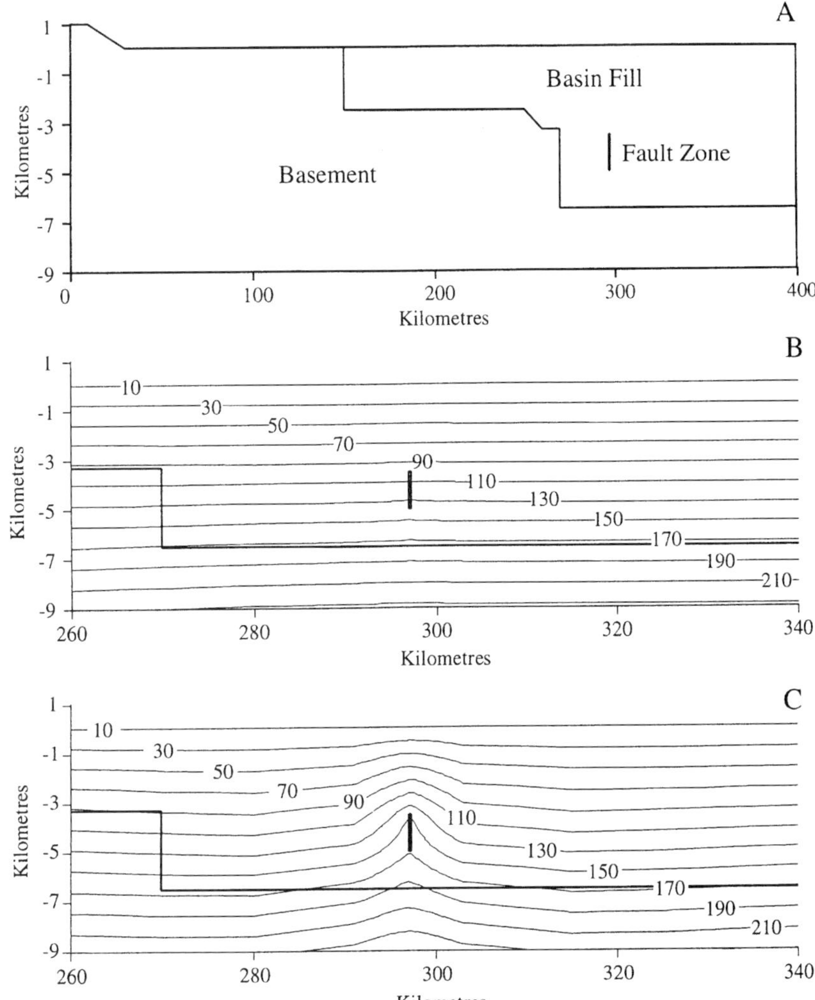

Fig. 3. (**A**) Simplified steer's head geometry of mature-stage rift basin showing depth-limited, vertical fault zone within syn-rift sedimentary rocks. (**B**), (**C**) Temperature contours (in °C) produced by coupled conduction/advection model in vicinity of fault zone. Fault $k_v = 50$ mD. Homogeneous sedimentary pile (syn-rift and post-rift rocks are identical), with sedimentary fill $k_h = 5$ mD in (B) and 10 mD in (C). Basement $k_h = 10^{-4}$ mD in each model. Note approximately 5°C temperature disturbance in (B) and 40°C disturbance in (C).

this set of models are caused by convective circulation located predominantly within the fault zone (Fig. 5a), whereas the larger anomalies are caused by whole-system convection that uses the fault zone as an updraught site (Fig. 5b).

As expected, there is an inverse trade-off between changes in fault permeability and fault zone width (i.e. the concept of transmissibility applies), such that a halving of the fault zone width is compensated by a doubling in the permeability. This relationship extends for more than two orders of magnitude on either side of the values noted above, so that a 3 m wide fault zone with k_v of 5000 mD gives the same results as those described above.

Of course, a real basin is more heterogeneous than is illustrated in the preceding models. A degree of complexity is gained by dividing the sedimentary pile into four units nominally patterned after the Tertiary ($k_h = 1$ mD), Palaeocene (100 mD), Cretaceous (0.1 mD), and pre-Cretaceous (10 mD) of the North Sea (Figs 6 & 7). This configuration behaves very much like the homogeneous basin case described above, with almost indiscernable thermal effects associated with minor within-fault circulation, and

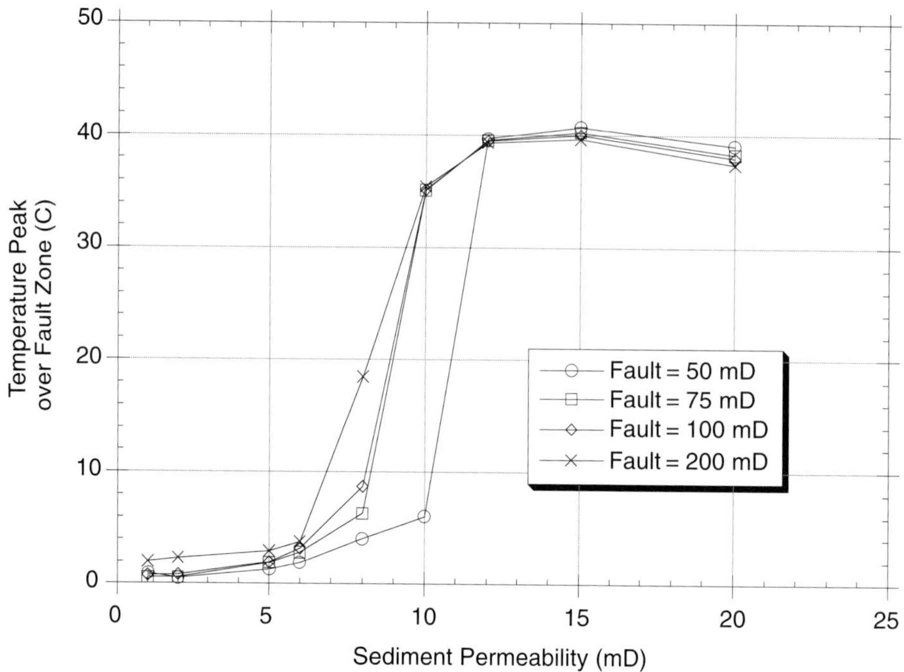

Fig. 4. Magnitude of temperature anomaly (temperature difference of advective state against conductive only state) at depth equal to top of fault zone, as a function of basin fill permeability (k_h), for a range of several fault zone permeabilities (k_v). In each case, fault zone $k_h = 0.01\,k_v$.

even more minor circulation in the country rocks (results not shown). Reductions in these sedimentary unit permeabilities have essentially no effect on the form of the flow system, but increases in them do have a large effect associated with what we might call 'synthetic' minor circulations within the sedimentary units concerned (Fig. 6). Minor increases in sedimentary unit permeabilities can once again create behavioural bifurcations, producing major convection within that particular unit (Fig. 7). The resulting temperature disturbances do not bear a straightforward relationship to fault zone properties, or the flow rate within the fault zone.

Central Graben example

The results described above suggest several possible phenomenological explanations for many of the short wavelength temperature anomalies observed in the Central Graben (Fig. 2) which cannot be explained via conductivity variations. Do these concepts remain valid in a more realistic model, and can they explain the full range of observations?

We here depict part of a regional scale model (Fleming 1996) of the North Sea Basin where it crosses the Central Graben (Fig. 8). The whole model is 850 km long, extending from northeast England to southern Norway. Along the margin of one of the rifted fault blocks, we assume the presence of a fault damage zone some 300 m wide extending from the upper part of the Jurassic to the lower Tertiary. We vary the fault zone permeability from 50 to 500 mD (k_v), but the resulting temperature disturbance above the fault is only about 15°C at the maximum fault zone permeability (Fig. 9). In contrast to this model result, the subsurface temperature data (Fig. 2) indicate extreme anomalies (up to 50°C) in similar situations. Something else is needed in order to explain these larger observed anomalies.

The only alteration of the model which we find 'works' (in the sense of producing the larger temperature anomalies) is the addition of an 'outlet zone' at the top of the fault. These outlet zones are some 500 m thick and +/− 3 km wide, and their petrophysical properties need to be similar to those of the fault zone. Interestingly, outlet zones centred across the fault have little effect, but those which are asymmetric (offset to one side; Fig. 8) can produce temperature anomalies of up to 50°C. The function of the outlet zones seems to be to allow the buoyant, upwelling

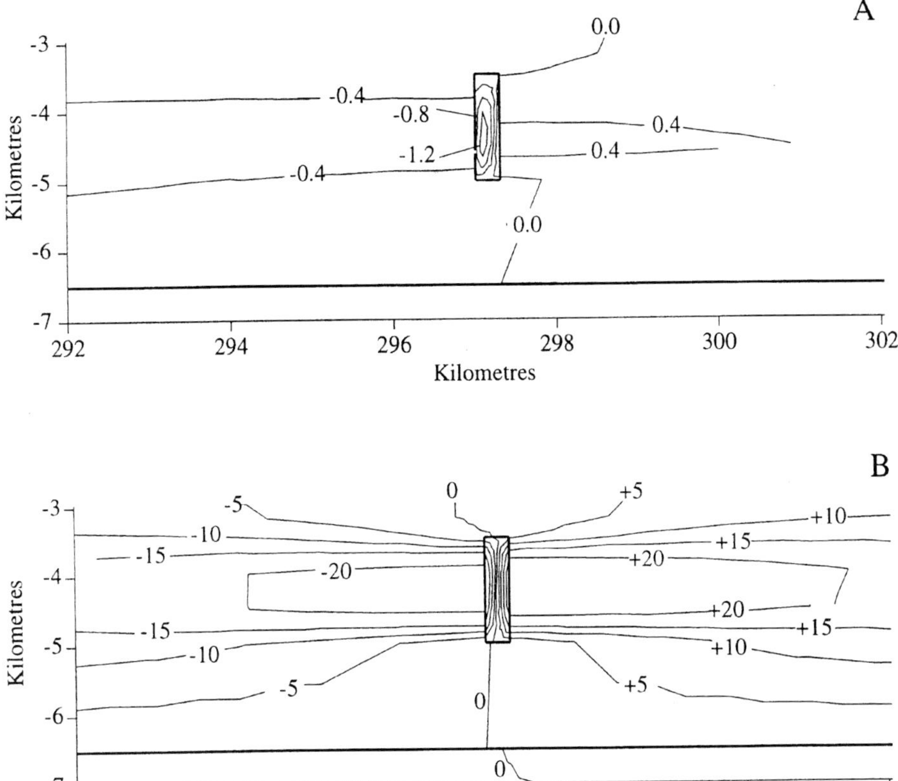

Fig. 5. (**A**) Stream Function (contour interval is $0.4\,m^3\,a^{-1}$) for model depicted in Fig. 3(B). Note anti-clockwise circulation concentrated within the fault zone, and limited sympathetic circulations in adjacent regions. (**B**) Stream Function (contour interval is $5.0\,m^3\,a^{-1}$) for model depicted in Fig. 3(C). Note that flow in fault zone is not circulatory, but is instead an updraught common to major fluid circulations lying to the left and right of the fault zone.

fluid to disperse into the surrounding materials across a wide interface, and therefore to permit a greater flux of fluid through the fault zone itself than would otherwise occur. The required asymmetry of the outlet zones is reminiscent of the typical dog-bone shape of damage zones as described by Knipe *et al.* (1997). The configuration is also compatible with the notion that fault tips should have asymmetric dilatant and compactant zones. Here, of course, we are possibly only seeing the dilatant part in terms of the relevant petrophysical impact.

Discussion

In the simple grid models, the pronounced change in temperature associated with an increase in sedimentary rock permeability (Fig. 4) represents a mode shift between within-fault zone circulation (Fig. 5a) and whole-system convection which uses the fault as an updraught (Fig. 5b). The layering heterogeneity of a realistic sedimentary pile means that similar whole system circulations are unlikely in a typical basin (cf. Figs 6 & 7), and the large magnitude thermal disturbance that is produced by the unrealistic modelled flow system in Fig. 5b is not an acceptable explanation for the temperature anomalies observed in the North Sea.

The occurrence of convective circulation concentrated within the fault zone is an unexpected result, made even more surprising by the observation that it continues to occur in the simulations even under conservative conditions (fault zone k_v as low as 1 mD; results not shown). At the lower end of the range of studied fault zone permeabilities, the calculated temperature disturbances due to this convective flow within the fault zone become numerically indiscernable. The small temperature anomalies produced under

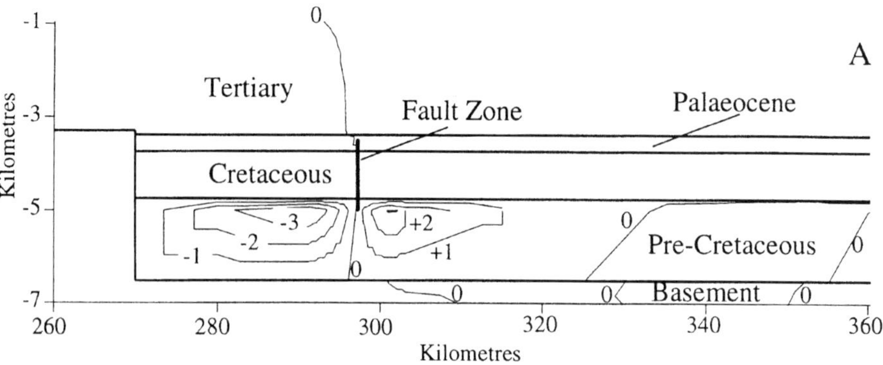

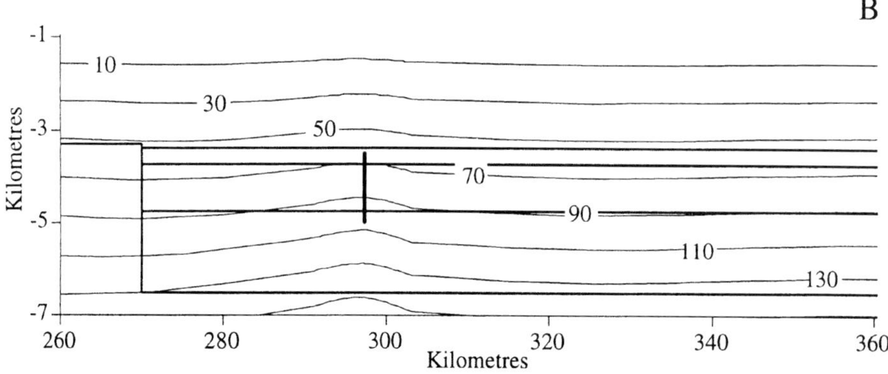

Fig. 6. 'Minor' sympathetic circulations (anticlockwise to lower left of fault zone) produced by increasing permeability of part of pre-Cretaceous layer to 30 mD. (**A**). Stream Function (contour interval 1.0 m^3 a^{-1}). (**B**). Temperature (contour interval 20°C).

these conditions would surely not be discovered in nature, since the uncertainty associated with bottom hole temperatures is, at best, +/− 3°C (Carstens & Finstad 1981) and is often taken to be +/− 8 to 10°C (Hermanrud *et al.* 1990). Although convective flows associated with lower permeabilities may not be detectable by temperature measurements, it is important to recognize the potential for them to occur, since they could operate for long time periods, with consequent effects on diagenesis, fault mineralization, hydrocarbon migration, and thermal history.

Salt piercement structures occur in the Central Graben (Fig. 1), but we have not explicitly included them in the models shown here. Fleming (1996) considers the thermal effects of such salt features and concludes that this type of conductivity anomaly is inadequate to explain the largest of the observed temperature variations. However, the deformation caused by a piercement structure could resemble a fault damage zone as modelled here, and the hydrogeological results we describe might apply equally well to such a case (there could be an additional effect on the fluid flow related to the thermal anomaly produced by the salt).

The boreholes which supply the subsurface temperature data for our study lie close to the line of our profile, with most lying within 5 km, although a few are projected 8–10 km. Additional wells positioned farther away were not used in constructing the profile because they indicate a distinct three-dimensional character to the temperature distribution. This observation makes it necessary to comment briefly on our choice of two dimensional models. There are two reasons why this approach is selected: firstly, the along-strike temperature variation seems to be less pronounced than the variability we observe in the 2D depth profile; and secondly, our purpose is to learn about first-order phenomena associated with fault zone flow systems, and this is best achieved by (initially) avoiding the extra complications brought by the third dimension. Nevertheless, as recently shown by Ondrak (1997) for a similar (but hypothetical) situation,

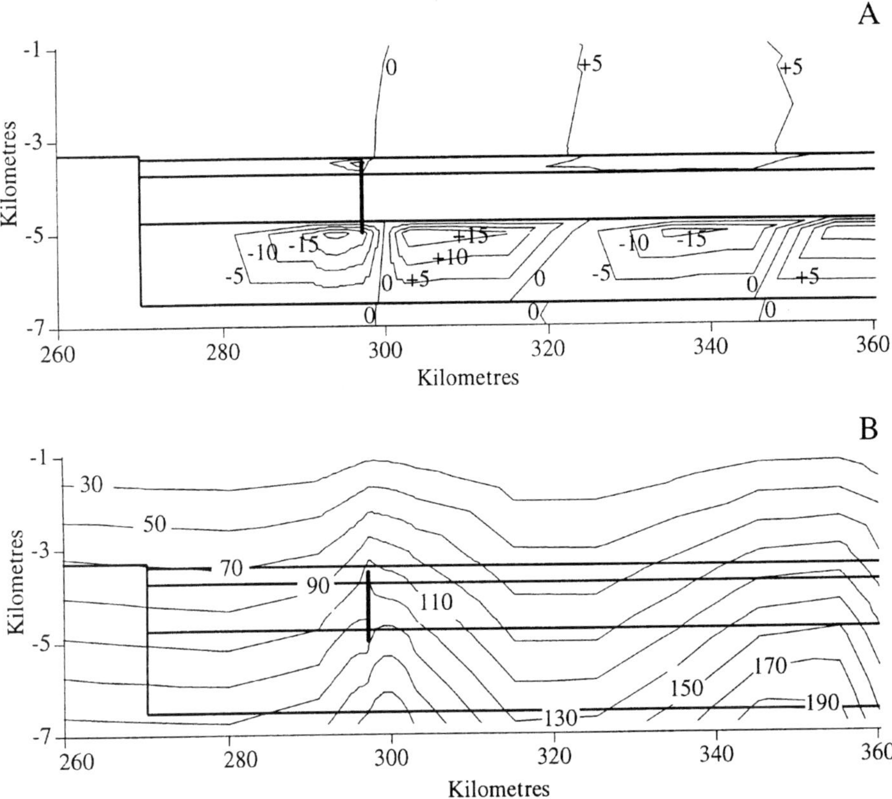

Fig. 7. 'Major' sympathetic circulations (anticlockwise to lower left of fault zone) produced by increasing permeability of part of pre-Cretaceous layer to 40 mD. Locations of rock units as given in Fig. 6. (A). Stream Function (contour interval 5.0 $m^3 a^{-1}$). (B). Temperature (contour interval 20°C). Although temperature anomalies are large, they have a long half wavelength.

a fully three dimensional approach is likely to yield a more thorough understanding of the heat transfer processes operative in the setting we have studied here.

Another aspect of the North Sea Basin which we have not addressed here is the widespread occurrence of overpressure in the deep Graben (Gaarenstroom *et al.* 1993). The various cells of

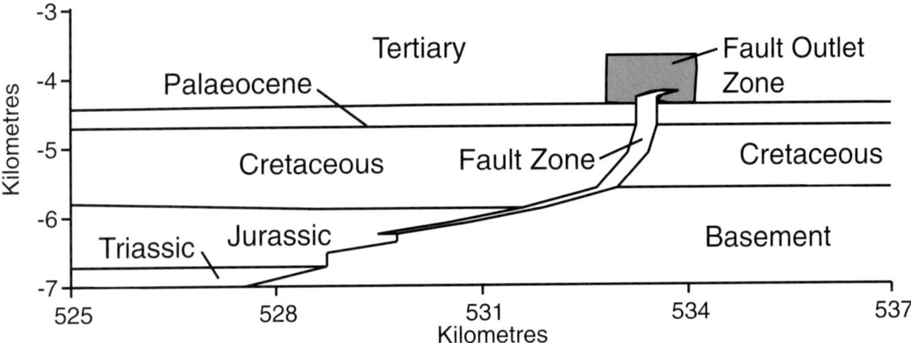

Fig. 8. Part of regional-scale North Sea model showing fault at margin of rift fault block (approximately true scale). Note geometry of 'outlet zone' (offset to left of fault zone top) referred to in text. Models with such an asymmetric outlet zone configuration produce 50°C temperature anomaly (at fault zone vertical permeabilities greater than 50 mD).

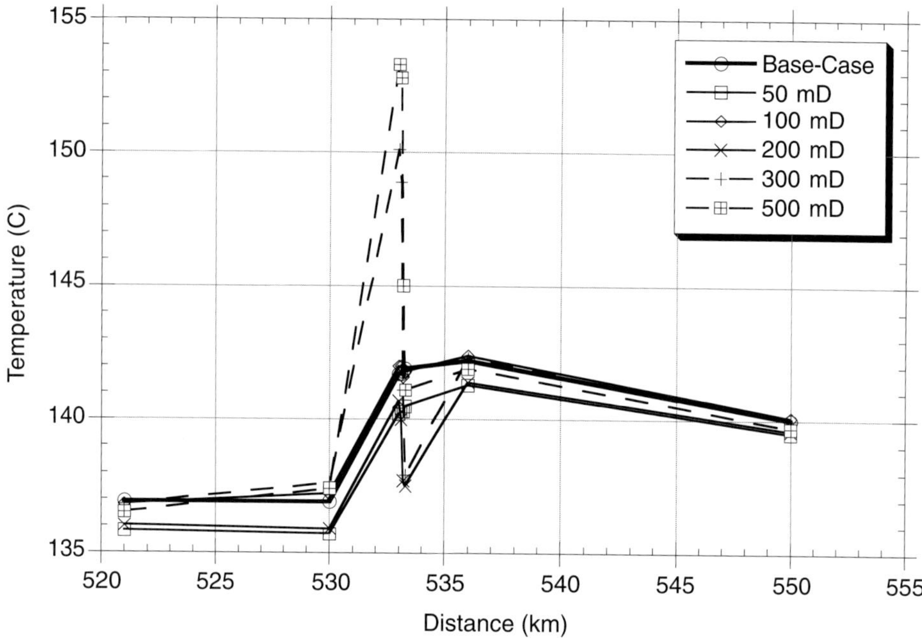

Fig. 9. Temperature profiles in the Lower Tertiary caused by fluid flow in and around a fault zone at the margin of a North Sea faulted block (as illustrated in Fig. 8, but without an outlet zone). Fault zone permeabilities (k_v) given in inset. Note approximately 15°C anomaly related to cases with fault permeability of 300 and 500 mD.

overpressure which are known (Darby *et al.* 1996) pose a problem for a model such as ours which operates at nearly hydrostatic conditions. However, the scale of overpressure cells (a few to perhaps tens of km) is small in comparison with the flow systems studied, and we believe that our models are nevertheless reasonable ones for the purpose of revealing the thermal effects of fault-focused fluid flow.

Another aspect of overpressure may be more significant: is it possible that venting of overpressured fluids ('leaks') is responsible for upwards flow along some fault zones (Roberts & Nunn 1995; Darby *et al.* 1996, 1998; Wilkinson *et al.* 1997), and the resulting temperature anomalies? If the pressure cell boundaries are dynamic, leaking barriers (Neuzil 1995), then the pressure of the released fluid will decrease markedly, and the heat transport processes we simulate here (buoyancy driven flow in normally pressured conditions) may be directly relevant to those parts of the hydrogeological system where the escaping fluids have experienced a pressure drop.

Conclusions

If steeply dipping fault damage zones are characterized by permeability enhancement parallel to the fault, such zones can have a significant impact on the hydrogeological system of the basin. Even under low driving energies, buoyancy driven fluid motion occurs in such zones, with a dominance of convective flow within the fault zone itself. Larger scale convective flow systems within the surrounding sedimentary rocks, which become significant as the bulk permeability of the system increases, can use the fault zone to localize an updraught, but such situations are unlikely to be common in a typical basin due to the permeability heterogeneity. Simple, up-the-fault flow does not readily occur in our simulations, but modifications of the fault damage zone, to include a 'dog-bone' end, enhance this behaviour. The thermal effects of fault-associated flow in normally pressured basins can range from undetectable anomalies (<5°C) to quite major disturbances (50°C). Fluid flow systems which involve fault damage zones need to be considered in a range of investigations concerning basin processes.

We thank Professor G. Garven (Johns Hopkins) for allowing us the use of his simulation code OILGEN, and for numerous helpful discussions. We also thank Robertson Research for allowing access to their North Sea data, and the Natural Environment Research

Council for financial support to CGF. J. Iliffe, M. Russell, H. Lewis (to whom further thanks is due for assistance with graphics), A. McCaig, and two anonymous reviewers provided constructive comments on this work.

References

ANDERSON, R. N., FLEMINGS, P. B., LOSH, S., WHELAN, J., BILLEAUD, L. B., AUSTIN, J. & WOODHAMS, R. 1995. The Pathfinder drilling program into a major growth fault in Eugene Island 330: Implications for behavior of hydrocarbon migration pathways. *In*: ANDERSON, R., BILLEAUD, L. B., FLEMINGS, P. B., LOSH, S. & WHELAN, J. (eds) *Results of the Pathfinder Drilling Program into a Major Growth Fault*, Lamont Doherty Earth Observatory Press, Palisades, NY, 4–22.

ANDREWS-SPEED, C. P., OXBURGH, E. R. & COOPER, B. A. 1984. Temperatures and depth-dependent heat flow in western North Sea. *American Association of Petroleum Geologists Bulletin*, **68** 1764–1781.

AUDET, J. M. & MCCONNELL, J. D. C. 1992. Forward modelling of porosity and pore pressure evolution in sedimentary basins. *Basin Research*, **4**, 147–162.

BETHKE, C. M. 1985. A numerical model of compaction driven groundwater flow and heat transfer and its application to the paleohydrology of intracratonic sedimentary basins. *Journal of Geophysical Research*, **90**, 6817–6828.

BJØRLYKKE, K. 1994. Fluid flow processes and diagenesis in sedimentary basins. *In*: PARNELL, J. (ed.) *Geofluids: Origin, Migration and Evolution of Fluids in Sedimentary Basins*. Geological Society, London, Special Publication, **78**, 127–140.

CACAS, M. C., LEDOUX, E., DE MARSILY, G., TILLIE, B., BARBREAU, A., DURAND, E., FEUGA, B. & PEAUDECERF, P. 1990a. Modelling fracture flow with a stochastic discrete fracture network: calibration and validation: 1. The flow model. *Water Resources Research*, **26**, 479–489.

——, ——, ——, ——, ——, ——, —— & —— 1990b. Modelling fracture flow with a stochastic discrete fracture network: calibration and validation: 2. The transport model. *Water Resources Research*, **26**, 491–500.

CARSTENS, H. & FINSTAD, K. G. 1981. Geothermal gradients of the northern North Sea basin: 59–62°N. *In*: ILLING, L. V. & HOBSON, G. D. (eds) *Petroleum Geology of the Continental Shelf of North-West Europe*. London, Institution of Petroleum, 152–161.

DARBY, D., HASZELDINE, R. S., & COUPLES, G. D., 1996. Pressure cells and pressure seals in the UK Central Graben. *Marine and Petroleum Geology*, **13**, 865–878.

——, —— & —— 1998, Central North Sea overpressure: insights into fluid flow from one- and two-dimensional basin modelling. *In*: DUPPENBECKER, S. P. & ILIFFE, J. (eds) *Basin Modelling*. Geological Society, London, Special Publications, **141**.

de MARSILY, G. 1986. *Quantitative Hydrogeology*. Academic Press, San Diego.

D'ONFRO, P., RIZER, W. D., QUEEN, J. H., MAJER, E. L., PETERSON, J. E., DALEY, T. M., VASCO, D. W., DATTA-GUPTA, A. & LONG, J. C. S. 1998. An integrated approach for characterizing fractured reservoirs. *This volume*.

ELLEVSET, S. O., KNIPE, R. J., OLSEN, T. S., FISHER, Q. & JONES, G. 1998. Fault controlled communication in the Sliepner West Field, Norwegian continental shelf; detailed, quantitative input for reservoir simulation and well planning. *This volume*.

FLEMING, C. G. 1996. Modern day temperature field, Central Graben, North Sea: Investigation of conduction and fluid advection. PhD Thesis, University of Glasgow.

——, COUPLES, G. D. & HASZELDINE, R. S. 1997. Modern temperature state of the central North Sea: necessity for advection of heat by moving porefluids. *Terra Nova*, **9**, 181.

FOLEY, L., DALTABAN, T. S. & WANG, J. T. 1998. Numerical simulation of fluid flow in complex faulted regions. *In*: COWARD, M. P., DALTABAN, T. S. & JOHNSON, H. (eds) *Structural Geology in Reservoir Characterization*. Geological Society, London, Special Publications, **127**, 121–132.

FOWLES, J. & BURLEY, S. 1994. Textural and permeability characteristics of faulted, high-porosity sandstones. *Marine and Petroleum Geology*, **11**, 608–623.

FOXFORD, K. A., WALSH, J. J., WATTERSON, J., GARDEN, I. R., GUSCOTT, S. C. & BURLEY, S. D. 1998. Structure and content of the Moab fault zone, Utah, USA. *This volume*.

GAARENSTROOM, L., TROMP, R. A., DE JONG, M. C. & BRANDENBERG, A. M. 1993. Overpressure in the Central North Sea: Implications for trap integrity and drilling safety. *In*: PARKER, J. (ed.) *Petroleum Geology of Northwest Europe, Proceedings of the Fourth Conference*, Geological Society, London, 1305–1313.

GARVEN, G. 1989. A hydrogeologic model for the formation of the giant oil sands deposits of the Western Canada sedimentary basin. *Americal Journal of Science*, **289**, 105–166.

—— 1995. Continental-scale groudwater flow and geologic processes. *Annual Reviews of Earth and Planetary Sciences*, **24**, 89–117.

GE, S. & GARVEN, G. 1994. A theoretical model for thrust-induced deep groundwater expulsion with application to the Canadian Rocky Mountains. *Journal of Geophysical Research*, **99**, 13851–13868.

GRAULS, D. J. & BALIEX, J. M. 1994. Role of overpressures and in-situ stresses in fault-controlled hydrocarbon migration: a case study. *Marine and Petroleum Geology*, **11**, 734–742.

HASZELDINE, R. S., BRINT, J., FALLICK, A. E., HAMILTON, P. & BROWN, S. 1992. Open and restricted hydrologies in Brent Group diagenesis. *In*: MORTON, A., HASZELDINE, R. S., GILES, M. & BROWN, S. (eds) *Geology of the Brent Group*. Geological Society, London, Special Publications, **61**, 401–419.

—— & MCKEOWN, C. 1995. A model approach to radioactive waste disposal at Sellafield UK. *Terra Nova*, **7**, 87–95.

HERMANRUD, C., CAO, S. & LERCHE, I. 1990. Estimates of virgin rock temperature derived from BHT measurements: bias and error. *Geophysics*, **55**, 924–931.

JESSOP, A. M. & MAJOROWICZ, J. A. 1994. Fluid flow and heat transfer in sedimentary basins. *In*: PARNELL, J. (ed.) *Geofluids: Origin, Migration and Evolution of Fluids in Sedimentary Basins*, Geological Society, London, Special Publications, **78**, 43–54.

KNAI, T. A. & KNIPE, R. J. 1998. Fault impact on fluid flow in the Heidrun field. *This volume*.

KNIPE, R. J. 1993. The influence of fault zone processes and diagenesis on fluid flow. *In*: ROBINSON, A. D. (ed.) *Diagenesis and Basin Development*, American Association of Petroleum Geologists Studies in Geology, **36**, 135–154.

——, FISHER, Q. J., JONES, G., CLENNELL, M. B., FARMER, A. B., HARRISON, A., KIDD, B., MCALLISTER, E., PORTER, J. R. & WHITE, E. A. 1997. Fault seal analysis: successful methodologies, applications, and future directions. *In*: MÖLLER-PEDERSEN, P. & KOESTLER, A. G. (eds) *Hydrocarbon Seals: Importance for Exploration and Production*. NPF Special Publications, 7, Elsevier, Singapore, 15–40.

KÜMPEL, H.-J. 1991. Poroelasticity: parameters reviewed. *Geophysical Journal International*, **105**, 783–799.

LECARLIER, C., ROYER, J.-J. & FLORES, E. L. 1994. Convective heat transfer at the Soultz-sous-Forets geothermal site: implications for oil potential. *First Break*, **12**, 553–562.

LEWIS, H. 1996. Characterization of the fluid flow systems for Irish lead–zinc deposits. PhD thesis. University of Glasgow.

MACAULAY, C. I., BOYCE, A. J., FALLICK, A. E. & HASZELDINE, R. S. 1997. Quartz veins record vertical flow at a graben edge: Fulmar oil field, Central North Sea. *American Association of Petroleum Geologists Bulletin*, **81**, 2024–2042.

MANZOCCHI, T., RINGROSE, P. S. & UNDERHILL, J. R. 1998. Flow through fault systems in high-porosity sandstones. *In*: COWARD, M. P., DALTABAN, T. S. & JOHNSON, H. (eds) *Structural Geology in Reservoir Characterization*. Geological Society, London, Special Publications, **127**, 65–82.

MATTHÄI, S. K., AYDIN, A., POLLARD, D. D. & ROBERTS, S. G. 1998. Numerical simulation of departures from radial drawdown in a faulted sandstone reservoir with joints and deformation bands. *This volume*.

MOURZENKO, V. V., THOVERT, J.-F. & ADLER, P. M. 1996. Geometry of simulated fractures. *Physical Reviews, E* **53**, 5606–5626.

MUIR-WOOD, R. & KING, G. C. P. 1993. Hydrologic signatures of earthquake strain. *Journal of Geophysical Research*, **98**, 22,035–22,068.

MULLIS, A. M. & HASZELDINE, R. S. 1995. Numerical modelling of diagenetic hydrogeology at a graben edge: Brent oilfields, North Sea. *Journal of Petroleum Geology*, **18**, 421–438.

NEUZIL, C. E. 1995. Abnormal pressures as hydrodynamic phenomena. *American Journal of Science*, **295**, 742–786.

ONDRAK, R. 1997. Numerical modelling of heat and reactive mass transport in hydrothermal systems. *Terra Nova*, **9**, 544.

PEDERSEN, T. & BJÖRLYKKE, K. 1994. Fluid flow in sedimentary basins: model of porewater flow in a vertical fracture. *Basin Research*, **6**, 1–16.

PERSON, M. & GARVEN, G. 1994. A sensitivity study of the driving forces on fluid flow during continental–rift basin evolution. *Geological Society of America Bulletin*, **106**, 461–475.

——, RAFFENSPERGER, J. P., GE, S. & GARVEN, G. 1996. Basin-scale hydrogeologic modelling. American Geophysical Union, *Reviews of Geophysics*, **34**, 61–87.

ROBERTS, S. J. & NUNN, J. A. 1995. Episodic fluid flow from geopressured sediments. *Marine and Petroleum Geology*, **12**, 195–204.

ROJSTACZER, S. & WOLF, S. 1992. Permeability changes associated with large earthquakes: an example from Loma Prieta, California. *Geology*, **20**, 211–214.

RUSSELL, M. J., COUPLES, G. D., & LEWIS, H. 1995. SEDEX genesis and super-deep boreholes: can hydrostatic pore pressures exist down to the brittle–ductile boundary? *In*: PASAVA, J., KRIBEK, B. & ZAK, K (eds) *Mineral Deposits: From Their Origin to Their Environmental Impacts*. A. A. Balkema Publishers, Rotterdam, 315–318.

SIBSON, R. H. 1994. Crustal stress, faulting and fluid flow. *In*: PARNELL, J. (ed.) *Geofluids: Origin, Migration and Evolution of Fluids in Sedimentary Basins*, Geological Society, London, Special Publications, **78**, 69–84.

SMITH, L. & CHAPMAN, D. S. 1983. On the thermal effects of groudwater flow. 1. Regional scale systems. *Journal of Geophysical Research*, **88**, 593–608.

STEEN, O., SVERDRUP, E. & HANSEN, T. H. 1998. Predicting the distribution of small faults in hydrocarbon reservoirs by combining outcrop, seismic and well data. *This volume*.

THOMAS, A. L. & LAPOINTE, P. R. 1995. Conductive fracture identification using neural networks. *Proceedings, 35th US Symposium on Rock Mechanics*, Lake Tahoe, 627–632.

TOWNSEND, C., FIRTH, I. R., WESTERMAN, R., KIRKEVOLLEN, L., HÄRDE, M. & ANDERSON, T. 1998. Small seismic-scale fault identification and mapping. *This volume*.

WALSH, J. J., WATTERSON, J., HEATH, A., GILLESPIE, P. A. & CHILDS, C. 1998. Assessment of the effects of subseismic faults on bulk permeabilities of reservoir sequences. *In*: COWARD, M. P., DALTABAN, T. S. & JOHNSON, H. (eds) *Structural Geology in Reservoir Characterization*. Geological Society, London, Special Publications, **127**, 99–114.

WAPLES, D. W. & COUPLES, G. D. 1998, Some thoughts on porosity reduction—rock mechanics, overpressure, and fluid flow. *In*: DUPPENBECKER, S. P., & ILIFFE, J. (eds) *Basin Modelling*. Geological Society, London, Special Publications, **141**.

WHELAN, J. K., KENNICUTT, M. C., BROOKS, J. M., SCHUMACHER, D. & EGLINTON, L. B. 1993. Organic geochemical indicators of dynamic fluid flow processes in sedimentary basins. *Organic Geochemistry*, **22**, 587–615.

WIECK, J. M., PERSON, M. & STRAYER, L. 1995. A finite element method for simulating fault block motion and hydrothermal fluid flow within rifting basins. *Water Resources Research*, **31**, 3241–3258.

WILKINSON, M. W., DARBY, D., HASZELDINE, R. S., & COUPLES, G. D., 1997. Secondary porosity generation during deep burial associated with overpressure leak-off: Fulmar Formation, UKCS. *American Association of Petroleum Geologists Bulletin*, **81**, 803–813.

WOOD, J. R. & HEWITT, T. A. 1984. Reservoir diagenesis and convective fluid flow. *In*: MCDONALD, D. A. & SURDAM, R. C. (eds) *Clastic Diagenesis*. American Association of Petroleum Geologists Memoirs, **37**, 3–13.

ZIEGLER, P. A. 1982. *Geological Atlas of Western and Central Europe*. Shell International Petroleum and Elsevier, Amsterdam.

The influence of fault compaction on fault zone evolution

J. R. HENDERSON

Department of Geological Sciences, University of Durham, Durham DH1 3LE, UK

Abstract: A numerical simulation of coupled fluid flow and failure in a fault zone is presented, that shows the impact of the dynamics of fault growth on the relationship between seismic and sub-seismic fault scaling. A tabular fault zone is modelled which undergoes non-uniform compaction, leading to development of pressure gradients in the fault zone. Increased pore fluid pressures result in a reduction of the fault-normal stresses, and hence frictional failure. The model suggests a possible physical mechanism for the development of non-power-law scaling of fault displacements, implying that observations of such behaviour are real, rather than the result of poor sampling. The fault displacement scaling law appears to result from the dynamics of the fault evolution, which depend upon the relative rates of compaction of fault material and fluid flow within the fault zone. When fault compaction is slow, event sizes show a power-law scaling relationship. When compaction is fast, pressure gradients are unable to dissipate and the distribution of these event sizes is non-power-law. It is suggested that the dynamics of fault evolution may be inferred from geological evidence, with the relics of repeated small events (e.g. aligned calcite fibres) implying a relatively high compaction rate, whereas cataclasites would suggest relatively low compaction rates.

In order to estimate the number and size of faults and fractures falling below a measurement limit, it is necessary to employ some sort of fault-scaling relationship. Such procedures have practical applications in, for example, the study of hydrocarbon reservoirs, in which the larger scale faults may be adequately mapped using seismic data, whereas smaller faults, which play an important role in determining the reservoir characteristics (seal integrity, permeability etc.), may not. Similarly, fault populations will be important in determining suitability of sites for storage of hazardous materials.

The presence of small faults may act to increase or decrease reservoir permeability. The faults act either as conduits (e.g. Caine *et al.* 1993) or seals (e.g. Antonellini & Aydin 1994), depending on the characteristics of the fault zone material. In a dynamic fault system it is to be anticipated that these properties will evolve over the lifetime of the fault system. In order to make a successful prediction of the distribution and fluid transport properties of small scale faults, it is therefore necessary to understand the processes occurring during faulting, their impact on the material properties of the fault zone material, and the consequent fault scaling relationships.

The belief that fault systems have a number of fractal properties, including power-law distributions of displacements, fractal patterns of fault traces etc., leads to an appealing method of predicting the size of populations of small faults. According to this hypothesis, one would simply measure the fractal dimension of a fault size distribution on the basis of the available data, and extrapolate it to small fault sizes. Unfortunately, real datasets frequently do not lend themselves to this type of procedure, showing a non-power-law distribution of sizes over the observed scale range (Nicol *et al.* 1996) and offering no clue as to how to extrapolate the data into the small-scale range.

The non-fractal nature of real fault systems, for example the distribution of displacements, suggests that the processes controlling their formation impose a characteristic length scale upon the system. One possible candidate for such a process may be fluid flow in the fault zone (e.g. Byerlee 1993). A number of workers have addressed the question of how pore fluids may influence the mechanics of faulting: it has been suggested that fluid-related processes may account for the anomalous weakness of major faults such as the San Andreas Fault (Lachenbruch & Sass 1980, Zoback & Healy 1992) and the Cascadia subduction fault (Wang *et al.* 1995). This observation suggests that it may be valuable to investigate further the consequences of the presence of overpressured fluids within the fault zone. It appears that fault size distributions may be influenced by pore fluid pressures in a number of ways. For example, Hill (1977) suggested that a fracture 'mesh' may be formed as a result of the influence of fluid pressure on sequences of rock units with different material properties.

A simple model of coupled fluid flow in a fault zone and frictional failure arising from compaction of the fault material is presented here.

HENDERSON, J. R. 1998. The influence of fault compaction on fault zone evolution. *In*: JONES, G., FISHER, Q. J. & KNIPE, R. J. (eds) *Faulting, Fault Sealing and Fluid Flow in Hydrocarbon Reservoirs*. Geological Society, London, Special Publications, **147**, 231–242.

Compaction of fault material results in a loss of porosity due to a number of processes possibly including pressure solution and ductile flow, and consequently fluid pressure increases. The model suggests a possible physical mechanism for the development of non-power-law scaling of fault sizes, as measured by slip areas, thereby supporting the conclusions of Nicol *et al.* (1996) that such behaviour is real, rather than the result of poor sampling discussed by Needham *et al.* (1996) and Fossen & Rørnes (1996). The fault scaling law appears to result from the dynamics of the fault evolution, which depends upon the relative rates of compaction of fault material and fluid flow within the fault zone.

The analysis presented suggests that the dynamics of fault evolution may be inferred from geological evidence, with the relics of repeated small events (e.g. aligned calcite fibres) implying a relatively high compaction rate whereas cataclasites would suggest relatively low compaction rates.

The model

Relationship to earlier work

Earlier work has focussed on the influence of pore-fluid pressure in fault zones on sequences of seismicity. Sleep & Blanpied (1992) and Byerlee (1993) present models in which pore-pressure variations result in episodic seismicity. The aim of the present work is to look instead at the way the diffusion of fluids which occurs during compaction may have influenced the characteristics of faults observed in the geological record.

Numerical simulations of fluid flow in geological contexts have employed Lattice Boltzmann methods in finding solutions to the Navier–Stokes equation. Rothman (1990) modelled two-phase flow in porous media, and Wilson *et al.* (1996) modelled viscous flow inside a fracture in a coupled model of fluid flow and fracture. Fluid transport in porous media is often considered to be diffusive and governed by Darcy's law. In this work, the lattice BGK technique (a variant of the lattice Boltzmann method) is used to simulate fluid diffusion in a heterogeneous medium (Maillot & Main 1996). This technique is well suited to the study of diffusive processes which characterize fluid flow in rocks, and has the advantage of being straightforward to implement on simple lattices.

Rock failure is modelled using a simple cellular automaton approach similar to those described by Lomnitz-Adler (1993), formulated in terms of frictional failure. In the absence of pore fluids, the model resembles simulations of seismicity such as the cellular automaton model of Bak *et al.* (1987), but differs from the elastic network model of fault growth presented by Cowie *et al.* (1993) in that only local stress interactions are considered.

The model differs from other earlier works (e.g. Segall & Rice 1995) in a number of respects, principal amongst which is that the evolution of a large lattice of sliding elements is modelled and used to calculate the effect of fluid flow in the fault plane, rather than the effect of fluid flowing between the fault and the country rock. In addition, a simple model of frictional failure is employed, described by a single friction coefficient for each fault element.

The present model

A simple 2D model of a fault is considered, which consists of a network of $N \times N$ elements forming a square grid. Each element represents a 'patch' of the fault with uniform physical properties. The fault material is porous and it is isolated from the surrounding country rock by an impermeable seal. A shear traction is deemed to be applied to the fault resulting from, for example, distant plate motions (Fig. 1). An element i fails when the traction, τ^i, exceeds a critical value τ_c^i

$$\tau_c^i = \mu^i(\tau_n - P_j^i) \quad (1)$$

where τ_n is the normal stress, P_f^i is the fluid pressure in the element and μ^i is the friction coefficient appropriate to the element. The value of τ_n is held constant throughout the simulation, and P_f^i is calculated for each element during the deformation, and depends upon the porosity and fluid diffusivity of the element. The model of frictional failure is characterized by a single frictional coefficient, μ, for each element. Laboratory measurements of frictional coefficients range from around 0.5 for core samples of fault gouge (Scott *et al.* 1994) up to 0.9 (Byerlee 1978). These values may represent lower bounds, since surfaces might be expected to be 'locked' by mating of roughness.

The values of the friction coefficients μ_i are initially selected at random from the range 0.5 to 1.0. When an element fails, the applied traction is instantaneously relaxed by an amount $\Delta\tau^i = \beta\tau_c^i$, while the traction on the four neighbouring elements is increased by $\Delta\tau/4$.

After failure, the strength of the element is changed to a new random value. This implies that the failure process completely changes the

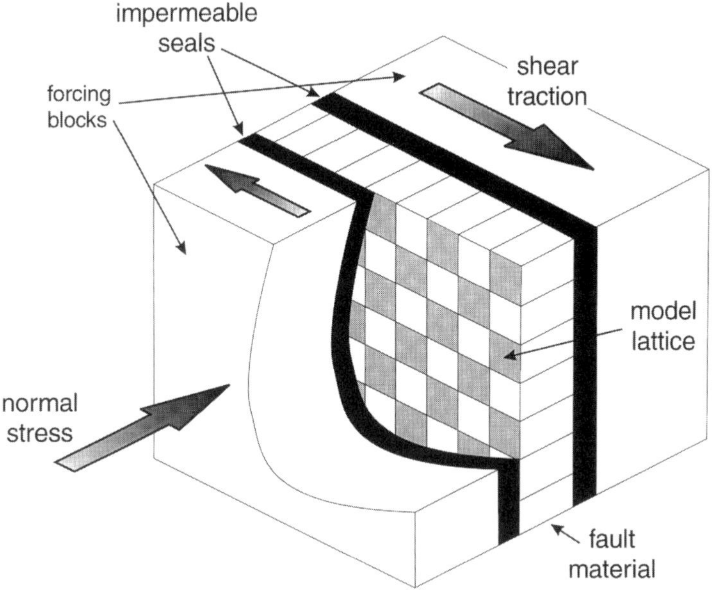

Fig. 1. Diagram showing the physical situation modelled. The fault material is separated from the country rock by an impermeable seal.

mechanical properties of the fault material, disrupting, for example, any existing fabric that may influence its frictional strength. The parameter β represents the efficiency with which a slip event relaxes the stress on an element. The model assumes that the process conserves energy, implying that the energy dissipated in seismic radiation is small (Spottiswoode & McGarr 1975). The value of β depends on the details of the slip process, which are not well constrained. The case where $\beta < 1$ represents a partial stress relaxation, whereas $\beta > 1$ implies a stress-drop 'overshoot'. For all the simulations described here, in the absence of good evidence placing strict bounds on the size of β, a value of $\beta = 1.0$, corresponding to a system in which a slip event transfers all of the stress to the four neighbouring elements, has been chosen. Exploration of the case where $\beta < 1$ suggests that these conclusions remain valid for smaller values of β.

An important point to be noted is that it is the increase in fluid pressure due to compaction, rather than an increase in externally applied traction, that is responsible for driving the system towards failure. Fluid pressure is initially set to approximately 75% of the fault-normal stress, representing the result of sealing a volume of fluid within a fault zone which is subsequently compacted. There are two principal ways in which fluid pressure varies during the simulations. Firstly, compaction results in a reduction of porosity and hence an increase in fluid pressure, whereas faulting causes an increase in the porosity of the fault material. Secondly, the gradients of fluid pressure resulting from changes in porosity cause a diffusion of pore fluid pressure.

Compaction

Porosity reduction takes place reversibly, due to poroelastic compression of the rock by external forces, and irreversibly due to the dissolution of the material (e.g during pressure solution). Porosity reduction resulting from pressure solution has been estimated as taking place at a rate of around $10^{-16}\,s^{-1}$ (Walder & Nur 1984). Sleep & Blanpied (1992) model the processes of compaction in terms of ductile flow in the fault zone and suggest that greater rates of porosity reduction, of the order $10^{-11}\,s^{-1}$, may be feasible.

Rock permeability k depends on the geometry and interconnectivity of the porosity ϕ, which in turn depends, in general, on the effective pressure. Hence no simple relation between the permeability, porosity and effective pressure exists. For the purposes of this work, the model considers a case in which porosity in a rock is made up of pores of a wide variety of aspect ratios, and that the flatter pores are rapidly closed, while the more equant pores remain open. This implies that porosity loss will occur at a rate which depends on the effective pressure, and will also vary during compaction, being more rapid during the initial stages. This type of behaviour (Type I compaction) appears to be typical of crystalline rocks, and is observed in the laboratory (David *et al.* 1994). An exponential dependence of permeability on confining pressure is widely used (e.g. Rice 1992). The actual sensitivity of this relationship varies considerably, however, and a number of different micromechanisms appear to be involved.

These processes are modelled by a simple law for the evolution of porosity:

$$\phi = \phi_0 \exp(\gamma\phi_0(\tau_n - P_f)) \quad (2)$$

(γ is a constant) which is physically reasonable, and at the same time sufficiently simple that the role of porosity reduction in the system will not be unnecessarily obscured.

The processes by which deformation influences porosity and permeability are complex. Field and laboratory observations suggest that porosity may be enhanced or reduced during deformation, and that the mechanisms involved may include poroelastic behaviour, thermal cycling, volume change during metamorphism and brittle pore collapse, as well as a host of chemical processes including cementation, and dissolution.

The laboratory experiments of Marone *et al.* (1990) show that simulated fault gouges increase in porosity up to a limit of around 10%, which may be analogous to the 'critical porosity' known in soil mechanics. In addition, Aydin & Johnson (1983) showed that cataclasis increases fault porosity only up to a value of about 10%. More recent laboratory experiments by Wong *et al.* (1997) and Zhu & Wong (1997) show that rocks may show a sequence of behaviours in which initial dilation in the brittle regime is followed by cataclastic flow and porosity reduction. It has also been found that the relationship between permeability and porosity changes during deformation. In many cases permeability and porosity are positively correlated, but Menéndez *et al.* (1996) found that a negative correlation exists during some phases of deformation.

Rather than attempt to simulate the entire range of complexity shown by real rock deformation processes, the model examines the consequences of a simple porosity evolution of fault material in order that the results be readily interpretable. It is assumed in this model that the energy available for porosity production during the faulting process is sufficient to generate a local porosity increase of 10%. It is assumed that the faulting event also has the effect of rupturing the seal retaining fluid in the fault zone. Fluid pressure therefore falls to a low value representing the fluid pressure in the country rock.

Fluid diffusion

During the simulation, porosity is reduced by compaction, increasing the pore fluid pressure, and created during element rupture, reducing the pore fluid pressure. These processes produce fluid pressure gradients, resulting in flow of fluid in the fault zone. It is supposed that fluid flow takes place only within the fault zone, except when failure of an element ruptures the seal and fluid pressure in the element falls to the level of the country rock.

The fluid pressure evolves throughout the simulation according to a diffusion equation. Diffusivity, D, is related to the permeability, k and the porosity, ϕ by:

$$D = \frac{kc^2}{\phi\nu} \quad (3)$$

where c is the speed of sound in water and υ is the dynamic viscosity of water (Phillips 1991).

The highly heterogeneous nature of rocks means that there is no simple relationship between porosity and permeability (Sahimi 1993). Brace (1977) found that using the formula:

$$k = \frac{m}{k_0}\phi^3 \quad (4)$$

along with values of hydraulic radius, m obtained from SEM studies of rock samples, and a value of 2.5 for the constant k_0, the calculated values of permeability compared well with measurements of a number of rock types. In the work presented here, the model also assumes that permeability is proportional to ϕ_3, but, as discussed above, this relationship does not take into account the uncertainty in the geometry of pore space (in particular the possibility that there may be large aligned fractures giving high permeabilities). A wide range of permeabilities are found in rocks. Brace (1994) reports permeabilities of crystalline rocks ranging from $10^{-20}\,m^2$ to $10^{-15}\,m^2$. Smith *et al.* (1990) measured permeabilities in fault core materials ranging from $10^{-22}\,m^2$ to $10^{-12}\,m^2$ and Morrow *et al.* (1984) found permeabilities of fault gouges from $10^{-22}\,m^2$ to $10^{-18}\,m^2$. This demands that a wide range of constants of proportionality be examined.

To simulate the diffusion of fluids the model employs the 'lattice BGK' method described by Maillot & Main (1996). This technique provides a fast and straightforward method for modelling fluid diffusion through materials with a heterogeneous distribution of fluid diffusivities. The technique permits the consideration of anisotropic diffusivities, although in the present simulation diffusivity is treated as a scalar quantity. The method is stable as long as the diffusivity is positive and anisotropy of the diffusion tensor is not excessive.

In the model of Maillot & Main (1996), $P(x, t)$, the fluid pressure at a lattice site $\mathbf{x}$ and at time t is the sum of the 'directional fluid pressures', i.e. the components of pressure moving in each of b lattice directions, α, represented by $P_\alpha(\mathbf{x}, t)$)

$$P_\alpha(\mathbf{x}, t) = \sum_{\alpha=1}^{b} P_\alpha(\mathbf{x}, t) \quad (5)$$

If the lattice unit vectors are represented by $\mathbf{e}_\alpha$ then, in the case of a two dimensional square lattice with isotropic diffusivities, the evolution of the model is represented by:

$$P_\alpha(\mathbf{x} + \mathbf{e}_\alpha, t + 1) = P_\alpha(\mathbf{x}, t) + \lambda(\mathbf{x}, t) \times (P_\alpha(\mathbf{x}, t) - P_\alpha^{\mathrm{eq}}(\mathbf{x}, t)) \quad (6)$$

where the equilibrium distribution P_α^{eq} is given by:

$$P_\alpha^{\mathrm{eq}}(\mathbf{x}\, t) = t_\alpha P(\mathbf{x}, t) \quad (7)$$

and the normalization factors t_α are simply 1/4. The relaxation parameters, $\lambda(\mathbf{x}, t)$) are related to the diffusion tensor, $D_{ij}(\mathbf{x}, t)$. General expressions for this relationship are given by Maillot & Main (1996). In the present case, the relation is:

$$\lambda = -\left(2D + \tfrac{1}{2}\right)^{-1} \quad (8)$$

Numerical simulations

The simulation proceeds as follows:

1. Model initialization. For each element of a 128×128 lattice, uniformly random values of friction coefficient (between 0.5 and 1.0) and porosity (between 0 and 0.1) are assigned. From these porosities, element diffusivities are calculated by combining equations (3) and (4):

$$D = \mathrm{d}\phi^2 \quad (9)$$

The applied shear stresses are set to a uniform value, as are the directional pore-fluid pressures P_α. A value for the confining pressure τ_n is set. In all the simulations described here, the confining pressure is set to 260 MPa, the initial directional pore fluid pressures are set to 50 MPa (i.e. a total pore-fluid pressure of 200 MPa) and the initial shear stresses are set to 20 MPa.

2. Faulting. The model is checked for element failure following equation (1). If an element has failed, its shear stresses are redistributed as described above, porosity is reset to 10%, and directional pore fluid pressures are reduced to hydrostatic levels. Friction coefficients are reassigned new random values. This step is repeated until no further failure occurs.

3. Compaction. Porosity (and diffusivity and pore fluid pressure) for the entire lattice are updated according to the compaction law (equation 2).

4. Fluid flow. The fluid pressures are updated using iteration of the fluid diffusion equation (6) over a large number of timesteps, typically 1000 timesteps each representing fluid pressure diffusion over about 2 days.

5. Iteration. Return to step (2).

The steps 2 to 5 represent a single time-step, which, in the simulations presented here, corresponds to a period of five years. In reality the compaction and fluid pressure diffusion would occur continuously but those processes are modelled here as occurring in sequential increments. The time-step for the model must therefore be small enough such that this approximation does not cause inaccuracies. Each simulation lasts for 500 time-steps; i.e. 2500 a. Step 2 is considered to occur instantaneously, whereas the compaction and fluid flow occur over the entire time-step. In this work, although the model considers the influence of fluid pressure on the rate of compaction, compaction and diffusion is modelled as occurring sequentially, the fluid diffusivities are not updated during the fluid diffusion process. The diffusivity is itself a function of the fluid pressure, leading to a non-linear diffusion equation. Consideration of that more complex model is deferred to a later time.

Although modelling a diffusive process, where the model is expressed in terms of fluid diffusivities, it may be useful to examine the results using permeabilities, rather than diffusivities. The relation between the two has already been described (equation 3). The behaviour of the model depends upon the relative importance of compaction and permeability, and hence it is possible to parameterize the model in terms of a non-dimensional quantity, Ψ:

$$\Psi = \frac{k^0}{\dot{\phi}^0} \times L^{-2} \quad (10)$$

where k^0 and $\dot{\phi}^0$ represent the permeability and compaction rate of an element under a defined set of standard conditions, and L is the size of a lattice element. The chosen standard conditions are a porosity of 0.05, and effective confining pressure of 40 MPa.

As an example, consider a situation in which, in the standard state, the compaction rate were $8 \times 10^{-12}\,s^{-1}$ (implying a porosity reduction, over 100 a, of 1.7%), and the permeability $6.0 \times 10^{-18}\,m^2$ (6.0×10^{-3} md), and letting each lattice element represent a square of side 100 m, then Ψ would be approximately 10^{-10}. Simulations were performed for a range of values of Ψ, ranging from a value of 0 (corresponding to a situation in which no flow of fluids occurs) up to $\Psi = 10^{-8}$.

Results

At each time-step, a number of iterations are required before the system settles in to a stable configuration. At each iteration one or more elements fail. These failed elements may fail again, at the same time-step, during subsequent iterations, or they may remain intact for the rest of the time-step. The failure history of the lattice may, therefore, be complex.

In this study, characterization of the failure history is achieved by examining the distribution of elements which have failed during a single time-step, regardless of the iteration at which this failure occurs. As each iteration is instantaneous, a group of contiguous failed elements may be considered to represent a single failure event. Furthermore, it is possible to identify the number of contiguous failed elements, the failure cluster size, with the area of the fault plane participating in the failure event. By analogy with seismic failure, an event magnitude, m, which is related to fault area, can be defined as:

$$m \propto \log A \qquad (11)$$

(Kanamori & Anderson 1975). Extending the analogy with the seismic case, the frequency of occurrence can be plotted against the cluster magnitude to define a b-value as the negative slope of the resulting graph.

Figure 2 provides an illustration of these principles. It shows a snapshot of the broken elements after several time-steps for the case where $\Psi = 0.0$ i.e there is no flow of fluid. There are few clusters with many elements, and these clusters are evenly distributed. Multi-element clusters may form when the stress from the broken element is transferred to neighbours and their critical stress level is exceeded, but since there is no spatial correlation of critical stress level, this does not occur in an organized fashion. Once an element has broken it re-heals, resetting its 'strength'. In this situation, the element remains relatively strong for an extended period of the simulation since the porosity production and seal rupture accompanying failure produces a low pore fluid pressure which cannot be increased by flow from other elements. Broken elements, therefore, tend not to break again for some time.

Figure 3 shows the frequency magnitude distribution for this simulation. An estimate for b may be defined for this distribution over a small range of event sizes from 10 to 40, but over the whole range of sizes displayed there is significant curvature of the line, indicating that no power law relationship between magnitude and frequency can be confidently defined.

A contrasting behaviour is displayed using a value of $\Psi = 10^{-8}$. In this case fluid flow is rapid relative to compaction, and once an element has broken, its fluid pressure is quickly restored to close to average levels. Perhaps more importantly, however, an element in which the fluid pressure has been increased greatly by compaction may be stabilized against failure by the rapid diffusion of the fluid pressure. As a result compaction may occur without failure resulting for a long period of time, as an area of the lattice undergoes 'dilatant hardening'. After a time, a large area of the lattice may be on the brink of stability, and failure of a small part of this may result in large failure events.

Figure 4 shows a snapshot of the distributions of broken elements for a case where fluid pressure diffusion is faster. In this case, organized 'islands' of broken elements have formed. Figure 5 shows fluid pressure in a detail of part of the lattice. Since no fluid transport takes place except within the fault zone, dilatant hardening occurs most effectively when fluid diffusivities are high. Segall & Rice (1975) consider the case in which fluid flow between the fault zone and country rock permits recovery of the fluid pressures within the fault zone, and hence high fluid diffusivity in their model prevents dilatant hardening. When a large cluster of elements fail simultaneously, large amounts of stress are transferred to neighbouring regions of the lattice which may themselves have been 'conditioned' to a state bordering on failure by the diffusion of fluid from areas of high fluid pressure. Consequent failure of these areas results in an avalanche of further failure, necessitating a large number of iterations of algorithm step 2 (see earlier). This behaviour is an important feature of the models, 'avalanches' of failure are particularly important, punctuating periods of quiescence. In contrast, when fluid diffusivity is low (low values of Ψ), a stable

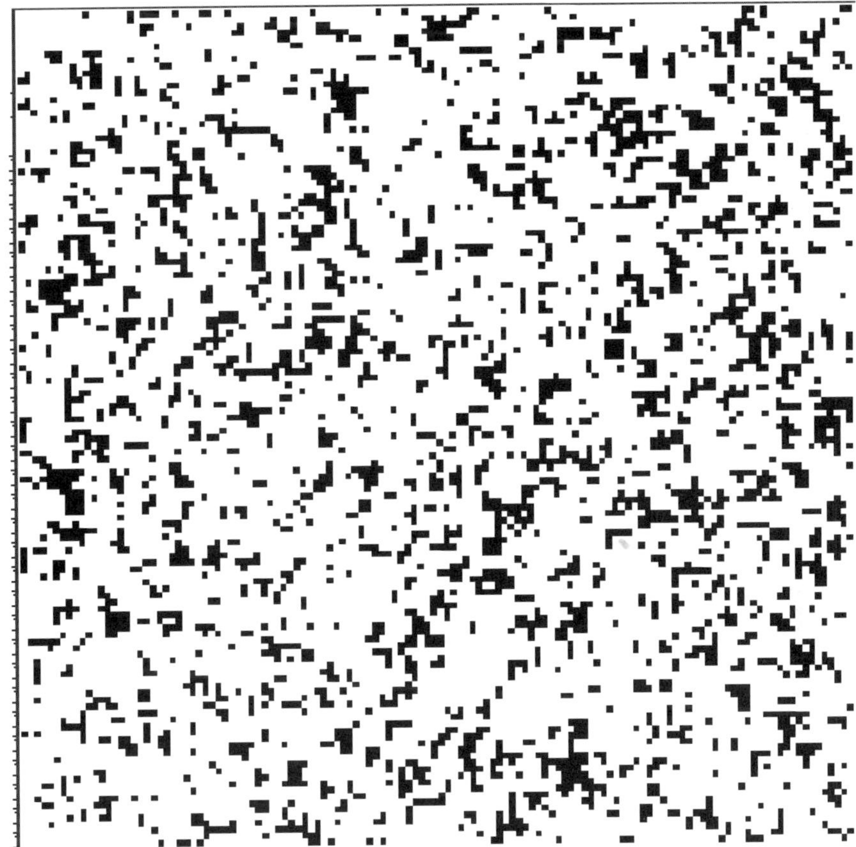

Fig. 2. Snapshot of the lattice during a simulation using a value of $\Psi = 0$, i.e. no fluid flow takes place. Broken elements are black, intact areas are white.

configuration is reached quickly, and activity occurs at a constant low level.

Figure 3 shows the frequency–magnitude distribution for all the simulations. The frequency–magnitude distribution for the simulation with $\Psi = 10^{-8}$ shows firstly a pronounced linear nature, and secondly a lower value of b than that which might be obtained for the case of $\Psi = 0.0$. Although a linear trend is observed over earthquake sizes from 1 to over 200, the trend is interrupted at high sizes. This is a feature of most of the simulations, and is a result of the finite size of the model. Note that this occurs not just at cluster sizes comparable to the whole lattice size, but for all clusters which have significant intersection with the boundaries of the model.

The frequency magnitude distributions (Fig. 3) for the range of values of Ψ investigated show a spectrum of properties. The cases of $\Psi = 0.0$ and $\Psi = 10^{-8}$ have already been discussed. Intermediate values of Ψ show intermediate behaviours, with an increasingly plausible straight line fit to the data being possible as the fluid diffusivity increases. The gradients of the line fit are approximately the same in the cases of $\Psi = 10^{-10}$ and 10^{-8}. For lower values of Ψ the distribution is markedly non power-law.

Geological implications

The results presented above demonstrate that the statistics of failure in a fault zone are strongly determined by the relationship between fault zone compaction and fluid flow. In this section, the implications of these findings will be examined in the context of elucidating the behaviour of natural faults.

The results show that when permeabilities are high relative to compaction rates, fluid pressures are able to equilibrate during the deformation cycle, and the resulting faults show a power-law distribution of sizes similar to those obtained in simple dry sliding-block cellular automaton

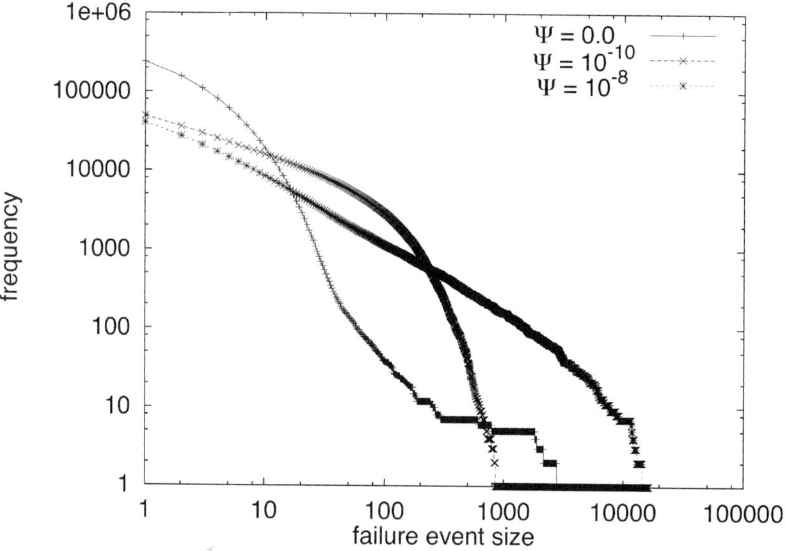

Fig. 3. Frequency–magnitude distribution for clusters of failed elements during simulations with different values of Ψ.

Fig. 4. Snapshots of the lattice during a simulation using a value of $\Psi = 10^{-10}$. Broken elements are black, intact areas are white.

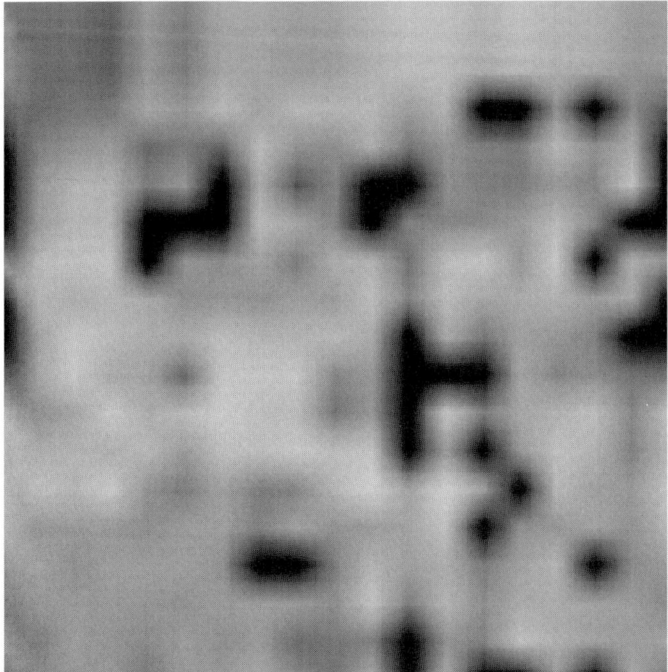

Fig. 5. Detail of part of the lattice showing broken elements in black, and fluid pressure in shades of grey from low (light grey) to high (dark grey). Notice that fluid pressure is low near broken elements, as fluid pressure has diffused into the broken element; and in places where elements have broken earlier, leaving 'ghosts' of low pressure areas.

models (Bak *et al.* 1987). However, when permeabilities are low, fluid pressure disequilibria may persist, and the distribution of sizes shows a distinctly non-power-law character. The latter case resembles the gamma distribution of earthquake sizes proposed by Main & Burton (1984).

Specifically, the results show that when the value of Ψ is greater than 10^{-10}, a power-law scaling will occur, and for values less than this, a gamma-law scaling will result. This result can be put into context by examining the geological parameters controlling the transition from power-law to gamma-distribution scaling. Figure 6 shows, for representative compaction rates, the boundary between power-law and gamma-distribution scaling. Faults with higher values of Ψ will plot in the field above the line, and those with lower values will fall in the region below the line. As has already been noted, for a given length scale, high permeability is likely to result in power-law fault scaling.

Limitations of the model

The model presented here simulates the evolution of a single fault zone. In a basin setting, however, it is usual for faulting to take place on systems of faults, which may interact with each other, and this interaction may significantly influence their development. In addition, a fault is not a simple planar feature, but is found in association with a host of subsidiary structures, including antithetic and synthetic faults, jogs, kinks, sub-parallel strands, forks, fault wall topographies etc. In this model, simulation of these effects is achieved by using an inherently discrete cellular automaton model of fault failure. This approach models fault heterogeneity in a general way, but it is possible that specific structures may influence the results in a more systematic way. Results from the present work should therefore be viewed in the light of these *caveats*. Once the processes described here are properly understood, then it may be appropriate to move to more realistic and complex fault models.

Permeability and scale length

Figure 6 gives an insight into the way in which changes in rock properties with length scale may influence the type of fault scaling behaviour observed. For example, although a rock may

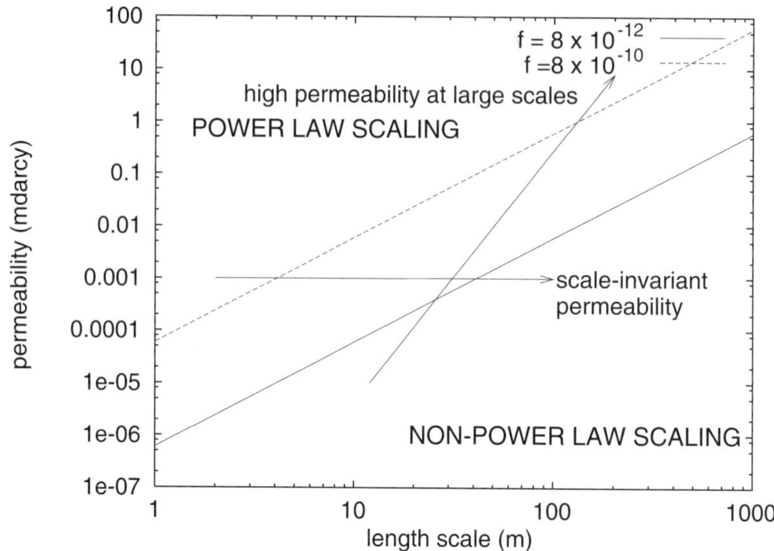

Fig. 6. Diagram showing the type of fault-scaling predicted by the model for different values of permeability and scale length. The position of the boundary between power-law and non-power-law behaviour is fixed by the compaction rate. This diagram is for the case $\Psi = 10^{-10}$, and the compaction rates shown are $8 \times 10^{-12}\,s^{-1}$ (solid line) and $8 \times 10^{-10}\,s^{-1}$ (dashed line).

have high permeability at large scales, resulting from large scale fractures etc., and hence show a power-law scaling at seismic resolution, at smaller scales it is likely to have a lower permeability (Brace 1984), and hence show gamma-distribution scaling at those scales which controls reservoir permeability. Conversely, if permeability remains constant over all length scales, then at large scales the faulting may show gamma-distribution characteristics, whereas at smaller scales it may exhibit power-law behaviour. The simulations presented here provide a rationale for predicting the smaller scale fault behaviour based on observations of large scale faulting, coupled with a judgement as to the general trend of permeability changes in a target area. Our model predicts, therefore, that the type of fault-size distribution may change depending on the scale at which they are observed. This suggests that attempting to predict fault patterns at small scales simply by extrapolating a power-law curve into small length scales may not be successful.

The effect of compaction rate

The boundary between power-law and gamma-law behaviours is not, however, fixed. The key element in determining the position of the boundary between power-law and non-power-law behaviour is the compaction rate (Fig. 6). Thus, one must know the rock permeability and also the dynamics of fault evolution in order to predict the type of scaling.

Conversely, it may be possible to deduce the compaction history of a fault given the failure history inferred from textural evidence. Continuous small-scale fault slip, forming part of a non-power-law fault size distribution, may result in, for example, aligned calcite fibres within the fault zone. This may imply that the compaction rate is rapid. Slow compaction, however, may permit failure in a power-law regime, and be indicated in the geological record by cataclasites.

Many authors have presented scaling data for faults which shows the general 'convex-upward' gamma distribution shape that is produced by some of our simulations. Needham *et al.* (1996) ascribe this shape to sampling bias. Fossen & Rørnes (1996) describe similar data in terms of a number of overlapping power laws. This model suggests a mechanism by which non-power-law fault size distributions may be generated, supporting the conclusion of Nicol *et al.* (1996) that observations of such distributions may be accurate, and not simply the result of poor sampling.

Finally, this work has implications for other fault models which have been used to predict fault-scaling relationships. Cowie & Scholz (1992*a*,*b*) present a model of fault growth in which a fault grows by failing along its entire length. Our model suggests that in many circumstances faults may grow by a process of incremental failure, constrained by the diffusion of fluids. This would imply that the fault scaling laws predicted by Cowie & Scholz (1992*a*,*b*) are not universally applicable.

This work was supported by grants from Durham University Research Fund. I am grateful for the assistance and advice of B. Maillot, and the comments of I. Main and P. Cowie.

References

ANTONELLINI, M. & A. AYDIN, A.. 1994. Effect of faulting on fluid flow in porous sandstones: Petrophysical properties. *American Association of Petroleum Geologists Bulletin*, **78**, 355–377.

AYDIN, A. & A. M. JOHNSON, A. M. 1983. Analysis of faulting in porous sandstones. *Journal of Structural Geology*, **5** 19–31.

BAK, P., TANG, C. & WIESENFELD, K. 1987. Self organized criticality: An explanation of 1/*f* noise. *Physical Review Letters*, **59**, 381–384.

BRACE, W. F. 1977. Permeability from resistivity and pore shape. *Journal of Geophysical. Research*, **82**, 3343–3349.

—— 1984. Permeability of crystalline rocks: New in situ measurements. *Journal Geophysical Research*, **89 (B6)**, 4327–4330.

BYERLEE, J. 1978. Friction of rocks. *Pure & Applied Geophysics*, **116**, 615–626.

—— 1993. Model for episodic flow of high-pressure water in fault zones before earthquakes. *Geology*, **21**, 303–306.

CAINE, J. S., FORSTER, C. B. & EVANS, J. P. 1993. A classification scheme for permeability structures in fault zones. *EOS*, **74**, 677.

COWIE, P. A. SCHOLZ, C. H. 1992*a*. Displacement–length relationship for faults: data synthesis and discussion. *Journal of Structural Geology*, **14**, 1149–1156.

——, & —— 1992*b*. Physical explanation for the displacement–length relationship using a postyield fracture mechanics model. *Journal of Structural Geology*, **14**, 1133–1148.

——, VANNESTE, C. & SORNETTE, D. 1993. Statistical physics model for the spatiotemporal evolution of faults. *Journal of Geophysical Research*, **98**, 21809–21821.

DAVID, C., WONG T., W. ZHU, W. & ZHANG, J. 1994. Laboratory measurements of compaction-induced permeability change in porous rocks: Implications for the generation and maintenance of pore pressure excess in the crust. *Pure & Applied Geophysics*, **143**, 425–456.

FOSSEN, H. & RØRNES, A. 1996. Properties of fault populations in the Gullfaks Field, northern North Sea. *Journal of Structural Geology*, **18**, 179–190.

HILL, D. P. 1977. A model for earthquake swarms. *Journal of Geophysical Research*, **82**, 1347–1352.

KANAMORI, H., ANDERSON, D. L. 1975. Theoretical basis of some empirical relations in seismology. *Bulletin Seismology Society America*, **65**, 1073–1095.

LACHENBRUCH, SASS, J. H. 1980. Heat flow and energetics of the San Andreas fault zone. *Journal of Geophysics Research*, **85**, 6185–6223.

LOMNITZ–ADLER, J. 1993. Automaton models of seismic fracture: Constraints imposed by the magnitude–frequency relation. *Journal of Geophysics Research*, **98**, 17745–17756.

MAILLOT, B., MAIN, I. G. 1996. A lattice BGK model for the diffusion of pore fluid pressure, including anisotropy, heterogeneity, and gravity effects. *Geophysical Research Letters*, **23**, 13–16.

MAIN, I. G., BURTON, P. W. 1984. Information theory and the earthquake frequency–magnitude distribution. *Bulletin Seismology Society America*, **74(4)**, 1409–1426.

MARONE, C., RALEIGH, C. B. & SCHOLZ, C. H. 1990. Frictional behavior and constitutive modeling of simulated fault gouge. *Journal of Geophysical Research*, **95**, 7007–7025.

MENÉNDEZ B., ZHU, W. & WONG, T. 1996. Micromechanics of brittle faulting and cataclastic flow in Berea sandstone. *Journal of Structural Geology*, **18**, 1–16.

MORROW, C. A., SHI, L. Q. & BYERLEE, J. D. 1984. Permeability of fault gouge under confining pressure and shear stress. *Journal of Geophysical Research*, **89**, 3193–3200.

NEEDHAM, T., YIELDING, G. & FOX, R. 1996. Fault population description and prediction using examples from the offshore U.K. *Journal of Structuiral Geology*, **18**, 155–167.

NICOL, A., WALSH, J. J., WATTERSON, J. & GILLESPIE, P. A. 1996. Fault size distributions – are they really power-law? *Journal Structural Geology*, **18**, 191–197.

PHILLIPS, O. M. 1991. Flow and Reactions in Permeable Rocks. Cambridge University Press, New York.

RICE, J. R. 1992. Fault stress states, pore pressure distributions, and the weakness of the San Andreas Fault. *In*: EVANS, B. & WONG, T.-F. (eds) *Fault Mechanics and Transport Properties in Rocks* Academic Press, San Diego, California, 475–503.

ROTHMAN, D. H. 1990. Macroscopic laws for immiscible two-phase flow in porous media: results from numerical experiments. *Journal of Geophysical Research*, **95**, 8663–8674.

SAHIMI, M. 1993. Flow phenomena in rocks: From continuum models to fractals, percolation, cellular automata and simulated annealing. *Reviews in Modern Physics*, **65**, 1393–1534.

SCOTT, D. R., LOCKNER, D. A., BYERLEE, J. D. & SAMMIS, C. G. 1994. Triaxial testing of Lopez fault gouge at 150 MPa mean effective stress. *Pure & Applied Geophysics*, **142**, 749–775.

SEGALL, P., & RICE, J. R. 1995. Dilatancy, compaction and slip instability of a fluid-infiltrated fault. *Journal of Geophysical Research*, **100**, 22,155–22,171.

SLEEP, N. H., & BLANPIED, M. L. 1992. Creep, compaction and the weak rheology of major faults. *Nature*, **359**, 687–692.

Smith, L., Forster, C. B. & Evans, J. P. 1990. Interaction of fault zones, fluid flow and heat transfer at the basin scale. *In*: Hydrogeology of permeability environments, Vol. **2**, 41–67. International Association of Hydrogeologists.

Spottiswoode, S. M., & McGarr, A. 1975. Source parameters of tremors in a deep level gold mine. *Bulletin of Seismology Society America*, **65**, 93–112.

Walder, J., & Nur, A. 1984. Porosity and crustal pore pressure development. *Journal of Geophysical Research*, **89**, 11539–11548.

Wang, K., Mulder, T., Rogers, G. F. & Hyndman, R. D. 1995. Case for very low coupling stress on the Cascadia subduction fault. *Journal of Geophysical Research*, **100**, 12907–12919.

Wilson, S. A., Henderson, J. R. & Main, I. G. 1996. The coupled evolution of fracture populations and fluid flow using a cellular automaton method. *Journal of Structural Geology*, **18**, 343–349.

Wong, T., David, C. & Zhu, T. 1997. The transition from brittle faulting to cataclastic flow: Mechanical deformation. *Journal of Geophysical Research*, **102**, 3009–3025.

Zhu, W., & Wong, T. 1997. The transition from brittle faulting to cataclastic flow: Permeability evolution. *Journal of Geophysical Research*, **102**, 3027–3041.

Zoback, M. D., & Healy, J. H. 1992. In situ stress measurements to 3.5 km depth in the Cajon Pass Scientific Research Borehole: Implications for the mechanics of crustal faulting. *Journal of Geophysical Research*, **97**, 5039–5057.

Relating microscale rock–fluid interaction to macroscale fluid flow structure

P. C. LEARY

Department of Geology & Geophysics, University of Edinburgh, Grant Institute, West Mains Road, Edinburgh EH9 3JW, UK

Abstract: Borehole logs of rock-property spatial fluctuations in the metre to kilometre scale range have power-law Fourier power-spectra that scale inversely with spatial frequency to a power near unity, $S(k) \propto 1/k^{\alpha}$. The spectral scaling is 'universal' in the sense that a narrow range of scaling exponents, $\alpha_v \approx 1.1 \pm 0.1$ for vertical wells and $\alpha_h \approx 1.34 \pm 0.1$ for horizontal wells, describes rock density, elastic modulus, porosity and lithology fluctuation spectra of a sample of 50 well logs; for 35 vertical logs the bounds on α_v hold for both sedimentary and crystalline rock types. The power-law nature of rock-property fluctuation spectra are modelled as long range spatial correlations arising from short range (grain scale) uncorrelated random fluctuations. Such long range random spatial correlations occur in thermodynamic order–disorder phase transitions. Applying to rock the statistical physics of thermodynamic order–disorder transitions, rock heterogeneity observed in borehole logs formally emerges from grain-scale elastic interactions and long range spatial organization of finite strain induced grain-scale defects associated with fluid percolation. If fluid flow paths in rock are significantly influenced by long range correlated random structures, reservoir management cannot be accurately conducted from flow models constrained by small-scale sampling of the reservoir rock; macro scale measurements of site-specific long range random correlation structures are needed.

A powerful rationale for investigating the details of generic and/or specific rock–fluid interactions is to understand the structure of *in situ* rock–fluid systems. How, for instance, do fluid species, pressure, temperature, differential stress, lithology or grain size and sorting affect large scale fluid flow and transport structure in rock? Since advance determination of detailed fluid flow and transport behaviour in large volume hydrocarbon reservoirs, water resource aquifers, waste isolation and excavation sites or whole crust sections is impractical, it is necessary to create models that encapsulate the small-scale physics and chemistry of rock–fluid interactions in order to extrapolate or 'upscale' data from limited sample volumes to reservoir-scale volumes.

Attempts to encapsulate the large scale effects of small-scale rock–fluid chemistry and rock fabric are likely to be legitimate exercises for a rock mass that is accurately characterized by a set of mean properties – mean density, porosity, permeability, cementation chemistry, aqueous/non-aqueous fluid content, and so forth. Mean rock-property values as an accurate and extrapolatable feature of a medium are possible if the medium is moderately to strongly ordered, or if the medium is moderately to strongly disordered. In the former case, uniform or quasi-uniform rock units can produce spatially and temporally steady-state mean flows which effectively average over small-scale, small-amplitude fluctuations in rock properties. For ordered material, the small-scale fabric established in, say, thin-section or in drill-core samples establishes a range of spatial statistical fluctuation that is maintained more or less throughout the rockmass.

In the disordered case, rock can be regarded as 'so heterogeneous it is homogeneous' (Tyler 1988). Small-scale rock samples may not capture the essence of rock fabric or flow character, but spatially diverse small scale samples will, it is hoped, establish a useful estimate of rock-property fluctuations. Here the spatial fluctuations are seen as too complicated to understand in terms of a specific fluid history or fabric pattern, but effective bounds on the spatial variability of rock property can in principle yield a sufficiently accurate estimate of the large-scale flow behaviour to manage the geofluid resource.

The 'order or disorder' approach to rock-property fluctuation has its roots in the assumption that spatial variation in rock properties is in some functional sense uncorrelated. Uncorrelated fluctuations about a mean value is the usual assumption in treating measurement or estimate uncertainty. Within a collection of physical property measurements, deviations from the mean in one sample or location are assumed to be balanced elsewhere by a similar deviation in the opposite direction. If several uncorrelated processes are at work in determining the flow properties of a rock unit, then the uncertainty

LEARY, P. C. 1998. Relating microscale rock–fluid interaction to macroscale fluid flow structure. *In*: JONES, G., FISHER, Q. J. & KNIPE, R. J. (eds) *Faulting, Fault Sealing and Fluid Flow in Hydrocarbon Reservoirs*. Geological Society, London, Special Publications, **147**, 243–260.

caused by all processes is the sum over the uncertainty caused by the individual process. The summation property of uncorrelated statistical events is an important feature unique to Gaussian or normal statistics usually employed to describe uncertainty in the description of rock.

The assumption of uncorrelated processes at work in shaping the flow properties of rock is convenient but it is not plausible. A more plausible assumption is that rock–fluid chemistry, fluid flow paths and finite-strain-induced brittle failure link to produce correlated rather than uncorrelated spatial variation. Rock–fluid interactions create considerable difficulties for investigators because of the formidable range of outcomes of combined physical and chemical processes influencing the rock flow character. Most attempts to determine the spatial distribution of *in situ* rock flow structures are made for their expedience rather than their accuracy. Failure of these investigations to produce accurate forecasts of spatial and temporal flow behaviour is often conceded as the rule rather than the exception. However, in spite of the recognized and often costly deficiencies of many statistical treatments of rock, there has been no physically-based assessment of how badly suited to rock the assumption of uncorrelated statistical events is.

This paper seeks to show that applying the statistics of uncorrelated events to the problem of spatial heterogeneity in fluid flow in rock violates an important, if not the important, physical process responsible for the heterogeneity. The spatial distribution of rock properties is complex, but the complexity does not have its origin solely in interacting physical and chemical processes that combine to create spatially correlated flow structures. Rather, extensive quantitative evidence from borehole logs indicates that a single, simple physical process tied to rock granularity plays a vital role in generating a spatial heterogeneity in rock that is correlated at all scale lengths. The resulting correlated spatial rock heterogeneity generated by this physical process is effectively unpredictable on the basis of small-scale sampling of the rockmass. It follows that using normal or Gaussian statistical tools for managing this component of spatially heterogeneous geofluid resources is fundamentally inadequate.

Two physical concepts, one the effect and one the cause, characterize the spatial heterogeneity of rock revealed by borehole logs. The effect is long range correlated statistical fluctuations that violate the normal or Gaussian statistics assumption at every scale length. The cause is the statistical physics process by which the long range correlations arise from the free energy budget of generic grain-scale elastic interaction and the energy associated with generic irreversible inelastic grain-scale defect formation promoting micro scale fluid percolation. The statistical physics concept of generic micro scale physical random processes that generate macro scale correlated random spatial complexity can explain the 'universal' systematics of spatial fluctuations in rock properties recorded in borehole logs.

The structure of borehole log fluctuations seems to be 'universal' in the sense that the Fourier power-spectra of spatial fluctuations maintain a particular form independent of rock property or rock type. A closely similar form of 'universality' in spatial heterogeneity is well established in a class of thermodynamic states known as critical point order–disorder phase transitions. A close physical analogy exists between the spatial heterogeneity of rock and the spatial heterogeneity seen in thermodynamic critical point order–disorder transition states.

The analogy between rock and thermodynamic critical point order–disorder transition states provides an intermediate case between rock as an ordered medium and rock as a disordered medium. Where conventional statistical thinking is rooted in the assumption of uncorrelated random events, rock heterogeneity is rooted in correlated randomness at all scale lengths. The correlated spatial fluctuations in rock involve arbitrarily large fluctuations in rock-properties on arbitrarily large scale lengths. The assumption that rock-property fluctuations are bounded is invalid. In the intermediate or 'critical' state between order and disorder, rock sample mean values and statistical variances are poor guides to rock structure and flow behaviour. On the evidence of borehole logs, a great deal of rock appears in the intermediate or critical state statistical category. This feature of rock can account for much of the difficulty in adequately managing geofluid resources.

Borehole logs indicate that spatial variation in rock properties measured over three decades of scale length obeys a specific scaling law independent of rock type or physical property. The quantitative expression of this scaling property is that the Fourier power-spectra of borehole logs have a power-law form $S(k) \propto 1/k^{\alpha}$. For a sample of 50 well logs recorded in a variety of crystalline and sedimentary rock types, the scaling exponents α cluster about 1.1 for vertical well data and about 1.34 for horizontal well logs. The power-law scaling property is not consistent with the statistics of uncorrelated random fluctuations. The tight constraints on the exponents cannot be regarded as accidental or incidental, and their origin should be sought in a physical process that is common to all rock types.

In order to understand the observed power-law scaling nature of borehole logs, this paper treats rock–fluid interactions from a generic statistical physics perspective rather than the perspective of specific geochemical or geological processes. To develop the statistical physics perspective on rock, the following three sections consider the statistics of borehole log data from increasingly abstract points of view. The first section introduces borehole logs and shows that *in situ* rock fluctuation amplitudes systematically increase with the dimensions of the system, so that the largest fluctuations affect the largest rock volumes. From a reservoir management point of view, this immediately suggests that small-scale rock samples are unreliable guides to large-scale rock-property distributions.

The second section interprets the $1/k^{\alpha}$ power-law scaling for $\alpha \approx 1$ as evidence that rock has evolved to a statistically near-stationary state. Since many types of rock–fluid interactions appear to lead to the same near-stationary statistics, it follows that the details of internal rock–fluid interactions are not essential in determining the large scale flow and transport structures of interest to reservoir management or to crustal scale tectonics.

The third section develops the analogy between rock fluctuations and fluctuations observed in many-body systems undergoing a thermodynamic order-disorder transition. The analogy supplies a mechanism by which familiar and tractable uncorrelated short range micro scale fluctuations inflate to the less familiar large scale correlated fluctuations recorded in borehole logs. Linking uncorrelated micro scale structure to correlated macro scale structure in this manner involves loss of information of small-scale processes. The exception is that micro scale stress anisotropy appears to affect the scaling exponent (the Fourier spectral power-law exponent is 1.1 ± 0.1 for vertical wells and 1.34 ± 0.1 for horizontal wells). The apparent anisotropy effect on power-law scaling exponents indicates how and to what extent micro scale physics and chemistry can be expected to influence macroscopic rock structure.

At each of these levels of abstraction, the observed structure of rock-property fluctuations implies that detailed knowledge of micro scale rock–fluid interactions is unsuited to predicting the macro scale configuration of rock–fluid structures. If this lack of predictive power of micro scale models is an accurate assessment of the problem facing reservoir management, then for better geofluid management it appears necessary to measure *in situ* systems at the scale for which structure information is needed rather than attempt to upscale model results based on small scale sampling of the rock–fluid system.

Power-law scaling statistics of borehole log fluctuations

Borehole logs provide a direct, simple and compact basis for investigating the nature of rock-property fluctuations. Rock-property fluctuations probably occur over scale lengths from grains to crustal sections, but no physical instrument or technique is currently capable of establishing this statement. Borehole logs do, however, establish that fluctuation correlations exist from about 1 m to about 1 km, a scale range relevant to hydrocarbon reservoirs, groundwater aquifers, toxic-waste isolation volumes, mines and excavations, and crustal tectonics. Other forms of evidence suggest that where larger and smaller scale measurements of rock are possible, correlated fluctuations and scaling are observed on correspondingly larger (Bak & Tang 1989; Abercrombie & Leary 1993; Leary 1995) and smaller scales (Hirata *et al.* 1987; Main *et al.* 1990; Barton & Zoback 1992; Dunnicliff 1993) or at both larger and smaller scales (Heffer & Bevan 1990; Yielding *et al.* 1992; Cowie *et al.* 1995).

Figures 1–3 illustrate how scaling emerges from borehole-log data. The three figures analyse log data for, respectively, P-wave sonic velocity (related to elastic modulus, fracture content and porosity), rock mass density (related to porosity and lithology), and gamma ray activity (related to lithology). In each figure panel, log data are shown for vertical wells in North Sea sediments, Irish Sea sediments, and granitic rock, and for a horizontal well in North Sea sediments.

Each data panel has three sub-panels. From left to right, the sub-panels show the fluctuating log of rock properties as a function of position along the borehole, the Fourier power-spectrum of the log, and the histogram of log fluctuation amplitudes. Straight-line (power-law) trends are fit to the spectral power magnitudes between spatial frequencies of *c.* 1 cycle/km to *c.* 256 cycles/km; the slope of the power-law fit is noted for each spectrum (Table 1 lists the power-law scaling exponents for these logs and for the remaining sample of velocity, density and gamma activity logs). Gaussian curves fit to the fluctuation-amplitude number distributions show that the fluctuations have a normal (Gaussian) distribution; the mean value and

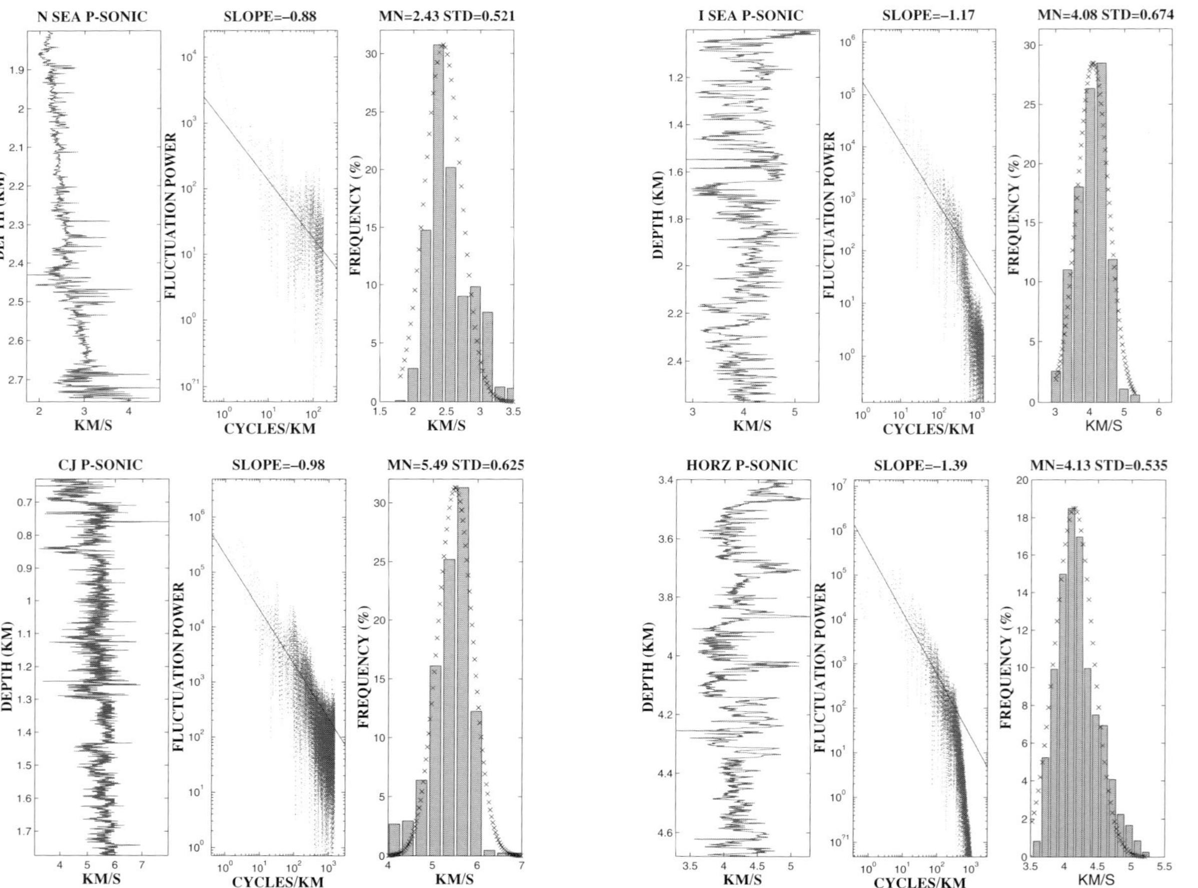

Fig. 1. P-wave sonic velocity borehole log data for three vertical wells (upper two panels and lower left panel) and one horizontal well (lower right panel). Each panel contains three sub-panels showing, from left to right, the log type and well location, the log power-spectrum labelled with the exponent of the power-law fit to the spectrum, and a histogram of the log amplitude distribution labeled with the mean and standard deviation of the Gaussian curve fitted to the histogram. Log data are given in km s^{-1}; well trajectory distances in km. Well locations for vertical wells are the North Sea (N Sea), the Irish Sea (I Sea), and the Cajon Pass scientific drill site in southern California (CJ); the horizontal well is located in the North Sea.

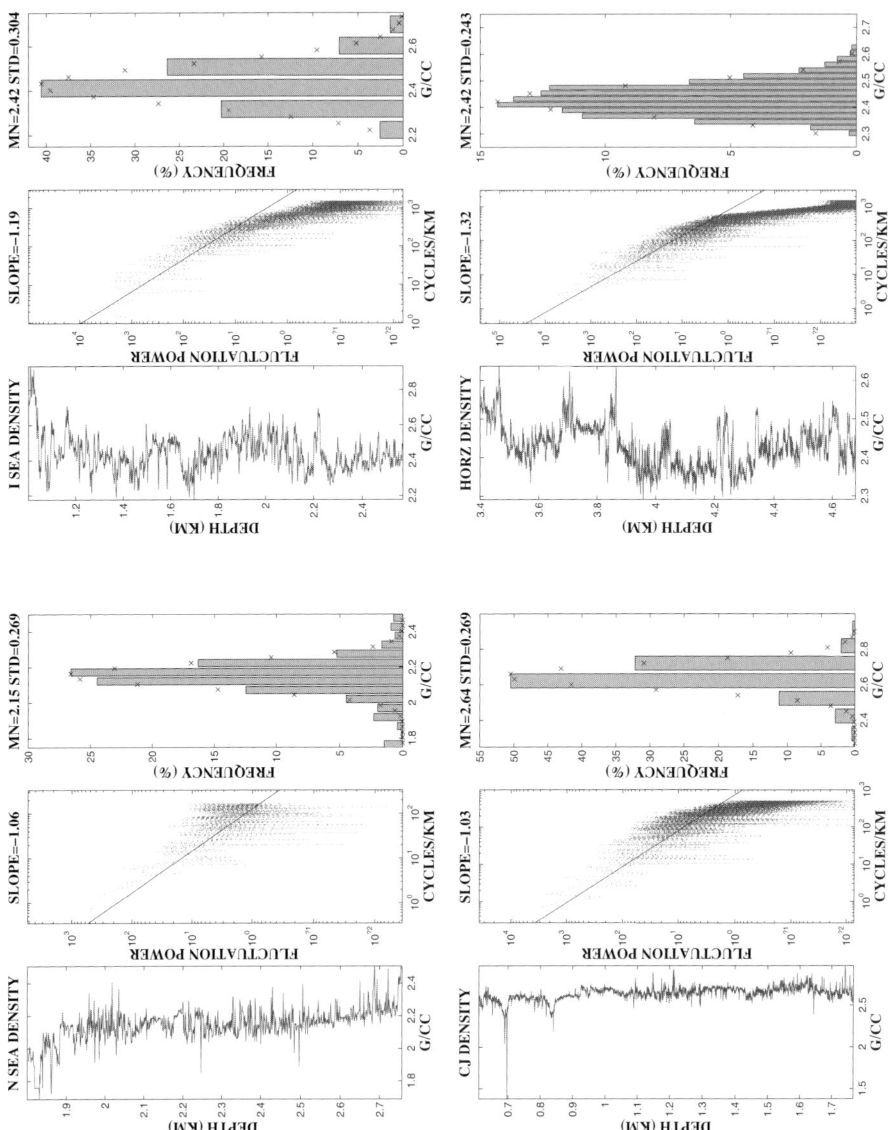

Fig. 2. Mass density borehole log data for the same wells as Fig. 1. Log data are given in $g\,cm^{-3}$.

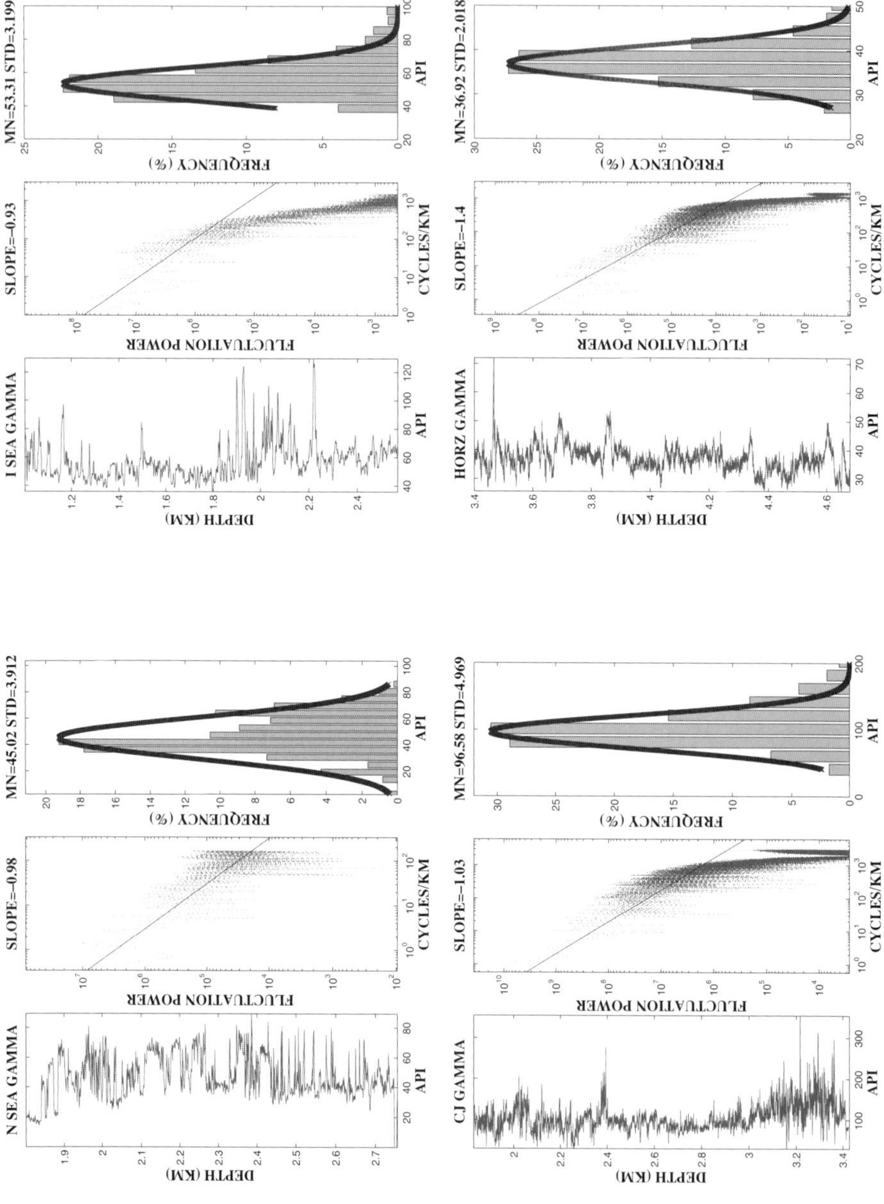

Fig. 3. Gamma activity borehole log data for the same wells as Fig. 1. Log data are given in standard API units.

Table 1. *Absolute value of power-law exponent of power-spectra of sample density, gamma activity and sonic velocity borehole logs*

SITE		DEN	GAM	VEL
VERT. WELL, XTL ROCK				
California	1	1.03	1.03	0.98
	2		0.96	1.10
	3			1.20
Sweden		0.96	1.2	1.14
Germany	1	1.08	1.2	1.10
	2			0.90
		1.02 ± 0.06	1.1 ± 0.1	1.07 ± 0.11
SED. ROCK				
Japan	1	1.1		1.1
	2	1.3		1.1
	3	1.2		1.1
North Sea	1	1.0		1.0
	2			0.90
	3	1.06	0.98	0.88
Irish Sea	1	1.19	0.93	1.18
	2	1.15	1.17	1.21
		1.14 ± 0.1	1.03 ± 0.1	1.06 ± 0.15
HORZ. WELL, SED. ROCK				
Well	1	1.26	1.37	
	2	1.43	1.39	1.43
	3	1.31	1.40	1.39
	4	1.47	1.49	1.47
	5	1.17	1.12	1.24
		1.32 ± 0.12	1.37 ± 0.15	1.38 ± 0.1

standard deviation for each log are noted. Since the fluctuations have a normal distribution, the power-law form of their Fourier power-spectrum is a result of spatial organization rather than anomalous composition of the population.

Like most logs, those shown in Figs 1–3 give readings of rock properties at 15 cm (1/2 foot) intervals along the borehole. Because of the tool length, successive log readings at 15 cm intervals are not independent. Independent rock-property readings are achieved at 2 m to 3 m intervals. The finite length of the logging tool produces an averaging effect over the data at the shortest sample interval (highest spatial frequencies). Averaging causes a change of power-law slope or 'droop' in some of the power-spectra at high spatial frequency. As data averaging over the finite tool interval reduces the high-frequency fluctuation power, the droop in power-law spectra is an artifact of measurement rather than a physical feature of rock.

Km-long logs provide 250–500 independent (unbiased by finite tool aperture) samples of rock properties at 2–3 m intervals of *in situ* rock. Each borehole measurement averages over a rock volume around the borehole. The sampled rock volume lateral dimension ranges from a few tens of centimetres to about 1–2 m. The physical measurement process is stable in a borehole. Trends and fluctuations in rock properties in the scale range from meters to kilometers can be established by examining spatial correlations in the sequence of instrumentally-independent logging values. When, as in the middle sub-panels of Figs 1–3, the logarithm of the borehole-log spectral power is plotted against the logarithm of spatial frequency, a straight-line amplitude trend is evident amidst the otherwise random fluctuations of rock property recorded by the logging tool. The slope of the straight line in the log–log plot is the scaling exponent α in the spectral power-law $1/k^{\alpha}$. For any log, an exponent α can be resolved to about ± 0.1, depending on the log interval (here typically 90–95% of the total interval) and the number of spectral magnitudes fit (256 spatial frequencies in all cases shown in Figs 1–3). The power-law slope is estimated for spatial frequencies smaller than those affected by the spectral 'droop' due to tool averaging. Later, the relationship between log fluctuations and their power-spectra is discussed.

Power-law exponents

The vertical logs of Figs 1–3 have, in common with numerous other vertical logs (Hewitt 1986; Todeschuk *et al.* 1990; Leary 1990, 1991; Turcotte 1992; Wu, *et al.* 1994; Bean 1996; Holliger 1996; Shiomi *et al.* 1997), power-law spectral exponent ≈ 1. Figure 4 summarizes the distribution of 35 power-law scaling exponents for vertical logs in crystalline and sedimentary rock; the mean and standard deviation are 1.1 and 0.1, respectively. In contrast, the spectral exponents of the horizontal logs, illustrated in Figs 1–3 and summarized in Fig. 4, range from 1.1 to 1.5 with mean 1.34 and standard deviation 0.12. We may note that, in comparison with the narrower cluster of vertical log exponents, the more broadly clustered horizontal log exponents may reflect a range of well azimuths relative to the local horizontal principal stress axes.

There appear to be no systematics within the vertical and horizontal well log categories of the well log sample of Fig. 4. Table 1 details the power-law spectral exponent distributions for rock type and rock property. While sediments have a slightly wider distribution about the mean than do crystalline rocks, this is to be expected

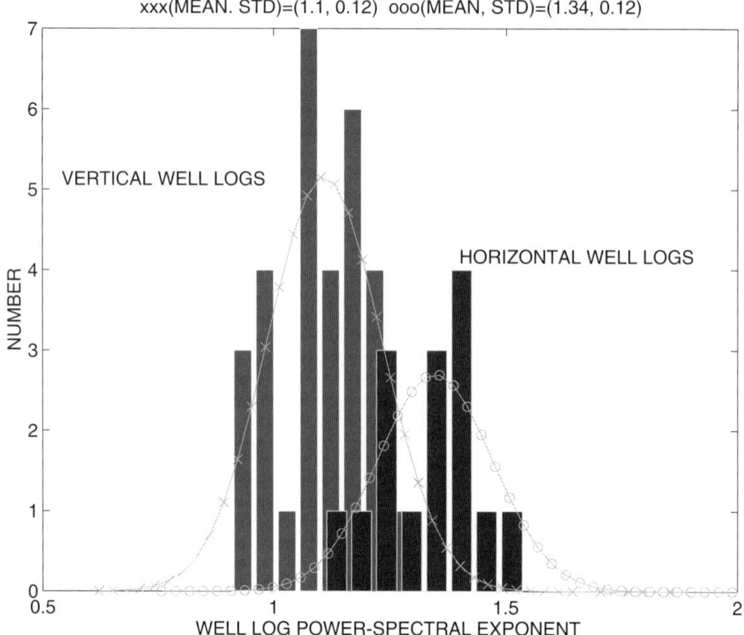

Fig. 4. Histograms of power-spectral scaling exponents for 35 vertical well logs (lighter shading) and 15 horizontal well logs (darker shading) with Gaussian curve fits. The mean and standard deviation of the fitted curves are keyed to the curve symbol (× for vertical logs, ○ for horizontal logs). The vertical logs are from deep crustal drill holes in California, Sweden and Germany, from three boreholes in sedimentary rock in Japan, and from four wells in three UKCS hydrocarbon fields. The horizontal logs are from five wells in a single UKCS hydrocarbon field.

because spatial fluctuations in crystalline rock are more likely to be fracture dominated than are the spatial fluctuations in sedimentary rocks. This feature of the well log data again emphasizes that the essential ingredient of well log systematics in the 1 m to 1 km scale range is the granularity of rock rather than the specific rock types or rock properties.

Scaling

The principal points to made from Figs 1–4 are:

(i) rock structure is a power-law distribution and is therefore inherently scale-independent;
(ii) a narrow range of power-law exponents applies to a variety of logs over three-decades of scale length and a range of rock properties and rock environments;
(iii) in all cases, bigger fluctuation amplitudes are associated with bigger spatial dimensions; and
(iv) log fluctuation amplitudes are normally distributed.

As there is no attributable scale length in the physical mechanisms that generate fluctuations in various rock-property distributions in either the vertical or horizontal direction, the details of micro scale physical and chemical interactions cannot be of first order importance in defining the rock structure encountered by the logging tools. Moreover, since the frequency distribution of fluctuation amplitudes is normal, there is no evidence for a class of physical property structures outside the class of scaling structures. As the logs show that these scaling fluctuations define the spatial distribution of rock properties, it seems unlikely that fluid flow structures are associated with another set of rock-physical attributes not manifested in the logs. Where a specific rock volume has a fluid flow structure, it is likely that the structure obeys a similar scaling law. In particular, observed scaling implies that the larger the rock volume, the greater is the likelihood of a high amplitude (high permeability) flow structure. This aspect of rock permeability is generally acknowledged if not well understood.

Rock as $1/k^{\alpha}$ scaling noise of near-stationary long range correlation structures

The Fourier power-spectral scaling property of borehole logs, $S(k) \propto 1/k^{\alpha}$, is the same for mass density (sensitive to porosity and fracture content), elastic modulus (sensitive to fracture content), and gamma activity (e.g. sensitive to mafic mineral content), and the scaling property is not greatly different between sedimentary rock types nor between crystalline and sedimentary rock. It might not be expected that different physical properties scale in the same manner, particularly if the micro scale physics and chemistry of rock were of paramount importance. There is, therefore, reason to consider what underlying common features unite the upper crustal rock volumes sampled by borehole logs. Features common to most competent rock are:

(i) granularity, with grains having greater elastic competence and physical property uniformity than the rock composite;
(ii) weakness in tension;
(iii) a history of finite strain after having been once buried at elevated temperatures and pressures;
(iv) presence of water.

Seen from this perspective, rock is principally a mechanical entity in which brittle fracture and tensile damage to weak inter-granular bounds during finite strain in the presence of water could be a source of the 'universality' of fluctuation scaling recorded by borehole logs.

We have seen in the right-hand sub-panels of Figs 1–3 that rock-property fluctuations are normally distributed. The structure of the well log power spectra cannot therefore be attributed to the population of rock-property fluctuations. It is instead how the fluctuations are organized in space that gives rise to the power-law spectra. The normally distributed fluctuations are spatially correlated in a manner such that the spatially largest correlation structures are also the highest fluctuation amplitude structures. The link between power-spectra and spatial correlations can be made as a formal mathematical statement.

Study of fluid–mechanical systems gave rise to the understanding of spatial organization of random processes. In particular, the study of lengthy temporal and spatial sequences of fluid dynamical behaviour in liquids and gases, comprising many decades of temporal and spatial frequencies, indicates that a range of power-law scaling exponents are associated with different degrees of stability in time and space (Davis *et al.* 1994). Borehole logs may be considered in relation to spatial sequences of turbulent flow fluid–mechanical fluctuation structures. To gain a better understanding of the rock physical system revealed by Figs 1–3, let us look in more detail at how random fluctuations can be classified in terms of spatial correlations. Familiarity with correlation structure can be useful when considering rock from a statistical mechanics point of view.

Correlated and uncorrelated random structures

The simplest, most stable, and most familiar spatial or temporal sequence is that of the Gaussian uncorrelated random numbers. Gaussian sequences are simple because, being uncorrelated, they sum to a finite mean over any interval and do not build up arbitrarily large trends over such intervals. The uncorrelated nature of successive elements of a Gaussian sequence means that sooner or later every positive trend in the sequence is balanced by a negative trend of equal magnitude and duration. As such, Gaussian sequences are 'stationary' in time or space; that is, the underlying lack of correlation between numbers at any spacing guarantees that no unexpected trends or bursts of activity develop in the sequence. Gaussian sequences are also tractable because two uncorrelated Gaussian sequences add to give a third uncorrelated Gaussian sequence. Finally, a Gaussian sequence has a flat or 'white' power spectrum. This follows because, as there is no correlation between numbers at any spacing, there is no scale-dependent structure in the sequence; a Gaussian sequence does not have any way of knowing what is a 'big' or a 'small' interval or scale length.

As the values of any two neighbouring numbers of a Gaussian sequence are independent, any realizable physical or numerical Gaussian sequence has an arbitrarily short correlation length ξ. This feature in Gaussian sequences is in sharp contrast to a number of physical systems which are characterized by arbitrarily long correlation lengths. In particular, consider a sequence of N numbers constructed by Fourier transforming an uncorrelated Gaussian sequence F_i, $i = 1 \ldots N$, to obtain the sequence of complex spectral amplitudes f_k, $k = 1 \ldots N/2$. Weighting each member of the sequence f_k by the square root of the spatial frequency k, $h_k = f_k/\sqrt{k}$, and computing the inverse Fourier transform, H_i, $i = 1 \ldots N$, gives a real number sequence H_i with a power-spectrum that scales as $1/k$. Figure 5 illustrates that the sequence H_i has the

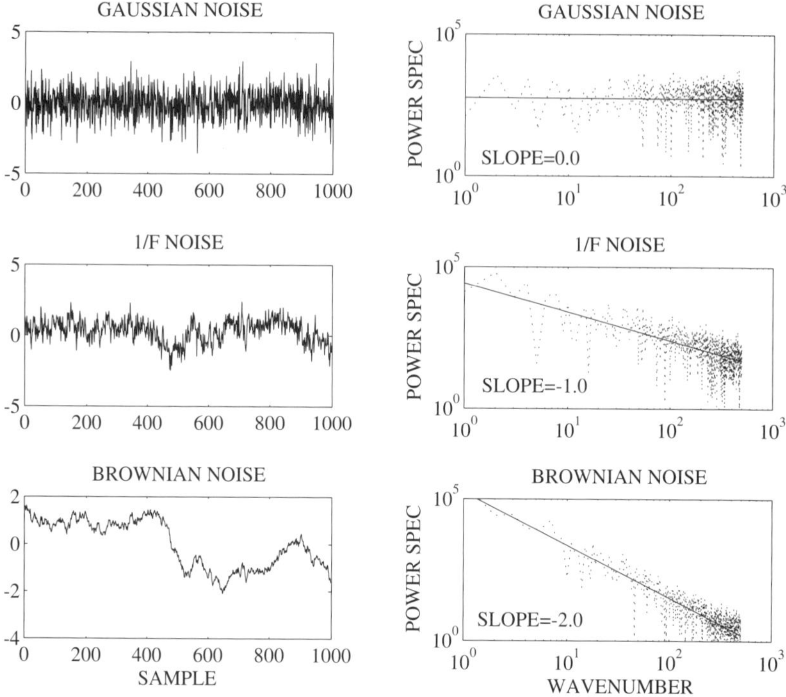

Fig. 5. Uncorrelated and correlated noise sequences and their power spectra. The uncorrelated random numbers (Gaussian noise, upper left trace) were used to generate the correlated ($1/F$ and Brownian) noises by power-law spectral filtering. $1/F$ ($1/k$ in the text) noise was generated by weighting the amplitude spectrum of the uncorrelated noise inversely by the square root of the spatial frequency $\sqrt{k}$, and Brownian noise inversely by the spatial frequency k. The $1/F$ noise sequence resembles the logs of Figs 1–3 while the Gaussian and Brownian noise sequences do not.

spectral scaling properties of a vertical borehole log (cf. Figs 1–3). Note the qualitative difference between the uncorrelated random (Gaussian) sequence and the sequence with power-spectrum scaling as $1/k$. The latter resembles the logs of Figs 1–3 while the former does not. Note also that the droop in well log power-spectra due to averaging over the finite aperture of the logging tools is absent in the synthetic power-spectra.

Now consider the spatial correlation structure of the sequence H_i. The autocorrelation of sequence H_i is $G_r = \langle H_i H_{i+r} \rangle$ where r is an arbitrary distance between points in the sequence and the angled brackets denote summation over the repeated index i. If there is a correlation distance ξ that characterizes sequence H_i, it can be estimated by the summing the autocorrelation function,

$$\xi = \int G(r)\,\mathrm{d}r, \tag{1}$$

since summing over $G(r)$ will produce finite values for intervals r less than the correlation length and will tend to produce zero values for intervals r that exceed the correlation length. Since the autocorrelation function $G(r)$ is the cosine transform of the power-spectrum of a sequence H_i which scales as $1/k$, and the transform of the 1D sequence $1/k^{\alpha}$ is $1/r^{1-\alpha}$, we have the result that in the case of H_i, where $\alpha = 1$, the correlation length $\xi \approx \int \mathrm{d}r/r^0 = \int \mathrm{d}r$ is arbitrarily large. Thus, for sequences with power-spectra that scale as $1/k^{\alpha}, \alpha \approx 1$, the correlation length is large. Conversely, when $\alpha \approx 0$ the correlation length defined above approaches ln (r), which is effectively zero since the logarithm grows very slowly with interval r.

An arbitrarily large correlation length for statistical correlation structures is thus a defining property of rock. This recognition supports a unifying understanding of rock in two related ways. First, the statistical mechanics of critical-point thermodynamics systems identifies infinite correlation length as the key defining concept. Second, infinite correlation lengths emerge as the key concept in defining percolation structures in physical systems. We shall now see how the statistical physics of thermodynamic critical-point systems can be joined to the theory of percolation structures to provide a simple but inclusive description of rock heterogeneity.

Rock structure as a critical state fluctuation phenomenon

The scale-free $1/k$ fluctuation spectra of various rock physical properties expressed in 1 D borehole log data can be restated in terms of spatial fluctuation correlations in three dimensions. In systems of geometric dimension d, a $1/k^{\alpha}$ power-spectrum of spatial noise implies that the spatial correlation relation is power law $G(r) \propto 1/r^{d-\alpha}$ (Falconner 1990). A scale-free power-law correlation function is very different from the usual concept of correlation function, often expressed by an exponential $G(r) \propto \exp(-r/\xi)$, ξ being the characteristic correlation length and r being the distance between two points in the medium. If the correlation length for rock structures is indefinitely large, $\xi \rightarrow \infty$ (i.e. effectively the size of the rock system), then an exponential distribution is void. Power-law correlation functions can thus be regarded as expressions of long range correlations. Expressing $1/k^{\alpha}$ fluctuation scaling in terms of correlation functions makes contact with the theory of long range thermodynamic critical point fluctuations in many-body systems and provides a physical basis for correlation structures that do not permit accurate treatment by the statistics of sample means and variances.

Defect order and disorder

The long range correlation picture of rock does not indicate what causes variation in rock properties, or how and at what rate these variations are generated. Spatial correlations simply yield information about rock structure organization. However, borehole logs indicate that rock attains a 'universal' scaling structure, suggesting that granularity, weakness in tension, and a history of finite strain in the presence of water common to most rock may account for the universality. We may suppose that micro scale defects arising from finite strain are an important component of rock structure expressed in borehole logs. Stress-aligned micro cracks and similar defects offer an attractive analogue to micro scale interactive elements observed in thermodynamic critical phenomena.

The scaling symmetry illustrated by borehole logs has been noted in a wide range of thermodynamic critical phenomena such as ferromagnetism, liquid–gas phase transitions, binary fluid and alloy transitions, helium I/helium II transition, and conductor/superconductor transition (Binney *et al.* 1995; Cardy 1996). These phenomena are independent of the details of the micro scale interactions and depend essentially on the interplay between system free energy (energy above the heat energy that defines the temperature or mean kinetic energy of the micro scale 'molecular' interaction) and the entropy (probability of various spatial configurations of the system). The experimental observation of critical phenomena requires that the systems are in close thermal equilibrium somewhere between a state of extreme order and a state of extreme disorder (Binney *et al.* 1995; Cardy 1996). If a physical system is in mechanical equilibrium, has an analogue of thermodynamic free-energy expressed for mobile population of interacting elements, and has a well-defined probability for any particular spatial distribution of interacting elements, observation and simulation indicate that the system can exhibit anomalous long range, high amplitude random correlation structures at a point between the ordered state and disordered state.

Where equilibrium thermodynamic systems of elastically interacting 'molecules' vary with free energy at a given level of heat energy that defines the system temperature, a scaling theory for rock fluctuation structures must use analogues for thermodynamic free energy, heat energy and temperature. For rock these analogues are, respectively, elastic strain energy, damage induced by finite strain, and strain induced defect probability. Damage inducing strain is stored in the rock as quasi-irreversible anelastic defects. However, unlike thermodynamic systems with systematic access to a wide range of heat energy (i.e. a wide range of temperature), rock is maintained at its critical point by self-regulating the defect population induced by finite strain. The 'heat-energy' or critical defect population of brittle rock is regulated by adding defects through quasi-irreversible damage induced by tectonics and by subtracting defects through long term chemical healing processes. If there is too much strain without healing, rock disaggregates into a state of disorder; if there is too little strain, rock heals through recrystalization into a state of order. It appears from borehole log evidence that most crustal rock resides in an intermediate state between order and disorder.

The fracturing and healing processes that balance rock between order and disorder are closely associated with the water content of rock. When rock fails, elastic strain energy is released by folding or faulting and excess fluids percolate out of the system. When rock is unstrained, available water is consumed by recrystallization reactions to damage sites. Critical defect density and structural failure are thus conceptually closely related to the percolation threshold defect density at which a defect pathway is statistically likely to span the rock volume (Deutscher *et al.* 1983; Feder 1988; Falconner 1990; Bunde & Havlin 1991; Stauffer & Aharony 1994; Binney *et al.* 1995). Below the percolation threshold, the medium is more or less intact (ordered) and above the percolation threshold the medium is more or less disaggregated (disordered). Rock is observed to rest at the order–disorder threshold (in the 'self-organized critical' state of Bak *et al.* 1988) where the rock-property fluctuation correlation length approaches the size of the body (becomes 'infinite') and fluctuation structures exist at all scale lengths as recorded in borehole logs.

Long range defect correlation structures

A mathematical picture of rock structure as correlated scaling fluctuations at a critical point order–disorder transition can be sketched following Kadanoff *et al.* (1967). The operative result is given as equation (5) below. Details of the development appear in the Appendix. Accessible sources for the theory and extensive later developments are Landau & Lifshitz (1958), Lifshitz & Pitaevskii (1980), Kadanoff (1966, 1976) and Binney *et al.* (1995).

In the present analogy for order–disorder transitions in rock, a spatially varying defect density $D(\mathbf{r})$ is the measure of order in the order–disorder spectrum. A defect is thought of as a crushed, sheared or chemically altered grain, a broken cement bond, or whatever weakness changes the bulk elastic properties, affects the porosity or density of the rockmass, and/or allows fluid to flow where otherwise fluids would not flow. If the rock is uniform, $D(\mathbf{r})$ is constant or nearly constant and variations in D are small and spatially uncorrelated (normally distributed in space). If the rock is disaggregated, the amplitude variation of D is large but again the variations are spatially uncorrelated. In both the ordered or disordered state, traditional Gaussian means and variances are adequate to describe the rock.

The 'temperature' of the rock state is the probability $P, 0 \leq P \leq 1$, that a defect exists at a grain site; P is the normalized mean defect density. To make this definition of a temperature analogue easier to accept, it may be noted that it is customary in thermodynamics to use 'reduced temperature', $t = |T - T_c|/T_c$, for which $0 \leq t \leq 1$ (e.g. Binney *et al.* 1995; Cardy 1996). As stress and strain in the rock increase, the overall defect probability increases towards a critical value P_c. At the critical point, a defect pathway is statistically likely to span the rockmass and the rock system will tend to fail either by releasing stored elastic strain energy or by letting fluids percolate through the rock mass to relieve excess pore pressure, or both.

The spatial configuration state of a rock system is described with the spatial correlation function $G(\mathbf{r}, \mathbf{r}')$, spatial correlation range ξ and several scaling exponents. In equation (5) we will see that as the percolation threshold is reached, the correlation range becomes singular as $\xi \propto 1/\sqrt{|P_c - P|}$ and the correlation function has a Fourier transform $G(k) \propto 1/k^{2-\eta}$ where η in thermodynamics is a parameter typically near zero. In principle, the order of singularity ν in correlation length at $P_c = P$, $\xi \propto 1/|P_c - P|^{\nu}$, can have a value $\nu \neq 1/2$ and can possibly be estimated from numerical simulation or from laboratory observation of acoustic emissions (e.g. Hirata *et al.* 1987). Other properties of rock may have critical point power-law dependencies, as, say, bulk modulus. However, at present the correlation exponent η is the only theoretical parameter that can be straightforwardly observed for rock, where on the evidence of borehole-log spectra it has a value between ≈ 0.65 and ≈ 0.9.

From the Appendix, the thermodynamically stable state of an isotropic rock medium is described by the condition

$$[2A(P) + 4B(P)D(\mathbf{r})^2 + 2C(P))\Delta^2]D(\mathbf{r}) = \sigma(\mathbf{r}) \tag{2}$$

where Δ^2 is the radial Laplacian operator, and theory assumes that $A(P) \approx (P_c - P)$ vanishes at critical probability P_c and $C(P)$ approaches non-zero constant value at P_c. Standard theory also assumes that the spatially varying term $4B(P)D(\mathbf{r})$ can be ignored but the term is retained here (see Appendix) in an imposed phenomenological form to represent fluctuation interactions that produce a screening effect parameter η to broaden the fluctuation spectrum observed in borehole logs.

The fluctuations embodied in the spatial variation $D(\mathbf{r})$ are introduced by defining the spatial correlation function

$$G(\mathbf{r}, \mathbf{r}') \approx \langle D(\mathbf{r})D(\mathbf{r}')\rangle \tag{3}$$

and applying a variational principle in combination with equation (2). The result is that $G(\mathbf{r}, \mathbf{r}')$

satisfies the condition

$$[2A(P) + \eta B'\langle D(\mathbf{r})\rangle^2 - 2C(P))\Delta^2]G(\mathbf{r},\mathbf{r}') \approx \delta(\mathbf{r}-\mathbf{r}') \quad (4)$$

where $\eta B'\langle D(\mathbf{r})\rangle$ is retained as a phenomenological term accounting for radially-dependent defect–defect interactions and η is the scaling exponent describing the effect of the defect–defect interaction. In the absence of a distortion effect, $\eta = 0$ and the interaction term vanishes to yield the standard thermodynamic expression.

The solution to condition equation (4) gives the basic theoretical result that the correlation distance *xi* is singular at the critical defect probability, $\xi \propto 1/\sqrt{|P_c - P|}$, and hence at $P_c \approx P$ the correlation function is defined entirely in terms of a power law

$$G(\mathbf{r},\mathbf{r}') \approx 1/|\mathbf{r}-\mathbf{r}'|^{1+\eta}\exp[-|\mathbf{r}-\mathbf{r}'|/\xi] \rightarrow 1/|\mathbf{r}-\mathbf{r}'|^{1+\eta}. \quad (5)$$

Equation (5) is the desired physical description of rock structure. From this we can deduce the behaviour of borehole logs. As noted earlier, the 3D expression $G(\mathbf{r},\mathbf{r}') \approx 1/|\mathbf{r}-\mathbf{r}'|^{1+\eta}$ has the Fourier power-spectrum $G(k) \approx 1/k^{2-\eta}$. When the phenomenological parameter η is small, as in thermodynamic systems, the correlation spectrum assumes the appropriate $1/k^2$ scaling (see lowest panel of Fig. 5). When micro scale interactions induce η to take values between 0.6 and 1.0 the observed borehole log power-law spectral scaling is attained (as in middle panel of Fig. 5).

Discussion

Critical opalescence

To help consolidate the statistical physics model of rock as a medium characterized by long range random correlated fluctuation structures, it may be useful to consider the visual analogy between rock and the phenomenon of critical opalescence in fluid-gas mixtures (Binney *et al.* 1995). It is observed for some fluid systems that at a thermodynamic critical point in temperature and pressure, an otherwise clear, uniform medium becomes cloudy or opalescent. The opalescence is due to light scattered at the large scale density fluctuations as the fluid–gas mixture sits on the edge of a fluid-gas order–disorder phase change. In this thermodynamic phase, the range and amplitude of fluctuations vastly exceeds that of ordinary medium (gas or liquid) uncorrelated random fluctuations. The conventional measure of opalescence (and similar broadband spatial scaling correlation structures) is the intensity of light (or X-ray or neutron) scattering versus the fluctuation scale length (Kjems 1991). Light scattering data for liquid–gas transitions show that the intensity/scale-length relation is a power-law that scales inversely with spatial frequency with exponent $\alpha \approx 2$, $I(k) \propto 1/k^\alpha \propto 1/k^2$. In rock, the well-known 'coda wave' of extended seismic oscillations is generated by scattering analogous with producing critical opalescence. Leary (1995) shows that the spectrum of seismic coda waves is precisely that which is expected from rock elastic heterogeneity deduced from borehole logs. On the evidence of coda wave spectra, rock heterogeneity seen in borehole logs is characteristic of crustal rock in volumes of tens of thousands of cubic kilometres.

The power-law scaling exponent $\alpha \approx 2$ for critical opalescence is characteristic of molecular interactions and other thermodynamic systems in which the micro scale interaction is isotropic and relatively simple. Opalescence fluctuations occur in isotropic medium with relatively uncomplicated micro scale thermal kinetic interactions between molecules. In rock, the equivalent spectral scaling exponents $\alpha_v \approx 1.1$ and $\alpha_h \approx 1.34$ indicate that equivalent micro scale interactions are not quasi-ideal molecules in isotropic conditions but are anisotropic and probably spatially complex. This can be expected if micro scale interactions in rock involve quasi-planar, quasi-elastic defects or defect clusters, e.g. grain scale stress-aligned microfractures and microfracture structures, generated by finite strain in an anisotropic stress field and interacting with an anisotropic fluid permeability field. The effect of complex micro scale interactions in rock is evidently to shorten the range of structural correlations and to broaden the corresponding power-law spectrum, $1/k^2 \rightarrow 1/k^\alpha$, $\approx 1.1 < \alpha <\approx 1.35$, depending on the orientation of the sample relative to the vertical.

Rock type and the critical state

The statistical mechanical picture of rock emphasizes rock as a granular medium. However, treating the generic properties of rock as a granular medium undergoing an order–disorder transition due to finite strain brittle fracturing in the presence of percolating water, does not mean that geological rock classifications are destroyed. Rocks can exist in an order–disorder transition state while retaining their geological identity and the general properties associated with their geological classification.

Thermodynamic systems such as binary fluids, fluid–gas mixtures, ferromagnets, superconducting and superfluid systems exhibit critical-point behaviour at a critical temperature but the systems retain their physical nature. All physical systems exhibiting critical-point behaviour have a basic element of free energy interaction (grains in rock, magnetic spins in ferromagnets, molecules in gas–fluid and binary–fluid systems) and have states of complete order (uniform rock, crystal state or uniform fluid) and states of complete disorder (disaggregated rock, thermally disordered magnetic domains, or gas). Within the particular physics of each system, the order–disorder transition state links small scale uncorrelated random fluctuations to large scale correlated fluctuation structures.

This 'universality' applies equally to the variations of rock type. Sandstones may have a depositional fabric, complex cementation history, and aqueous fluid transport history that differs profoundly from a history of igneous melt crystal growth, loss of volatiles, and plastic deformation characteristic of granitic plutons. If undisturbed by tectonic forces, rock units of different compositions and process histories might manifest themselves in distinct statistical fluctuation profiles. However, rock accessible as geofluid reservoirs has usually undergone extensive episodes of finite strain and brittle fracture in the presence of abundant water. In the same way that injecting heat into a thermodynamic systems to bring it to the threshold of an order–disorder transition, finite strain induces percolation structures to grow in rock of all types.

The order–disorder critical state does not remove from granite or sandstone their particular characters as a rock type any more than the fact that binary fluids showing critical opalescence, or iron domains aligning to become ferromagnetic, means that the fluid and iron lose their physical character. Sandstones and granites will continue to exhibit very different ranges of porosity, permeability, and mineral composition and chemistry. While one may say that no rock type is immune to the process leading to an order–disorder transition state, the physical rock character of virgin state grain–grain interaction, and its related porosity and permeability, is always likely to influence how useful that rock unit is as a geofluid reservoir.

Interconnection of physical property fluctuations

Borehole logs measure physical properties of rock such as mass density, electrical resistivity, sonic velocity, porosity, and gamma ray activity. The Fourier power-spectra of the logs are observed to have the same spatial frequency scaling property regardless of rock property and rock type. This 'universality' is initially surprising but it has a natural explanation in terms of rock as a granular-medium analogue to critical-point thermodynamic systems. There is, however, a logical reason why the physical properties of rock measured by borehole logs appear to be interconnected.

First, the similarity of power-spectrum is not equivalent to point-by-point similarity in borehole logs. The power-spectrum retains only amplitude information, and eliminates all spatial phase information. There is, therefore, no constraint on logs of different rock properties to have more than a rough spatial correlation. Within this limit, an interconnection between density, resistivity, porosity, and sonic velocity is to be expected if 'defect density' is the main physical feature of rock heterogeneity. For macroscopic defect density we may understand macroscopic fracture density. It is plausible that there is a physical relation between mass density, porosity, sonic velocity and resistivity in terms of fractures. Denser, less porous rock intervals will be less likely to fail in finite strain than will less dense and more porous rock intervals. Sonic velocity is well known to be controlled by fracture density, as is resistivity in the presence of water. Finally, it is understandable that fluctuations in gamma ray activity share a broad interconnection with the other physical variables since fracture systems carry a large percentage of the fluids passing through the rock. Radionuclides such as potassium, thorium and uranium are water-soluble chemical elements that can be dissolved out of damaged rock grains and transported throughout the rock mass. As large volumes of water pass through crustal rock over geological time, it is not surprising that gamma activity can be distributed in the percolation pathways generated by finite-strain-induced defect populations.

Consequences for geofluid management

As a result of the finite-strain order–disorder transition state, rock fracture and percolation fabric appears, within any rock type, to be relatively insensitive to micro scale interactions of fluid chemistry, grain geometry, local stress field, local permeability and so forth. Quantitatively, the relation between micro scale processes and macro scale structures appears to be defined through the value of the borehole-log power-spectral scaling exponent. Variation in the exponent seems to be mostly determined by the orientation of the stress field relative to the well axis. The universal feature of the borehole log scaling is that the largest size defect structures are those with the largest magnitude of rock-property fluctuation. As such, large scale system-defining structures tend to be uncoupled from local scale processes, it seems to be more important to determine the effective rock structure through direct measurement of the *in situ* system rather than trying to infer 'probable' structures through simulations based on micro scale process and small scale rock samples. The physical dimension and spatial unpredictability of large scale critical-state fluctuations also mean that the rock is likely to be partitioned into large scale structures which can be thought of as cost-effective targets for remote sensing.

Probably the most effective way to sense these target volumes in a commercial geofluid system is through measurement of the fluid flow structure itself. It is likely that time-lapse seismic sensing of migrating oil–water, gas–oil, and gas–water substitution fronts can provide useful data for establishing the large scale fluctuation structure of hydrocarbon reservoirs. Similar geophysical methods may have to be developed to address the more subtle problem of determining the flow structure of aqueous rock–fluid systems.

Maintaining the critical state

Unlike thermal critical processes which require close thermal equilibrium and heat control, rock is maintained in the state of 'self-organised criticality' (Bak *et al.* 1988; Zatsepin & Crampin 1997) probably by the fact that order–disorder criticality exists when the rock defect population creates a through-going percolation structure. At that point, fluids are not retained by the rock-mass and strain can be relieved through finite strain displacement. Once drained of excess fluids and relieved of excess strain, the rockmass undergoes a renewed cycle of tectonic strain and approaches the percolation threshold.

This inferred state of crustal rock is consistent with much of what we know about rock in tectonically active and passive areas. On the time scale of brittle fracture healing, almost all rock is tectonically active. Evidence for this is the on-going in intra-cratonic regions of infrequent natural seismicity such as Fennoscandia, central North America and Canada, South Africa and Brazil. Despite the low intensity, intra-cratonic earthquake activity is indistinguishable in seismic properties from that of tectonically active areas. In a large number of tectonically active and passive regions, seismic shear-wave polarization data show that rock micro scale fracture fabric is aligned with the regional principal stress axes (Crampin 1994; Zatsepin & Crampin 1997). The stress alignment occurs equally in sedimentary and in crystalline rock. The seismic evidence indicates that a stress-induced critical state characterizes rock in most if not all continental crust.

It appears safe to say that crustal rock hosting geofluids of interest to commercial activity will be influenced by order–disorder transition state long-range correlation structures. In consequence of the long range correlation structures, fluid flow is in many cases likely to be spatially unpredictable on the basis of small scale sampling and attempts to 'up-scale' the flow character through models. The appropriate management technique for critical-state rock is to measure the site-specific flow structure at the scale lengths of interest.

Summary

A 'universal' borehole log fluctuation power spectrum of rock, $S(k) \propto 1/k^{\alpha}$ with $\alpha \approx 1$, is observed for velocity, mass density, gamma activity and chemical abundance fluctuations in both crystalline and sedimentary rock. The power spectrum can be viewed as evidence that rock heterogeneity is a stable or stationary form of random spatial correlation noise. Rock seems to be in a stable state that is neither ordered, rock is too flawed to be uniform, or disordered, *in situ* rock is not disaggregated. Intermediate states between order and disorder have been comprehensively investigated in the statistical mechanics of order–disorder phase transitions in thermodynamic systems. The statistical mechanics order–disorder transition model can be applied to rock if the role of order parameter is taken by a defect density, and percolation probability, anelastic strain and elastic strain assume the roles of thermodynamic variables temperature, heat energy and free energy, respectively. By focusing on the granularity of rock and finite-strain defects related to fluid percolation, the statistical mechanics analogy explains, in a straightforward way, why the power-spectral scaling of borehole logs is 'universal'. The 'universality' of rock heterogeneity indicates that detailed investigations of complex inter-related micro scale chemical and physical rock–fluid processes may not be necessary to understand large scale aspects of fluid flow in rock.

The stationary critical-point order–disorder transition state of rock shows semi-quantitatively that borehole log fluctuation distributions arise from a micro scale defect density related to the fluid percolation threshold. The chemico–physical details of defect production are probably of secondary importance. The long range fluctuations that underlie the borehole log power-spectra occur because at a critical defect density, the free energy state of the solid hovers between order and disorder and a large number of spatial configurations of equal energy and probability occur. In this stability point between order and disorder, it is possible for micro scale fluctuations to inflate to essentially infinite correlation length without violating either energy or probability constraints.

The intimate role that percolation structures play in generating the observed heterogeneity of rock implies that fluid flow paths in reservoirs are strongly influenced by long range correlated random flow and transport paths. Such paths cannot be predicted on the basis of small scale sampling of the reservoir volume. Effective management of reservoir fluids requires measuring *in situ* flow structures at the scale of the reservoir.

The horizontal well log data were supplied by Mobil North Sea Ltd through the courtesy of A. Tweedie of Heriot Watt University. Vertical well log data in sediments were supplied by industry through the courtesy of J. Underhill, A. Ziolkowski and R. Johnston of the University of Edinburgh. Vertical well data in crystalline rock were made available by C. Juhlin (Siljan well, Sweden), the KTB project (Germany) and the Cajon Pass project (California). Numerous conversations on these matters with S. Crampin, S. Zatsepin, R. Peveraro, K. Sorbie and R. Knipe are recalled with pleasure. Two referees provided cogent remarks for improving the manuscript. Support for this work was provided by NERC and PSTI.

References

ABERCROMBIE, R. & LEARY, P. 1993. Source parameters of small earthquakes recorded at 2.5 km depth, Cajon Pass, southern California: implications for earthquake scaling. *Geophysical Research Letters*, **21**, 971–974.

BAK, P., TANG, C., & WIESENFELD, K. 1988. Self-organized criticality *Physical Review*, **38**, 364–374.

—— & TANG, C. 1989. Earthquakes as a self-organised critical phenomenon. *Journal of Geophysical Research*, **94**, 15,635–15,638.

BARTON, C., & ZOBACK, M. 1992. Self-similar distribution and properties of macroscopic fractures at depth in crystalline rock in the Cajon Pass scientific drill hole. *Journal of Geophysical Research*, **97**, 5181–5200.

BEAN, C. J. 1996. On the cause of $1/f$ power spectral scaling in borehole logs. *Geophysical Research Letters*, **23**, 3119–3122.

BINNEY, J. J., DOWRICK, N. J., FISHER, A. J., & NEWMAN, M. E. J. 1995. *The Theory of Critical Phenomena*, Oxford Science Publications, Oxford University Press, Oxford.

BUNDE, A., & HAVLIN, S. 1991. Percolation. *In*: BUNDE, A. & HAVLIN, S. (eds). *Fractal and Disordered Systems*, Springer-Verlag, Berlin.

CARDY, J., 1996. *Scaling and Renormalization in Statistical Physics*. Cambridge University Press, Cambridge.

COWIE, P., SORNETT, E, D. & VANNESTE, C. 1995. Multifractal scaling properties of a growing fault population. *Geophysical Journal International*, **122**, 457–469.

CRAMPIN, S. 1994. The fracture criticality of rock. *Geophysical Journal International*, **118**, 428–438.

DAVIS, A., MARSHAK, A. & WISCOMBE, W. 1994. Wavelet-based multifractal analysis of non-stationary and/or intermittent geophysical signals. *In*: EFOUFOULA-GEORGIOU, E. & KUMAR, K. (eds). *Wavelets in Geophysics*. Academic Press, San Diego.

DEUTSCHER, G., ZALLEN, R. & ADLER, J. 1983. *Percolation Structures and Processes*. Annals of the Israel Physical Society, Vol 5.

DUNNICLIFF, J. 1993. *Geotechnical Instrumentation for Monitoring Field Performance*. Wiley & Sons, New York.

FALCONNER, K. 1990. *Fractal Geometry*. J. Wiley & Sons, Chichester.

FEDER, J. 1988. *Fractals*. Plenum Press, New York.

HEFFER, K. & BEVAN, T. 1990. Scaling relationships in natural fractures: data, theory and application. *Society of Petroleum Engineers Europec90*, SPE20981, The Hague, Netherlands.

HEWETT, T. 1986. Fractal distributions of reservoir heterogeneity and their influence on fluid transport. *Society Petroleum Engineers* SPE15386.

HIRATA, T., SATOH, T. & ITO, K. 1987. Fractal structure of spatial distribution of microfracturing in rock. *Geophysical Journal Royal Astronomical Society*, **90**, 369–374.

HOLLIGER, K. 1996. Seismic velocity heterogeneity of the upper crystalline crust as derived from P-wave sonic logs. *Geophysical Journal International*, **125**, 813–829.

KADANOFF, L. P. 1966. Scaling laws for Ising models near T_c. *In*: KADANOFF, L. P. (ed.) *From Order to Chaos*. Nonlinear Science, Series A, Vol. 1, World Scientific, Singapore.

—— 1976. Scaling, universality and operator algebras. *In*: KADANOFF, L. P. (ed.) *From Order to Chaos*. Nonlinear Science, Series A, Vol. 1, World Scientific, Singapore.

——, GÖTZE, W. & HAMBLEN, D. 1967. Static phenomena near critical points: theory and experiment. *In*: KADANOFF, L. P. (ed.) *From Order to Chaos*. Nonlinear Science, Series A, Vol. 1, World Scientific, Singapore.

KJEMS, J. 1991. Fractals and experiments. *In*: BUNDE, A. & HAVLIN, S. (eds) *Fractal and Disordered Systems*, Springer-Verlag, Berlin.

LANDAU, L. D. & LIFSHITZ, E. M. 1958. *Statistical Physics*. Pergammon Press, Oxford.

LEARY, P. 1990. Basement rock fracture structure from Cajon Pass, California, and Siljan Ring, Sweden, borehole geophysical logs. 60th Meeting, Society of Exploration Geophysicists, Expanded Abstracts, 153–155.

—— 1991. Deep borehole log evidence for fractal distribution of fractures in crystalline rock. *Geophysical Journal International*, **107**, 615–628.

—— 1995. Quantifying crustal fracture heterogeneity by seismic scattering. *Geophysical Journal International*, **122**, 125–142.

LIFSHITZ, E. M. & PITAEVSKII, L. P. 1980. *Statistical Physics, Part 1*, Pergamon Press, Oxford.

MAIN, I., MEREDITH, P., SAMMONDS, P. & JONES, C. 1990. Influence of fractal flaw distribution on rock deformation in the brittle field. *In*: KNIPE, R. & RUTTER, E. (eds). *Deformation Mechanisms, Rheology and Tectonics*. Geological Society, London, Special Publications **54**, 71–79.

SHIOMI, K., SATO, H. & OHTAKE, M., 1997. Broad-band power-law spectra of well-log data in Japan. *Geophysical Journal International*, **130**, 57–64.

STAUFFER, D. & AHARONY, A. 1994. *Introduction to Percolation Theory*. Taylor & Francis, London.

TODESCHUK, J. P., JENSEN, O. G. & LABONTE, S. 1990. Gaussian scaling noise model of seismic reflection sequences: Evidence from borehole logs. *Geophysics*, **55**, 480–484.

TURCOTTE, D. 1992. *Fractals and Chaos in Geology & Geophysics*. Cambridge University Press.

TYLER, N. 1988. New oil from old fields. *Geotimes*, **33(7)**, 8–10.

WU, R.-S., XU, Z. & LI, X.-P. 1994. Heterogeneity spectrum and scale-anisotropy in the upper crust revealed by the German continental deep-drilling (KTB) holes. *Geophysical Research Letters*, **21**, 911–914.

YIELDING, G., WALSH, J. & WATTERSON, J. 1992. The prediction of small-scale faulting in reservoirs. *First Break*, **10**, 449–460.

ZATSEPIN, S. & CRAMPIN, S. 1997. Modelling the compliance of crustal rock – I. Response of shear-wave splitting to differential stress. *Geophysical Journal International*, **129**, 477–494.

Appendix

Equations (2) and (4) in the main text embody the Landau theory of second order phase transitions devised to explain the long range spatial correlations at critical-point transitions observed in many-body systems such as ferromagnetism, super-conductivity, and critical opalescent binary fluids. Landau theory is presented in Chapter 14 of Landau & Lifshitz (1958) and in the later, expanded edition, Lifshitz & Pitaevskii (1980). Binney *et al.* (1995) give the modern version of Landau theory. Kadanoff (1966) recognized the 'universality' of critical state behaviour, and Kadanoff *et al.* (1967) discussed universality and the Landau theory in a particularly accessible manner. This appendix gives the Kadanoff *et al.* presentation in greater detail than in the main text and makes a slight ad hoc adaptation to allow for larger values of the parameter η than proceed naturally from standard critical-point theory (Binney *et al.* 1995).

Thermodynamic systems can be thought of as 2D or 3D lattice networks with an interacting element at each node. The element can interact with an external field, say a magnetic, electric or stress field, or the element can interact with its neighbours, or with both the external field and neighbours. In each case, the interaction consumes energy, and contributes to the thermodynamic 'free energy' budget of the system.

Lattice node interacting elements that contribute to the system free energy can in principle assume arbitrary spatial patterns. If element thermal energies are small, the spatial patterns are likely to be short range, uncorrelated fluctuations that reflect the order of the regular support lattice. If element thermal energies are high, interactions are likely to be overwhelmed by thermal fluctuations and the spatial patterns disordered. At an intermediate thermal energy, interaction and thermal energies are comparable and the pattern of interacting elements can assume a wide range of structures. In order to describe this intermediate state, the system free energy is parameterized in the fewest number of terms that can capture the physical observations.

Let $D(\mathbf{r})$ and $\sigma(\mathbf{r})$ denote, respectively, the state of the interaction element and value of the externally applied elastic stress field at lattice point $\mathbf{r}$. For rock $D(\mathbf{r})$ can represent the density of inter-granular defects at position $\mathbf{r}$. $D(\mathbf{r})$ can be regarded as either 0 or 1 to represent the absence or presence of a unit defect at site $\mathbf{r}$, or it could represent a continuum of defect densities given by averages over a small volume of rock centered on $\mathbf{r}$. The necessary quality of the density $D(\mathbf{r})$ is that it represent a physical interaction energy between lattice points $\mathbf{r}$ and $\mathbf{r}'$, and/or interaction energy with the external field σ at $\mathbf{r}$.

The total interaction energy $F(\mathbf{r})$ at position $\mathbf{r}$ is then parameterized in terms of density $D(\mathbf{r})$ by

$$F(\mathbf{r}) = -\sigma(\mathbf{r})D(\mathbf{r}) + A(P)D(\mathbf{r})^2 + B(P)D(\mathbf{r})^4 + C(P)\partial D(\mathbf{r})\cdot\partial D(\mathbf{r}) \quad \text{(A1)}$$

In equation (A1) the factors $A(P)$, $B(P)$ and $C(P)$ are unknown position-independent functions of the mean defect density expressed as a probability P that site $\mathbf{r}$ is occupied by a defect. If $P = 0$, there are no defects at any site; if $P = 1$, all lattice sites are occupied by defects. Probability P plays the role of temperature in thermodynamic systems. At low 'temperatures', order prevails since there are few or no defects in a regular granular material; at high 'temperatures', disorder prevails because there are so many defects that the material is disaggregated; at an intermediate critical 'temperature', defect populations interact to create large range correlation defect fluctuation structures.

The first term in equation (A1) represents the free energy interaction between the defect population and the applied stress field. The terms quadratic and quartic in $D(\mathbf{r})$ represent first and second order interactions between a defect or defect cluster at $\mathbf{r}$ and the defects or defect clusters elsewhere in the material. The gradient term $\partial D(\mathbf{r})$ is included to penalize spatial variations of defect density in the medium (the gradient is zero in a uniform medium, indicating that a smooth defect distribution costs the system less free energy).

The basis of statistical mechanics is that the probable state of the system is the lowest free-energy state. The spatial function $D(\mathbf{r})$ must be such that equation (A1) is essentially zero in the presence of small changes in the distribution $D(\mathbf{r})$. This global minimum energy condition is expressed by integrating equation (A1) over the volume of the rock mass V and setting to zero the integral over the free energy subject to small but finite variation in $D(\mathbf{r})$, $\delta D(\mathbf{r})$,

$$0 = \int \mathrm{d}V \delta D(\mathbf{r})[-\sigma(\mathbf{r}) + 2AD(\mathbf{r}) + 4BD(\mathbf{r})^3 - 2C\partial^2 D(\mathbf{r}). \qquad \text{(A2)}$$

Stationarity condition equation (A2) holds for arbitrary $\delta D(\mathbf{r})$ when the integrand is zero, i.e. when the spatial variation of $D(\mathbf{r})$ obeys the equation

$$[2A + 4BD(\mathbf{r})^2 - 2C\partial^2]D(\mathbf{r}) = \sigma(\mathbf{r}). \qquad \text{(A3)}$$

Observations of random systems are usually more easily treated in terms of mean spatial correlations between physical variables than by attempting to construct a complete spatial function $D(\mathbf{r})$. Expressing equation (A3) in terms of the spatial correlation function

$$G(\mathbf{r}, \mathbf{r}') = \langle [D(\mathbf{r}) - \langle D(\mathbf{r})\rangle][(D(\mathbf{r}') - \langle D(r')\rangle]\rangle$$

allows equation (A3) to be written in terms of the defect density fluctuation field $D(\mathbf{r}) - \langle D(\mathbf{r})\rangle$. Stationarity in the presence of arbitrary small changes in the stress field $\sigma(\mathbf{r})$ is given by Kadanoff *et al.* (1976) as

$$[2A + 12B\langle D(\mathbf{r})\rangle^2 - 2C\partial^2]G(\mathbf{r}, \mathbf{r}') \approx \delta(\mathbf{r} - \mathbf{r}'). \qquad \text{(A4)}$$

If the function B is zero and the function $A(P)$ vanishes at a critical defect probability P_c, $A(P) \propto P_c - P$, then equation (A4) has the solution

$$G(\mathbf{r}, \mathbf{r}') \approx \exp[-|\mathbf{r} - \mathbf{r}'|]/\xi)]/|\mathbf{r} - \mathbf{r}'|, \qquad \text{(A5)}$$

where $\xi \propto 1/\sqrt{(P_c - P)}$ as noted in the text. At $P \approx P_c$, ξ becomes arbitrarily large and yields the basic desired observation that rock exhibits arbitrary long range spatial correlations in its physical property distribution.

We make *ad hoc* use of the second order defect interaction term $B\langle D(\mathbf{r})\rangle^2$ to generate a phenomenological local radial dependence on defect interaction energy that gives an additional degree of freedom in the description of the long-range macro scale spatial correlation function. An empirical correlation function with power-law exponent appropriate to borehole logs,

$$G(\mathbf{r}, \mathbf{r}') \approx \exp[-|\mathbf{r} - \mathbf{r}'|]/\xi)]/|\mathbf{r} - \mathbf{r}'|^{1+\eta}, \qquad \text{(A6)}$$

satisfies equation (A4) if the intermediate term $B'\langle D(\mathbf{r})\rangle^2$ has an effective radial dependence,

$$B'\langle D(\mathbf{r})\rangle^2 \approx [\eta/|\mathbf{r} - \mathbf{r}'|][1/\xi + (1+\eta)/|\mathbf{r} - \mathbf{r}'|], \qquad \text{(A7)}$$

where $|\mathbf{r} - \mathbf{r}'|$ is the distance from the observation lattice site $\mathbf{r}$ to an arbitrary correlation point $\mathbf{r}'$. Term (A7) vanishes when $\eta = 0$. When $\eta > 0$, equation (A7) expresses the effect of a general distortion of the local stress field caused by the presence of defects in the vicinity of the observation point $\mathbf{r}$. Near the critical point P_c, $1/\xi$ vanishes so that the radial distortion in interaction energy is proportional to $\eta/|\mathbf{r} - \mathbf{r}'|^2$.

In thermodynamic systems η is assumed to vanish in the absence of an external body force field σ (Kadanoff *et al.* 1967). In the first order theoretical correction η is computed to be small, $\eta \approx 0.1$ (Binney *et al.* 1995). However, in rock, the stress field is never absent, so the intermediate term $B'\langle D(\mathbf{r})\rangle^2$ cannot be arbitrarily ignored, and phenomenological distortion parameter is comparatively large, $0.5 < \eta < 1$, and can vary with stress direction. The stress-related parameter η suggests that a general statistical physics approach to rock structure is suitable for describing the energetics of rock–defect interaction. Connections between the value of η and micro scale rock, fluid and stress conditions can in principle be observed in laboratory experiments. Numerical simulations developed for critical state thermodynamic systems may be adaptable to explore the macro scale consequences of general micro scale rock–fluid interactions.

An inverse problem to determine the piecewise homogeneous hydraulic conductivity within rocks

D. LESNIC[1], L. ELLIOTT[1], D. B. INGHAM[1], R. J. KNIPE[2] & B. CLENNELL[2]

[1] *Department of Applied Mathematical Studies, University of Leeds, Leeds LS2 9JT, UK*
[2] *Department of Earth Sciences, University of Leeds, Leeds LS2 9JT, UK*

Abstract: The experimental measurement of the piecewise homogeneous hydraulic conductivity of a rock sample in which the location of the discontinuities is unknown appears difficult, even in a one-dimensional situation. However, the values of these quantities can be obtained numerically using additional boundary and/or interior measurements of the pressure from transient hydraulic experiments. In the direct problem, the hydraulic conductivity is known, and only a solution for the pressure field is sought. However, in the inverse formulation of the problem, both the hydraulic conductivity and the pressure are unknown and have to be determined using additional pressure measurements. The numerical method employed for solving the diffusion equation is based on the boundary element method as a direct solution procedure, combined with an ordinary least-squares technique. The sensitivity coefficients are calculated for the unspecified boundary conditions and for interior pressures, and clearly show the need for interior measurement information to be imposed on the solution of the inverse ill-posed problem. The uniqueness of the solution of the inverse problem is thoroughly numerically investigated using additional pressure measurements at several prescribed times from various numbers of wells and different well locations. For a material presenting a single discontinuity in the hydraulic conductivity, and subject to certain experimental conditions, it was found that when the number of time measurements is limited then two interior well pressure measurements are necessary and sufficient in order to render a good estimate of the exact solution. Otherwise, it is necessary to increase the number of time measurements at a single interior well location to accomplish the same result.

The problem of accurately quantifying the flow properties of geological discontinuities is central to reservoir simulation and modelling of hydrocarbon reservoirs (King 1994). A large range of heterogeneities created by sediment as well as fault and fracture architectures can cause local flow conduits, baffles or barriers (Pickup *et al.* 1994; Knipe *et al.* 1997; Foley *et al.* 1998). The success of future predictions of fluid flow behaviour will depend upon the ability to upscale from local observations, as well as being able to interpret detailed rock properties and behaviour from large-scale flow measurements. The aim of this paper is to calculate the piecewise homogeneous hydraulic conductivity of rocks, where the locations of the discontinuities are unknown, using additional boundary and/or interior measurements of the pressure in transient hydraulic experiments. This formulation models the presence of geological discontinuities within rocks, such as sedimentary layering and faults, which are defined herein as discontinuity points of the hydraulic conductivity, due to the one-dimensional nature of the problem. Practically, this formulation models a hydraulic experiment in which two or more homogeneous rocks are butted together and the experiment performed over the whole of the linked material, (Fig. 1).

The situation where the location of the discontinuity is unknown, but where hydraulic conductivity is known, has been investigated theoretically by Cannon (1964). The case of unknown piecewise homogeneous hydraulic conductivity, but with known discontinuity location, has been briefly investigated numerically by Carrera & Neuman (1986). However, the simultaneous determination of the piecewise homogeneity of the hydraulic conductivity and the locations of the unknown discontinuities has not been previously described in the literature and therefore this study aims to investigate such a situation. The lack of knowledge of these quantities is compensated for in this analysis by measuring the pressure on the boundary and/or at internal positions; i.e. with simulated wells.

The boundary element method (BEM) introduced below combined with an ordinary, nonlinear least-squares method (LSM) are employed for obtaining an inverse numerical solution. The LSM minimizes the gap between the computed values and the measured readings, subject to physically sound constraints. The effect on the uniqueness of the numerical solution of the number and the location of the wells recording time measurements of the pressure is thoroughly

LESNIC, D., ELLIOTT, L., INGHAM, D. B. *et al.* 1998. An inverse problem to determine the piecewise homogeneous hydraulic conductivity within rocks. *In*: JONES, G., FISHER, Q. J. & KNIPE, R. J. (eds) *Faulting, Fault Sealing and Fluid Flow in Hydrocarbon Reservoirs*. Geological Society, London, Special Publications, **147**, 261–268.

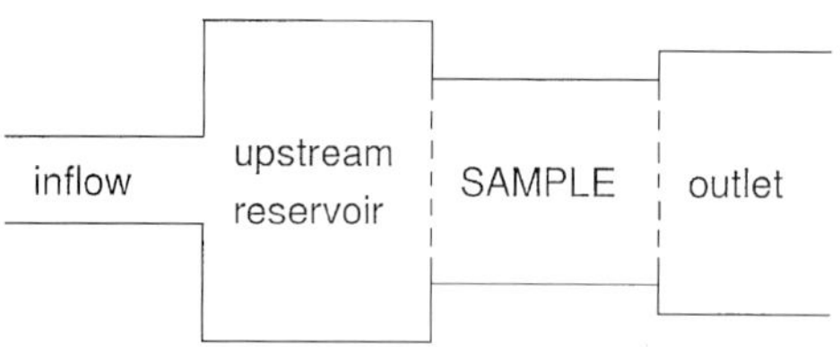

Fig. 1. Schematic diagram showing the hydraulic experiment arrangement of a pump flow test.

investigated. The inverse numerical results obtained using the BEM as a direct solution procedure, combined with the ordinary least-squares approach, are discussed and compared with their exact values for numerically simulated noisy input data.

Experimental set-up

The hydraulic experiment arrangement of a pump flow test is shown in Fig. 1. An inflow of mass is applied at an upstream reservoir which is in hydraulic contact with the rock sample and the induced pressure is measured with a differential transducer connected between the upstream and the downstream face of the sample which is buffered at a constant pressure. After the differential pressure has reached its steady-state, or even before that, the pump may be switched off.

Mathematical model

The flow of fluid of constant compressibility in a saturated, one-dimensional, isotropic, porous medium is governed by the equation (Atkinson *et al.* 1986):

$$S_s \frac{\partial p}{\partial t}(x,t) = \frac{\partial}{\partial x}\left(k(x)\frac{\partial p}{\partial x}(x,t)\right), \quad 0 < x < L, \quad 0 < t < \infty \quad (1)$$

where p is the hydraulic pressure, S_s is the specific storage, $k(x)$ is the hydraulic conductivity and L is the length of the sample. For composite homogeneous rocks the hydraulic conductivity is assumed to be piecewise constant, i.e.

$$k(x) = k_i \quad \text{for} \quad x^s_{i-1} \le x \le x^s_i, \quad i = \overline{1, N_s} \quad (2)$$

where $N_s \ge 2$ is the number of layers of different homogeneous materials butted together in the hydraulic experiment, $x^s_0 = 0$, $x^s_{N_s} = L$ and $0 < x^s_i < L$ for $i = 1, (N_s - 1)$ are the unknown discontinuity locations. As the model is constrained to be one-dimensional, the interfaces of the butted materials are simply the discontinuity points of the hydraulic conductivity.

The initial and boundary conditions are given by (Lesnic *et al.* 1997):

$$p(x,0) = 0 \quad \text{for} \quad 0 \le x \le L \quad (3)$$

$$p(L,t) = 0,$$

$$S_u \frac{\partial p}{\partial t}(0,t) = k(0)A\frac{\partial p}{\partial x}(0,t) + Q_u, \quad 0 < t < \infty \quad (4)$$

where S_u is the compressive storage of the upstream reservoir, A is the cross-sectional area of the sample and Q_u is the constant inflow rate of mass at the entrance of the upstream reservoir which is zero when the pump is off. At the interfaces $x = x^s_i$ for $i = \overline{1, (N_s - 1)}$, the continuity of the pressure and the hydraulic flux is applied, namely,

$$\lim_{x \uparrow x^s_i} p(x,t) = \lim_{x \downarrow x^s_i} p(x,t), \quad 0 \le t < \infty \quad (5)$$

$$\lim_{x \uparrow x^s_i} k_i \frac{\partial p}{\partial x}(x,t) = \lim_{x \downarrow x^s_i} k_{i+1} \frac{\partial p}{\partial x}(x,t), \quad 0 \le t < \infty \quad (6)$$

The inverse problem requires one to determine, in addition to the solution for the pressure field $p(x,t)$, the piecewise homogeneous hydraulic conductivity values k_i for $i = \overline{1, N_s}$ and the unknown discontinuity locations x^s_i for $i = \overline{1, (N_s - 1)}$. The lack of information is compensated for by additional pressure measurements at N_w well locations x_i within the medium $[0, L)$, recorded in time at N_T prescribed instants t_j, namely,

$$p(x_i, t_j) = [p(x_i, t_j)]^{(e)}, \quad i = \overline{1, N_w}, \quad j = \overline{1, N_T} \quad (7)$$

where the symbol (e) stands for simulated known measured data which may be exact or subject to noise.

Boundary element solution of the direct problem

The first step in the inverse analysis which will be finally undertaken is the development of the corresponding direct solution for the problem; i.e. equations (1)–(6), when k_i for $i = \overline{1, N_s}$, and x^s_i for $i = \overline{1, (N_s - 1)}$, are assumed to be known. In this study, the numerical method adopted for solving the direct initial boundary value problem, given by equations (1)–(6), is the boundary element method (BEM).

If;

$$\alpha(x) = \alpha_i = k_i/S_s \quad \text{for} \quad x^s_{i-1} \le x \le x^s_i, \quad i = \overline{1, N_s} \quad (8)$$

is defined to be the piecewise constant hydraulic diffusivity of the medium then the governing time-dependent diffusion equation (1), in which the hydraulic conductivity is given by equation (2) over each zonation $[x^s_{i-1}, x^s_i]$, possesses a fundamental solution, namely,

$$FS_i(x, t; \xi, \tau) = \frac{H(t-\tau)}{(4\pi\alpha_i(t-\tau))^{1/2}} \exp\left(-\frac{(x-\xi)^2}{4\alpha_i(t-\tau)}\right),$$
$$x^s_{i-1} \le x, \xi \le x^s_i, \quad t, \tau > 0, \quad i = \overline{1, N_s} \quad (9)$$

where H is the Heaviside function. Infinite series expressing Green's functions may also be employed, but then computational difficulties may arise with respect to choosing the level of truncation of the series. The use of the fundamental solution given by equation (9) enables one to reformulate the partial differential equation (1) over each zonation $[x^s_{i-1}, x^s_i]$ as a boundary integral equation (Brebbia *et al.* 1984),

$$\eta(x)p(x,t) = \int_0^{t_f} \alpha_i p'(x^s_{i-1}, \tau) FS_i(x, t; x^s_{i-1}, \tau)\, d\tau + \int_0^{t_f} \alpha_i p'(x^s_i, \tau) FS_i(x, t; x^s_i, \tau)\, d\tau - \int_0^{t_f} \alpha_i p(x^s_{i-1}, \tau) FS'_i(x, t; x^s_{i-1}, \tau)\, d\tau - \int_0^{t_f} \alpha_i p(x^s_i, \tau) FS'_i(x, t; x^s_i, \tau)\, d\tau + \int_{x^s_{i-1}}^{x^s_i} p_0(y) FS_i(x, t; y, 0)\, dy,$$
$$(x, t) \in [x^s_{i-1}, x^s_i] \times (0, t_f] \quad (10)$$

where t_f is a final time of interest over which duration the experiment is performed, $\eta(x)$ is a coefficient function equal to 1 if $x \in (x^s_{i-1}, x^s_i)$ and 0.5 if $x \in \{x^s_{i-1}, x^s_i\}$, the prime denotes the differentiation with respect to the outward normal n at the boundaries $x \in \{x^s_{i-1}, x^s_i\}$ and $p_0(y)$ is the initial pressure data which is zero, see equation (3), if the pump is on and which is equal to $p(y, t_f)$ (obtained by first solving the pump-on flow test), when the pump is switched off. Therefore, when the pump is on, the BEM requires the discretization of the boundary space solution domain, $x \in \{0, 1\}$, only, and it fully reduces the dimensionality of the problem by one. When the pump is off, the BEM requires an additional space solution domain discretization at $t = t_f$, but the line integral involved in equation (10) can easily be evaluated numerically without much computational cost.

The BEM discretization of the boundary integral equation (10) is performed by subdividing the time interval $[0, t_f]$ into N equal time intervals $[t_{j-1}, t_j]$, for $j = \overline{1, N}$, and, for simplicity, we assume that the boundary pressure and its normal derivative are constant over each time step and take their values at the midpoint, $\bar{t}_j = (t_{j-1} + t_j)/2$, namely:

$$p(x, t) = p(x^s_i, t)^{\mp} \simeq p(x^s_i, \bar{t}_j)^{\mp} = (p^i_j)^{\mp}, \quad t \in [t_{j-1}, t_j], \quad j = \overline{1, N}, \quad i = \overline{0, N_s} \quad (11)$$

$$p'(x, t) = p'(x^s_i, t)^{\mp} \simeq p'(x^s_i, \bar{t}_j)^{\mp} = (p'^i_j)^{\mp}, \quad t \in [t_{j-1}, t_j], \quad j = \overline{1, N}, \quad i = \overline{0, N_s} \quad (12)$$

where the symbol $\mp$ denotes the limits of the functions involved from the left and right at the interface $x = x^s_i$ for $i = \overline{1, (N_s - 1)}$, respectively. When the pump is switched off we also need the discretization of the space intervals $[x^s_{i-1}, x^s_i]$ into N^i_0 cells, namely, $[y^i_{k-1}, y^i_k]$, for $k = \overline{1, N^i_0}$, $i = \overline{1, N_s}$, and this is accomplished by a constant space cell approximation at $y = \bar{y}^i_k = (y^i_{k-1} + y^i_k)/2$, namely:

$$p_0(y) = p(y, t_f) \simeq p(\bar{y}^i_k, t_f) = p^i_{0k}, \quad y \in [y^i_{k-1}, y^i_k], \quad k = \overline{1, N^i_0}, \quad i = \overline{1, N_s} \quad (13)$$

Based on the constant BEM approximations as given by equations (11)–(13), the discretization of the boundary integral equation (10) results in:

$$\eta(x)p(x,t) = \sum_{j=1}^{N} \left[\alpha_i p'^{i-1}_j \int_{t_{j-1}}^{t_j} FS_i(x, t; x^s_{i-1}, \tau)\, d\tau + \alpha_i p'^i_j \int_{t_{j-1}}^{t_j} FS_i(x, t; x^s_i, \tau)\, d\tau - \alpha_i p^{i-1}_j \int_{t_{j-1}}^{t_j} FS'_i(x, t; x^s_{i-1}, \tau)\, d\tau - \alpha_i p^i_j \int_{t_{j-1}}^{t_j} FS'_i(x, t; x^s_i, \tau)\, d\tau \right] + \sum_{k=1}^{N^i_0} p^i_{0k} \int_{y^i_{k-1}}^{y^i_k} FS_i(x, t; y, 0)\, dy,$$
$$(x, t) \in [x^s_{i-1}, x^s_i] \times (0, t_f], \quad i = \overline{1, N_s} \quad (14)$$

where, in the first instance, the symbol $\mp$ has been dropped for simplicity. The integrals that appear in equation (14) can be evaluated analytically (Wrobel, 1983). Letting x in equation (14) tend to the interface locations x_i^s for $i = \overline{0, N_s}$, we obtain a system of linear equations involving boundary and interface unknowns only. This system of equations is completed with the boundary conditions (4)–(6), which in discretized form yield,

$$p_j^{N_s} = 0, \qquad j = \overline{1, N} \tag{15}$$

$$S_u\left(\frac{p_1^0 - p_0(0)}{t_f/N}\right) + k_1 A p_1'^0 = Q_u \tag{16}$$

$$S_u\left(\frac{p_{j+1}^0 - p_{j-1}^0}{2t_f/N}\right) + k_1 A p_j'^0 = Q_u, \qquad j = \overline{2, (N-1)} \tag{17}$$

$$S_u\left(\frac{p_N^0 - p_{N-1}^0}{t_f/N}\right) + k_1 A p_N'^0 = Q_u \tag{18}$$

$$(p_j^i)^- = (p_j^i)^+, \qquad i = \overline{1, (N_s - 1)} \tag{19}$$

$$k_i (p_j'^i)^- = -k_{i+1}(p_j'^i)^+, \qquad i = \overline{1, (N_s - 1)} \tag{20}$$

where $p_0(0) = 0$ if the pump is on and $p_0(0) = p_N^0$ (an already calculated value) if the pump is off.

Sensitivity coefficients

Sensitivity coefficients are the first derivatives of the pressure at well measurement locations x_i for $i = \overline{1, N_w}$, with respect to the unknowns, namely, k_i for $i = \overline{1, N_s}$, and x_i^s for $i = \overline{1, (N_s - 1)}$. They provide indicators of how well-designed the experiment is. In general, the sensitivity coefficients are desired to be large and uncorrelated, i.e. linearly independent (Beck and Arnold 1977). A sense of the magnitude of the sensitivity coefficients is gained through normalisation by multiplying them with their corresponding unknown differentiation variable, resulting in units of pressure for the normalized sensitivity coefficients. The degree of uncorrelation of these coefficients can then be illustrated by the departure of their ratios from a constant function. Based on this criterion, we can determine optimal well locations and time instant measurements to be imposed or recorded in equation (7) in order to reduce the ill-posedness of the inverse formulated problem. Therefore, a study of the sensitivity coefficients, prior to performing experiments, can lead to better experimental designs.

The (normalized) sensitivity coefficients when $N_s = 2$ are defined by

$$S(x_1^s; x, t) = x_1^s \frac{\partial p}{\partial x_1^s}, \quad S(k_1; x, t) = k_1 \frac{\partial p}{\partial k_1}, \quad S(k_2; x, t) = k_2 \frac{\partial p}{\partial k_2} \tag{21}$$

A much enhanced characterisation of the degree of correlation of the sensitivity coefficients can be viewed by plotting, as shown in Fig. 2, the ratios between the sensitivity coefficients, namely:

$$R_1(x, t) = S(x_1^s)/S(k_1), \quad R_2(x, t) = S(k_1)/S(k_2), \quad R_3(x, t) = S(k_2)/S(x_1^s). \tag{22}$$

From Fig. 2a it can be seen that, except for very short times, the sensitivity coefficients for the pressure at $x = 0$ possess two degrees of correlation, namely:

$$R_1(0, t) \approx 1; \quad R_2(0, t) \approx \tfrac{2}{3}; \quad R_3(0, t) \approx \tfrac{3}{2} \tag{23}$$

Therefore, prior to performing any inverse analysis, one may conclude from equation (23) that if additional pressure measurements, imposed by equation (7), are recorded only at $x = 0$ then only one of the three unknowns x_1^s, k_1 or k_2 can be identified, whilst assuming the other two parameters to be known.

Based on the sensitivity analysis corresponding to the information supplied by boundary pressure measurements only, it is concluded, and further numerically shown in Table 1, that interior well measurements are necessary in order to reduce the non-uniqueness of the ill-posed problem.

The ratios $R_i(x, t)$ for $i \in \{1, 2, 3\}$, as a function of time, between the sensitivity coefficients for the pressure taken at the interior well locations $x = x_1^s/2$ and $x = 3x_1^s/2$ plotted in Figs 2b & 2d, respectively, show single degrees of correlations, namely:

$$R_3\left(\frac{x_1^s}{2}, t\right) \approx \frac{3}{2}; \quad R_1\left(\frac{3x_1^s}{2}, t\right) \approx 1 \tag{24}$$

Therefore, one may conclude from equation (24) that both x_1^s and k_2 cannot be identified simultaneously from only pressure measurements at $x = x_1^s/2$ imposed in equation (7), whilst from only pressure measurements at $x = 3x_1^s/2$ imposed in equation (7) we cannot identify, simultaneously, both x_1^s and k_1. Therefore, in such situations, in general, one needs to prescribe one of the unknowns.

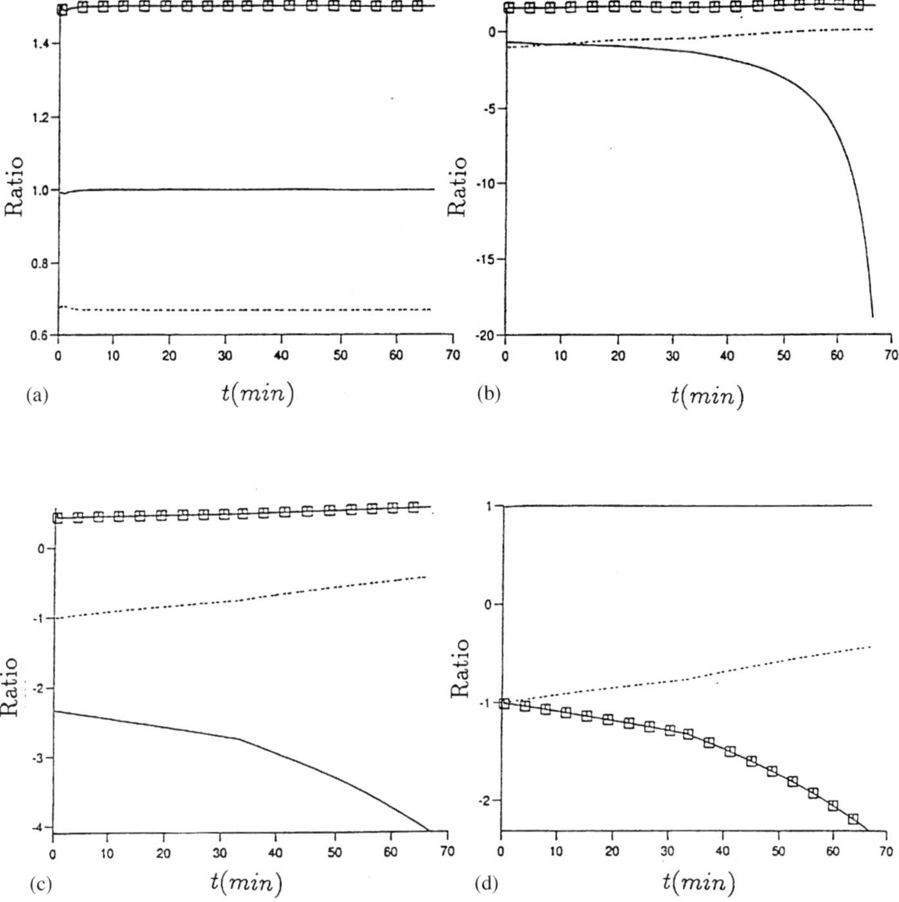

Fig. 2. The ratios $R_1(x,t)$ (——), $R_2(x,t)$ (– – –) and $R_3(x,t)$ (–□–), of the sensitivity coefficients, as a function of time, at the space locations, **(a)** $x = 0$; **(b)** $x = 10L/35 = x_1^s/2$; **(c)** $x = 20L/35 = x_1^s$; **(d)** $x = 30L/35 = 3x_1^s/2$.

In order to demonstrate the usefulness of the sensitivity analysis, although practically unavailable for the inverse problem, Fig. 2c shows the ratios of the sensitivity coefficients for the pressure measurements when the well is located at the unknown discontinuity, $x = x_1^s$. From Fig. 2c it can be observed that there is no degree of correlation between the sensitivity coefficients, although because of the distorted scale used for the y-axis it may appear that $S(x_1^s; x_1^s, t)$ and

Table 1. *The retrieval of a piecewise homogeneous hydraulic conductivity and its discontinuity location for $p\% = 1\%$ noisy data when $N_w \in \{1, 2, 3\}$ and $N_T = 3$*

BEM	x_1^s (cm)	$k_1 \times 10^7$ (cm min^{-1})	$k_2 \times 10^7$ (cm min^{-1})	LS (MPa)	Iter
Exact	1.95428	8.88000	4.44000	0	0
$\lambda_1 = 1$	2.26652	5.99770	5.99572	0.36594	50
$\lambda_2 = 1$	2.02447	9.07363	4.26596	0.10E – 8	312
$\lambda_3 = 1$	1.67942	10.0016	4.15429	0.12E – 2	50
$\lambda_1 = \lambda_2 = 1$	2.16605	8.71372	4.52733	0.36870	312
$\lambda_2 = \lambda_3 = 1$	1.92733	9.19323	4.46385	0.17E – 3	312
$\lambda_1 = \lambda_3 = 1$	2.07543	8.12811	4.26855	0.36782	47
$\lambda_1 = \lambda_2 = \lambda_3 = 1$	1.95169	8.66375	4.25567	0.40819	462

$S(k_2; x_1^s, t)$ are slightly correlated. However, one may conclude from Fig. 2c that if it is possible to obtain a measurement 'near' the discontinuity then this will produce a unique solution for the inverse problem. The term 'near' should be understood as implying that a prior estimate, x_1^{s*}, of the discontinuity location x_1^s is known. In many practical situations, this prior estimate can be supplied by the experimentalist and, in the inverse problem treated in this section, a large tolerance for the prior estimate of 50% error, is allowed. If smaller errors for the prior estimate are available from the experimentalist, then the degree of ill-posed problem is reduced. In addition, this prior estimate may be then used in other experiments in order to improve the convexity of the least-squares minimization functional.

So far, the sensitivity analysis performed in this subsection highlights the high level of ill-posedness of the problem that one has formulated when the number of well instalments assumed available is restricted to $N_w = 1$. However, when $N_w \in \{2, 3\}$, as will be shown numerically later, it might be possible to gather useful information supplied by two or three well measurements from the assumed available set of sensor locations $W = \{0, x_1^s/2, 3x_1^s/2\}$.

The least-squares minimization technique

In the inverse problem given by equations (1)–(7) in which $\underline{k} = (k_i)$ for $i = \overline{1, N_s}$, and $\underline{x}^s = (x_i^s)$ for $i = \overline{1, (N_s - 1)}$, are unknown, additional pressure measurements, as given by equation (7), are imposed. Then we can compare the measured, (e), and the BEM computed, (c), data in equation (7) in the least-squares sense, namely:

$$LS(\underline{k}, \underline{x}^s) = \sum_{i=1}^{N_w} \sum_{j=1}^{N_T} \lambda_i |p(x_i, t_j'; \underline{k}, \underline{x}^s)^{(e)} - p(x_i, t_j'; \underline{k}, \underline{x}^s)^{(c)}|^2 \quad (25)$$

where the parameters λ_i are set to be 0 or 1 according to whether or not we measure in time at the well locations, $x = x_i$, the multiplying quantities, i.e. the pressure values $p(x_i, t_j'; \underline{k}, \underline{x}^s)^{(e)}$. The variables λ_i were included in equation (25) only for a convenient presentation of some later results (Tables 1 and 2). The least-squares functional, given by equation (25), is minimized over some set of constraints, namely:

$$M = \{(\underline{k}, \underline{x}^s) | \underline{bu}^k > \underline{k} > \underline{bl}^k, \underline{bu}^s > \underline{x}^s > \underline{bl}^s, k_i \neq k_j, \forall i \neq j\} \quad (26)$$

where $\underline{bu}^k$, $\underline{bu}^s$, $\underline{bl}^k$ and $\underline{bl}^s$ are vectors of upper and lower bounds, respectively, to which the physical variables are subjected. Therefore, one has to deal with the constraint minimization problem described by the following expression:

$$(\underline{k}, \underline{x}^s)_{\text{optimal}} = \min_{(\underline{k}, \underline{x}^s) \in M} \{LS(\underline{k}, \underline{x}^s)\} \quad (27)$$

which is numerically solved using the NAG routine E04UCF (Gill *et al.* 1986). This routine is designed to minimize an arbitrary smooth function subject to certain constraints which may include simple bounds on the variables and linear and non-linear constraints. In summary, the method of E04UCF first determines a point that satisfies the bounds and linear constraints. Thereafter, each iteration includes:

Table 2. *The retrieval of a piecewise homogeneous hydraulic conductivity and its discontinuity location for* $p\% = 1\%$ *noisy data when* $\lambda_2 = 1$ *and* $N_T \in \{3, 4, 5, 6, 7, 8\}$ *and the pump is on and on and off*

BEM	x_1^s (cm)	$k_1 \times 10^7$ (cm min^{-1})	$k_2 \times 10^7$ (cm min^{-1})	LS (MPa)	Iter
Exact	1.95428	8.88000	4.44000	0	0
$N_T = 3$ (on)	2.02447	9.07363	4.26596	0.10E − 8	312
(on and off)	1.85337	9.07478	4.52840	0.83E − 8	283
$N_T = 4$ (on)	2.03323	9.07396	4.25186	0.27E − 8	272
(on and off)	1.84759	9.07471	4.53674	0.19E − 7	242
$N_T = 5$ (on)	2.04002	9.07415	4.24084	0.51E − 8	312
(on and off)	1.84969	9.07475	4.53370	0.20E − 7	218
$N_T = 6$ (on)	2.03432	9.07431	4.25020	0.65E − 8	162
(on and off)	1.85762	9.07473	4.52222	0.18E − 7	199
$N_T = 7$ (on)	2.06037	9.07432	4.20735	0.12E − 7	174
(on and off)	1.87048	9.07471	4.50347	0.13E − 7	189
$N_T = 8$ (on)	2.08003	9.07435	4.17453	0.20E − 7	404
(on and off)	1.88889	9.07470	4.47635	0.65E − 8	183

(a) the solution of a quadratic programming subproblem;
(b) a line search with an augmented Lagrangian merit function;
(c) a quasi-Newton update of the approximate Hessian of the Lagrangian function.

Finally, it should be noted that with prior estimates k_i^* and x_i^{s*} for the exact values k_i and x_i^s available, together with known values of their standard deviations σ_{k_i} and $\sigma_{x_i^s}$, respectively, then investigations have been undertaken by minimizing the modified least-squares functional given by equation (25) (Carrera & Neuman, 1986), namely:

$$(\underline{k}, \underline{x}^s)_{\text{optimal}} = \min_{(\underline{k}, \underline{x}^s) \in M} \left\{ LS(\underline{k}, \underline{x}^s) + \sum_{i=1}^{N_s} \frac{(k_i - k_i^*)^2}{\sigma_{k_i}^2} + \sum_{i=1}^{N_s - 1} \frac{(x_i^s - x_i^{s*})^2}{\sigma_{x_i^s}^2} \right\} \quad (28)$$

but these *a priori* additional assumptions did not improve significantly on the uniqueness of the inverse numerical solution of the ill-posed problem.

Numerical results and discussion

The numerical analysis performed in the previous sections has been tested for a rock material presenting a single fault, i.e. $N_s = 2$. The multiple faults situation is deferred to a future study.

Table 1 shows the recovery of the unknowns x_1^s, k_1 and k_2 for a piecewise homogeneous hydraulic conductivity material presenting a single discontinuity when $p\% = 1\%$ noisy pressure measurements are used. Good estimates of the exact solution are shown in Table 1 when the inverse problem is identifiable; i.e. when $N_w \geq 2$. In such situations, the relative errors between the exact and numerically obtained values are reasonably small and comparable to the amount of noise $p\%$. Also, all the results obtained for a single well case, i.e. $N_w = 1$, are improved when more measurement information is gathered (see the cases corresponding to $N_w \in \{2, 3\}$ in Table 1). However, it should be noted that the best stable estimate of the solution is obtained by measuring the pressure at $N_w = 2$, interior wells located at $x_1 = x_1^s/2$ and $x_2 = 3x_1^s/2$. The amount of redundant noisy data which may have been included in the overall number of imposed conditions in equation (7) by increasing to $N_w \times N_T \in \{6, 9\}$, when $N_w \in \{2, 3\}$, is negligible in comparison with both the extra information supplied and the reduction in the degree of the ill-posedness of the problem.

Next, we investigate the effect on the numerical inverse solution of increasing the number of time measurements N_T. Table 2 shows the recovery of the unknowns x_1^s, k_1 and k_2, when $p\% = 1\%$ noisy measurements, respectively, of the pressure at the single well location $x_1 = x_1^s/2$, are recorded at various $N_T \in \{3, 4, 5, 6, 7, 8\}$ times when the pump is on and on and off. As expected from Fig. 2b, in Table 2 only the value of k_1 can be retrieved very accurately. However, there is a significant improvement in the prediction of the values of x_1^s and k_2 when the number of time measurements increases from $N_T = 3$ to the interval 4 to 8. When the pump is on, from Table 2 it can be seen that there is little difference between the results obtained when N_T increases from 4 to 8 and therefore in this case it can be concluded that $N_T = 4$ is the optimal minimal number of time measurements. Better estimates of the exact values can be obtained by recording pressure measurements during the time when the pump is in both the on and off settings. In this case, from Table 2 it can be seen that this measurement information significantly improves the results, with the best accuracy obtained when N_T increases to about 7 or 8 time measurements.

Overall, from Table 2 it can be concluded that the measurement at the single well location $x_1 = x_1^s/2$ can be used successfully for retrieving the unknowns, provided that about $N_T \in \{7, 8\}$ time measurements when the pump is both on and off are imposed in equation (7). Furthermore, the inclusion of noise in the data did not produce any large or oscillatory deviations of the numerical results from their exact values, showing that the numerical solution is also stable.

Conclusions

Overall, from Tables 1 and 2 it can be concluded that for the single fault case investigated in this paper, in general, $N_w = 2$ interior well measurements located on each side of the fault are needed to be imposed for an accurate and stable estimation of the unknown discontinuity location x_1^s and the composite hydraulic conductivities k_1 and k_2, when the number of time measurements is limited to $N_T = 3$. In contrast, if the number of wells N_w is limited to one, then further improvement in retrieving the unknowns may be achieved by increasing the number of time measurements to about $N_T \in \{7, 8\}$ and also by recording pressure data when the pump is both on and off. This improvement is significantly increased when the single well is located at the left-hand side of the fault in the region where the hydraulic-contact boundary condition (4) is applied.

In addition some preliminary studies of pump flow tests on rock samples presenting multiple faults, i.e. $N_s > 2$, indicate that, in general, $N_w \in \{(N_s - 1), N_s\}$ interior well pressure measurements recorded at $N_T \geq (2N_s - 1)$ instants need to be inverted in order to ensure an identifiable and good retrieval of the piecewise homogeneous hydraulic conductivity and the locations of its discontinuities.

The authors wish to express their thanks to the referees A. Maltman and I. Main for their help in producing this paper.

References

ATKINSON, C., HAMMOND, P. S., SHEPPARD, M. & SOBEY, I. J. 1986. Some mathematical problems from the oil service industry, *In*: *Proceedings of the London Mathematical Society Symposium*. KNOPS, R. J. & LACEY, A. A. (eds) Cambridge University Press, Cambridge, UK.

BECK, J. V. & ARNOLD, K. J. 1977. *Parameter Estimation in Engineering and Science*, John Wiley, New York, USA.

BREBBIA, C. A., TELLES, J. C. F. & WROBEL, L. C. 1984. *Boundary Element Techniques: Theory and Application in Engineering*. Springer-Verlag, Berlin, Germany.

CANNON, J. R. 1964. Determination of certain parameters in heat conduction problems. *Journal of Mathematical Analysis and its Applications*, **8**, 188-201.

CARRERA, J. & NEUMAN, S. P. 1986. Estimation of aquifer under transient and steady state conditions: 2. Uniqueness, stability, and solution algorithms. *Water Resources Research*, **22**, 211-227.

FOLEY, L., DALTABAN, T. S. & WANG, J. T. 1998. Numerical simulation of fluid flow in complex faulted regions. COWARD, M. P., DALTABAN, T. S. & JOHNSON, H. (eds) *In*: *Structural Geology in Reservoir Characterization* Geological Society, London, Special Publications, London, **127**, 121–132.

GILL, P. E., HAMMARLING, S. J., MURRAY, W., SAUNDERS, M. A. & WRIGHT, M. H. 1986. *User's Guide for LSSOL*, (Version 1.0), Report SOL 86-1, Stanford University.

KING, P. R. 1994. Rescaling of flow parameters using normalisation. *North Sea Oil and Gas Reservoirs*, **III**, 265–271.

KNIPE, R. J., FISHER, Q. J., JONES, G., CLENNELL, B., FARMER, A. B., HARRISON, A., KIDD, B., MCALLISTER, E., PORTER, J. R., & WHITE, E. A. 1997. Fault seal analysis: successful methodologies, application and future directions. MOLLER-PEDERSEN, P. & KOESTLER, A. G. (eds) *Hydrocarbon Seals Importance for Exploration and Production*, Norwegian Petroleum Society (NPF), Special Publications, **7**, Elsevier, Singapore, 15–40.

LESNIC, D., ELLIOTT, L., INGHAM, D. B., CLENNELL, B. & KNIPE, R. J. 1997. A mathematical model and numerical investigation for determining the hydraulic conductivity of rocks. *International Journal of Rock Mechanics and Mining Sciences* **34**, 741–759.

PICKUP, G. E., RINGROSE, P. S., JENSEN, J. L. & SORBIE, K. S. 1994. Permeability tensors for sedimentary structures. *Mathematical Geology* **26**, 227–250.

WROBEL, L. C. 1983. A boundary element solution to Stefan's problem. *In*: BREBBIA, C. A., FUTAGAMI, T. & TANAKA, M. (eds) *Boundary Elements V*. Springer-Verlag, Berlin, Germany, 173–182.

The impact of faults on fluid flow in the Heidrun Field

T. A. KNAI[1] & R. J. KNIPE[2]

[1] *Statoil, Heidrun PETEK, PO Box 273, N-7501 Stjordal, Norway*
[2] *Rock Deformation Research, Leeds University, Leeds LS2 9JT, UK*

Abstract: A quantitative analysis of fault seal properties has been used in reservoir simulation modelling in the Heidrun Field. The inclusion of microstructural and petrophysical data from core analysis has resulted in a better prediction of reservoir communication and drainage patterns during production. Three main fault rock types were observed in the Heidrun cores, namely; *cataclasites* developed from clean sandstones, *phyllosilicate framework fault rocks*, created from impure sandstones and *clay smears*. The clay content of the host sediment is the controlling factor in determining which fault rock type dominates in the fault zones. Fault plane geometries were assessed from seismic-based juxtaposition analysis for input into the reservoir simulation model. The fault planes were divided into areas following the reservoir zonation, and transmissibility multipliers were assigned to each subarea of the fault planes in the model. The transmissibility multipliers are a function of fault rock permeability and fault rock width, as well as the matrix (host rock) permeability and the dimensions of the grid blocks used in the simulation model. Introducing a quantitative description of the faults has had a significant effect on the results of the reservoir simulation runs and has played an important role in the successful modelling and prediction of the observed gas breakthrough and pressure evolution. After actively using the transmissibilities during history matching of the simulation model, a remarkably good match to the observed well rates and pressures (flowing and shut-in) for the Jurassic Fangst and Upper Tilje reservoir units was achieved.

Analysis of the impact of faulting on hydrocarbon flow is now recognized as an important component of reservoir management (Bouvier *et al.* 1989; Jev *et al.* 1993; Møller-Pederson & Koestler 1997). The success of future prediction will depend upon testing and validation of the different methodologies being used for the evaluation of faults as retarders to flow in hydrocarbon reservoirs (Berg 1975; Schowalter 1979; Knipe 1992; Knott 1993; Knipe *et al.* 1997; Yielding *et al.* 1997). It is essential to determine what input parameters are critical and what level of quantification is required for the successful prediction of the role of faults in flow patterns. This short paper describes an example of fault analysis from the Heidrun field in the Norwegian Sea where testing has been possible.

The fluid system of the Heidrun field is heterogeneous with the interpretation of pressure data, PVT analysis and geochemical information indicating different fluid contacts and types within the different reservoirs and fault compartments present. Heum (1996) describes the general sealing and fluid system in Heidrun and concludes that the minor faults in the field cause a hydraulic resistance of approximately 2 bars. Welbon *et al.* (1997) have assessed the hydrocarbon migration and conducted a regional fault seal and pressure analysis. These authors recognized the possibility of dynamic fault seals in the Heidrun area.

Eight different OWCs have been mapped in Heidrun, with ~85 m difference between the deepest and shallowest contacts. These data clearly indicate the potential role of faults as flow retarders or barriers and a primary objective of the study reported here was to quantify the impact of faults on flow. The paper describes inclusion of a quantitative fault description, obtained from the detailed analysis of fault zones and fault rock properties based on samples recovered during coring, into a reservoir simulation model and the testing of this model against gas breakthrough results and down-hole pressure data.

The paper is divided into four more sections which:

(a) review the background to the geology of the Heidrun field;
(b) present the results of fault zone property analysis in the field;
(c) outline the basis of the reservoir models constructed; and
(d) compare the observed reservoir behaviour with the model results.

Heidrun background geology and reservoir description

The Heidrun Field was discovered in 1985 in block 6507/7 in the Haltenbank area, *c.* 200 km

KNAI, T. A. & KNIPE, R. J. 1998. The impact of faults on fluid flow in the Heidrun Field. *In*: JONES, G., FISHER, Q. J. & KNIPE, R. J. (eds) *Faulting, Fault Sealing and Fluid Flow in Hydrocarbon Reservoirs*. Geological Society, London, Special Publications, **147**, 269–282.

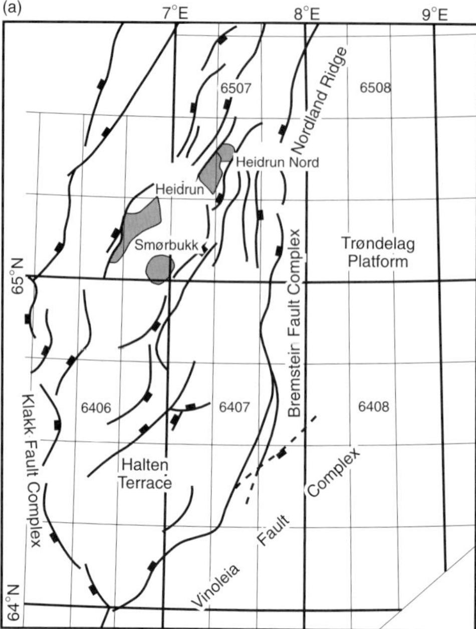

Fig. 1. (**a**) Location of the Heidrun Field, offshore Norway (modified fromWelbon *et al.* 1997).

from the Norwegian coast (Fig. 1a). The hydrocarbons (acidic oil and free gas) are present in three reservoirs of Jurassic age; the high quality Fangst Group and the more heterogeneous Tilje and Åre Formations (Hemmens *et al.* 1994; Reid *et al.* 1996; Welbon *et al.* 1997). Field production started in October 1995 and along with water injection, regular injection of gas into the Fangst gas cap has occurred since December 1995. The field drains a large area and is linked to the same hydrocarbon migration system as Smørbukk, Smørbukk South and Trestakk (Heum *et al.* 1986). In spite of well rates above 5000 Sm 3 day^{-1}, the Fangst wells produce with drawdown pressures of only 10–20 bars. Consequently, there is limited drive (differential pressure) to move fluid across flow retarders or seals. Faults with moderate permeability reductions and threshold pressures were therefore expected to influence the drainage pattern on Heidrun. This provided an important justification for understanding the impact of faulting on well behaviour in Heidrun.

The Heidrun Field is positioned at the transition between the Halten Terrace and the southwest trending Nordland ridge. The structure is dominated by a highly faulted and tilted fault block created during the Late Jurassic–Early Cretaceous extension (Schmidt 1993; Hemmens *et al.* 1994; Koch and Heum 1995). Three main fault trends are present in the field: NNE–SSW, ESE–WNW and N–S (Fig. 1b). Most faults have throws in the range of <30 m, although fault throws up to ~80 m are present. Seismic resolution allows the detection of faults with throws down to ~10 m (Reid *et al.* 1996). The top seal on the field is provided by the Viking Group Upper Jurassic shales and the Upper Cretaceous Springar Formation shales.

The sedimentological and reservoir properties of the Heidrun Field are reviewed in Fig. 2. A brief overview of the character of the main reservoirs is presented below. A more detailed account is presented in Whitley (1992), Hemmens *et al.* (1994) and Reid *et al.* (1996).

The Fangst Group is up to ~100 m thick and can be divided into three units. The Lower Fangst Group, equivalent to the Ile Formation was deposited in a shallow marine setting, and is approximately 50 m thick with permeabilities ranging from 10 to 5000 mD. The Upper Fangst Group is composed of a basal shallow marine deposit between 1 m and ~20 m which contains a thin (<2 m) claystone at the base equivalent to the Not Formation. The upper section of the Upper Fangst Group is a fluvially dominated sequence with excellent reservoir qualities (permeabilities 7–10,000 mD) and is ~13 m to 30 m thick.

The Tilje Formation is a shallow marine to paralic deposit up to ~130 m thick. The formation is heterogeneous and contains clay drapes which are known to be vertical flow barriers. The Tilje Formation can be divided into five zones and 12 sub-zones. The best quality reservoir unit within this formation is Tilje 3, which is dominated by tidal channel deposits, with permeabilities between 200–1500 mD.

The Åre Formation is divided into two units (Åre 1 and 2), with differing lithologies and reservoir properties. The fluvial to deltaic Åre 1 unit is greater than 300 m thick and composed of valley fills, channels and flood plane deposits. The Åre 2 member is a shallow marine deposit, with tidal influences increasing up the section.

Fault analysis in Heidrun

The overall objective of the fault seal studies conducted was to provide a quantitative description of the fault seal properties for the Heidrun Field reservoir simulation models and to improve prediction of reservoir communication and drainage pattern during production. The more detailed aims were to:

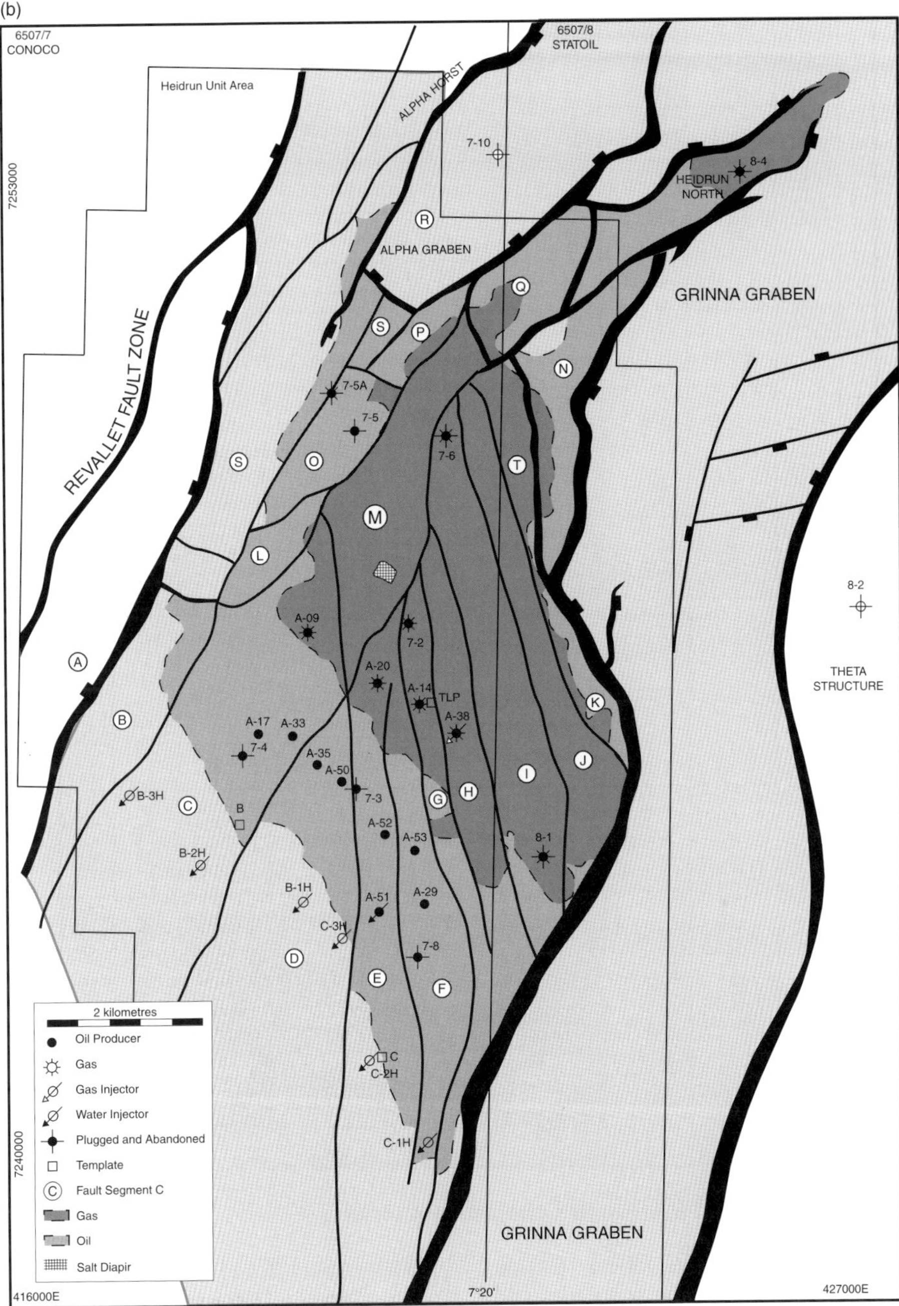

Fig. 1. (**b**) Structure of the Heidrun Field. The main fault trends (NNE–SSW, ESE–WNW and N–S) together with the distribution of hydrocarbons, wells and fault compartments are shown.

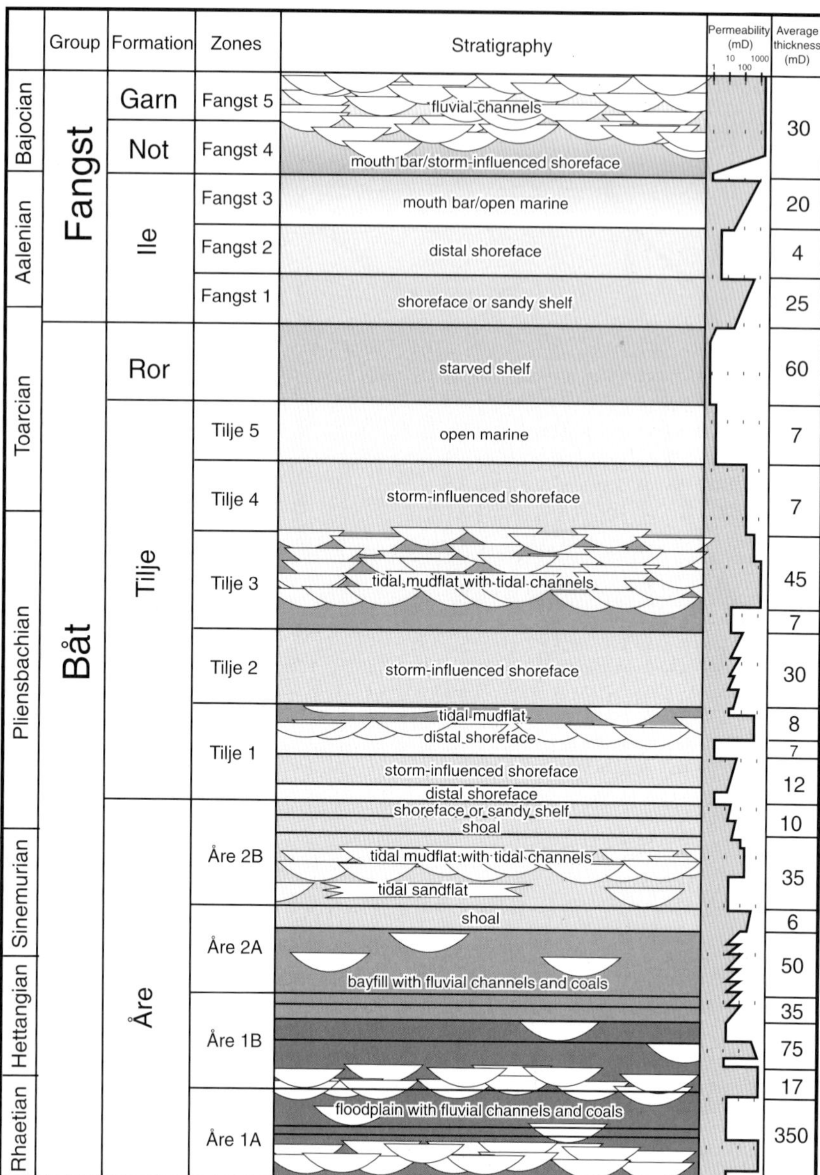

Fig. 2. Review of the Heidrun Field reservoir sedimentology and stratigraphy.

(a) Establish a fault seal model for the Heidrun Field, i.e. to determine the properties and distribution of the different fault seal types;
(b) Implement effective threshold pressures and permeabilities for fluid flow across the significant faults in the reservoir simulation model;
(c) To provide a fault description which could be adjusted in a geologically robust way as the model was history matched.

A fault can act as a restriction to fluid flow by creating either a juxtaposition seal (i.e. reservoir against non-reservoir rock across a fault), or a fault rock seal which is a zone of deformed, low permeability rock along fault zones (Knott 1993; Gibson 1994). Juxtaposition seals are traditionally handled well in reservoir simulation models, since the non-reservoir zones are assigned low permeabilities or set as inactive. The fault rock seals can have an equally important effect on the fluid flow, and the challenge of the current study was to provide input to model the fault seals in a robust way (Knipe *et al.* 1997).

It is important to recognize that a fault may not be complete or open to fluid flow in an absolute sense (Berg 1975; Schowalter 1979). A fault can be sealing for pressure differences up to a certain value (threshold pressure) across the fault plane, and fluid will start flowing only when this threshold pressure is exceeded. The flow rate is a function of the differential pressure across the fault, as well as the permeability and width of the deformation zone. It is the combination of the threshold pressure and the fault rock permeability and width which controls the sealing or retardation capacity of a fault.

The sealing capacity of a fault is not constant over the whole area of a fault plane or within the volume defined by a fault damage zone. The fault rock, and hence the sealing capacity, will typically vary as the juxtaposed stratigraphy changes laterally and vertically along a fault plane. In addition, the distribution of retardation effects will depend upon the history of juxtaposition (displacement) for different points, and on the connectivity of the structures within the fault damage zone (Knipe 1997; Yielding *et al.* 1997; Knipe *et al.* 1997). The same fault can thus have a high sealing capacity in some areas and be open to fluid flow elsewhere.

Fault rock properties and fault zone structures in Heidrun

A microstructural and physical property analysis of fault rock from cores was conducted to assess the deformation mechanisms and to quantify the petrophysical properties of Heidrun faults. Figure 3 presents a review of the core scale deformation features present in Heidrun. The Heidrun Field fault rock database is summarized in Table 1. Three main fault rock types were observed in the Heidrun cores, namely cataclasites (developed from clean sandstones), phyllosilicate framework fault rocks, (created from impure sandstones) and clay smears. The phyllosilicate framework fault rock was further grouped into two categories, based on their clay contents (Table 1). A more detailed account of the microstructure and evolution of these different fault rock types is presented in Fisher & Knipe (this volume).

Basis of the reservoir model input

The reservoir zonations used in the fault seal analysis reported here are reviewed in Table 2 and are simplified compared to the reservoir division in the simulation model, which contains up to 29 layers. This simplification was necessary for the fault seal model, to reduce the labour involved in defining and describing faults in the reservoir simulation model. The main reservoir units were treated as single zones (i.e. F5–F4A, T4–T3A) with the significant stratigraphic barriers (i.e. F4A, T3A) defining the zone boundaries. Some lower quality reservoir zones in Åre were grouped together, while individual cleaner sandstone units (i.e. T1C, Å1B1) were kept separate, as they have the potential to form leaky points on the fault plane. The resulting zonation is shown in the far left column of Table 2. The average undeformed sediment permeabilities listed are arithmetic mean values from the reservoir simulation model, i.e. arithmetic means summed over the whole area of the reservoir simulation grids for each zone.

The different fault rock types were grouped into the four categories and a representative permeability was assigned to each category (Table 1). Clay content of the host sediment appears to be the key factor controlling which fault rock dominates within the fault zone. An important observation from the fault rock analysis on Heidrun cores was that layers with phyllosilicate contents in the range 10–50% did not always form continuous clay smears, even when faulted by small offsets (compare with Lindsay *et al.* 1998). Consequently, it was decided to only take into account the composition of the juxtaposed reservoir units, when deciding which fault rock type was controlling the sealing capacity in a particular area of a fault plane. By doing so, holes or discontinuities in any clay smears created when the hanging wall stratigraphy is moved past this area during fault movements is accounted for. Only the Ror Formation (Fig. 2) with an estimated clay content of 57% is expected to form continuous clay smears close to the hanging wall and footwall cutoffs. We used a conservative estimate of the clay smearing, where the seal was assumed to be effectively continuous for a distance up and down the fault plane equal to the clay unit thickness. Table 3 shows the fault rock types and permeabilities which are expected to control the sealing capacity of a fault plane in Heidrun, for each combination of juxtaposed reservoir units. Figure 4 illustrates these properties on a juxtaposition/fault seal diagram (Knipe 1997) constructed for one of the wells on the field. The figure shows the size and location of high permeability windows expected along faults in the Heidrun reservoir. Figure 4 also highlights the impact of smearing of the Ror Formation on

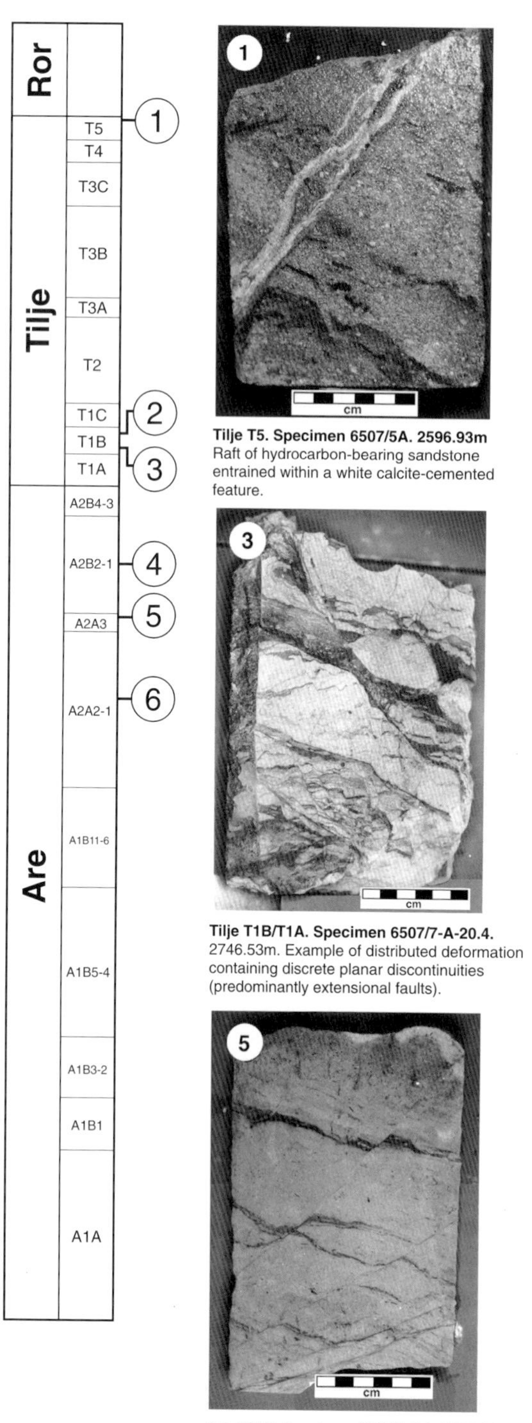

Tilje T5. Specimen 6507/5A. 2596.93m
Raft of hydrocarbon-bearing sandstone entrained within a white calcite-cemented feature.

Tilje T1B. Specimen 6507/7-A-20.1. 2741.71m
Set of anastomosing, predominantly extensional faults which are a mixture of phyllosilicate smear and disaggregtion zone.

Tilje T1B/T1A. Specimen 6507/7-A-20.4. 2746.53m. Example of distributed deformation containing discrete planar discontinuities (predominantly extensional faults).

Are A2B2-1. Specimen 6507/7-3.2. 2642.74m
Laminated sandstone with abundant normal faults with displacement magnitudes of 1-20mm.

Are A2A3. Specimen 6507/7-5.1. 2617.14m
Moderately-sorted sandstone containing four normal faults with displacement magnitudes of 5-20mm.

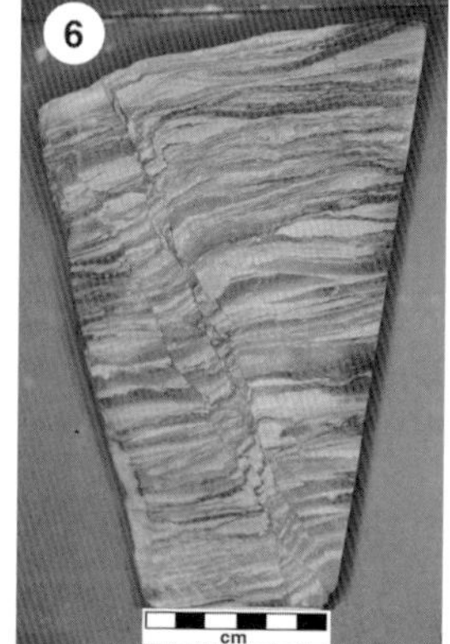

Are A2A2-1. Specimen 6507/7-6.2. 2339.96m
Well-sorted siltstone with abundant phyllosilicate-rich laminae. Note the two multiple faults at the centre.

Fig. 3. Review of the core scale deformation features present in Heidrun.

Table 1. *Review of fault rock properties in Heidrun*

Fault rock type	Clay content	Permeability (mD) Initial model	Average	Low	High	Th. Pressure oil-brine (Bars)
Cataclasite	<10%	0.3	0.14	0.006	0.3	0.4
Phyllo. Framework (low clay)	10–15%	0.1	0.11	0.1	0.12	0.7
Phyllo. Framework (high clay)	15–50%	0.02	0.018	<0.001	0.07	1.5
Clay smear	>50%	0.0015	0.0014	<0.001	<0.003	5

the permeability associated with Fangst juxtapositions with lower reservoir units. In the realization shown, the smearing in the Ror Formation is limited (only effective within less than the Ror thickness); increasing the smear distribution would increase the area of light brown on Fig. 4 and obscure more of the higher permeability Fangst/lower reservoir juxtapositions.

Table 2. *Review of reservoir zonation used for the fault seal model. Also shown are the clay contents of the more detailed reservoir zonation scheme which has been simplified for modelling*

Zonation for fault seal model		Average clay %	Av. Perm (mD)
Melke		82	Inactive
F5–F4A	F5–4B	1	9200
	F4A	52	Inactive
F3C–F1A	F3	9	
	F2	28	1650
	F1	11	
Ror–T5		57	Inactive
T4–T3A	T4	14	
	T3C	23	1000
	T3B	14	
	T3A	33	Inactive
T2		26	200
T1C		8	350
T1B–A2B1	T1B–A	32	
	A2B4–3	26	160
	A2B2–1	24	
A2A3		15	900
A2A2–A1B2	A2A2–1	26	
	A1B11–6	20	480
	A1B5–4	30	
	A1B3–2	44	
A1B1		10	2000
A1A		34	610

To describe the faults in the reservoir simulation model, the fault planes were divided into sections (or grids) following the modified reservoir zonation for the fault seal model (Table 3). Threshold pressures were calculated for each fault rock type, based on standard Hg porosimetry results. Fault rock permeabilities were input as transmissibility multipliers, where the conversion from permeability to transmissibility included the impact of fault rock thickness, and the dimensions of the model cells used. Finally, permeability multipliers were assigned to each subarea of the fault planes with different throw ranges in the model. Such a layered fault description enables the stratigraphical controls on fault rock distribution to be implemented, and allows for closing or opening parts of the fault planes for history matching purposes. The average permeability for flow between two neighbour grid blocks is:

$$k_{\text{no fault}} = L/[(0.5L/k_1) + (0.5L/k_2)]$$

$$k_{\text{with fault}} = L/[(0.5L/k_1) + (0.5L/k_2) + (L_f/k_f)]$$

where L = average grid block length; k_1 and k_2 are the permeabilities of the neighbouring grid blocks; L_f is the fault rock width and k_f is the fault rock permeability. The transmissibility multiplier is simply the ratio between the two calculated permeabilities: $(k_{\text{with fault}}/k_{\text{no fault}})$. The average grid block length in the Heidrun simulation grid is 120 m. Data on fault zone width is sparse on Heidrun. Only two faults of significant size were cored successfully; well 6507/-8, where a 25 m throw fault has an estimated total fault rock thickness of ~1 m (based on addition of individual small fault thicknesses) and well 6507/7-A-38, where a fault with a throw of 65 m has a total fault rock thickness of 1–2 m.

The initial transmissibility values used in the modelling were based on assigning the large faults (throws >30 m) an initial cumulative fault rock thickness of 2 m, and the smaller faults (throws <30 m) given a fault rock thickness value of 1 m. Characteristic permeabilities for each fault rock type (Tables 1 & 3) used in the transmissivity estimates were chosen from the high side of the measured range rather than

Table 3. *Fault rock permeabilities used for different juxtapositions in the fault seal model*

Reservior zone	F5–F4A	F3C–F1A	Ror	T4–T3A	T2	T1C	T1B–A2B1	A2A3	A2A2–A1B2	A1B1	A1A
F5–F4A	0.3										
F3C–F1A	0.3	0.3									
Ror	0.0015	0.0015	0.0015								
T4–T3A	0.0015	0.0015	0.0015	0.1							
T2	0.02	0.0015	0.0015	0.02	0.02						
T1C	0.3	0.0015	0.0015	0.1	0.02	0.3					
T1B–A2B1	0.02	0.0015	0.0015	0.02	0.02	0.02	0.02				
A2A3	0.1	0.0015	0.0015	0.1	0.02	0.1	0.02	0.1			
A2A2–A1B2	0.02	0.0015	0.0015	0.02	0.02	0.02	0.02	0.02	0.02		
A1B1	0.3	0.0015	0.0015	0.1	0.02	0.3	0.02	0.1	0.02	0.3	
A1A	0.02	0.0015	0.0015	0.02	0.02	0.02	0.02	0.02	0.02	0.02	0.02

taking an average value. It should also be noted that fault rocks associated with very high permeability undeformed clean sandstones in Heidrun were not analysed in this study, which was aimed at assessing the Tilje and Åre Formation, where faults were believed to have the largest impact on reservoir communication. For the initial modelling conducted, the permeability values measured from the Heidrun fault rock specimens were used. The validity of this assumption is discussed later in the paper.

Fault juxtaposition diagrams were constructed using FAPS 2 software (Badley Earth Sciences) and used in combination with the matrices presented in Tables 1–3 to establish how the sealing capacity varies laterally and vertically on the fault planes. The faults incorporated in this study are highlighted in Fig. 5. This figure also provides an example of how the transmissibility multipliers were distributed at different reservoir levels along the faults using variations in the fault throw as a discriminator. The map shown is for the transmissivity factors applied initially to the Upper Tilje Formation (T4–T3A).

Discussion and comparison of the observed breakthrough behaviour with the model results

In this section the parameters defining fault sealing capacity used in the reservoir simulation model (transmissibility multipliers) are discussed, and the associated uncertainties are assessed. This analysis is based on the results of the history matching and from an evaluation of the fault rock database for the Heidrun Field.

Introducing a quantitative description of the faults had a significant effect on the results from the reservoir simulation runs. In the Upper Tilje, the proposed transmissibilties gave a good match to the rise in reservoir pressure in water injector A-51 and to the GOR of produced oil from A-14, but the production history is still limited from this reservoir. No production history was available from Lower Tilje and Åre at the time of the study. A good first order match to the observed well rates and pressures (flowing and shut-in) was also achieved for the Fangst reservoir. However, in this case modelling the behaviour of the reservoir unit containing the highest host rock permeability values, was improved after locally opening (increasing the transmissibility) the faults by a factor 10. Figure 6a illustrates the almost perfect model match of the observed gradual pressure build-up in well A-50 during the shut-in in September 1996, while an almost immediate pressure build-up in the well would be the result of the simulation model run with open faults (Fig. 6b).

The new model created from the detailed fault rock property data, i.e. reduced flow across faults, is used in Fig. 7 to compare model and observed data for gas breakthrough in two oil production wells A-52 and A-53. The new model predicted breakthrough in July 1996, one month after the actual break through. A model with open faults predicts an even and gravity stable expansion of the gas cap with the gas break through occurring as late as 2000/2001 (see Fig. 7). This demonstrates the remarkably good match achieved by adding detailed fault properties as input. Two mechanisms contribute to this match: firstly the fault restrictions cause lower reservoir pressure within the fault segments around the wells in question and secondly the approaching gas can flow unrestrained along the fault segments without spreading across to neighbouring segments/compartments to the extent that completely open faults would allow.

Although the new reduced fault flow model prediction was successful, the moderate adjustment (approximately a factor of 5–10) needed to

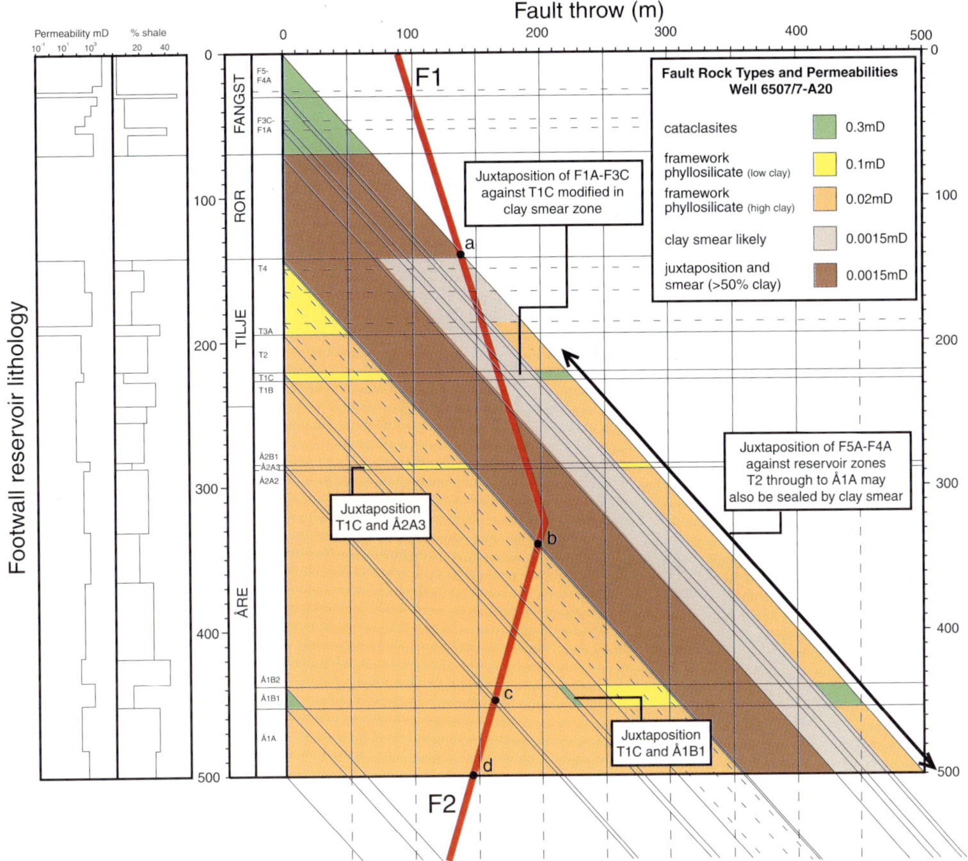

Fig. 4. Juxtaposition and seal diagram for the Heidrun reservoir based on the 6507/7-A-20 well stratigraphy. The diagram shows the triangle of reservoir juxtapositions possible. The permeabilities used in the reservoir modelling are colour coded. Note the importance of the extent of clay smearing from the Ror (shown in light brown) on the permeability windows associated with the Fangst Group. Increasing the effectiveness of the clay smear (clay smear factor) would reduce the size of the medium/high permeability fault rock windows. The figure also illustrates the juxtapositions along a fault with a maximum throw of 200 m. This fault would be sealed by clay smears between 'a' and 'b' and be dominated by the phyllosilicate framework fault rocks (orange), developed from impure sandstones, between 'b' and d; a small window of high permeability (A1B1 to A2A3) is shown at 'c'.

improve the flow where Fangst/Fangst self-juxtaposition existed indicates that the transmissibility multipliers for self-juxtaposition of the Fangst Group based on Tables 2 and 3 were too low for this reservoir unit. This can be corrected for by either increasing the fault rock permeability or reducing the fault rock thickness used in the Fangst. These factors are discussed further.

The sealing capacity of a fault is controlled by the highest permeability present on the fault plane. This was kept in mind when establishing the fault seal model and representative permeabilities for Fangst fault rocks were chosen from the high side of the measured range from the sample analysed. However, this may have under-estimated the permeabilities of cataclasites generated from the very high permeability (~2000 mD) host rocks present in the undeformed Fangst. Such cataclasites, from the very high permeability Fangst unit 5 were not sampled by the coring. From the history matching results it can be inferred that the cataclasite permeability should be higher than the measured upper limit of 0.3 mD. A small adjustment to a value of 1 mD would be compatible with the observed production history. This value was outside the range of measurements for Heidrun but is well within the range of permeabilities measured on cataclasites from other fields (see later discussion). A value of 1 mD rather than 0.3 mD for cataclasites

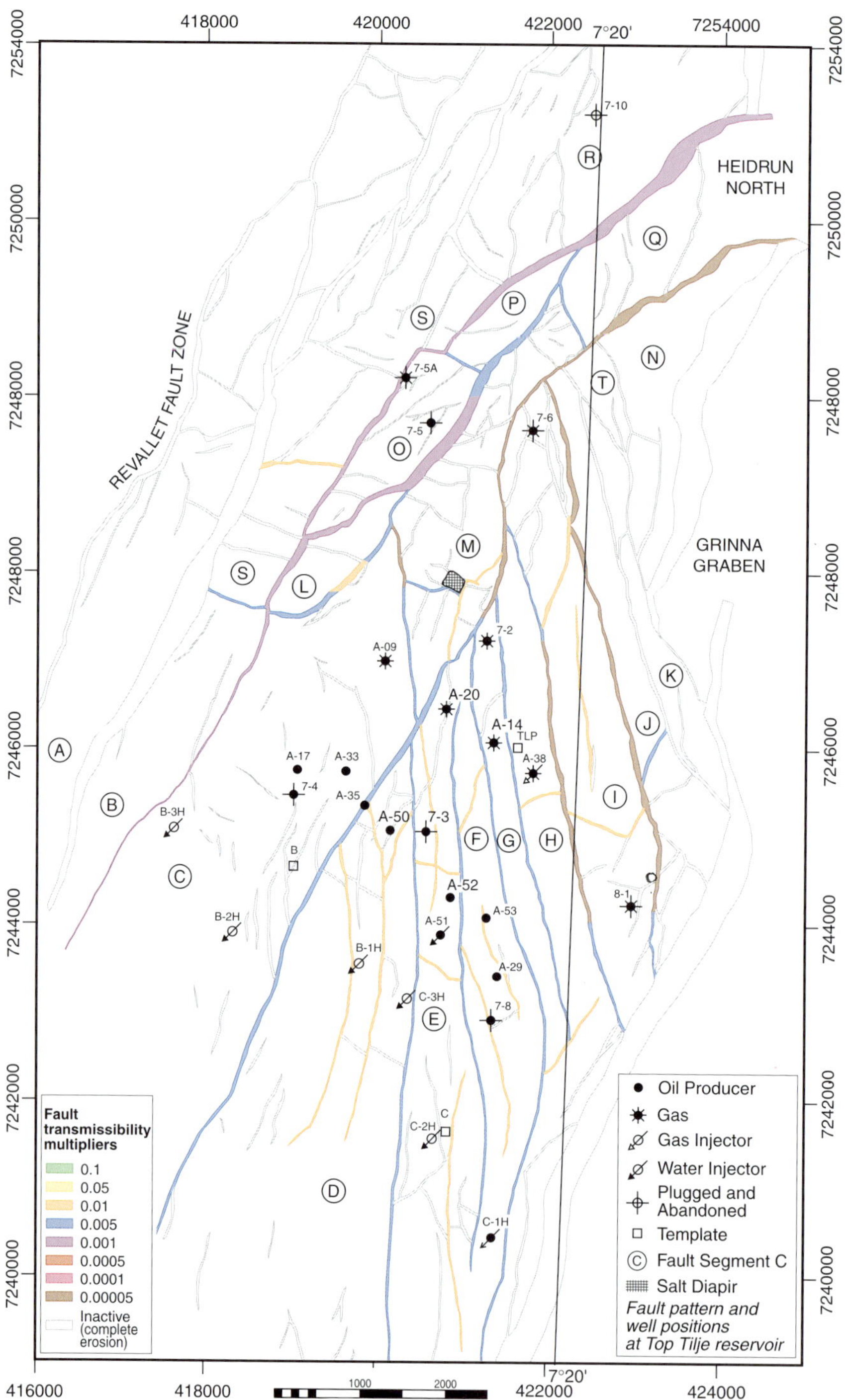

Fig. 5. An example of how the transmissibility multipliers are distributed along the faults in the reservoir model for the Upper Tilje Formation. Fault throw and juxtaposition/seal types are used to define each section/subarea of the fault planes.

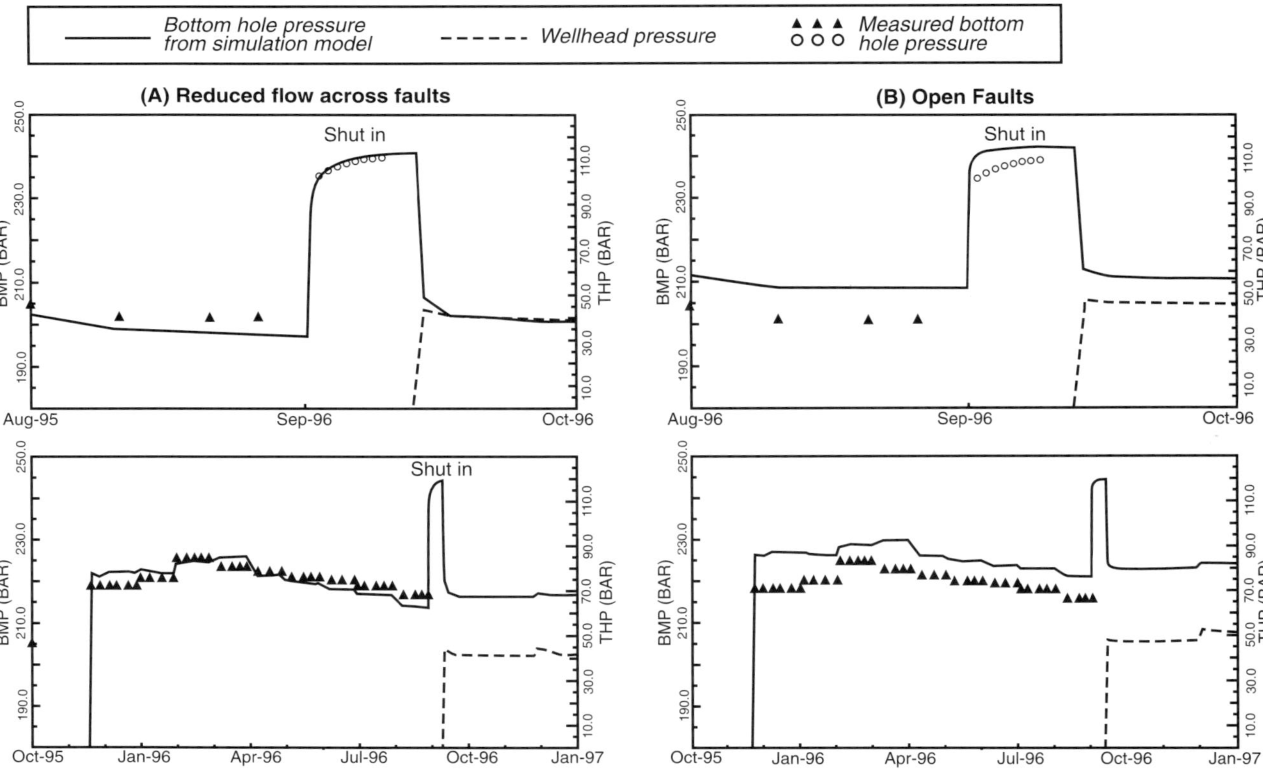

Fig. 6. Comparison of the bottom hole measured pressures for well 6507/7-A35 with the simulated pressures. Two simulation models are shown; one for the reduced fault flow model (**a**) and secondly for the model with open faults (**b**) Note the increased fit of the reduced flow model.

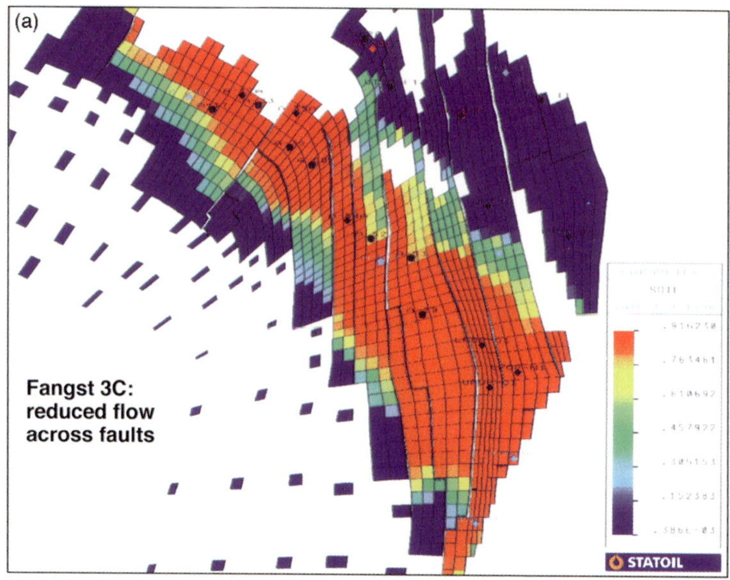

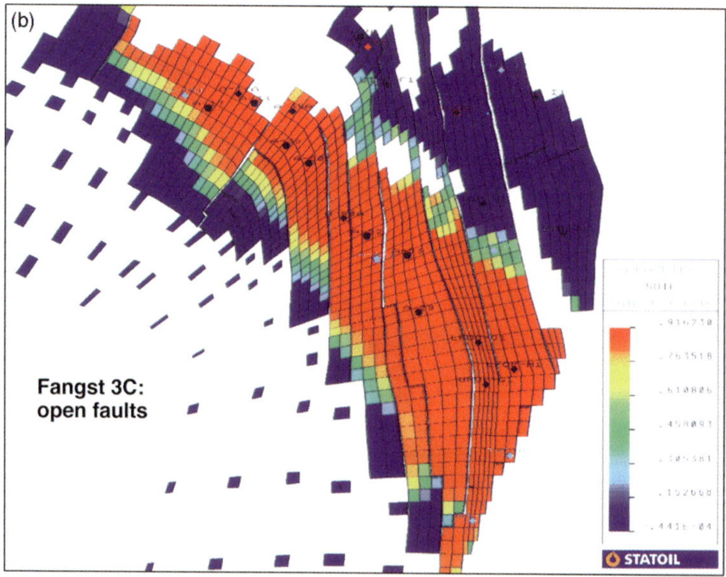

Fig. 7. Comparison of oil saturation in the simulation model for (**a**) reduced fault flow and (**b**) open faults by 1.9.96. Note the predicted gas breakthrough in well 65070-A-52 and A53 in the reduced flow model shown in (a).

derived from very high permeability host rocks was used in later model runs. The values were unchanged for the other fault rocks.

The other factor which may increase the transmissibility of fault rocks developed from the high permeability sandstones in the Fangst is a reduced fault rock thickness. It is not straightforward to determine the correct effective fault zone width or flow path length for calculating transmissibility multipliers. A fault zone is commonly made of several slip planes with lenses of undeformed rock in-between the deformed rock. A simple, first order solution, is to use the cumulative fault rock thickness from all the faults along a straight line flow-path across the fault damage zone. However, the length of the fluid flow path across a fault zone can be significantly greater than that measured along a straight line pathway through the damage zone. The fluid will probably follow a high tortuosity, zig-zag route around complex fault/fracture arrays following higher permeability windows on the fault planes within the array. This tortuosity factor will be higher for the clay-rich reservoir units, as these create lower permeability fault rocks and are likely to have wider spaced high-permeability windows on the fault planes (see Knipe *et al.* 1997). For these reasons, the fault rock thickness for cataclastic faults was reduced to 0.4 m and 0.75 m for small and large faults, respectively, which was compatible with the observed production history for the Fangst reservoir. The fault rock thicknesses were kept unchanged for faults dominated by the more clay-rich fault rocks (see later discussion).

The applicability of the threshold pressures listed in Table 1 can only be validated at present by comparison to the observed differences in fluid contacts across faults. A threshold pressure of approximately 0.2 bar is required to be consistent with the observed difference of 7 m in Fangst OWC between segments D and E (see Fig. 1b). A permeability of 1 mD suggested for the cataclasites developed from clean sandstones in Heidrun corresponds to a threshold pressure of approximately 0.25 bar, which is close to the required value. The maximum OWC difference of 33 m (based on extrapolation of FMT pressure gradients) for upper Tilje between segments G and H, corresponds to a threshold pressure of approximately 0.8 bar, which compares well with the value for phyllosilicate rich fault rocks listed in Table 1.

Conclusions

The additional constraints for reservoir simulation provided by structural and petrophysical property analysis of faults from core material dramatically impact on the accuracy of the modelling. Reservoir simulation modelling using a restricted fault flow model was able to predict the observed behaviour remarkably well after only a slight adjustment of the initial fault rock properties. This reflects a number of factors which highlight the present limitations of modelling fluid flow through faulted reservoirs in general and some specific restrictions related to the data available for Heidrun in particular. These restrictions and limitations are listed below:

(1) There is a lack of data needed for understanding, modelling and up-scaling the effective flow through complex fault zones (composed of intersecting clusters of deformation features). This results in the relative impact of three potentially important factors being unknown. These are: (a) the tortuosity factor (flow path length) for the array; (b) the cumulative fault rock thickness along the dominant flow-paths through the array; and (c) the fault-rock permeabilities encountered along these dominant flow paths. At present it is possible to choose a range of combinations of these factors which describe the large scale flow behaviour. Future refinements to fault related flow will require separation and improved modelling of the impact of these variables.

(2) In the case of Heidrun, the data input available from core for fault rock characterization did not allow sampling of deformation features from all the critical reservoir units. This arises from the 'random' sampling provided by the cores and emphasizes the importance of building databases from tightly constrained examples for use in less well-constrained situations. In the study reported here, a key element for the flow modelling, the cataclasites created from the high permeability Fangst units, were not sampled. The transfer and application of data from other examples will improve as future studies allow collation of fault rock and fault zone properties from a range of geohistories as was the case in this example.

Despite these limitations and restrictions, the study reported has influenced the future development strategies for Heidrun. Firstly, the simulation model with restricted fault related flow predicts that well drainage volumes will be limited in the Tilje and Åre reservoir units to within fault segments. Several extra wells are now planned for further development of these reservoirs. In addition, the new modelling predicts earlier and more severe gas cusping into the Fangst oil producers, compared to earlier models run with open faults. This is to be addressed in planning the location and character of new wells.

We would like to thank the following for discussions and comments on early drafts of this paper: R. Davies, J. Cosgrove, T. Needham, N. Porter, Q. Fisher, and G. Jones. The support from colleagues in Heidrun PETEK is also acknowledged and in particular the contribution from K. Christoffensen and Ø. Lund was valuable. Thanks to A. Roberts (Badley Earth Sciences) for great co-operation and help in carrying out juxtaposition analysis. We would also like to acknowledge the co-operation of the Heidrun Partners in allowing publication.

References.

BERG, R. R. 1975. Capillary pressure in stratigraphic traps. *American Association of Petroleum Geologists Bulletin*, **59**, 939–956.

BOUVIER, J. D., SIJPESTEIJN, K., KLEUSNER, D. F., ONYEJEKWE, C. C. & VAN DER PAL, R. C. 1989. Three-dimensional seismic interpretation and fault sealing investigations, Nun River field, Nigeria. *American Association of Petroleum Geologists Bulletin*, **73**, 1397–1414.

FISHER, Q. J. & KNIPE, R. J., 1998. Microstructural controls on the petrophysical properties of deformation features. *This volume*.

GIBSON, R. G., 1994. Fault zone seals in siliclastic strata of the Columbus Basin, Offshore Trinidad. *American Association of Petroleum Geologists Bulletin*, **78**, 1372–1385.

HEMMENS, P. D., HOLE, A., REID, B. E., LEACH, P. R. L. & LANDRUM, W. R. 1994. The Heidrun Field. In, North Sea Oil and Gas Reservoirs III. Norwegian Inst. of Technology (NTH). Kluwer Acad. publ. Dordreet/Boston/ London , 1–23.

HEUM, O. R., DALLAND, A. & MEISINGSET, K. K. 1986. Habitat of hydrocarbons at Haltenbanken (PVT-modelling or a predictive tool in hydrocarbon exploration). *In*: SPENCER, A. M. *et al.* (eds) *Habitat of Hydrocarbons on the Norwegian Continental Shelf*, Norwegian Petroleum Society, Graham and Trotman, 259–274.

HEUM, O. R. 1996 A fluid dynamic classification of hydrocarbon entrapment. *Petoleum Geoscience*, **2**. 145–158.

JEV, B. I., KARS-SIJPESTEIJN, C. H., PETERS, M. P. A. M., WATTS, N. L. & WHITE, J. T. 1993. Akaso field, Nigeria: Use of integrated 3-D seismic, fault slicing, clay smearing, and RFT pressure data on fault trapping and dynamic leakage. *American Association of Petroleum Geologists Bulletin*, **77**, 1389–1404.

KNIPE, R. J. 1992. Faulting processes and fault seal. *In*: LARSEN, R. M., BREKKE, H., LARSEN, B. T. & TALLERAS, E. (eds) *Structural and Tectonic Modelling and its application to Petroleum Geology*. Norsk Petroleumsforening (NPF) Special Publications, **1**, 325–342. Elsevier.

——, R. J. 1997. Juxtaposition and seal diagrams to help analyze fault seals in hydrocarbon reservoirs. *American Association of Petroleum Geologists Bulletin*, **81**. 187–195.

——, R. J., FISHER, Q. J., JONES, G., CLENNELL, M. B., FARMER, A. B. HARRISON, A., KIDD, B., MCALLISTER, E., PORTER, J. R. & WHITE, E. A. 1997. Fault seal analysis: successful methodologies, application and future directions. *In*: MØLLER-PEDERSON, P. & KOESTLER, A. G. (eds) *Hydrocarbon Seals – Importance for Exploration and Production*. Norsk Petroleums-forening/NPF (Norwegian Petroleum Society) Special Publications. **7**, 15–38.

KNOTT, S. D., 1993. Fault seal analysis in the North Sea. *American Association of Petroleum Geologists Bulletin*, **77**, 778–792.

KOCH, J. O. & HEUM, O. R. 1995. Exploration trends of the halten terrace. *In*: HANSLEIN, S. (ed.) *Petroleum Exploration and Exploitation in Norway*. NPF Special Publications **4**, pp 235–241.

LINDSAY, N. G., MURPHY, F. C., WALSH, J. J. & WATTERSON, J. 1993. Outcrop studies of shale smears on fault surfaces. Special Publication. *International Association Sedimentologists*, **25**, 113–123.

MØLLER-PEDERSON, P. & KOESTLER, A. G. 1997. Hydrocarbon Seals: *Importance for Exploration and Production*. NPF Special Publications **7**, 250 pp.

REID, B. E., HØYLAND, L. A., OLSEN, S. R. & PETTERSON, O. 1996. The Heidrun Field – Challanges in Reservoir Development and Production. *Offshore Technology Conference*, OTC 8084 Houston.

SCHMIDT, W. J. 1992. Structure of the mid-Norway Heidrun field and its regional implications. *In*: LARSEN, R. M. *et al.* (eds) *Structural Modelling and its Application to Petroleum Geology*, Special Publication, **1**, Norwegian Petroleum Society, 381–395.

SCHOWALTER, T. T. 1979. Mechanisms of secondary hydrocarbon migration and entrapment. *American Association of Petroleum Geologists Bulletin*, **63**, 723–760.

WELBON, A. L., BEACH, A., BROCKBANK, P. J., FJELD, O., KNOTT, S. D., PEDERSON, T. & THOMAS, S. 1997. Fault seal analysis in hydrocarbon exploration and appraisal: examples from offshore mid-Norway. *In*: MØLLER-PEDERSON, P. & KOESTLER, A. G. (eds) *Hydrocarbon Seals—Importance for Exploration and Production*. Norsk Petroleums-forening/NPF (Norwegian Petroleum Society) Special Publications. **7**, 125–138.

WHITLEY, P. K. 1992. The Geology of Heidrun: A giant oil and gas field on the mid-Norwiegian self. *In*: AAPG Gaint Oil and Gas Fields of the Decade 1980–1990, Tulsa.

YIELDING, G., FREEMAN, B. & NEEDHAM, D. T. 1997. Quantitative fault seal prediction. *American Association of Petroleum Geologists Bulletin*, **81**, 897–917.

Fault controlled communication in the Sleipner Vest Field, Norwegian Continental Shelf; detailed, quantitative input for reservoir simulation and well planning

S. OTTESEN ELLEVSET[1], R. J. KNIPE[2], T. SVAVA OLSEN[1], Q. J. FISHER[2] & G. JONES[2]

[1] *Statoil, PO Box 300, 4001 Stavanger, Norway*

[2] *Rock Deformation Research Group, Department of Earth Sciences, University of Leeds, Leeds LS2 9JT, UK*

Abstract: The Sleipner Vest field, located in blocks 15/6 and 15/9 on the Norwegian Continental Shelf, contains hydrocarbons (mainly gas condensates) within the marginal marine units of the Middle Jurassic Hugin Formation. The field is segmented into fault-bounded compartments, which exhibit differences in gas–water contacts of between 10 and 100 m. Microstructural analysis of core samples has identified three principal fault types; cataclasites, developed from clean sandstones, framework phyllosilicate fault rocks created from impure sandstones and clay smears developed from phyllosilicate-rich units. The distribution of these has been linked to the phyllosilicate content of the undeformed Hugin reservoir at the time of deformation. Petrophysical property analysis has been used to quantify the representative permeabilities and threshold capillary pressures of the fault rocks and their undeformed equivalents. Juxtaposition/seal diagrams and detailed fault plane maps were also constructed and provide a basis for mapping sand–sand juxtapositions and the distributions of the different fault rock types with assigned permeabilities and capillary threshold pressures. The results provide an explanation of the mapped variations in gas–water contacts. An important correlation exists between the type of fault rock predicted to dominate the fault plane near the hydrocarbon–water interface and the difference in hydrocarbon depths across the faults. Compartments separated by faults with windows of juxtaposed clean sandstone cataclasites have small hydrocarbon–water level differences (7–11 m), suggesting that the low capillary threshold pressure (more permeable) cataclastic fault rocks control communication. Compartments separated by faults predicted to have extensive phyllosilicate-rich fault rocks (developed from impure sandstones and clay rich units) with high capillary threshold pressures correlate to larger hydrocarbon–water differences (>39 m), reflecting the reduced communication.

In recent years, fault seal analysis has become an increasingly important element in hydrocarbon exploration and production (Bouvier *et al.* 1989; Knipe 1992; Knott 1993; Knipe *et al.* 1997). Understanding fault behaviour is now recognized as critical to successfully locating hydrocarbon accumulations and for their efficient and cost-effective development. In this paper, we discuss the results of an integrated fault seal study of the Sleipner Vest (west) field in the Norwegian North Sea, which was initiated to assess the transmissibility of a number of mapped sealing faults as the field entered a programme of production and development.

Sleipner Vest is a large, structurally complex gas-condensate field which is located in the eastern South Viking Graben (blocks 15/6 and 15/9) in the central North Sea Basin (Fig. 1). Mapping of recently acquired 3D seismic data (1994) has shown that the field is extensively segmented by faults (Fig. 2). Differences in hydrocarbon contacts in the order of metres to tens of metres have been observed within the Middle Jurassic (Callovian) marginal marine reservoir sandstones of the Hugin Formation. Larger contact differences occur across faults which juxtapose reservoir against non-reservoir (sealing) lithologies. On the basis of these differences and the mapped fault pattern, a model for compartmentalization of the Sleipner Vest field has been proposed (Fig. 2). The successful appraisal and development of the field, which was brought onstream in 1995, was therefore considered to be strongly dependent on understanding the nature and degree of fluid communication across these compartmentalizing faults under production drawdown conditions.

The analysis of fault zone behaviour in Sleipner Vest reported here was based on a detailed evaluation of 3D seismic data (including extensive fault plane mapping), the characterization of deformation features present in cores and

Ottesen Ellevset, S., Knipe, R. J., Olsen, T. S. *et al.* 1998. Fault controlled communication in the Sleipner Vest Field, Norwegian Continental Shelf; detailed, quantitative input for reservoir simulation and well planning. *In*: Jones, G., Fisher, Q. J. & Knipe, R. J. (eds) *Faulting, Fault Sealing and Fluid Flow in Hydrocarbon Reservoirs*. Geological Society, London, Special Publications, **147**, 283–297.

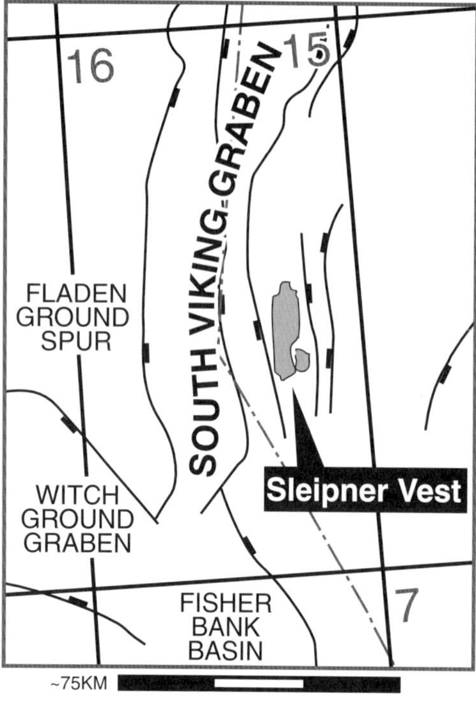

Fig. 1. Structural depth map of the Top Hugin Fm, contoured inside the GWC.

also the microstructural and physical property analysis of selected fault rocks. We also introduce the applied use of a new graphical technique for analysis of cross-fault lithological juxtapositions in fault zones (Knipe 1997). The overall objective was to combine data from these different scales and provide detailed input to well planning and reservoir simulation studies.

Fault seal assessment methodology

The sealing behaviour of faults has often been assessed by relating the probability of a fault plane sealing to a number of input parametres, which are then transferred to a map of the fault plane. Common parametres used include the displacement along the fault surface, the fault zone thickness (Knott 1993), the lithology either side of the fault, (often expressed as a net/gross ratio) and also some factor which estimates the effect of deforming and smearing clay units (e.g. the Smear Gouge Ratio, SGR; Bouvier *et al.* 1989; G. Skerlec unpublished report; Yielding *et al.* 1997). Such analysis is usually completed by mapping the connectivity across the fault using (vertical) fault plane diagrams (Allan 1989), in order to assess the transmissivity. Where known, hydrocarbon–water contacts and pressure data are used to calibrate the depths where transmissivity occur (Yielding *et al.* 1997).

Whilst these methods do help to reduce risk during exploration, they do not provide enough detail for reservoir evaluation and modelling. As pointed out in Knipe (1992), fault seal mechanisms, and thus petrophysical properties and sealing behaviour, will vary over a given fault surface and depend not only on the factors described above, but also as a result of the internal structure of fault zones and their geohistory. The aim of any fault seal study should be to map this variation, and ultimately, to quantify variations in transmissibility as a basis for understanding reservoir communication across the faults.

In this study, intra-reservoir communication across the faults separating the main compartments of the Sleipner Vest Field has been systematically analysed. Communication across a total of 61 selected faults or fault segments has been assessed. The analysis presented has involved the integration of structural geological data obtained from core to seismic scales, with detailed microstructural and petrophysical property analysis of fault rocks obtained from core. Initially, structural logging of cored wells highlighted the nature of the fault zones and the main fault rock and fault seal types present in Sleipner Vest. The diagenetic and microstructural evolution of selected fault rocks and their petrophysical properties were characterized using the techniques described in Fisher & Knipe (this volume).

The detailed fault rock characterization exercise described has been integrated with the mapping of cross-fault reservoir juxtapositions. This exercise was undertaken using two different methodologies. Firstly, graphical juxtaposition analysis (triangular Juxtaposition Diagrams of Knipe 1997) were used to assess the effects of deforming typical Hugin Formation reservoir stratigraphies. This technique has the advantage of providing rapid visualization of the distribution of unit juxtapositions and can also highlight the critical throws at which particular juxtaposition types (e.g. sand–shale) are developed. Also, the technique is not dependent on other factors such as the quality and resolution of the seismic data, the availability of seismic reflectors for mapping and interpreter preference.

The second methodology involved creating vertical fault plane diagrams (Allan 1989), based on the 3D seismic fault and horizon mapping, using the FAPS software (Badley Earth

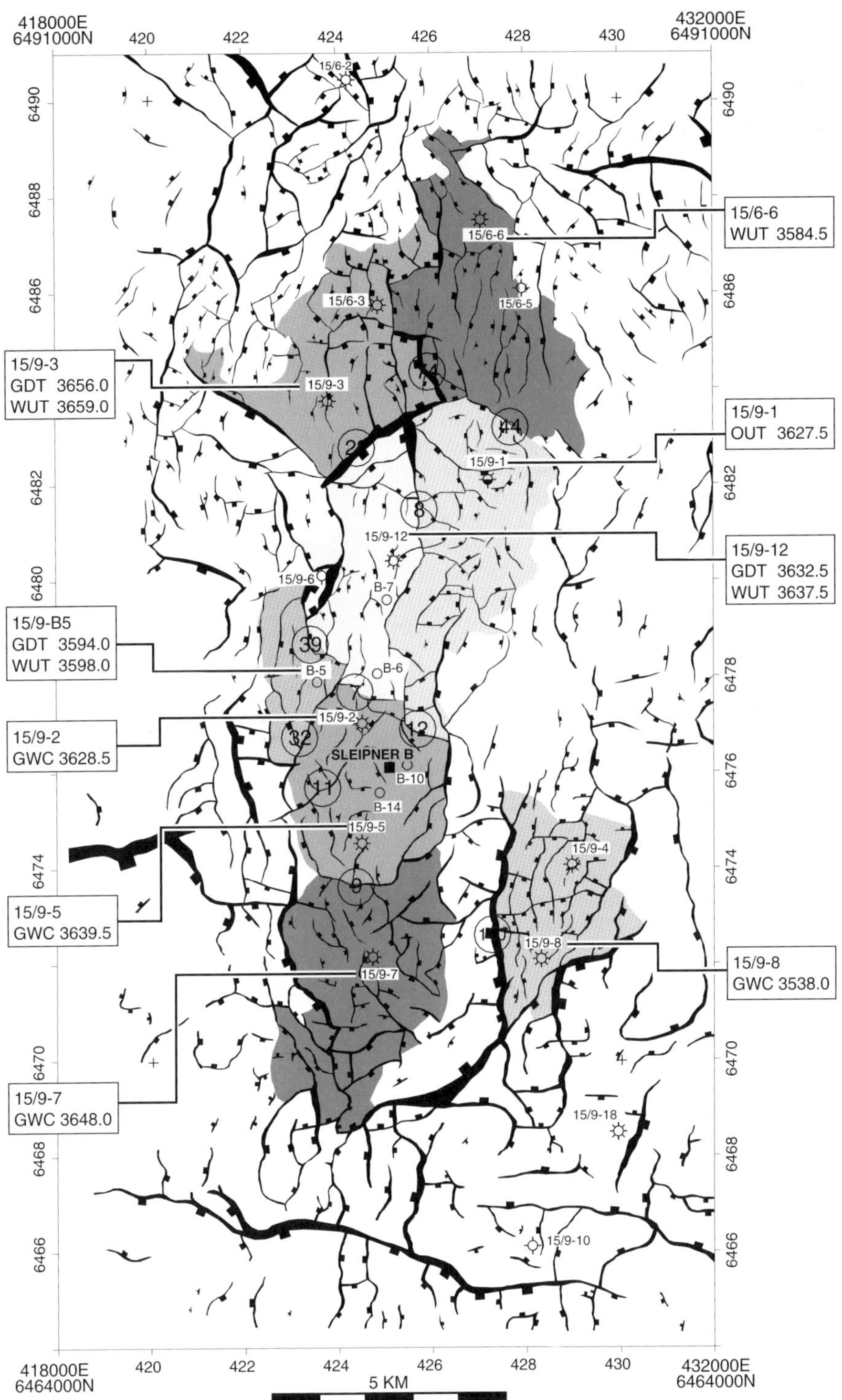

Fig. 2. Suggested compartmentalization of the field, based on the fault pattern and observed differences in GWC.

Sciences). This allows throw variations along the fault zone to be assessed and defines the approximate depths at which the critical juxtapositions identified occur in the subsurface. The top Hugin reservoir is only a moderately reliable reflector, and reservoir units have been isopached up from a strongly reflective coal unit which lies close to the base of the reservoir sequence. Vertical seismic resolution over Sleipner Vest is in the order of 30 m, which could generate some important possible variations in reservoir topography.

Tectonostratigraphic evolution

The Sleipner Vest field lies in a tectonically complex part of the central North Sea Basin that has experienced a long history of deformation. The fundamental structural elements in this area are probably of Early Palaeozoic age. These include deep-seated fault systems aligned along a WNW trend, developed as part of the Tornquist structural zone in northern Europe (Pegrum 1984). Early Mesozoic (Lower Triassic) extension was probably an important element in the initial development of the N–S trending South Viking Graben, possibly re-activating Devonian basins (Coward 1994). Subsequently, during the early Middle Jurassic (Toarcian–Aalenian), the structural evolution of the Sleipner area was markedly influenced by uplift and erosion resulting from thermal doming of the central North Sea area (Underhill & Partington 1993). This resulted in the removal of Lower Jurassic sediments.

The Sleipner area was characterized by an essentially W–E oriented coastline system during the deposition of the mainly Callovian age Hugin Formation reservoir, in which marginal marine to offshore sediments were deposited from south to north, respectively. Comparatively little tectonic activity is recorded at the seismic scale during reservoir deposition, and a progressively deepening marine environment is recorded throughout the Upper Jurassic. Widespread extension occurred on the Sleipner terrace area during the latest Jurassic and earliest Cretaceous, which generated both N–S faults and WNW cross-faults in the Hugin reservoir. The Sleipner terrace presently forms a down-thrown area, (*c.* 1000 m offset), along a NNE–SSW fault zone, relative to the Gamma High in the east. Extension was probably assisted by the presence of evaporites at depth. Although presently unproved by drilling, seismic data suggests fault-controlled salt pillows may be present. Most faults cut through the regionally diachronous 'base Cretaceous' reflector (Rattey & Hayward 1993) and appear to terminate in the Early Cretaceous shale units.

The Sleipner area shows some evidence of compressional tectonic activity during the Early Cretaceous. This is possibly related to re-activation of the Tornquist zone, which may have switched to dextral transpression from earlier regional transtension (Pegrum 1984; Pegrum & Ljones 1984). Additional, poorly understood compressional reactivation of normal faults also took place in the Late Cretaceous and Early Tertiary, as a result of intra-plate stresses resulting from both Alpine collision and North Atlantic opening.

Reservoir diagenesis

The diagenetic history of the Hugin Formation has proved remarkably consistent across the Sleipner Field (Knipe *et al.* unpublished report). Four diagenetic processes have affected the samples examined:

(i) quartz overgrowth development,
(ii) kaolin neoformation;
(iii) K-feldspar dissolution; and
(iv) illite precipitation.

In addition to these field extensive diagenetic processes, the Hugin sandstones have also experienced more localized precipitation of pyrite, calcite, siderite, dolomite and barite. This diagenetic history is summarized in Fig. 3. The timing of deformation, in relation to the diagenetic history, was established for each sample. This provides a framework for the assessment of fault rock development in the field. The deformation features preserved in the cores can be assigned to different combinations of a four phase history involving:

(i) early deformation of poorly consolidated sediments;

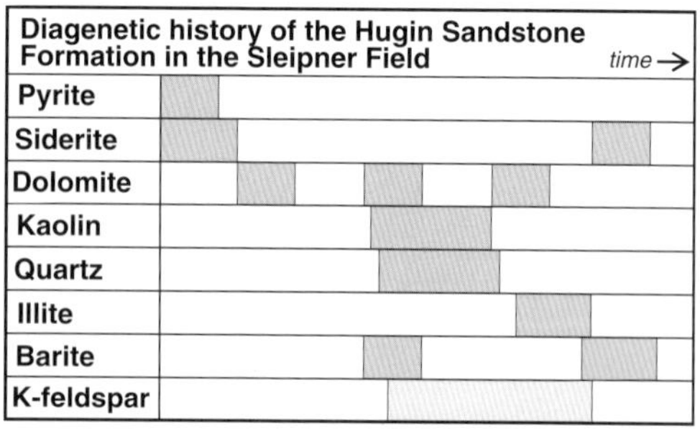

Fig. 3. The diagenetic history of the Hugin Formation in the Sleipner Vest Field. The dark coloured boxes refer to precipitation events. The light coloured box is a dissolution event.

(ii) intermediate stage faulting which pre-dates the main phase of kaolin and quartz precipitation;
(iii) late stage faulting which post-dates the main period of meso-diagenesis and
(iv) inversion.

Fault rocks and fault seal

Most faults are best considered as membranes (Watts 1987) or flow retarders with variable transmissibility properties. This view is in contrast to the sealing/non-sealing assessment of some recent studies (Knott 1993; Skerlec unpublished report). In the analysis reported here, the distribution of capillary threshold entry pressures is used to evaluate cross-fault communication, as the sealing capacity of the fault rocks can be related to the capillary threshold pressure (Watts 1987) and a permeability to threshold entry pressure relationship can also be established. No hydrocarbon flow through the fault zone will take place unless the threshold pressure is reached; flow is then controlled by the permeability. Microstructural and petrophysical analysis has allowed the fault rocks to be classified according to the deformation mechanisms involved in their formation as well as their present-day properties. The classification used is based on that presented in Knipe *et al.* (1997) and reviewed in Fig. 4. Seals present in the Sleipner Vest field cores include:

(i) phyllosilicate smears developed from clay-rich units;
(ii) phyllosilicate framework fault rocks developed from impure sandstones;
(iii) cataclastic collapse faults developed from clean sandstones;
(iv) cemented deformation features.

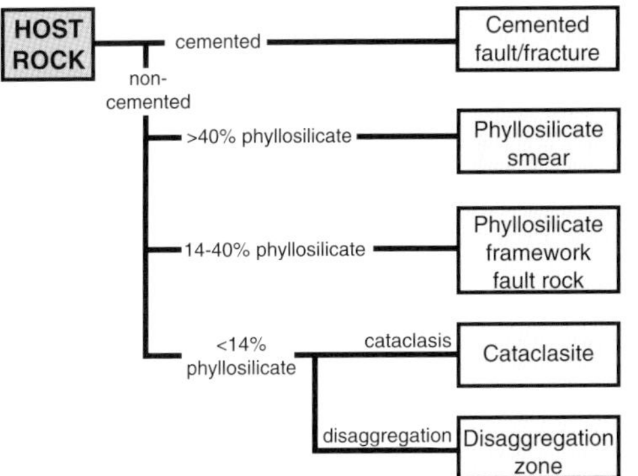

Fig. 4. Host rock lithology and fault rock type. The relation between phyllosilicate and microcrystalline quartz content of the sandstone and the fault rock development.

One of the most important aspects of this classification is the recognition of the relation between phyllosilicate content of the host sandstone and the seals which develop. A more detailed discussion on the classification and properties of fault rocks is given in Fisher & Knipe (this volume). In addition, regional compression (inversion faulting) events have in some cases induced a further decrease in the petrophysical properties of fault zones by localized pressure solution processes. There is also evidence for a late rupturing of early formed cataclastic fault rocks, which may be ascribed to this late inversion. Halite has been found in faults within the Sleipner Formation and the Skaggerak Formations but not in the overlying Hugin Formation reservoir. The salt sampled appears most likely to be a secondary feature associated with core recovery and drying. Each of the main fault rock classes and their role in the Sleipner Vest field is outlined below.

Cemented deformation features

Extensive development of cemented fault zones will invalidate the analysis of fault seals based on the simple prediction of fault rocks from the host undeformed lithologies (Knipe 1993; Knipe *et al.* 1997; Sverderup & Bjørlykke 1997). However, in many cases the continuity of cements may not be extensive enough to impact on communication, but may be restricted to parts of a fault plane. Cemented sandstone horizons, which contain cemented (calcite) fractures, are present in the Sleipner Vest field. This material may be redistributed and reprecipitated during deformation. However, the core data suggest that during faulting, this redistribution is likely to be restricted to an area close to the hanging wall and footwall cut-off regions of the cemented units. Given the patchy nature of the undeformed cement, such redistribution will probably be laterally discontinuous. In Sleipner Vest, cemented deformation features present have a range of permeabilities from 1 mD to >0.0001 mD, which is controlled by the micro-architecture of the cement. For our analysis of fault zone permeability in Sleipner Vest, we have assumed that the extent of any cementation on the fault plane does not exceed three times the thickness of the cemented sedimentary bed which acts as a source.

Phyllosilicate smear

In Sleipner, sedimentary units containing more than approximately 40% phyllosilicates (clay, micas) develop phyllosilicate smears when faulted. Physical property analysis of these rocks was not obtained directly from Sleipner Vest, but similar material sampled at equivalent depths usually has permeabilities well below 0.001 mD and capillary threshold pressures of $\gg$1500 psi. ($\sim$100 bars, or $\sim$10 MPa) for the Hg–air system (Fisher & Knipe this volume). For our analysis of the distribution of clay smears in Sleipner Vest, we have used the criterion that the smear is discontinuous when the effective fault throw exceeds three times the thickness of the clay-rich unit (see Knipe *et al.* 1997).

Phyllosilicate framework fault rocks (PFFR)

Faults in sedimentary rocks with 14–40% phyllosilicate contents develop phyllosilicate framework fault rocks (PFFRs) from impure sandstone lithologies and are composed of microsheared phyllosilicate domains (see Knipe 1992; Knipe *et al.* 1997; Fisher & Knipe this volume). The permeability of the measured phyllosilicate framework fault rocks developed in impure sandstones in the Sleipner Vest Field ranges from 0.3 to 0.004 mD and Hg threshold capillary pressures are above 120 psi ($\sim$1 MPa, $\sim$10 bars). There are two areas of a fault plane where these PFFR are likely to impact on sealing behaviour. The first of these is where impure sandstones are directly juxtaposed against the fault zone, and the second is along faults between the hanging wall and footwall cut-offs of the impure sandstone units. This latter situation is analogous to the formation of clay smears. The effectiveness of the PFFRs to form barriers to flow depends on their continuity, which is likely to decrease with increasing throw in the same manner as clay smears. There will be a zone near to the hanging wall and footwall cut-offs to the impure sandstones where there is an increase in the probability of these fault rocks forming seals. In Sleipner Vest, these zones of high potential for development of phyllosilicate framework fault rocks are taken to be effective for a distance (up or down fault planes) equal to the thickness of the impure sandstone units which create the fault rocks.

Cataclasites developed from clean sandstones

Sedimentary rocks containing less than 14% phyllosilicates develop either cataclasites or disaggregation zones during deformation. There is a very large variation in the permeability of this type of fault rock, depending upon the geohistory of the fault (Knipe *et al.* 1997; Fisher & Knipe this volume). In Sleipner Vest, these fault rocks have high permeabilities, with a range from 43–0.4 mD and they form the least effective flow barriers. The high permeability cataclasites are associated with early faulting at shallow depths of burial, and thus the lower permeability range (<1–5 mD) are more representative of the larger offset, seismically visible faults which were active at deeper levels. The fault plane windows dominated by cataclastic fault rocks developed from clean sandstones are likely to occupy the parts of fault planes rocks where sand–sand juxtapositions occur away from clay smears and phyllosilicate framework fault rocks.

Fault seal evaluation at sampled wells

Based on the detailed core, microstructural and petrophysical analysis, as well as the lithological description of the reservoir, a fault seal evaluation has been possible. The lithological description from the nearest well(s) can be used to produce a diagram showing the varying juxtapositions of the different reservoir stratigraphies with different fault throws. Using the results of core analysis, it is possible to predict which fault rocks will develop for a specific reservoir stratigraphy, and for a range of throw values. It is then possible to identify the throw windows that will develop local areas dominated by the three important fault rocks in Sleipner Vest; i.e. cataclasites, phyllosilicate framework fault rocks, and phyllosilicate smears (see Knipe 1997). The RDR Group has developed a software programme called 'Juxtaposition' that allows rapid visualization and assessment of juxtaposition and fault seal patterns generated by faulting of varying reservoir stratigraphies. An example of a juxtaposition and fault seal triangular plot from well 15/9-8 of the Sleipner Vest Field is given in Fig. 5.

Upscaling of physical properties

The hydrocarbon filling history of the Sleipner Vest Field is complex, with the possibility of several source areas contributing to the fluid pool. The timing of the main filling was from Tertiary until present and very little burial has occurred after filling of the structures. The measured capillary threshold pressures for the

(a) Juxtaposition Diagram for Well 15/9-8

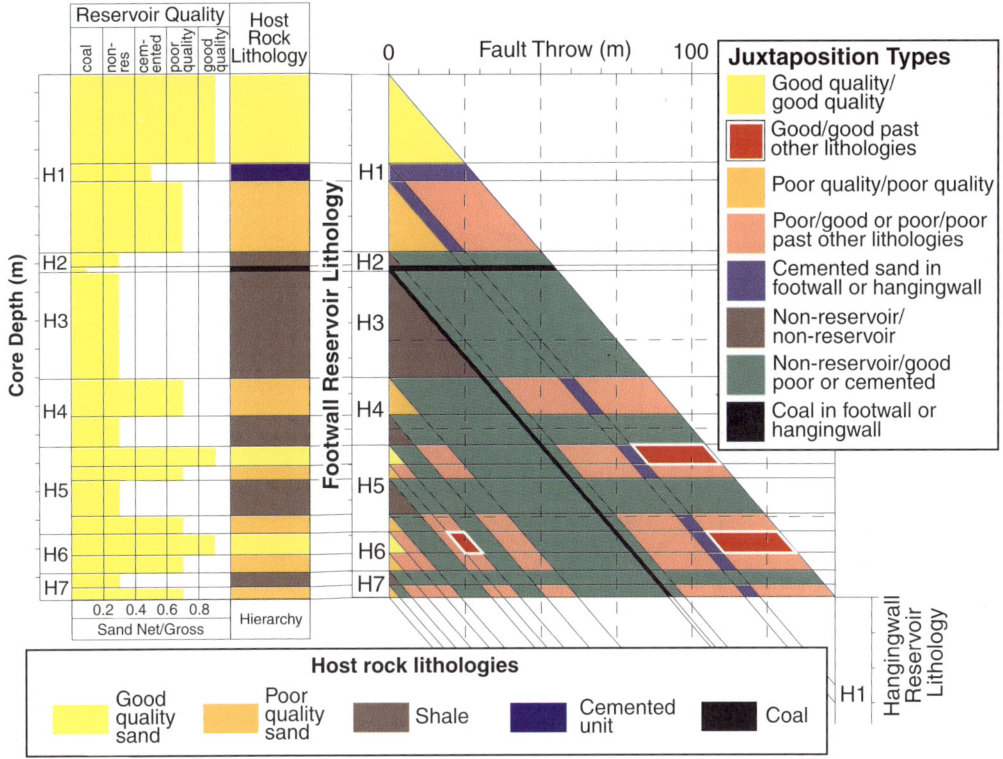

(b) Fault Seal Types for Well 15/9-8

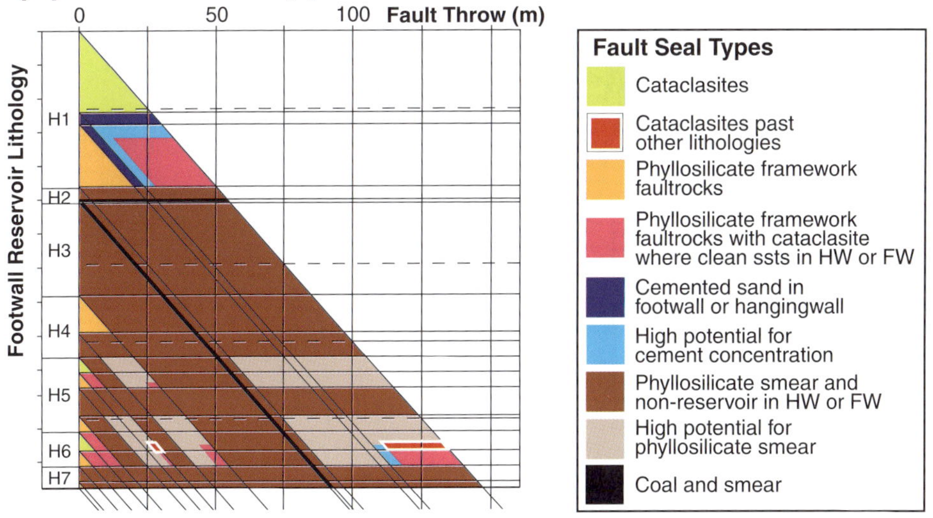

Fig. 5. Juxtaposition and fault seal patterns generated by faulting of the specific reservoir stratigraphy from well 15/9-2 of the Sleipner Vest Field.

faults are therefore expected to be representative of the fault rocks at the time of filling, apart from in the case of reactivated faults. The differences in hydrocarbon column heights can thus be used for a calibration and validation for the up-scaling of fault rock properties.

The distribution of fault rocks near the GWCs across the field was investigated using the results from the Juxtaposition software, together with simple Allan-type diagrams. The predicted fault rock distribution map is shown in Fig. 6. The cataclasites are the most leaky fault rocks,

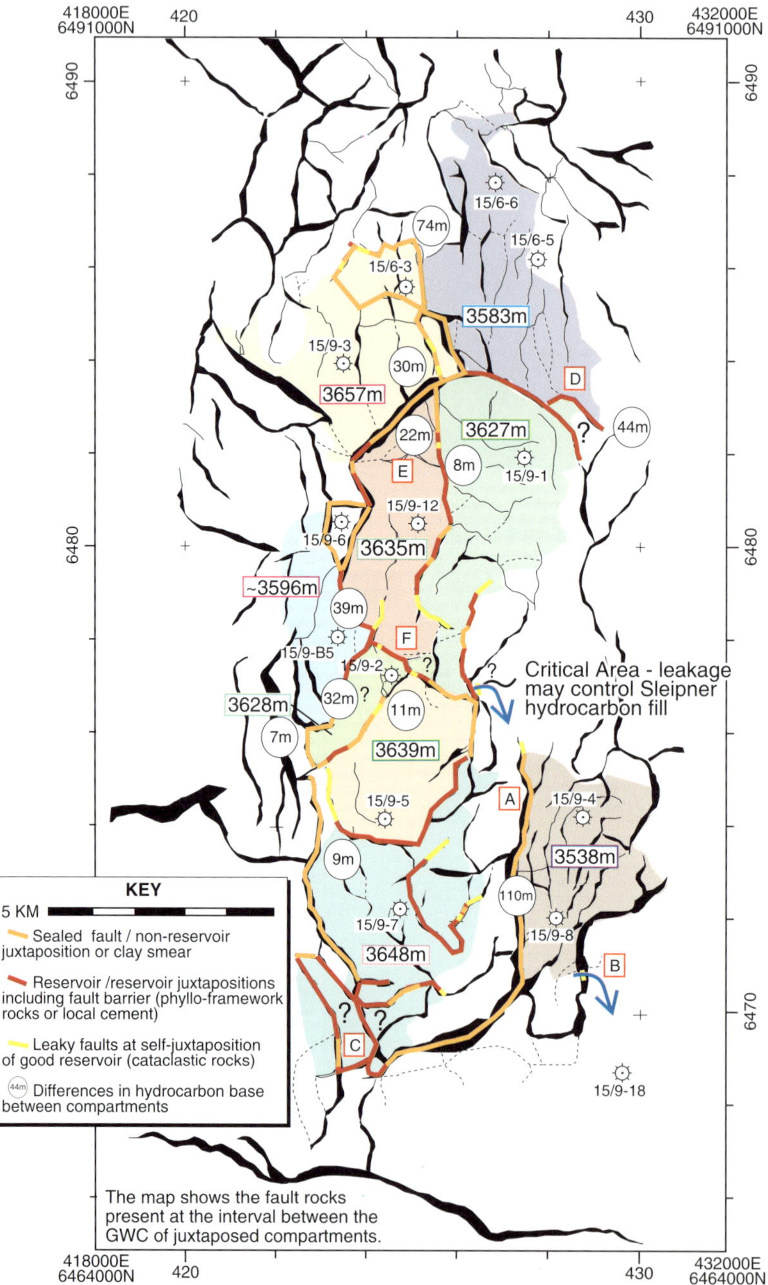

Fig. 6. Distribution of the most 'leaky' fault rocks at and near the GWC.

and are believed to control the differences in the hydrocarbon column heights present in the southern part of the Sleipner Vest Field. The map in Fig. 6 shows the distribution of the different types of fault rocks at the depth interval near the gas–water contacts. The figure illustrates that in areas where the fault zones are dominated by the cataclasites, they are associated with cross-fault differences of 7–11 m in the hydrocarbon column. A prediction of the sealing capacities (column heights of hydrocarbons which can be held below the seal) of these fault rocks can be made from the measured threshold capillary pressures and their equivalent permeabilities. The predicted sealing capacities correspond well to the observed GWC differences. For example, a 7 m difference in GWC is equivalent to a fault rock permeability ~1 mD.

A good correlation between the predicted sealing capacity of the phyllosilicate framework fault rocks and the values indicated from the field GWC data is also apparent. The measured permeabilities on small displacement phyllosilicate framework fault rocks vary from 0.3 to 0.004 mD. According to fault rock distribution mapping (Fig. 6), the sections of faults where phyllosilicate framework fault rocks dominate near the GWC depth intervals, create differences in the hydrocarbon base of 22–44 m. This corresponds to a range of laboratory permeabilities of ~0.05–~0.01 mD, which is nearer the upper, high permeability values for the measured range. This emphasizes, as expected, that the higher permeabilities in the measured range dominate the large scale sealing (membrane) behaviour of the fault zones. Smaller areas of the fault plane will be characterized by the lower permeabilities (higher threshold capillary pressures), as measured in selected specimens, but the localized distribution of these fault rocks will not allow them to control the large-scale effective fault permeability. To assess the wide range of physical properties associated with this class of fault rocks, permeabilities in the range of 0.3 mD–0.005 mD have been selected for input for the reservoir simulation.

Phyllosilicate smears

Phyllosilicate smears from Sleipner Vest are expected to have permeabilities of ≪0.001 mD. The phyllosilicate smear fault rocks will have such very low permeabilities, even for small faults, and up-scaling seems unnecessary as these faults probably act as seals during production. This is supported by the correspondence between the GWC differences (>44 m) in the areas of the field where smears dominate. The maximum sealing capacity possible associated with the low permeability end of these phyllosilicate smears is >300 m.

Calibration against production data and pressure history

The production history of the Sleipner Vest Field is too short for calibrating the detailed permeabilities of the distributed fault rocks with the pressure history. However, two wells from a neighbouring field, that is also producing from the Hugin Formation, provide an initial means of calibration. Faults are present between the two wells and are considered to represent membrane seals consisting of phyllosilicate framework fault rocks, as they are believed to fault an impure sandstone within the Hugin Formation. The pressure history of the two wells (A and B) is shown in Fig. 7. The data illustrate that there is some communication between the two wells, because the pressure dropped in well B before production started in this well in October 1995, inferring a pressure drawdown from another well. There was a pressure difference of about 10 bars when well B (Fig. 7) started production. A pressure of 10 bars for the Sleipner hydrocarbon–water system is equivalent to a 61 bar (~900 psi) capillary pressure for the Hg–air system. This falls within the range measured for the phyllosilicate framework fault rocks in Sleipner. A good history match was made using 3 bars for the hydrocarbon–water capillary pressure on a fault separating the two wells (H. Hansen, pers. comm).

Distribution of fault rocks on fault planes

The recognition that different phyllosilicate contents in sandstones create different fault rock types by faulting is important to seal evaluation (Knipe 1992; Gibson 1994; Yielding *et al.* 1997). We have used FAPS software (Freeman *et al.* 1989; Needham *et al.* 1996) to provide an estimate of the sealing potential distributions on faults, as the percentage of shale within the slipped interval changes. That is, the GOUGE RATIO (hereafter named GR) has been estimated for different parts of the same fault. The overall procedure was as follows:

(i) the depth model obtained from the IRAP™ model of the Sleipner Vest Field, with the subdivision of the Hugin Fm reservoir into 8 reservoir zones, was imported into the FAPS database;

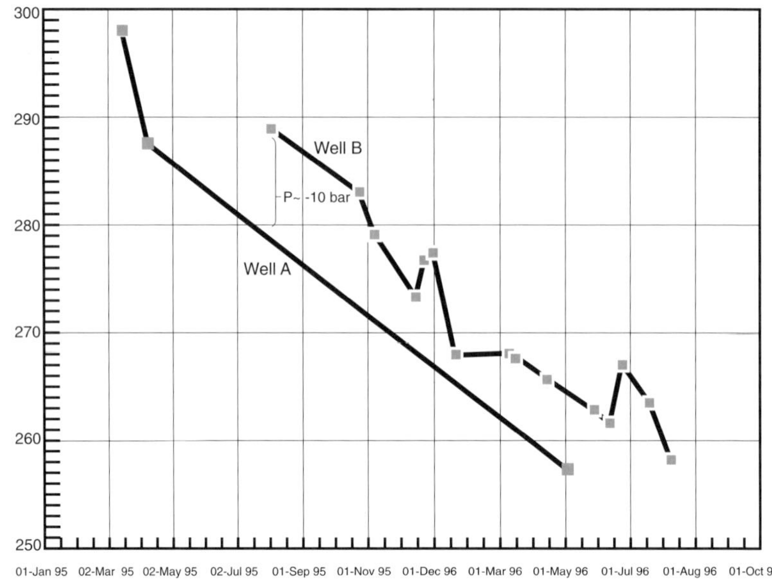

Fig. 7. Pressure history from two wells producing from the Hugin Formation on a neighbouring field.

(ii) A detailed stratigraphic subdivision of the reservoir was separated into three different classes (good, poor and non-reservoir), based on the clay-content of the rock in the nearest well;

(iii) the sealing potential (or GR) was calculated and contoured for a number of points on the mapped fault surface, using equation 1 below.

$$GR = \{(\Sigma V_{\mathrm{d}} \cdot \mathrm{Dz})/t \times 100\% \qquad (1)$$

V_{d} is the clay volume fraction in each interval of thickness, t is the fault throw at that point.

The assumption of a correlation between the volume of shale in the undeformed section and the volume of phyllosilicates in the fault is made. The GR values were divided into three groups (0–14, 14–40 and >40), and the expected distribution of the different fault rock types contoured on the whole fault plane, or only at the reservoir–reservoir juxtapositions as shown in Figs 8 a & b. Due to a large spread of permeabilities, the phyllosilicate framework fault rock class has been subdivided into three groups (14–20%, 20–30% and 30–40%) and decreasing permeabilities assigned to increasing phyllosilicate contents and increasing GR values. The contouring of the fault plane makes it possible to estimate the differences in communication for different juxtaposed sandstones. Figure 8b illustrates that lower permeabilities (i.e. higher entry pressures) are required for the thin sands juxtaposed in the lower part of the fault plane compared to the cleaner sands juxtaposed in the top of the fault plane.

Tectonic reactivation

Reactivation of faults is observed both in the microstructural data and on seismic data from the Sleipner Vest. A systematic pattern for the distribution of the reactivation has been difficult to map from seismic data. Microstructural analysis on Sleipner Vest has detected both an increase and decrease of the petrophysical properties related to late inversion. The fault orientation relative to the inversion stress axes are believed to control the permeability reduction or increase. The late inversion has a maximum horizontal compressional stress orientated approximately N–S (Pegrum & Ljones 1984). This will give compression on the E–W faults and extension on the N–S faults. Better transmissibilities may therefore be expected for the same fault rock types on N–S striking faults, compared to E–W striking faults.

Communication across faults in Sleipner Vest

The map in Fig. 9 shows the distribution of the most leaky fault rocks on the main faults of the Sleipner Vest Field. It is based on the type

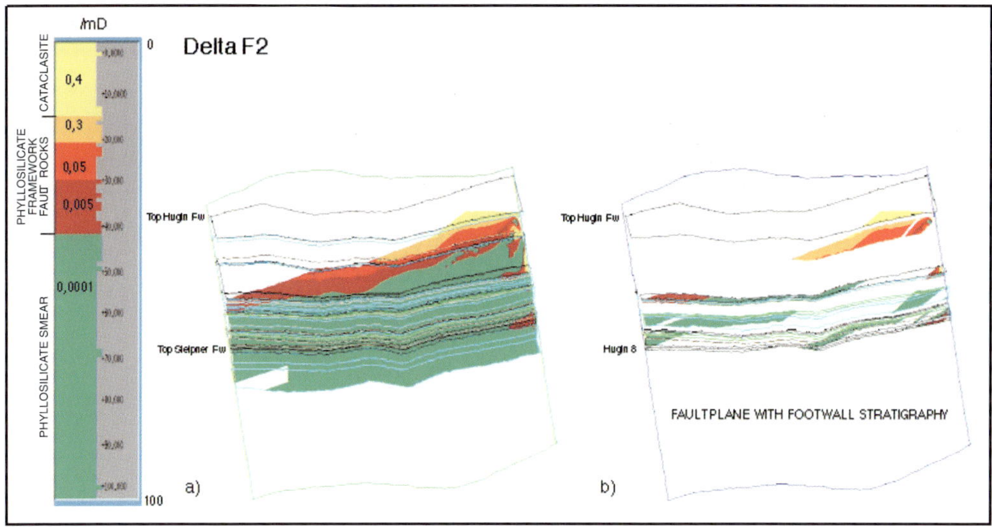

Fig. 8. Fault rock type distributions contoured on a fault plane.

of fault plane diagrams shown in Figs 8a & b. In the case of Sleipner Vest (primarily a gas-condensate field) a cut-off permeability value of 0.1 mD can be used as a general seal / non seal value in the reservoir model. For oil, a permeability cut off of 1 mD is used. A brief review of the more detailed controls on the communication in different parts of the field is presented.

Southern Sleipner Vest area

In the southern part of the field, the faults with reservoir–reservoir juxtaposition are dominated by cataclasites and relatively high permeability phyllosilicate framework fault rocks, with expected permeabilities and threshold capillary pressures of ~0.3 mD and ~2 bars, respectively, for the hydrocarbon–water system in Sleipner Vest. Good communication across intra-reservoir faults is expected, especially in the upper parts of the reservoir, where the cleanest sands are present. There are no pressure data indicating lithological controlled horizontal barriers in the field, but this cannot be excluded.

Central Sleipner Vest area

The faults limiting the fault block of well 15/9-B5 (Fig. 9) are considered to be dominated by phyllosilicate framework fault rocks with permeabilities of 0.05–0.005 mD and threshold capillary pressures of ~4 to ~10 bars for the hydrocarbon–water system in Sleipner. The difference in the hydrocarbon contacts of 39 m confirms the poorer communication across the fault blocks (e.g. in the central western part of the field).

For the southern/central part of the field, high permeability phyllosilicate framework fault rocks dominate and communication is expected to be good. Faults with throws exceeding 50 m are, however, likely to produce large areas with low permeability phyllosilicate framework fault rocks. Permeabilities and entry pressures of 0.05–0.005 mD and ~4 to ~10 bars for the hydrocarbon–water system in Sleipner, respectively, are predicted, which will retard communication.

Northern Sleipner Vest area

In the north of the Sleipner field, communication appears to be more difficult as impure sandstones and shale layers of the 'distal' facies are more common, which will create lower permeability phyllosilicate framework fault rocks and phyllosilicate smears. The expected permeabilities and capillary entry pressures are 0.3–0.005 mD and ~2 to ~10 bars and ≪0.001 mD and >56 bars (for the hydrocarbon–water system in Sleipner), respectively. Communication appears difficult between the east and west areas in the north. Communication is, however, good on the flanks of the northern part of the field and this indicates that water support will be effective.

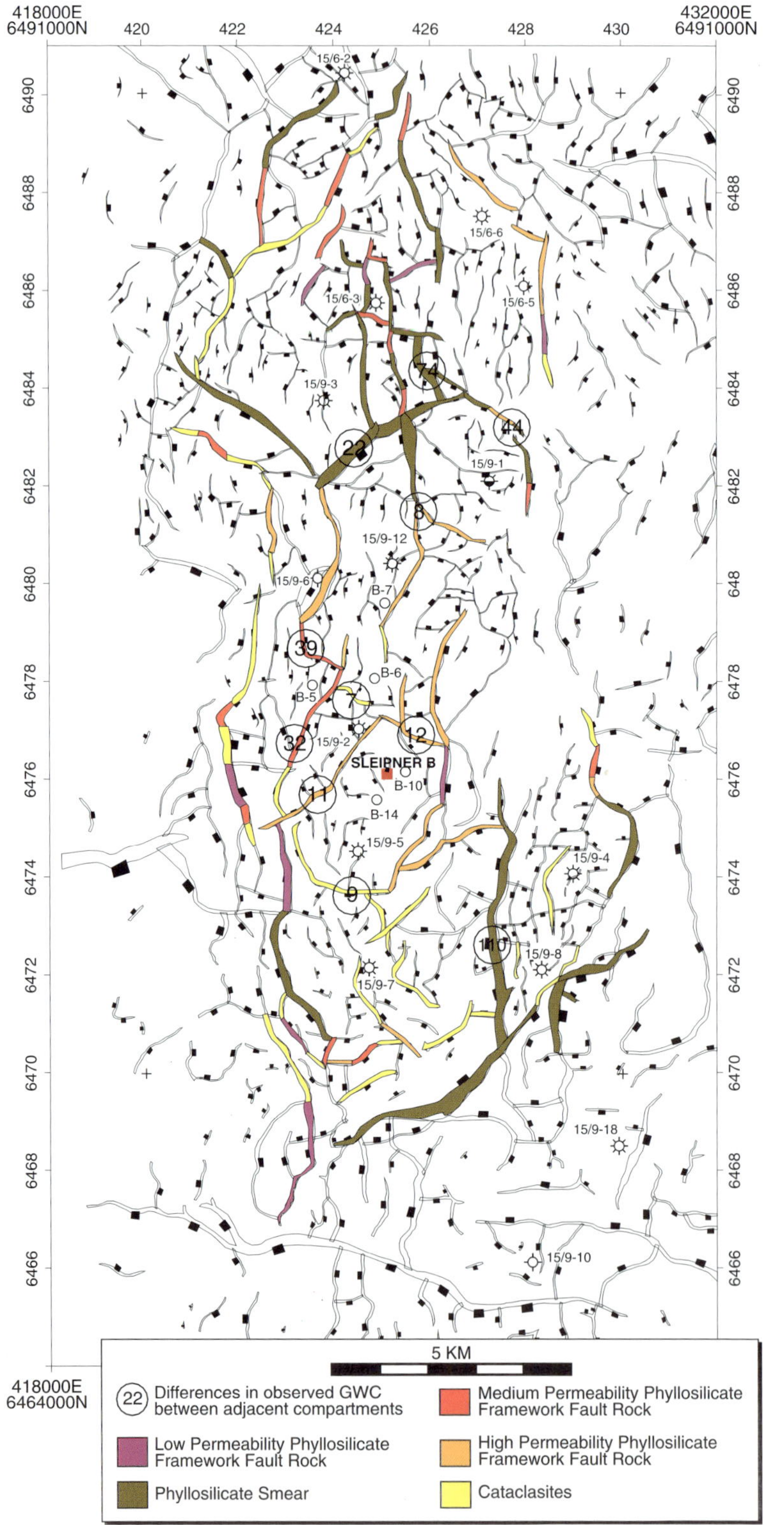
418000E
6491000N
432000E
6491000N
418000E
6464000N
15/6-2
15/6-3
15/6-5
15/6-6
15/9-1
15/9-2
15/9-3
15/9-4
15/9-5
15/9-6
15/9-7
15/9-8
15/9-10
15/9-12
15/9-18
B-5
B-6
B-7
B-10
B-14
SLEIPNER B
5 KM
Differences in observed GWC between adjacent compartments
Low Permeability Phyllosilicate Framework Fault Rock
Phyllosilicate Smear
Medium Permeability Phyllosilicate Framework Fault Rock
High Permeability Phyllosilicate Framework Fault Rock
Cataclasites

Although salt invasion along the fault zones has been recognized as a possibility for the northern part of the Sleipner Vest Field, salt invasion in the reservoir fault rocks is not required to explain the differences in GWC in this region. During the production of the field, a pressure drop in the range of 200 bars is to be expected. Given the conclusions from this study, this appears adequate to allow some communication across all the intra-reservoir faults with the possible exception of the phyllosilicate smear fault rocks. The initial modelling suggests that many months of production is required to induce the pressures needed for production through low permeability phyllosilicate framework fault rocks.

Discussion

The GR is similar to the shale/gouge ratio (SGR) of Skerlec and with the clay smear potential (CSP) of Bouvier *et al.* (1989) and Jev *et al.* (1993). Skerlec (unpublished report) studied known sealing and non-sealing faults affecting the Upper Jurassic in the Block 35/11 area in the North Viking Graben, and concluded that non-sealing faults had SGR <29 whereas SGR >40 implies sealing faults. These conclusions are very similar to those conclusions of Bouvier *et al.* (1989) who also calibrated the CPS against known sealing and non-sealing faults. CPS <28 have low sealing potential while CPS >36 have high sealing potential. Jev *et al.* (1993) used the same technique on the Akaso Field and quoted CPS <15 as non-sealing while CPS >30 as leaking. Fristad *et al.* (1997) have studied the Oseberg Syd area using similar methods to this study, and have found that SGR values of 15–18 are consistent with adjacent fault blocks having small pressure differentials (typically less than ~1–2 bars for the hydrocarbon–water system). Values of 18 or greater correspond to significant fault seal barriers, 8 bars of pressure differential or 100 m difference in oil–water contacts.

In the Sleipner Vest study reported here, a GR >40 is associated with clay smearing, and sealing in general. The data also imply that the low permeability PFFR (GR 30–40) can hold considerable hydrocarbon columns but may be transmissible for gas late in the proposed production history. The cataclastic fault rocks (GR 0–14) and the high permeability PFFR (GR 14–20) can in general be considered as being non-sealing for hydrocarbons. The medium permeability PFFR (GR 20–30) can also be considered as non-sealing for gas, but will probably be sealing for oil.

The results from the Sleipner Vest Field are in good agreement with these previous studies but contain more information and detail on the communication across faults. The GWC data allow validation of the core based petrophysical analysis of faults and sealing. In addition, the study has emphasized the need to consider the complexities of communication and the likely impact of the variation in transmissibility on different parts of the same fault plane. In the Sleipner Vest field, the overall hydrocarbon level is probably primarily controlled by a topographic spill point. In this situation, the differences in the GWCs recorded are likely to be related to the sealing properties of the faults near to the GWCs.

Despite the success of the integrated study reported here, and the demonstration that small-scale core studies can make a significant contribution to both the prediction and the modelling of fault sealing, there are still uncertainties. The most important ones are listed below:

(1) Seismic interpretation of fault throw. In Sleipner Vest, a large proportion of the intra-reservoir faults have interpreted throws in the range of seismic resolution (~30 m). This is a critical limitation in mapping fault geometries and estimating fault throws. The detailed analysis reported involved the evaluation of a number of possible/viable geometries to both the stratigraphy and the faults (including damage zone structures) in critical areas identified by the initial (triangular) Juxtaposition analysis.
(2) The VSH-log used to differentiate the phyllosilicate content of the host rock. The correlation with the true phyllosilicate content depends upon the resolution and ability to detect different phyllosilicate types. Given the critical link between the clay content and sealing properties in this field, careful calibration of this V-Shale data is essential.
(3) The GR calculation in the FAPS module assumes that the fault gouge is accumulating equal amounts of sand and shale and that equal amounts are accumulated from hanging-wall and footwall. This may be an oversimplification of the complex processes which take place in fault zones, but the pressure

Fig. 9. Distribution of the fault rocks on the main faults of the Sleipner Vest field. The map is based on fault plane diagrams as in Fig. 8 and shows the projected location of high permeability windows.

or inter-well permeability data from across a fault zone will incorporate this complexity.

(4) The strengths of the phyllosilicate smears are uncertain. Quantification of the strength of the different seals is important for the estimation of gas communication and water support during production. The first well planned on the adjacent Delta structure will provide an excellent opportunity to test the communication across these fault rocks, and will provide early information needed for back-calculating fault behaviour. With limited vertical communication in Sleipner Vest, and no water communication expected across the faults with phyllosilicate smears, the thin and poor quality sands in the deeper sections will be very difficult to produce, and possibly 1/3 of the volumes could be unproducible. Production testing from the first well on the Delta structure should be used to assess how much pressure build-up the low permeability phyllosilicate framework fault rocks can hold before leaking, either because the threshold capillary pressure has been reached, or because the seal experiences a mechanical failure.

Conclusions

The study has illustrated that prediction of the variations in hydrocarbon–water contact levels across Sleipner derived from careful evaluation of fault rock properties is possible. The validation of predictions of GWCs over the field support the integrated approach to fault seal analysis, which combines detailed microstructural and petrophysical characterization with seismic data on fault geometry. In addition, the amalgamation of the results with field information of hydrocarbon distribution creates an important platform for future modelling of the reservoir behaviour. The study has also identified the possible future directions which will continue this integration as the field development proceeds.

In summary:

(i) faults in the Sleipner Vest field do not have the ability to completely seal gas over much of the crest of the field. However, the fault rocks developed on intra-reservoir faults provide flow retarders or barriers which will restrict flow by variable degrees;

(ii) faults probably act as seals to gas in the central northern parts of the field, due to more extensive distribution of phyllosilicate smears on the faults in this lower reservoir net/gross region;

(iii) the study provides a consistent explanation for the observed differences in GWC in different fault blocks and highlights the importance of clay content in creating seals and flow retarders in this field.

The support of Statoil and the Sleipner Vest field partners is gratefully acknowledged by the Rock Deformation Research Group. Discussions with T. Harper and E. Lundin, as well as with other members of the RDR group are acknowledged. Constructive comments and reviewing by G. Yielding and S. Digert are also appreciated.

References

ALLAN, U. S. 1989. Model for hydrocarbon migration and entrapment within faulted structures. *American Association of Petroleum Geologists Bulletin*, **73**, 803–811.

BOUVIER, J. D., SIJPESTEIJN, K., KLEUSNER, D. F., ONYEJEKWE, C. C. & VAN DER PAL, R. C. 1989. Three-dimensional seismic interpretation and fault sealing investigations, Nun River field, Nigeria. *American Association of Petroleum Geologists Bulletin*, **73**, 1397–1414.

COWARD, M. P. 1994. Inversion Tectonics. *In*: HANCOCK, P. L. (ed.) *Continental Deformation*. Pergammon Press, 289–304.

FISHER, Q. J. & KNIPE, R. J. 1998. Fault sealing processes in sliliciclastic sediments. This volume.

FREEMAN, B., YIELDING, G. & BADLEY, M. 1989. Fault correlation during seismic interpretation. First Break, 8 (3), 87–95.

FRISTAD, T., GORTH., A., YEILDING, G. & FREEMAN, B. 1997. Quantitative fault seal prediction – a case study from Oseberg Syd. *In*: MOLLER-PEDERSON, P. & KOESTLER, A. G. *Hydrocarbon Seals – Importance for Exploration and Production*. NPF Special Publication **7**, Elsevier, Singapore, 107–124.

GIBSON, R. G. 1994. Fault zone seals in siliclastic strata of the Columbus Basin, Offshore Trinidad. *American Association of Petroleum Geologists Bulletin*, **78**, 1372–1385.

JEV, B. I., KARS-SIJPESTEIJN, C. H., PETERS, M. P. A. M., WATTS, N. L. & WHITE, J. T. 1993. Akaso field, Nigeria: Use of integrated 3D seismic, fault slicing, clay smearing, and RFT pressure data on fault trapping and dynamic leakage. *American Association of Petroleum Geologists Bulletin*, **77**, 1389–1404.

KNIPE, R. J., 1992. Faulting processes and fault seal. *In*: LARSEN, R. M., BREKKE, H., LARSEN, B. T. & TALLERAS, E. (eds) *Structural and Tectonic Modelling and its application to Petroleum Geology*. NPF Special Publications **1**, Elsevier, Amsterdam, 325–342.

—— 1993. The influence of fault zone processes and diagenesis on fluid flow. *In*: HORBURY, A. D. & ROBINSON, A. G. (eds) Diagenesis and Basin Development. *American Association of Petroleum Geologists Bulletin*, **36**, 135–154.

—— 1997. Juxtaposition and seal diagrams to help analyze fault seals in hydrocarbon reservoirs. *American Association of Petroleum Geologists Bulletin*, **81** (2) 187–195.

——, FISHER, Q. J. JONES, G., CLENNELL, M. B., FARMER, A. B., HARRISON, A. KIDD, B., MCALLISTER, E., PORTER, J. R. & WHITE, E. A. 1997. Fault seal analysis: successful methodologies, application and future directions. *In*: MOLLER-PEDERSON P. & KOESTLER, A. G. *Hydrocarbon Seals – Importance for Exploration and Production*. NPF Special Publications, **7**, Elsevier, Singapore, 15–38.

KNOTT, S. D. 1993. Fault seal analysis in the North Sea. *American Association of Petroleum Geologists Bulletin*, **77**, 778–792.

NEEDHAM, D. T, YIELDING, G. & FREEMAN, B. 1996. Analysis of fault geometry and displacement patterns. *In*: BUCHANAN, P. G. & NIEUWLAND, D. A. (eds) *Modern Developments in Structural Interpretation Validation and Modelling*. Geological Society, London, Special Publications, **99**, 189–200.

PEGRUM, R. M. 1984. The extent of the Tornquist Zone in the Norwegian North Sea. *Norsk Geologisk Tidsskrift*, **64**, 39–68.

—— & LJONES, T. E. 1984. 15/9 Gamma gas field offshore Norway, new trap type for the North Sea basin with regional structural implications. *American Association of Petroleum Geologists Bulletin*, **68**, 874–902.

RATTEY, R. P. & HAYWARD, A. B. 1993. Sequence stratigraphy of a failed rift system; the Middle Jurassic to Early Cretaceous basin evolution of the Central and Northern North Sea. *In*: PARKER J. R. (ed.) *Petroleum Geology of Northwest Europe: Proceedings of the 4th Conference*, Geological Society, London, 215–250.

SVERDRUP, E. & BJØRLYKKE, K. A. 1997. Fault properties and the development of cemented fault zones in sedimentary basins. Field examples and predictive models. *In*: MOLLER-PEDERSON, P. & KOESTLER, A. G. (eds) *Hydrocarbon Seals - Importance for Exploration and Production*. NPF Special Publications, **7**, Elsevier, Singapore, 91–106.

UNDERHILL, J. R. & PARTINGTON, M. A. 1993. Jurassic thermal doming and deflation in the North Sea: implications of the sequence stratigraphic evidence. *In*: PARKER, J. R. (ed.) *Petroleum Geology of Northwest Europe: Proceedings of the 4th Conference*. Geological Society, London, 337–326.

WATTS, N. L. 1987. Theoretical aspects of cap-rock and fault seals for single and two-phase hydrocarbon columns. *Marine and Petroleum Geology*, **4**, 274–307.

YIELDING, G., FREEMAN, B. & NEEDHAM, D. T. 1997. Quantitative fault seal prediction. *American Association of Petroleum Geologists Bulletin*, **81**, 897–917.

Facies and curvature controlled 3D fracture models in a Cretaceous carbonate reservoir, Arabian Gulf

J. B. ERICSSON[1], H. C. McKEAN[2] & R. J. HOOPER[3]

[1] *Conoco Inc., 10 Desta Dr-505E, Midland, TX 79707, USA*
[2] *McKean Petroleum Eng. Services, R.D.7., Dannevirke, New Zealand*
[3] *Conoco Inc., PO Box 2197, Houston, TX 77252, USA*

Abstract: The Late Cretaceous Ilam Formation of the Arabian Gulf's Fateh Field is a very heterogeneous, fractured carbonate reservoir. Porosity, matrix permeability and facies trends do not appear to be related, but fracture density exhibits a strong relationship to both facies distribution and structural curvature. Three-dimensional (3D) models of dip change, position with respect to the crest of the field, facies and porosity variation are combined to produce a qualitative 3D model of fracture density which matches empirical fracture density data from more than 16 750 m of interpreted image log. The fracture model aids in transmissibility prediction for reservoir simulation, producing a better history match and decreasing uncertainty for future drilling programmes.

The Fateh Field is a large elongate domal structure which is located in the offshore southern Arabian Gulf (Fig. 1). The Fateh dome is interpreted to be salt-cored and to have a similar developmental history to many of the other numerous salt-related domes and swells identified in the southeastern Arabian Gulf (Hooper *et al.* 1996). Salt-related structures in the southeastern Arabian Gulf were initiated and developed throughout the Jurassic and Early Cretaceous in an extensional setting on the trailing southern margin of Tethys. Throughout most of this time, the Arabian platform remained a broad, stable, shallow shelf (Dercourt *et al.* 1993; Grabowski & Norton 1994; Hooper *et al.* 1994*a*,*b*); and the thick carbonate successions that developed contain much of the region's hydrocarbon wealth (Beydoun 1991). The tectonic configuration of the Arabian platform changed irrevocably in the Late Cretaceous (early Santonian) as allochthons were emplaced onto the margin in Iran and Oman. There was a brief return to stability on the margin during which carbonate platform deposition resumed before the onset of the late Tertiary Zagros orogeny (Alavi 1980, 1994; Hooper *et al.* 1994*a*,*b*).

Salt-related structures in the southern Arabian Gulf display several of the characteristics of diapirs that have experienced contractional rejuvenation (as defined by Vendeville & Nilsen 1995). Two principal periods of accelerated growth can be recognized in the Fateh dome. These growth periods broadly correspond in time to contractional events associated with the late Mesozoic emplacement of the allochthons in Oman, and also the late Tertiary Zagros orogeny (Hooper *et al.* 1996). The fault and fracture array in the Fateh dome consequently reflects not only a dynamic response to the vertical and radial growth of the dome during contraction but also the enhancement of older, inherited, extensional fault trends present in the carbonates.

Reservoir geology

Oil within Fateh Field is produced from a series of Cretaceous carbonate reservoirs (Fig. 2), many of which exhibit performance that is strongly controlled by fracturing. The youngest commercially producing horizon in the field is the Late Cretaceous Ilam Formation, a reservoir in which both fracture- and facies distribution play a significant role. The form of the dome at the Ilam level is very gentle. Maximum structural dip is 7 degrees. However, structural dip is 4 degrees or less over most of the flanks of the dome. The Ilam is not one of the primary reservoirs in the field, but its commerciality has been revived by recent horizontal drilling. In this paper, we discuss the critical control on fluid flow in the Ilam Formation exerted by fractures and the consequent importance of accurate fracture prediction in the development of the field.

The Ilam Formation consists of 70–90 m of complex bedded carbonates. One third of this thickness is attributed to its two main reservoir zones. The upper zone is characterized by sheet-like accumulations of peloidal and bioclastic grainstones and packstones, while the thicker lower zone comprises broad aggradational bioclastic grainstone and packstone belts

Ericsson, J. B., McKean, H. C. & Hooper, R. J. 1998. Facies and curvature controlled 3D fracture models in a Cretaceous carbonate reservoir, Arabian Gulf. *In*: Jones, G., Fisher, Q. J. & Knipe, R. J. (eds) *Faulting, Fault Sealing and Fluid Flow in Hydrocarbon Reservoirs*. Geological Society, London, Special Publications, **147**, 299–312.

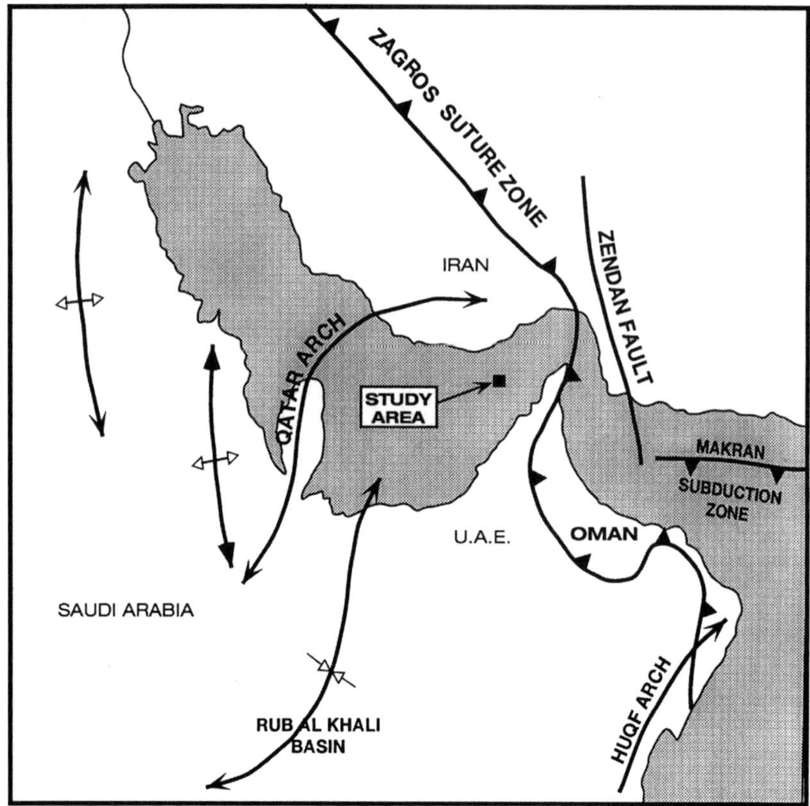

Fig. 1. Regional locator map. The location of the study area in the southern offshore Arabian Gulf.

with intervening packstones and wackestones. Non-reservoir zones that occur above, below and between the two reservoirs are low porosity wackestones and mudstones.

The reservoirs are characterized primarily by microporosity and intraparticle porosity that ranges from 6–20%. Porosity patterns are related to diagenetic processes and show little correlation to grain supported versus matrix supported facies distribution. Virtually no correlation exists between porosity and permeability. Matrix permeability from core is 0–10 mD, however, well behaviour shows that there is considerable enhancement to overall reservoir transmissibility due to the presence of a fracture system.

Database

Acquisition of a 3D seismic survey in 1994 has provided valuable information regarding the complexity of faulting in the field. The Ilam reservoir has 27 years of production data from 47 wells. However, more than 200 logged near-vertical well penetrations exist from drilling to deeper reservoirs (Fig. 3). This log coverage combined with over 900 m of core and 16 750 m of horizontal drilling in the reservoir have resulted in a vast database. Image logs have been acquired in the horizontal wells and interpreted over the entire horizontal section drilled in the reservoir.

Grain supported facies in the Fateh Ilam reservoir can be distinguished from matrix supported facies with 90% confidence using a normalized gamma log technique. Use of this method augments our existing database of facies type beyond just the core data, by including all of the logged well data. Facies mapping and statistical data presented are based on this normalized gamma log technique, with horizontal results supported by biostratigraphic wellsite analysis.

The large Fateh Ilam database allows for detailed interpretation of facies distributions, rock properties (porosity, fluid saturations, etc.) and fracture systems in the reservoir. This information is being used in a recently rejuvenated programme of reservoir simulation, horizontal

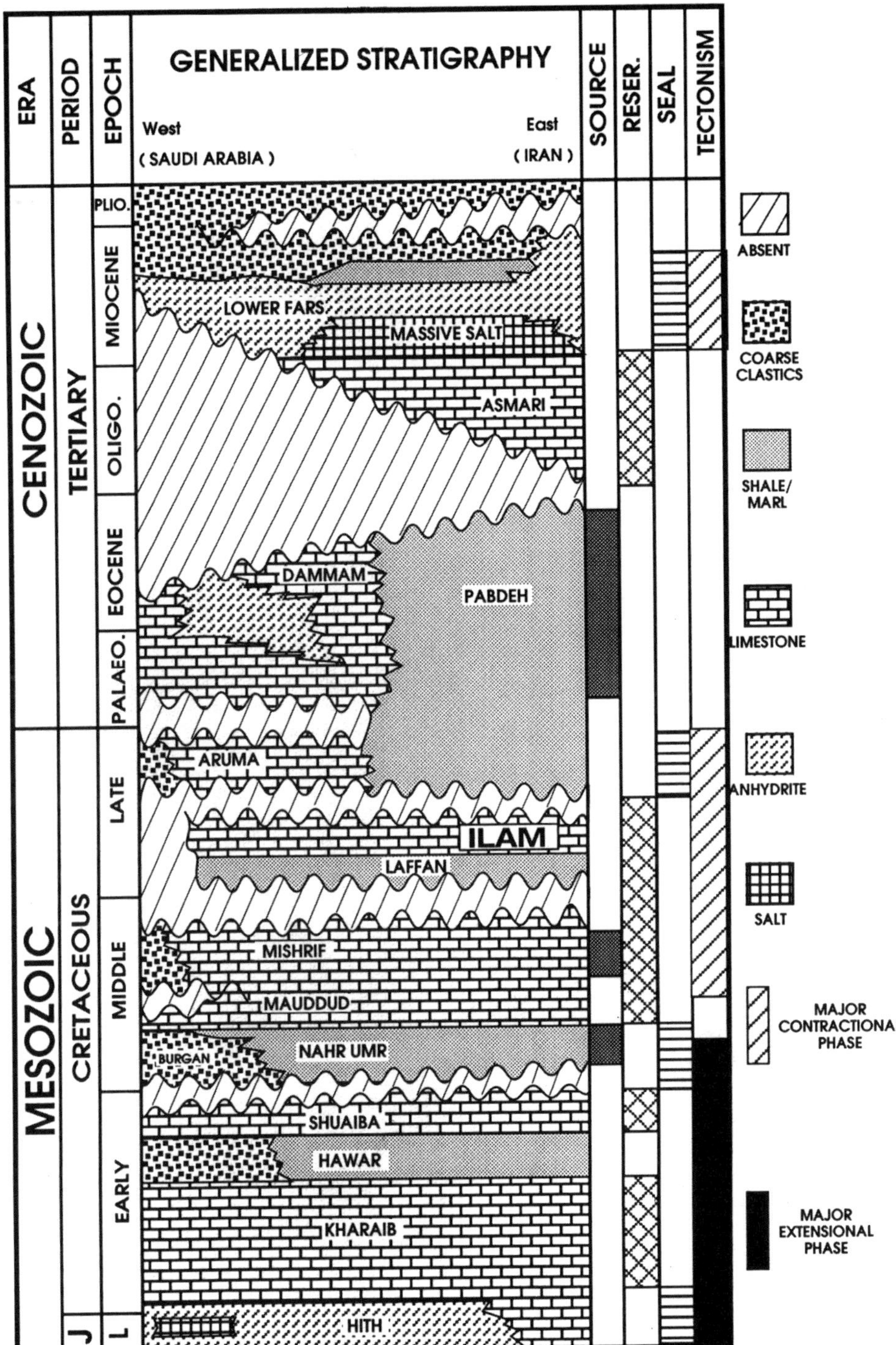

Fig. 2. Stratigraphic column. The Ilam Formation is of Late Cretaceous (Santonian) age.

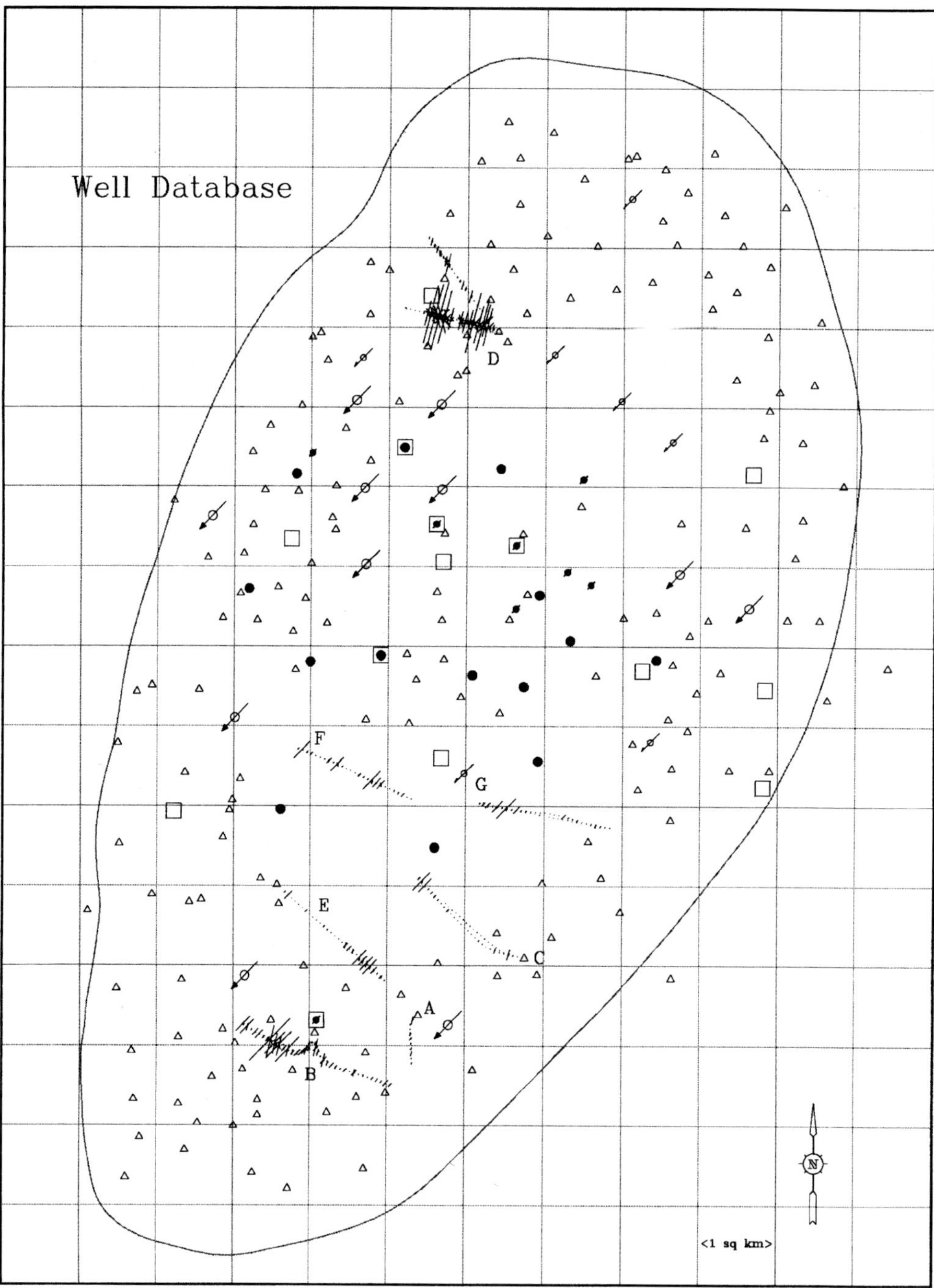

Fig. 3. Database map of the Ilam Formation. Logged wells to deeper reservoirs are shown as triangles; current producing wells are shown as solid circles (former producers are shown as smaller solid symbols); current injection wells are shown as open circles with arrows (past injectors are shown as smaller symbols); core coverage is shown by open squares; horizontal producers lettered in order of their drilling and are shown as lines along which fracture density roses are distributed at 60 m intervals. The longest line from the roses occurs in well D, and represents 750 fractures interpreted in that 60 m interval along the well track. The strike of the fractures is within 15 degrees of the fracture rose line orientations as displayed.

drilling, recompletions and pressure maintenance, to test ideas about fracture distribution and conductivity in the reservoir.

Results

The results presented are based on observations made in the course of ongoing horizontal reservoir development. A qualitative methodology that uses standard geoscience software was developed for use in 3D geological and reservoir models that has resulted in improved predictive capabilities for this reservoir. The principal achievement of the study is the incorporation of a geologically based framework of relative fracture transmissibility for simulation of dynamic flow behaviour. Application to other reservoirs requires specific study of the factors that influence that reservoir's fracture systems, and a customized model based on results of that study. Qualitative information or descriptions have been provided here rather than quantitative data or illustrations for cases in which proprietary data are concerned.

Fracture and facies characteristics

A strong relationship exists between facies and fracturing in the Ilam that is borne out by both core data and image log interpretation (Fig. 4). While the mineralogy of the different Ilam facies is virtually identical, their textures exhibit strong variation. Given comparable porosities, facies with grain supported textures in the reservoir exhibit enhanced fracture density and appear to be more brittle, while facies with matrix supported textures are less prone to fracturing.

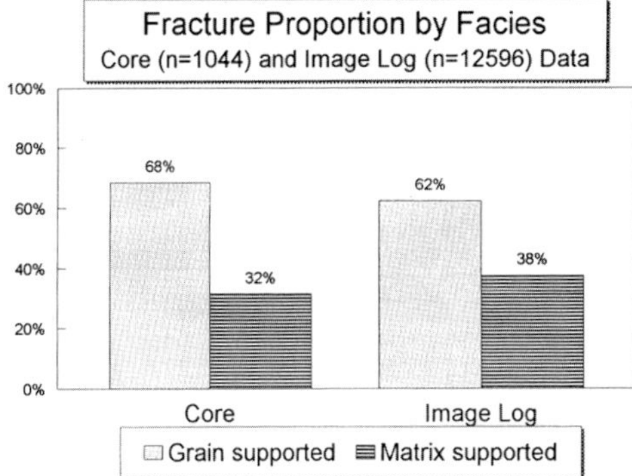

Fig. 4. Fracture proportion in core and image log by facies type. Both core and image log analyses show that fractures are twice as common in grain supported facies than in matrix-supported facies.

Three dominant types of fractures have been distinguished in the reservoir on the basis of core and microresistivity image log interpretation. Two of these types extend beyond the 21.5 cm wellbore diametres and are classified as 'macrofractures'. The third type of fractures are classed as 'mesofractures' because they are shorter than the wellbore diametres. This simplified nomenclature is appropriate for the level of detail interpreted from the image logs.

Hairline Macrofractures, are the first type of macrofractures and are associated with enhancement of transmissibility and reservoir flow. These fine, open fractures develop as pervasive, dense arrays and are prevalent in grain supported textures. They are less well developed in matrix supported facies and increase in abundance in all facies near faults. These features are usually represented by rubble zones in core but are very obvious on image logs. They commonly exhibit small apertures and densities of 25 fractures per metre in the grainstones and 3–10 fractures per metre in packstones. These type of fractures tend to terminate in large stylolites and bedding surfaces and exhibit strong directional orientation. Geological modelling efforts discussed later in this paper are focused primarily on the prediction of these fractures.

The second type of macrofractures, termed *Flow Disruptive Macrofractures*, are far less common but no less important than hairline fractures in terms of reservoir behaviour, as they are suspected of causing oil to be bypassed under water-flood conditions (Wood & East 1992). These features are both preserved in core and can be seen on image logs where they exhibit large open apertures, porosity occlusion in the adjacent matrix, and occasional crystalline cements. They have unpredictable occurrence, but are almost always found near the location of mapped seismic faults. Incomplete waterflood sweep and matrix bypass is thought to result from these fractures as a result of both their high fracture permeability and their tendency to develop adjacent cemented matrix which seals fluid in the fracture conduit from the matrix porosity.

The third type of fractures important to Ilam reservoir behaviour are the Mesofractures. For the most part, these features are short tension gashes that terminate into wispy pressure solution seams which concentrate the minor clay constituents of the unit. They occur primarily in the matrix supported facies. These are often preserved in core with random orientation and have lengths of 1–10 cm. At this scale, these small aperture mesofractures are at the limits of image log resolution, and appear as indistinct disruptions of image log character at best.

Influence of faulting and curvature on fracture distribution

Both hairline and flow disruptive macrofractures show clear relationships to tectonic features, whereas mesofractures appear to be related solely to the diagenetic processes that formed the wispy pressure solution seams from which they emanate when examined in core. All known occurrences of the rare flow disruptive macrofractures are spatially associated with faults; however, the occurrence and individual behavior of these macrofractures are difficult to predict. The hairline macrofractures are the most common fracture type interpreted from image logs and appear to be strongly influenced by their proximity to faults and flexures in the Ilam reservoir surface. As the most likely source of beneficial enhancement to Ilam reservoir transmissibility, the prediction of hairline macrofractures is a key to understanding dynamic reservoir response.

The strike of hairline macrofractures is primarily parallel to the present-day maximum principal horizontal stress, which has been inferred from bore-hole breakout data to be NNE–SSW. The alignment of the fractures with the maximum principal horizontal stress supports the open character of these features as seen on image logs and as demonstrated by production data. This direction is also parallel to the strike of many of the faults defined from a high-resolution 3D seismic survey over the field.

When most faults are crossed by horizontal wells in the reservoir, there is no single discontinuity image that stands out as the dominant fault plane. Instead, numerous hairline macrofractures are observed on the image logs in zones ranging from three m to nearly 100 m along the horizontal wellbore. The width and intensity of the zones of accentuated fracture density show some qualitative relationship to three observed characteristics: the orientation of the seismically resolvable faulting, the disparity in the dip directions of the footwall and hanging wall blocks and the textural facies of the faulted matrix.

Where seismic faulting deviates from the typical NNE–SSW orientation, additional fractures develop which parallel the fault direction, while fracture density in the NNE–SSW orientation is accentuated. The fractured zones in these cases are observed to be wider, which may indicate the existance of sub-seismic en echelon fault configurations in which the dominant individual fracture orientation parallels the *in situ* stress field while the overall trend of the zone roughly parallels the seismically mapped fault. High frequency 3D seismic data character also supports the idea that fault zones may be made up of numerous en echelon segments.

If the dip direction of the reservoir surface changes by more than 15 degrees across a fault, the fault zone is almost always characterized by a zone of higher fracture density. Conversely, faults that have similar hanging wall and footwall dips may be the only faults which exhibit single fracture planes on image logs. When compared to the strong impact of change in dip azimuth on the fracture system of this gently dipping structure, the effect of change in dip magnitude appears to be minimal.

Grain supported facies typically exhibit zones of increased fracture intensity that are more than 4 times wider in an areal sense than are seen in the matrix supported Ilam textures (Fig. 5a & b). This fracture accentuation is suspected to contribute to fluid conductivity both along and across fault zones, especially where grain supported facies are offset against each other.

Fluid flow characteristics

The differentially fractured nature of the Ilam Formation is observed in the production behaviour of wells completed in the reservoir. At both the wellbore and intra-reservoir scales, a high degree of variability in fluid flow behaviour is evident, as shown by well productivity, pressure gradients and water injection response.

Well productivity indices for the Ilam reservoir show wide variation in an areal sense. This index is a measure of the ability of a well to produce fluid for a given pressure drawdown, and in the Ilam reservoir it can vary from 0.3 $\text{bbl day}^{-1}\text{psi}^{-1}$ to more than 20 $\text{bbl day}^{-1}\text{psi}^{-1}$. The higher productivities observed are inconsistent with expectations for rock of such low matrix permeability and are therefore inferred to be evidence of enhanced permeability from the fracture system contributing to well production performance. An apparent broad qualitative correlation between well productivity index and the proportion of grain-supported facies present occurs in vertical wells (Fig. 6) suggesting that wells with higher productivity indices tend to have higher proportions of grain supported facies present. The inference is therefore drawn, that grain

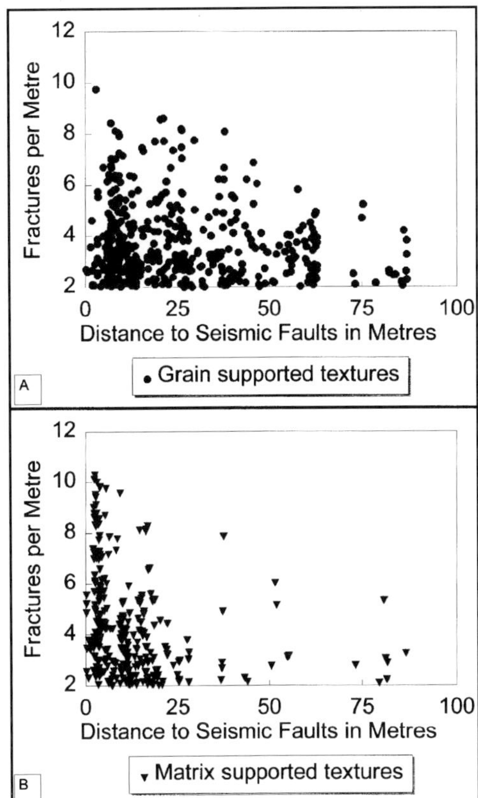

Fig. 5. Fracture density per metre interval along the horizontal wellbore versus distance to faults for (**a**) grain supported facies and (**b**) matrix-supported facies, $n = 6230$ data bins. Fractures in grain-supported textures commonly occur more than 4 times farther from seismic faults than fractures in matrix-supported textures. Faulting appears to create a wider zone of fractures in grain-supported facies, which leads to preferential enhancement of fluid conductivities in this facies near faults.

supported facies tend to have a higher proportion of fracture related permeability than the more matrix supported facies. Additional support for this observation occurs at a larger reservoir scale, as history matching of reservoir and well performance to actual production data with reservoir simulation models is often not possible without invoking fracture enhancement of reservoir transmissibility.

Pressure data collected from the reservoir are also consistent with a highly heterogeneous fracture system. Strong variations in interwell pressure gradient across the reservoir are indicative of the wide range in intra-reservoir transmissibility. While these variable gradients do not correlate specifically with the NNE–SSW

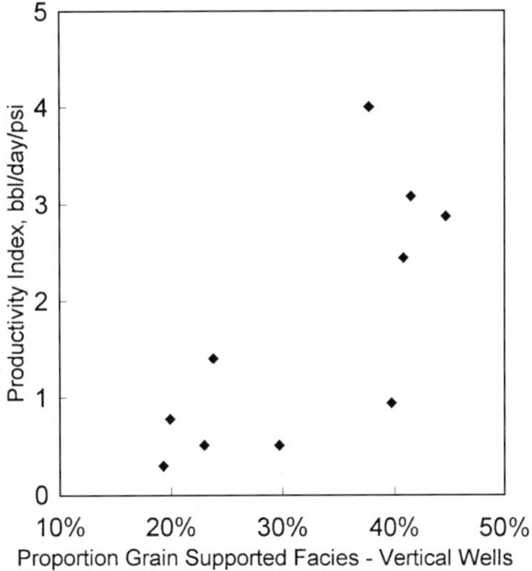

Fig. 6. Percent of grain supported facies in the total Ilam section versus productivity index (vertical wells only). Productivity index is a measure of the ability of a well to produce fluid for a given pressure drawdown, calculated in bbl day^{-1} psi^{-1}. Well productivity is enhanced when a greater proportion of grainstone facies is encountered in the well due to the higher degree of macrofracturing in this facies.

orientation of either the fractures or of the maximum principal horizontal stress direction, they often show a relationship to facies type.

The response of the reservoir to water injection is also found to be highly variable. In certain cases, the injection flood pattern appears to be as expected for a layered matrix system; however, in general, the flow of injection water through the reservoir is very irregular. In other cases, the flood front has migrated considerable distances in very limited periods of time in response to production. This is inferred to be due to water channelling along the fracture system toward production wells.

Discussion

In contrast to the deleterious effects of the flow disruptive macrofractures, the more diffuse hairline fracture arrays may be exploited as beneficial aids to reservoir flow if they can be understood. In an effort to understand their distribution in the reservoir, structural curvature mapping of the Ilam reservoir surface was employed. This method resulted in the striking qualitative empirical match between structural curvature and image log data (in rose diagram format) shown in Fig. 7.

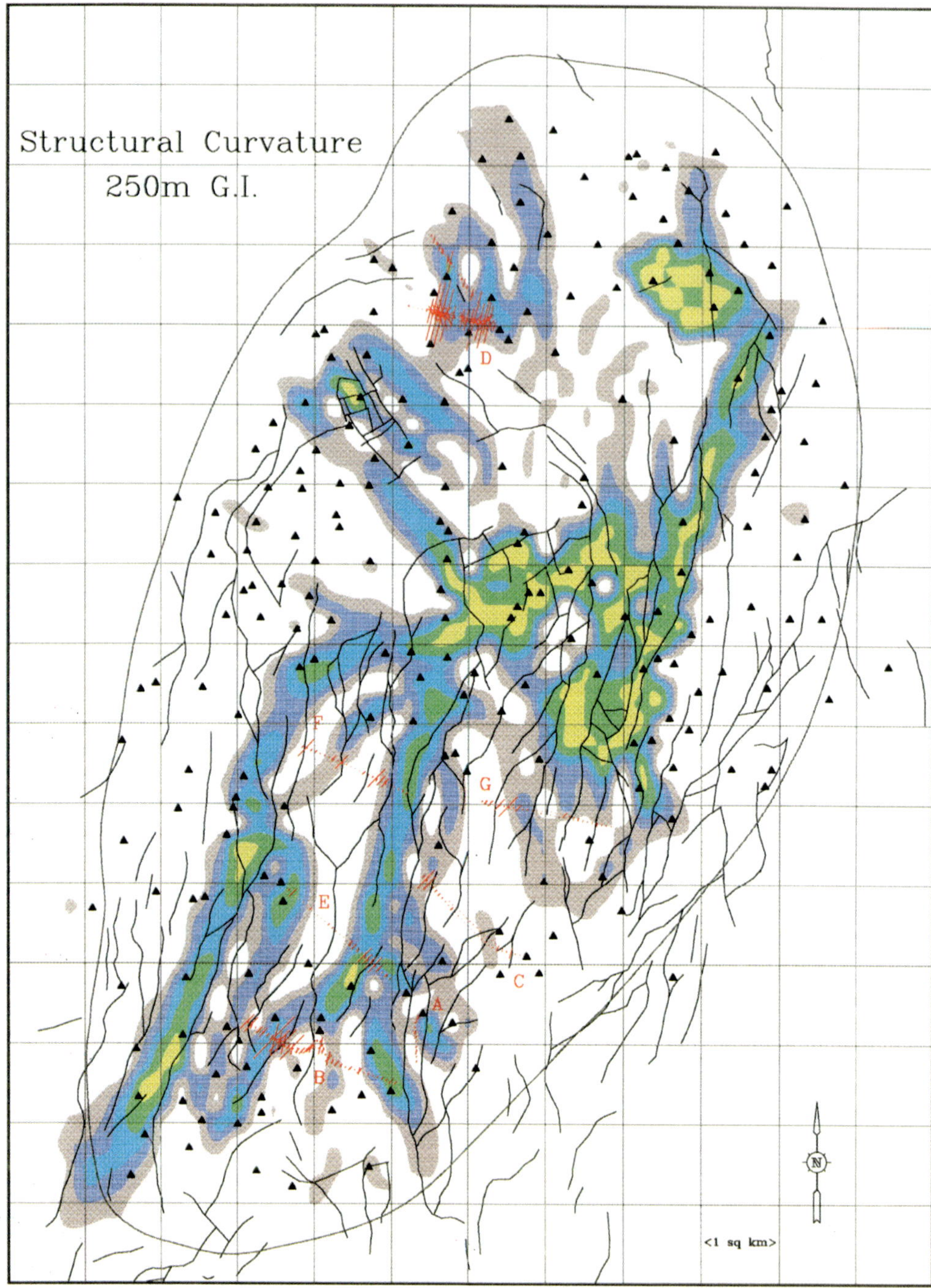

Fig. 7. Structural Curvature from dip azimuth change for the 250 m grid increment case is shown by shading. Yellow denotes areas of highest structural curvature while grey represents lower structural curvature regions. Note the qualitative match to fracture density from image log roses displayed every 60 m along the well paths for horizontal wells. The length of the lines in the roses are proportional to the number of fractures (0–750) in 30 degree 'bins' around the orientations shown by the direction of the six lines that constitute each rose.

3D geological and rock property modelling

Facies, porosity and other rock properties have been modelled in stochastic and deterministic three dimensional geological 'volumes' for the reservoir. These models preserve geological detail from log character as 1–3 m thick layers contained within correlated sequences. This results in 3D volumes made up of more than a million cells that are up-scaled to reservoir simulation models as required. The latest versions include fracture density models, an example of which is shown in Fig. 8.

One of the most functional aspects of the 3D geological and fracture models is the interactive capability to slice them both horizontally and vertically along a well path to optimally position a proposed well to intercept the highest porosity, the richest oil saturation, and the most advantageous fracture system in that area of the field.

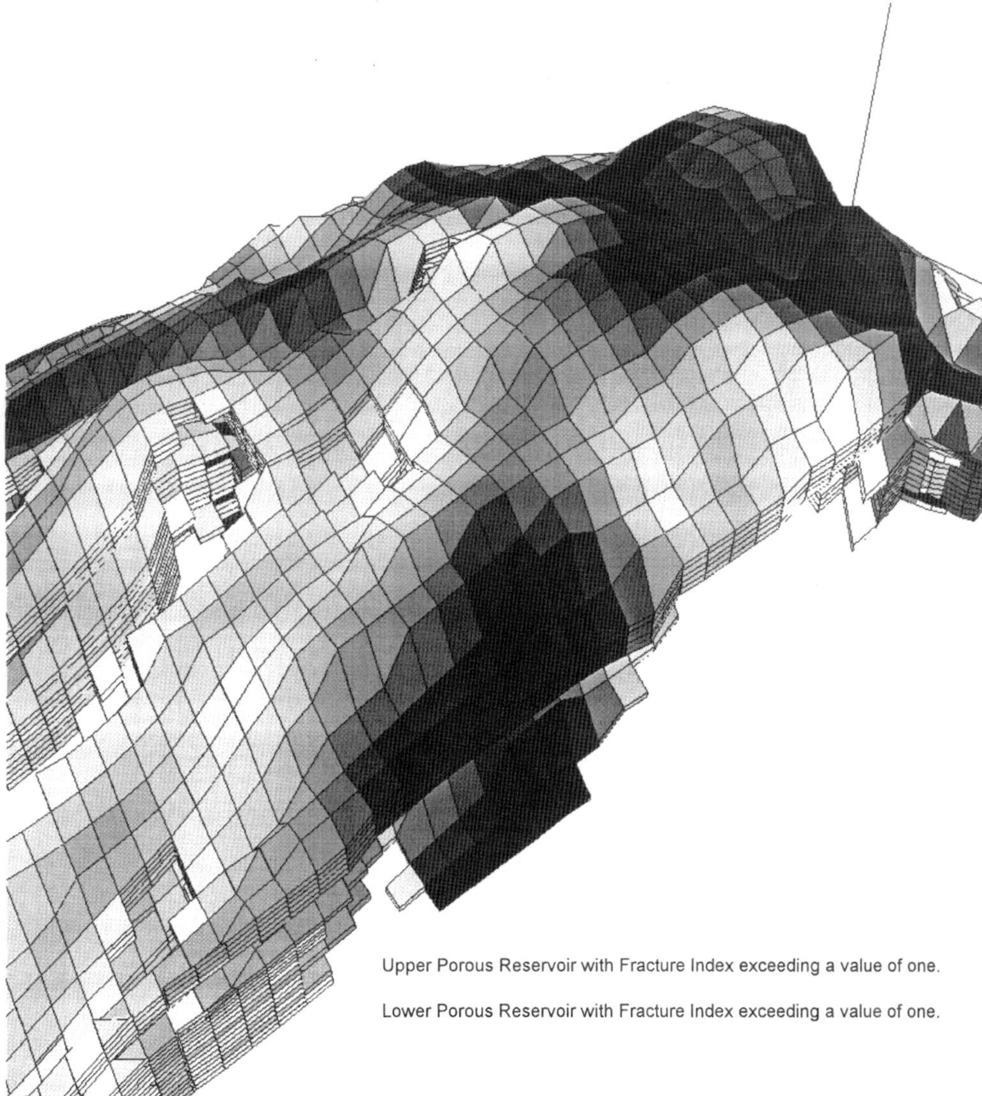

Fig. 8. Three dimensional view of a slice of the 'Fracture Density Index' model from the 3D geological model. Cells are shown only if they contain reservoir quality rock from the two reservoir zones that exceed a fracture density index value of one. Darker shading in the cells indicates higher fracture density index in this example. Cells are 100 m by 100 m by 1.5 to 3.0 m thick.

Fracture density prediction by 3D fracture density modelling

Based on observations of the fracture characteristics in the Ilam Formation, it is apparent that facies type and structural curvature are the primary keys to predicting its fracture density. Porosity and crestal position also have an impact on fracture development but to a much lesser degree than was originally expected. Prior to model development, analysis of more than 12 000 fractures logged in nearly 12 200 m of horizontal section, showed that 62% fall into regions characterized by a high proportion of grainstone facies, while 68% fall into areas distinguished by high structural curvature (Fig. 9). Using the union of these two aspects, 82% of the fractures identified fall into either high structural curvature areas or grainstone areas. These statistics are supported by continued drilling.

In order to take advantage of this excellent level of correlation, 3D fracture density models have been created to complement the existing 3D rock property model. To create the fracture model, four parametres are combined in proportions that provide the best fit to the empirical data, creating a *Fracture Density Index*. These four parametres are defined below for the Fateh Ilam reservoir as *Structural Curvature Index*, *Facies Index*, *Porosity Index* and *Crestal Position Index*.

The most important constituent is structural curvature, derived from dip change mapping. After experimenting with several different structural curvature methods, it was noted that changes in dip direction (dip azimuth) are more sensitive than changes in dip magnitude for this low relief structure, where most dips are less than 4 degrees. Initially, first derivative of dip azimuth is computed from a 3D seismic based structural map. Then, various grid increments are experimented with to emulate observed fracture zone widths from horizontal wells. Unusually high values which may occur as artifacts near larger faults are clipped to the field maximum, and normalized between values of zero and three, with higher values of this Structural Curvature Index indicating larger changes in dip direction. Finally, the results are compared to image log data using both continuous fracture density logs and rose diagrams computed for each 60 m along existing horizontal wells and best fit grid increments are selected for each facies type (Fig. 7 shows the result from the 250 m grid processing). In this case, a 250 m grid was used for grain supported facies and a 50 m grid was used for matrix-supported facies. Values from the two grids are imported to the 3D model conditioned upon the 3D facies model in order to produce a Structural Curvature Index volume that reflects wider fracture zone development in the grain supported facies, as is observed from the horizontal wells.

Even in areas distant from faulting and flexural effects the strong relationship of grain supported facies to fracture intensity makes three dimensional facies distribution in the reservoir another important component in the modelling process. The 3D facies model incorporates both facies map templates and log data. The established relationship between normalized gamma log values and cored facies designations for the Ilam reservoir allows for facies distinction as fine as the layer thickness of the model will allow. The facies values range continuously between zero and three, with higher values of

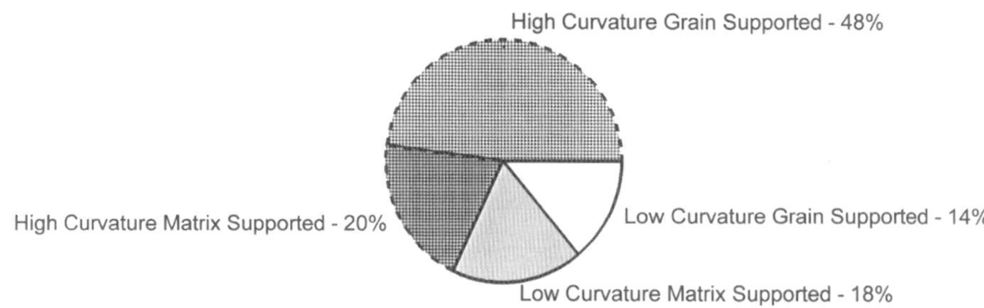

Fig. 9. Proportion of fractures by facies and structural curvature for the Ilam. This graph represents more than 12 000 fractures logged from image log for more than 12 250 m of horizontal section. Grain-supported textures were encountered in 48% of the logged section, while matrix dominated textures were found in 52% of the section. Of the fractures logged, 68% occur in high structural curvature zones, while 62% of fractures fall into the grainstone category. The union of these two characteristics accounts for 82% of the fractures logged. Fractures in orientations suggesting they might be induced have been discarded.

the Facies Index indicating higher proportions of grain-supported textures.

The third component is the three dimensional porosity model derived from petrophysical analysis of wireline logs. Low porosity materials tend to be more brittle than their high porosity equivalents because of the inverse relationship between porosity and rock strength. However, the effect of facies type is much stronger than the effect of porosity for the Ilam, which may reflect rock strength differences that are more strongly related to textural facies type than to porosity. Porosity values are normalized from zero to three, with higher Porosity Index numbers indicating lower porosity or higher propensity to fracturing.

The last factor used is crestal position, calculated from structural height in the reservoir. In the region, several low relief structures demonstrate increased apparent fracture-type flow characteristics on their crests relative to their flanks. This may reflect the influence of the more tensional crestal environment on the low tensile strength carbonates. Crestal position is simplistically represented as vertical distance from the crest and is ranked continuously from zero to three so that the crest of the field has the highest value of Crestal Position Index.

Processing of all four parametres results in four distinct 3D data volumes with values ranging from zero to three. These data volumes are arithmetically combined in different proportions and the result is compared to existing image log data. Iterations of different proportions are made until the best fit result is found (Figs 10 & 11).

Iterations resulted in a good qualitative match for the Ilam reservoir using a proportion of 85% for the combination of Structural Curvature Index and Facies Index, with the remaining 15% shared between Porosity Index and Crestal Position Index. The values used were intuitive based on observations and data, and were the first iteration attempted (later iterations did not improve the match). This demonstrates the importance of a thorough understanding of both geology and dynamic reservoir behaviour for the reservoir being studied. Each reservoir's correlation to these and other reservoir characteristics will most likely be unique. The key is to identify those properties which dominate each reservoir's behaviour.

Utilization of the fracture density model

Interactive manipulation of the 3D models provides a powerful multi-discipline team communication tool. Unfortunately, the visual impact of the full field match with empirical data is difficult to capture on paper as a 3D model snapshot. In order to display the results in two dimensions, averages of the 3D fracture model values for the multiple layers that comprise the two reservoir zones are shown in map view in Figs 10 & 11. Image log fracture density and orientation data are overlain where the horizontal wells penetrate each reservoir layer to show the quality of the empirical match. These maps show the average of the Fracture Density Index for 11 model layers in the lower reservoir (Fig. 10), and for 7 model layers in the upper reservoir (Fig. 11). When compared, these two maps clearly show the effects of reservoir layer facies variation in the lower zone's belt-like geometry and the upper zone's sheet-like character.

The 3D Fracture Density Index volume is transferred to the reservoir simulator where it is used as a framework to guide dual porosity behaviour or reservoir transmissibility. This qualitative index can be converted into a quantitative simulation parametre by creating transform functions based on iterations of best-fit history matches.

Areas of higher fracture densities can now be predicted prior to drilling. Since the development of the fracture model, two wells have been drilled, F (single lateral) and G (stacked dual lateral), adding 4500 m of horizontal section and image log to the database (Figs 3, 7, 10 & 11). Both wells show agreement between the model-predicted and actual fracture densities on a qualitative basis (within the error of the directional and seismic surveys) and their production performance has been as expected.

Conclusions

Implications for further development

The combination of a predictive fracture model with rock property characteristics in reservoir simulation modelling of the heterogeneous Ilam Formation enhances our ability to predict its dynamic reservoir responses.

The key to maintaining oil flow rates and ultimately increasing recovery in the Ilam reservoir lies in a program of pressure maintenance through water injection. Definition of the fracture system will allow for the optimal placement of water injection, such that water breakthrough times are extended and pressure maintenance is maximized. To date, the success of water injection has been limited, mainly due to an inability to predict where fractured areas of the reservoir can be expected.

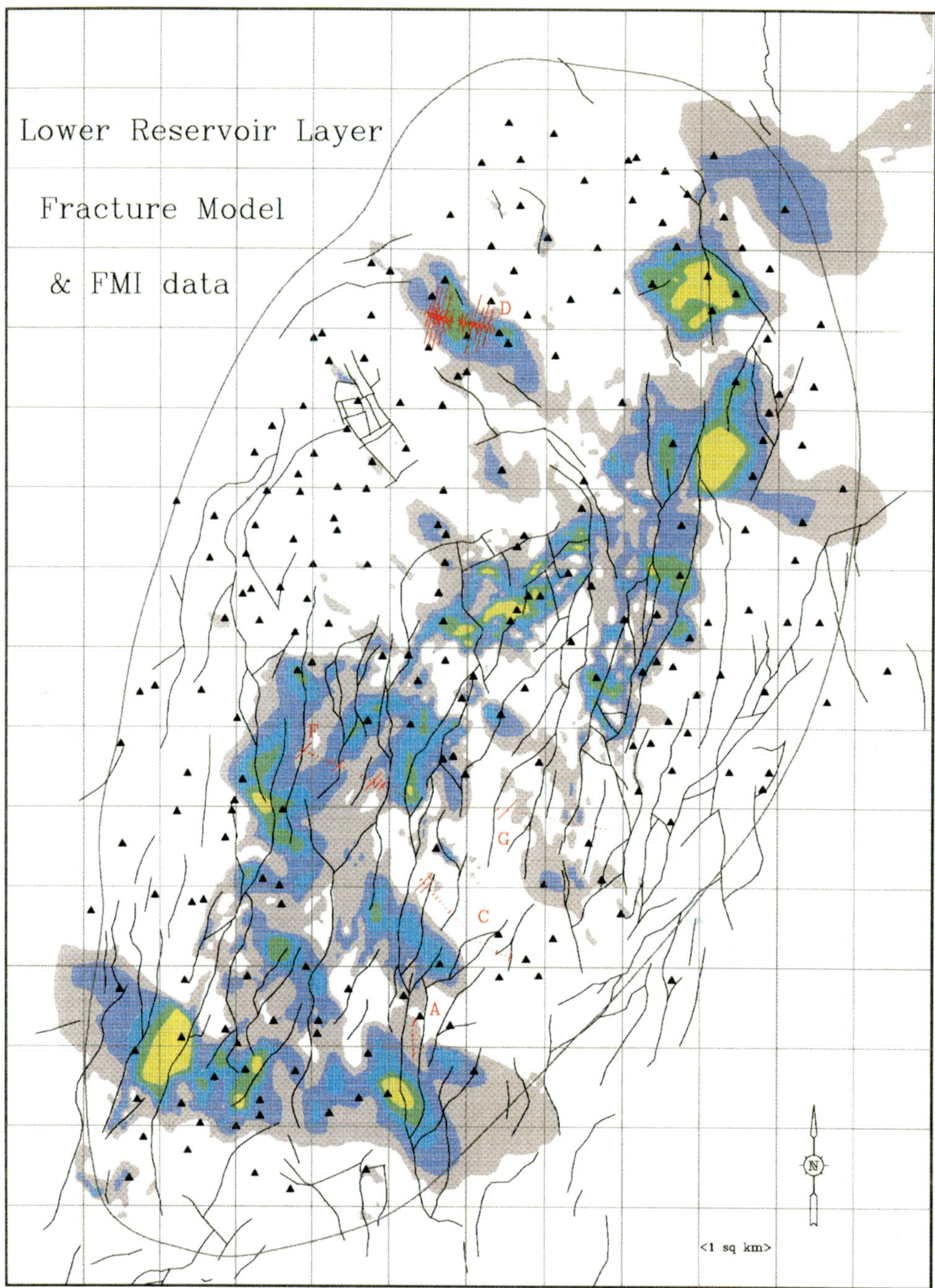

Fig. 10. Lower Reservoir Fracture Model averaged from 11 layers out of the 3D geological model. Yellow shading indicates highest predicted fracture densities, while grey shading denotes lower predicted fracture densities. Overlay of image log roses portraying fracture azimuth and density for 60 m increments along the wellbores drilled within each zone shows the strong correlation between the model and the actual fracture densities. (See captions for Figs 3, 7 & 11 for additional details).

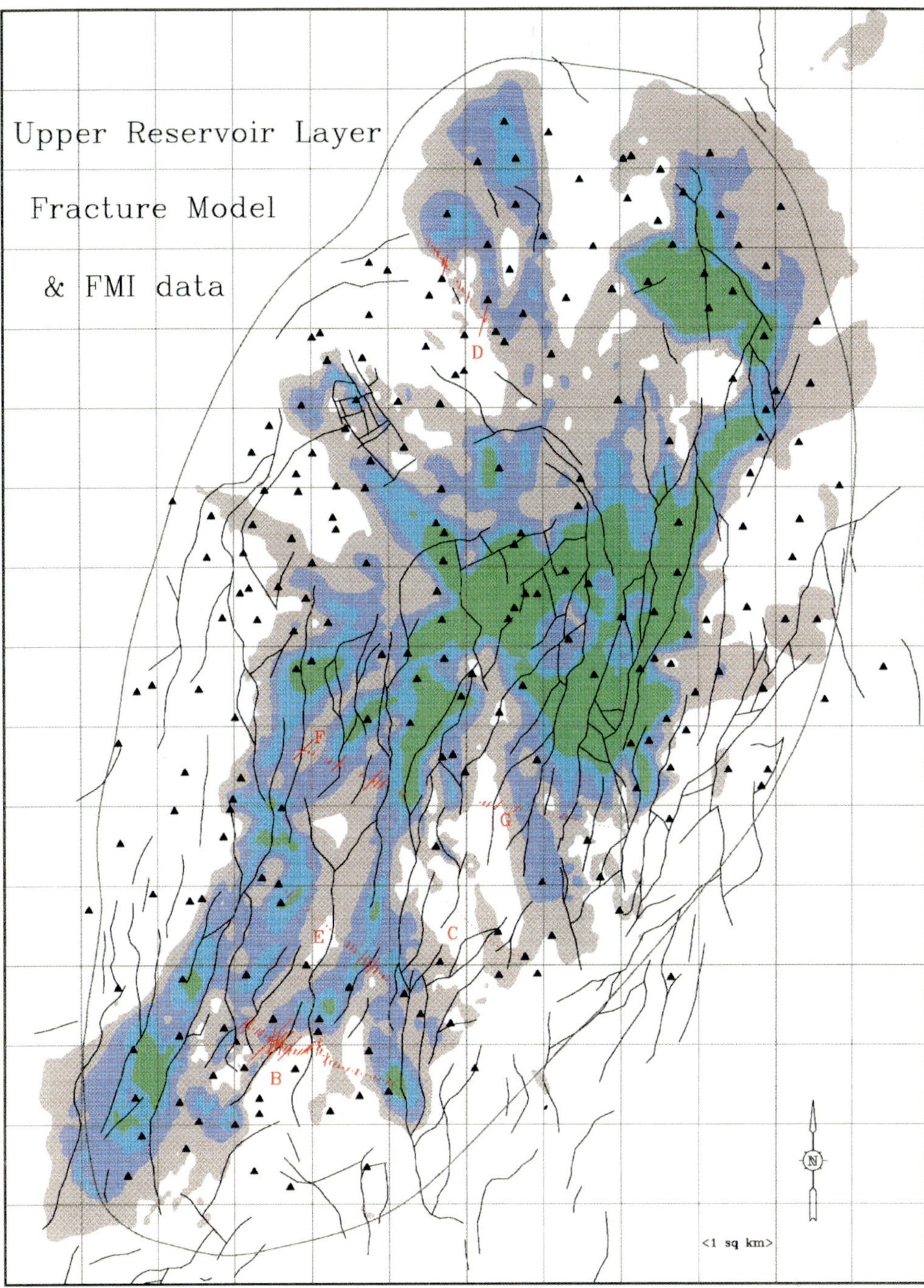

Fig. 11. Upper Reservoir Fracture Model averaged from 7 layers out of the 3D geological model (see captions for Figs 3, 7 & 10 for more details). Yellow shading indicates highest predicted fracture densities, while grey shading denotes lower predicted fracture densities. These maps are derived from 3D geocellular models of facies, structural curvature, porosity and crestal position of 55 layers in 8 sequences in the Ilam. The models have more than 1 million cells, and so preserve much of the geologic detail of the reservoir. These models are up-scaled for reservoir simulation at whatever level of detail is deemed necessary by the reservoir team.

As part of the planning for further production off-take and the implementation of additional water injection, reservoir simulation can now be undertaken whereby the fracture system is characterized to some predictive extent. This is a great advance over previous modelling undertaken, in which well behaviour was matched by property enhancement. Away from well control, the predictive capability of the models was poor.

The drilling and completion of horizontal wells in the Ilam reservoir presents unique problems when compared to other non-fractured reservoirs. A more definitive knowledge of the fracture system combined with detailed fault determination from 3D seismic allows for the optimization of well positions for productivity and injectivity, recovery and the avoidance of potential drilling problems.

The fracture model currently allows us to include qualitative values for influential geologic input parametres; the next obvious challenge is to define quantitative descriptions of fracture behaviour in the form of a 3D model. Additional methods, including utilization of 3D seismic attributes, investigation of rock mechanics characteristics, timing and genetic history of fracture development and other techniques are being explored. It is hoped that applications and future enhancements of the fracture modelling methods described herein will result in a decrease in reservoir uncertainty as exploitation continues.

The authors wish to express their thanks to the following contributors: M. Noble and H. Sheline for geophysical interpretation of the field; W. Esco for his review and application of the fracture model in simulation experiments; M. Akbar and K. Saxena of the Schlumberger Abu Dhabi Processing Center for documentation of the image log results; J. Wood and G. East for early studies leading to this work. Internal reports authored by T. Harland, W. Rizer and K. Soofi, J. Ericsson and H. McKean, and M. Noble, M. Levret, R. Hooper, J. Ericsson, H. Sheline and P. Williams were utilized in the study. Reviews of this work at various stages were undertaken by W. Rizer, P. D'Onfro, A. Cole, P. Williams, A. Al Awar (Al Bastaki), T. Sebire, B. Sager and R. Darr. Thoughtful reviews by E. Edwards, G. Howes, and volume editor G. Jones significantly improved the clarity and focus of the paper.

References

ALAVI, M. 1980. Tectonostratigraphic evolution of the Zagrosides in Iran. *Geology*, **8**, 144–149.

—— 1994. Tectonics of the Zagros orogenic belt of Iran: new data and interpretations. *Tectonophysics*, **229**, 211–238.

BEYDOUN, Z. R. 1991. Arabian plate hydrocarbon geology and potential – a plate tectonic approach. *American Association of Petroleum Geology, Studies in Geology*, **33**, 49–65.

DERCOURT, J., RICOU, L. E. & VRIELYNCK, B. (eds). 1993. *Atlas of Tethys Paleo-environmental Maps*. Gauthier-Villars, Paris, 307.

GRABOWSKI, G. J. & NORTON, I. O. 1994. Tectonic controls on the stratigraphic architecture and hydrocarbon systems of the Arabian Plate. *In*: AL-HUSSEINI, M. I. (ed.) *GEO '94. Selected Middle East Papers from The Middle East Geoscience Conference: Bahrain*, Gulf Petrolink, 413–430.

HOOPER, R. J., BARON I. R., AGAH, S. & HATCHER, R. D., Jr 1994*a*. The Cenomanian to Recent development of the southern Tethyan margin in Iran: *In*: AL-HUSSEINI, M. I. (ed.) *GEO '94. Selected Middle East Papers from The Middle East Geoscience Conference: Bahrain*, Gulf Petrolink, 505–516.

——, ——, HATCHER, R. D. Jr & AGAH, S. 1994*b*. The development of the southern Tethyan margin in Iran after the break-up of Gondwana – implications for the Zagros hydrocarbon province: *Geosciences, Scientific Quarterly Journal of the Geological Survey of Iran*, **4**, 913), 14.

——, ——, MURPHY, C. M. & PERKINS, S. M. 1996. The role of structural inversion in the development of diapirs in the southeast Arabian Gulf. *GeoArabia*, **1**, 148.

VENDEVILLE, B. C. & NILSEN, K. T. 1995. Episodic growth of salt diapirs driven by horizontal shortening. *In*: TRAVIS, C. J. (ed.) *Salt, Sediment and Hydrocarbons*, GCSEPM Foundation 16th Annual Research Conference, 285–295.

WOOD, J. R. & EAST, G. E. 1992. An integrated approach to the recognition and characterisation of a fractured reservoir in the Ilam Formation, Fateh Field, Dubai, UAE. *Proceedings of the 5th Annual Abu Dhabi Petroleum Exhibition and Conference*, 313–339.

Index

Page numbers in italics refer to Figures and Tables